AF844013

VIABILITY, INVARIANCE AND APPLICATIONS

NORTH-HOLLAND MATHEMATICS STUDIES 207

(Continuation of the Notas de Matemática)

Editor: Jan van Mill
Faculteit der Exacte Wetenschappen
Amsterdam, The Netherlands

ELSEVIER

Amsterdam – Boston – Heidelberg – London – New York – Oxford
Paris – San Diego – San Francisco – Singapore – Sydney – Tokyo

VIABILITY, INVARIANCE AND APPLICATIONS

OVIDIU CÂRJĂ
MIHAI NECULA
IOAN I. VRABIE
"Al. I. Cuza" University Iaşi, Romania
(for all authors)
&
"O. Mayer" Mathematics Institute Iaşi, Romania
(for the first and the third author)

ELSEVIER

Amsterdam – Boston – Heidelberg – London – New York – Oxford
Paris – San Diego – San Francisco – Singapore – Sydney – Tokyo

Elsevier
Radarweg 29, PO Box 211, 1000 AE Amsterdam, The Netherlands
The Boulevard, Langford Lane, Kidlington, Oxford OX5 1GB, UK

First edition 2007

Copyright © 2007 Elsevier B.V. All rights reserved.

No part of this publication may be reproduced, stored in a retrieval system or transmitted in any form or by any means electronic, mechanical, photocopying, recording or otherwise without the prior written permission of the publisher.

Permissions may be sought directly from Elsevier's Science & Technology Rights Department in Oxford, UK: phone (+44) (0) 1865 843830; fax (+44) (0) 1865 853333; email: permissions@elsevier.com. Alternatively you can submit your request online by visiting the Elsevier web site at http://elsevier.com/locate/permissions, and selecting *Obtaining permission to use Elsevier material.*

Notice
No responsibility is assumed by the publisher for any injury and/or damage to persons or property as a matter of products liability, negligence or otherwise, or from any use or operation of any methods, products, instructions or ideas contained in the material herein. Because of rapid advances in the medical sciences, in particular, independent verification of diagnoses and drug dosages should be made.

Library of Congress Cataloging-in-Publication Data
A catalog record for this book is available from the Library of Congress.

British Library Cataloguing in Publication Data
A catalogue record for this book is available from the British Library.

ISBN-13: 978-0-444-52761-5
ISSN (Series): 0304-0208

For information on all Elsevier publications
visit our website at books.elsevier.com

Transferred to Digital Printing 2008.

Contents

Preface

The book collects some of the main mathematical concepts, results and applications in the particularly flourishing field of differential equations whose solutions are constrained to live in a given set. In other words, the problem is to find the proper theoretical conditions in order that certain phenomena evolve within some bounds imposed by the objective we intend to reach. The cases of an aircraft which has to stay in a pre-assigned air-tunnel, or of the concentration of a substance which, on the long term of its evolution, must obey some requirements, are specific examples of this sort of problems. We may also think at two competing species living within the same region and whose densities have to be kept within some given limits.

The monograph is intended as an almost self-contained presentation of the most important concepts and results in viability and invariance. The *viability* of a set K with respect to a given function or multi-function, F, defined on it, describes the property of the differential equation driven by that function or multi-function, i.e., $u'(t) \in F(u(t))$, that for each initial data in K there exists at least one solution. The *invariance* of a set K with respect to a function or multi-function, F, defined on a set D strictly larger that K, is the property which asserts that each solution (if there is any), of the differential equation or inclusion driven by F, which starts in K remains in K, at least for a short time.

The book includes the most important necessary and sufficient conditions for viability starting with Nagumo's Viability Theorem for ordinary differential equations with continuous right-hand sides and continuing with the corresponding extensions either to differential inclusions or to semilinear or even fully nonlinear evolution equations, systems and inclusions. In the latter cases (i.e., multi-valued), the results, due to the authors, are very recent, completely new and extend significantly, in several directions, their well-known classical counterparts.

We present here a thorough study of viability problems involving multi-functions by means of some new tangency concepts, such that as the one of *tangent set*, or of *A-quasi-tangent set* to a set K at a point $\xi \in K$, with A

a given, possible nonlinear, operator. Recently introduced by the authors, these concepts are multi-valued extensions of the Bouligand–Severi tangent vectors. More than this, they prove to be more natural and appropriate, when handling with evolution problems with multi-valued right-hand sides, than the just mentioned Bouligand–Severi tangent vectors. We notice that the tangency conditions based on tangent, or A-quasi tangent sets are necessary and sufficient for viability in more general settings than the usual tangency conditions expressed in terms of vectors. Furthermore, the use of A-quasi-tangent sets, in the study of a general controllability problem for semilinear evolutions, is almost imposed by the problem, and proves very elegant and efficient. See Section 13.7.

Apart from the general abstract viability and invariance results, we include various applications showing the power of the abstract developed theory. For instance, we prove several comparison results referring to linear, or even nonlinear partial differential equations and systems of parabolic type. See Sections 13.3~13.6. We also prove an existence result of positive solutions for a pseudo-parabolic problem by reducing it to an ordinary differential equation which is under the incidence of a simple extension to infinite dimensions of the Nagumo Viability Theorem. See Section 7.5.

The book is divided into two parts. In Part 1 we confined ourselves merely to the study of viability and invariance problems for ordinary differential equations and inclusions, while in Part 2 we focused our attention on the more general situation of differential evolution equations and inclusions. We include below a short presentation of both parts.

Chapter 1 has an introductory character and it may be skipped by anyone with a reasonable training in Linear Functional Analysis and Evolution Equations. Chapter 2, also introductory, differs from the preceding one in that it contains several general results very precisely circumvented to the topic of the book. Therefore, its careful reading would be of great help for anyone wishing to make the reading of the whole book run smoothly. We insist on the importance of Sections 2.1, 2.3 (containing completely new concepts and results on tangent sets) and 2.7 in the construction and, of course, in the easy understanding of the whole book. Chapter 3 contains the main viability results referring to both autonomous and nonautonomous ordinary differential equations, driven by continuous functions defined on locally closed subsets, in general Banach spaces. Chapter 4 is mainly devoted to general sufficient conditions of invariance, also in the specific case of ordinary differential equations in general Banach spaces. Chapter 5 reconsiders the viability problems studied in Chapter 3 in the more general frame of nonautonomous ordinary differential equations driven by Carathéodory functions defined on cylindrical domains. We notice that the constraint on the domain is somehow dictated by the simple observation that in the general, i.e., noncylindrical case, the a.e. tangency condition is not enough for viability no matter how smooth the right-hand side of the equation is. Chapter 6 contains an extension of the viability theory developed for ordinary differential equations to ordinary differential inclusions driven by upper semicontinuous multi-functions with nonempty convex and weakly compact values. Here, taking advantage of the notion of tangent set, we were able to rebuild the whole existing viability theory in the single-valued case to work similarly in the multi-valued u.s.c., nonempty, convex and weakly compact valued case as well. We did not touch upon the lower semicontinuous case simply because, thanks to the Michael Selection Theorem, this reduces to the single-valued continuous one. Also, in order to keep the book under reasonable lenght limits, we decided not to consider the general Carathéodory multi-valued case. We notice that the latter requires a sharp mathematical apparatus involving rather difficult and laborious technicalities. Chapter 7 (the last in the first part) collects several applications of the viability and invariance results referring to: viability of an epigraph, the existence of monotone solutions, the existence of positive solutions for some pseudo-parabolic partial differential equations, the Hukuhara Theorem, the Kneser Theorem, the existence of Lyapunov functions, the characteristics method for a class of first order partial differential equations.

Part 2 starts with Chapter 8 which is devoted to the study of the viability of a given set with respect to a single-valued continuous perturbation of an infinitesimal generator of a C_0-semigroup. We added here significant extensions of the viability theory built in Chapter 3 to this more general setting, as well as several viability results referring to semilinear reaction-diffusion systems. In Chapter 9, taking advantage of the concept of A-quasi tangent vector to a set K at a point $\xi \in K$, we extended the previous theory to the case of nonempty, convex and weakly compact-valued u.s.c. perturbations of infinitesimal generators of C_0-semigroups. Chapter 10 is mainly concerned with the viability of a given set K with respect to a vector field of the form $A + f$, with A an m-dissipative (possibly) nonlinear and (possibly) multi-valued operator and f a given continuous function. Here we also discuss several situations concerning reaction-diffusion nonlinear systems. Again, using the concept of A-quasi tangent vector (this time with A nonlinear), in Chapter 11, we develop a viability theory handling the general case of nonempty, convex and weakly compact-valued u.s.c. perturbations of m-dissipative operators. In order to give the reader an idea of what happens in the fully nonlinear Carathéodory case, in Chapter 12 we proved several viability results concerning nonlinear evolution equations driven by Carathéodory perturbations of m-dissipative operators. Like in Chapter 5, we confined ourselves merely to the particular case of a cylindrical domain. Chapter 13 includes more applications to concrete partial differential equations and systems. Here we included several new results: the existence of orthogonal solutions for a class of first-order hyperbolic systems, the existence of Lyapunov pairs ensuring the asymptotic stability of certain semilinear evolution systems, several comparison results for solutions of linear (and even nonlinear) partial differential equations and systems of parabolic (and even degenerate) type, sufficient conditions for null controllability of abstract semilinear parabolic equations and sufficient conditions for the existence of periodic solutions for fully nonlinear evolution equations.

Each chapter also contains a set of problems, with detailed solutions at the end of the book in the section entitled "Solutions to the proposed problems". The book ends with a Bibliographical Notes and Comments section, a Bibliography, a Name Index, a Subject Index and a Notation Index.

Acknowledgements. The research done during the writing of this monograph was partially supported by the Projects CERES 4-194/2004, CEx05-DE11-36/05.10.2005 and the CNCSIS Grant A 1159/2006.

We take this opportunity to thank Professor Mihai Turinici and our graduate students Monica Burlică and Daniela Roşu for the critical reading of large portions of the manuscript.

Iaşi, November 11th, 2006

Ovidiu Cârjă, Mihai Necula and Ioan I. Vrabie

CHAPTER 1

Generalities

The aim of this chapter is to put together most of the frequently used auxiliary notions and results which are needed for a good understanding of the whole book. Therefore, we included here basic facts about Banach spaces, the Bochner integral and usual function spaces, compactness theorems including the infinite dimensional version of Arzelà–Ascoli Theorem, C_0-semigroups, m-dissipative operators and the nonlinear evolutions governed by them, m-dissipative possibly nonlinear partial differential operators, differential and integral inequalities.

1.1. Basic facts on Banach spaces

Throughout this book, X is a real Banach space[1] with norm $\|\cdot\|$ and X^* is its *topological dual*, i.e., the vector space of all linear continuous functionals from X to $\mathbb{R}$, which, endowed with the *dual norm* $\|x^*\| = \sup_{\|x\|\leq 1} |(x, x^*)|$, for $x^* \in X^*$ is, in its turn, a real Banach space too. Here and thereafter, if $x \in X$ and $x^* \in X^*$, (x, x^*) denotes $x^*(x)$. We denote by $\operatorname{Fin}(X^*)$ the class of all finite subsets in X^*. Let $F \in \operatorname{Fin}(X^*)$. Then, the function $\|\cdot\|_F : X \to \mathbb{R}$, defined by

$$\|x\|_F = \max\{|(x, x^*)|;\ x^* \in F\}$$

for each $x \in X$, is a seminorm on X.

The family of seminorms $\{\|\cdot\|_F;\ F \in \operatorname{Fin}(X^*)\}$ defines the so-called *weak topology* and X, endowed with this topology, is a locally convex topological vector space.

Whenever we refer to weak topology concepts, we shall use the name of the concept in question preceded, or followed, by the word weak (weakly). For instance, a subset B in X is called weakly closed if it is closed in the weak topology. If B is norm or strongly closed, we simply say that B is closed.

---

[1]Sometimes, we will merely assume that X is a real vector space, but in all those cases we will clearly specify that.

Theorem 1.1.1. *The weak closure of a convex subset in a Banach space coincides with its strong closure.*

See Hille–Phillips [**107**], Theorem 2.9.3, p. 36.

Corollary 1.1.1. *If* $\lim_n x_n = x$ *weakly in* X*, then there exists* $(y_n)_n$*, with* $y_n \in \operatorname{conv}\{x_k;\ k \geq n\}$*, such that* $\lim_n y_n = x$*.*

See Hille–Phillips [**107**], Corollary to Theorem 2.9.3, p. 36.

Definition 1.1.1. A Banach space X is *uniformly convex* if for each $\varepsilon \in (0, 2]$ there exists $\delta(\varepsilon) > 0$ such that, for each $x, y \in X$ with $\|x\| \leq 1$, $\|y\| \leq 1$ and $\|x - y\| \geq \varepsilon$, we have $\|x + y\| \leq 2(1 - \delta(\varepsilon))$.

Definition 1.1.2. A Banach space X is *reflexive* if the natural mapping $x \mapsto x^{**}$ defined by $(x^*, x^{**}) = (x, x^*)$ for each $x^* \in X^*$ is an isomorphism between X and X^{**} — the topological dual of X^*.

Theorem 1.1.2. *A Banach space is reflexive if and only if its topological dual is reflexive.*

See Hille–Phillips [**107**], Corollary 2, p. 38.

Theorem 1.1.3. *Every uniformly convex space is reflexive.*

See Yosida [**185**], Theorem 2, p. 127. An immediate consequence of Theorems 1.1.2 and 1.1.3 is

Corollary 1.1.2. *A Banach space whose topological dual is uniformly convex is reflexive.*

We recall that the *duality mapping*[2] $J : X \rightsquigarrow X^*$ is defined by

$$J(x) = \left\{x^* \in X^*\,;\ (x, x^*) = \|x\|^2 = \|x^*\|^2\right\}$$

for each $x \in X$. In view of the Hahn–Banach Theorem, it follows that, for each $x \in X$, $J(x)$ is nonempty.

Theorem 1.1.4. *If the dual of* X *is uniformly convex, then the duality mapping* $J : X \rightsquigarrow X^*$ *is single-valued and uniformly continuous on bounded subsets in* X*.*

See Barbu [**10**], Proposition 1.5, p. 14.

---

[2]Whenever F is a multi-valued mapping from a set D to a set E, i.e., $F : D \to 2^E$, we denote this by $F : D \rightsquigarrow E$.

1.2. The Bochner integral and L^p spaces

A measure space (Ω, Σ, μ) is called *σ-finite* if there exists $\{\Omega_n;\ n \in \mathbb{N}\} \subseteq \Sigma$ such that $\mu(\Omega_n) < +\infty$ for each $n \in \mathbb{N}$ and $\Omega = \cup_{n\in\mathbb{N}}\Omega_n$. It is called *finite* if $\mu(\Omega) < \infty$. The measure space (Ω, Σ, μ) is called *complete* if the measure μ is *complete*, i.e., if each subset of a null μ-measure set is measurable, i.e., belongs to the σ-field Σ.

Let X be a Banach space with norm $\|\cdot\|$ and (Ω, Σ, μ) a measure space with a σ-finite and complete measure.

Definition 1.2.1. A function $x : \Omega \to X$ is called:

(i) *countably-valued* if there exist two families: $\{\Omega_n;\ n \in \mathbb{N}\} \subseteq \Sigma$ and $\{x_n;\ n \in \mathbb{N}\} \subseteq X$, with $\Omega_k \cap \Omega_p = \emptyset$ for each $k \neq p$, $\Omega = \cup_{n\geq 0}\Omega_n$, and such that $x(\theta) = x_n$ for all $\theta \in \Omega_n$;

(ii) *measurable* if there exists a sequence of countably-valued functions convergent to x μ-a.e. on Ω.

Theorem 1.2.1. *A function $x : \Omega \to X$ is measurable if and only if there exists a sequence of countably-valued functions from Ω to X which is uniformly μ-a.e. convergent on Ω to x.*

See Vrabie [**175**], Theorem 1.1.3, p. 3 and Remark 1.1.2, p. 4.

Since the two families $\{\Omega_n;\ n \in \mathbb{N}\}$ and $\{x_n;\ n \in \mathbb{N}\}$ in the definition of a countably-valued function are not unique, in the sequel, a pair of sets,

$$(\{\Omega_n;\ n \in \mathbb{N}\}, \{x_n;\ n \in \mathbb{N}\}),$$

with the above properties, is called *a representation of the countably-valued function x*. Since Ω has σ-finite measure, for each countably-valued function $x : \Omega \to X$ there exists at least one representation such that, for each $n \in \mathbb{N}$, $\mu(\Omega_n) < +\infty$. A representation of this sort is called *σ-finite representation.*

Definition 1.2.2. Let $x : \Omega \to X$ be a countably-valued function and let $\mathcal{R} = (\{\Omega_n;\ n \in \mathbb{N}\}, \{x_n;\ n \in \mathbb{N}\})$ be one of its σ-finite representations. We say that $\mathcal{R}$ is *Bochner integrable* (B-integrable) on Ω with respect to μ, if

$$\sum_{n=0}^{\infty} \mu(\Omega_n)\|x_n\| < +\infty.$$

Remark 1.2.1. If $\mathcal{R}$ and $\mathcal{R}'$ are two σ-finite representations of the very same countably-valued function $x : \Omega \to X$, the series $\sum_{n=0}^{\infty} \mu(\Omega_n)x_n$ and $\sum_{n=0}^{\infty} \mu(\Omega'_n)x'_n$ are either both convergent, or both divergent, and, in the former case, they have the same sum. So, $\mathcal{R}$ is B-integrable on Ω with respect to μ if and only if $\mathcal{R}'$ does.

This remark enables us to introduce:

Definition 1.2.3. The countably-valued function $x : \Omega \to X$ is called *Bochner integrable* on Ω with respect to μ if it has a σ-finite representation

$$\mathcal{R} = (\{\Omega_n;\ n \in \mathbb{N}\}, \{x_n;\ n \in \mathbb{N}\})$$

which is B-integrable on Ω with respect to μ. In this case, the vector

$$\sum_{n=0}^{\infty} \mu(\Omega_n)x_n = \int_\Omega x(\theta)\, d\mu(\theta) = \int_\Omega x\, d\mu,$$

which does not depend on the choice of $\mathcal{R}$ (see Remark 1.2.1), is called *the Bochner integral* on Ω of the function x with respect to μ.

Definition 1.2.4. A function $x : \Omega \to X$ is *Bochner integrable* on Ω with respect to μ if it is measurable and there exists a sequence of countably-valued functions $(x_k)_k$, Bochner integrable on Ω with respect to μ, such that

$$\lim_k \int_\Omega \|x(\theta) - x_k(\theta)\|\, d\mu(\theta) = 0.$$

Proposition 1.2.1. *If $x : \Omega \to X$ is Bochner integrable on Ω with respect to μ and $(x_k)_k$ is a sequence with the properties in Definition* 1.2.4, *then there exists*

$$\lim_k \int_\Omega x_k\, d\mu$$

in the norm topology of X. In addition, if $(y_k)_k$ is another sequence of countably-valued functions with the property that

$$\lim_k \int_\Omega \|x(\theta) - y_k(\theta)\|\, d\mu(\theta) = 0,$$

then

$$\lim_k \int_\Omega x_k(\theta)\, d\mu(\theta) = \lim_k \int_\Omega y_k(\theta)\, d\mu(\theta).$$

See Vrabie [**175**], Proposition 1.2.1, p. 5.

Definition 1.2.5. Let $x : \Omega \to X$ be a Bochner integrable function on Ω. The vector

$$\lim_k \int_\Omega x_k\, d\mu = \int_\Omega x(\theta)\, d\mu(\theta) = \int_\Omega x\, d\mu$$

which, according to Proposition 1.2.1, exists and does not depend on the choice of the sequence $(x_k)_k$ in Definition 1.2.4, is called *the Bochner integral* of the function x on Ω with respect to μ.

Theorem 1.2.2. *A function $x : \Omega \to X$ is Bochner integrable on Ω with respect to μ if and only if x is measurable and the real function $\|x\|$ is integrable on Ω with respect to μ.*

See Vrabie [**175**], Theorem 1.2.1, p. 6.

We denote by $\mathcal{L}^p(\Omega, \mu\,; X)$ the set of all functions $f : \Omega \to X$ which are measurable on Ω and $\|f\|^p$ is integrable on Ω with respect to μ. Let us define $\|\cdot\|_{\mathcal{L}^p(\Omega,\mu\,;X)} : \mathcal{L}^p(\Omega, \mu\,; X) \to \mathbb{R}_+$ by

$$\|f\|_{\mathcal{L}^p(\Omega,\mu\,;X)} = \left(\int_\Omega \|f\|^p\, d\mu\right)^{1/p}$$

for each $f \in \mathcal{L}^p(\Omega, \mu\,; X)$. This is a seminorm on $\mathcal{L}^p(\Omega, \mu\,; X)$. The relation "$\sim$" defined by $f \sim g$ if $f(\theta) = g(\theta)$ μ-a.e. for $\theta \in \Omega$ is an equivalence on $\mathcal{L}^p(\Omega, \mu\,; X)$. Let $L^p(\Omega, \mu\,; X)$ be the quotient space $\mathcal{L}^p(\Omega, \mu\,; X)/\sim$. If $f \sim g$, then $\|f\|_{\mathcal{L}^p(\Omega,\mu\,;X)} = \|g\|_{\mathcal{L}^p(\Omega,\mu\,;X)}$. As a consequence, the mapping $\|\cdot\|_{L^p(\Omega,\mu\,;X)} : L^p(\Omega, \mu\,; X) \to \mathbb{R}_+$, given by $\|\hat{f}\|_{L^p(\Omega,\mu\,;X)} = \|f\|_{\mathcal{L}^p(\Omega,\mu\,;X)}$ for each $\hat{f} \in L^p(\Omega, \mu\,; X)$, is well-defined (i.e. it does not depend on the choice of $f \in \hat{f}$) and is a norm on $L^p(\Omega, \mu\,; X)$. Endowed with this norm, $L^p(\Omega, \mu\,; X)$ is a Banach space.

Next, let $\mathcal{L}^\infty(\Omega, \mu\,; X)$ be the space of all functions $f : \Omega \to X$ satisfying

$$\|f\|_{\mathcal{L}^\infty(\Omega,\mu\,;X)} = \inf\{\alpha \in \overline{\mathbb{R}};\ \|f(\theta)\| \le \alpha, \quad \text{a.e.} \quad \theta \in \Omega\} < +\infty.$$

The mapping $\|\cdot\|_{\mathcal{L}^\infty(\Omega,\mu\,;X)} : \mathcal{L}^\infty(\Omega, \mu\,; X) \to \mathbb{R}_+$, defined as above, is a seminorm. Let $L^\infty(\Omega, \mu\,; X) = \mathcal{L}^\infty(\Omega, \mu\,; X)/\sim$, where "$\sim$" is the μ-a.e. equality on Ω and let $\|\cdot\|_{L^\infty(\Omega,\mu\,;X)} : L^\infty(\Omega, \mu\,; X) \to \mathbb{R}_+$, given by

$$\|\hat{f}\|_{L^\infty(\Omega,\mu\,;X)} = \|f\|_{\mathcal{L}^\infty(\Omega,\mu\,;X)}$$

for each $\hat{f} \in L^\infty(\Omega, \mu\,; X)$. Obviously $\|\cdot\|_{L^\infty(\Omega,\mu\,;X)}$ is well-defined and, in addition, is a norm on $L^\infty(\Omega, \mu\,; X)$, with respect to which this is a Banach space. For simplicity, we denote by f either a fixed element in $\mathcal{L}^p(\Omega, \mu\,; X)$ or its corresponding equivalence class in $L^p(\Omega, \mu\,; X)$.

Theorem 1.2.3. *Let $(f_n)_n$ be a sequence in $L^1(\Omega, \mu\,; X)$ with*

$$\lim_n f_n(\theta) = f(\theta)$$

μ-a.e. for $\theta \in \Omega$. If there exists $\ell \in L^1(\Omega, \mu\,; \mathbb{R})$ such that

$$\|f_n(\theta)\| \le \ell(\theta)$$

for $n = 1, 2, \ldots$ and μ-a.e. for $\theta \in \Omega$, then $f \in L^1(\Omega, \mu\,; X)$ and $\lim_n f_n = f$ in the norm of $L^1(\Omega, \mu\,; X)$.

For the proof of Theorem 1.2.3, known as the Lebesgue Dominated Convergence Theorem, see Dinculeanu [**85**].

We also need the following specific form of the Fatou Lemma.

Lemma 1.2.1. *Let (Ω, Σ, μ) be a σ-finite and complete measure space and let $(f_n)_n$ be a sequence of measurable functions from Ω to $\mathbb{R}_+$, but not necessarily μ-integrable. Then*

$$\limsup_k \int_\Omega f_k(\theta)\, d\mu(\theta) \leq \int_\Omega \limsup_k f_k(\theta)\, d\mu(\theta).$$

See Dunford–Schwartz [**90**], Theorem 19, p. 52.

The next result gives a simple but precise description of the topological dual of $L^p(\Omega, \mu\,; X)$ for certain classes of Banach spaces.

Theorem 1.2.4. *If either X is reflexive, or X^* is separable, then, for each $p \in [1, +\infty)$, $(L^p(\Omega, \mu\,; X))^*$ can be identified with $L^q(\Omega, \mu\,; X^*)$, where $\frac{1}{p} + \frac{1}{q} = 1$ if $p > 1$ and $q = \infty$ if $p = 1$.*

See Dinculeanu [**85**], Corollary 1, p. 252. Some extensions and variants of Theorem 1.2.4 can be found in Edwards [**91**], Theorem 8.18.2, p. 588, Remarks, p. 589 and Theorem 8.20.5, p. 607.

A remarkable consequence of Theorem 1.2.4 is stated below.

Corollary 1.2.1. *If X is reflexive and $p \in (1, +\infty)$, then $L^p(\Omega, \mu\,; X)$ is reflexive. If X is separable, then, for each $p \in [1, +\infty)$, $L^p(\Omega, \mu\,; X)$ is separable.*

Now, let $(\Omega_i, \Sigma_i, \mu_i)$, $i = 1, 2$, be two measure spaces and let us define *the product measure space* (Ω, Σ, μ) as the measure space for which $\Omega = \Omega_1 \times \Omega_2$, Σ is the smallest σ-field containing all the sets $E_1 \times E_2$ with $E_i \in \Sigma_i$, $i = 1, 2$, and such that $\mu(E_1 \times E_2) = \mu_1(E_1)\mu_2(E_2)$ for each $E_i \in \Sigma_i$, $i = 1, 2$.

Theorem 1.2.5. *Let $(\Omega_i, \Sigma_i, \mu_i)$, $i = 1, 2$, be finite measure spaces and let (Ω, Σ, μ) be their product space. Let X be a Banach space and let $f \in L^1(\Omega, \mu; X)$. Then, for μ_1-a.e. $s \in \Omega_1$, $t \mapsto f(s, t)$ belongs to $L^1(\Omega_2, \mu_2; X)$, the function $s \mapsto \int_{\Omega_2} f(s, t)\, d\mu_2(t)$ belongs to $L^1(\Omega_1, \mu_1; X)$ and*

$$\int_{\Omega_1} \int_{\Omega_2} f(s, t)\, d\mu_2(t)\, d\mu_1(s) = \int_\Omega f(\theta)\, d\mu(\theta).$$

For the proof of this result, known as the Fubini Theorem, see Dunford–Schwartz [**90**], Theorem 9, p. 190.

In the theorem below, μ is the Lebesgue measure on $\mathbb{R}^n$. In order to simplify the notation, whenever Ω is a Lebesgue measurable subset in $\mathbb{R}^n$ and μ is the Lebesgue measure on Ω, we denote by $L^p(\Omega; X) = L^p(\Omega, \mu\,; X)$.

If, in addition, $X = \mathbb{R}$, a further simplification is made, i.e., we denote by $L^p(\Omega) = L^p(\Omega;\mathbb{R})$. Finally, if $\Omega = [\tau, T]$, we simply write $L^p(\tau, T; X)$ instead of $L^p([\tau, T]; X)$.

Theorem 1.2.6. *If $\Omega \subseteq \mathbb{R}^n$ is nonempty, bounded and Lebesgue measurable, $n \geq 1$, and $p \in (1, +\infty)$, then the space $L^p(\Omega)$, endowed with its usual norm, is uniformly convex.*

See Cioränescu [**64**], Teorema 4.1, p. 113.

We conclude this section with a differentiability result.

Theorem 1.2.7. *Let X be a Banach space and let $f \in L^1(\tau, T; X)$. Then, for a.a. $t \in (\tau, T)$, we have*

$$\lim_{h\downarrow 0} \frac{1}{h} \int_t^{t+h} \|f(\theta) - f(t)\|\, d\theta = 0.$$

See Bochner [**20**] or Diestel–Uhl [**83**], Theorem 9, p. 49.

1.3. Compactness theorems

Here we gather several compactness results which will be used later.

Definition 1.3.1. A subset C of a topological space $(X, \mathfrak{T})$ is called:

(i) *relatively compact*, if each generalized sequence in C has at least one generalized convergent subsequence;
(ii) *compact*, if it is relatively compact and closed;
(iii) *sequentially relatively compact*, if each sequence in C has at least one convergent subsequence;
(iv) *sequentially compact*, if it is sequentially relatively compact and closed.

If (X, d) is a metric space, $C \subseteq X$ is called *precompact*, or *totally bounded*, if for each $\varepsilon > 0$ there exists a finite family of closed balls of radius ε whose union includes C.

Remark 1.3.1. As each metric space satisfies the First Axiom of Countability, i.e., each point has an at most countable fundamental system of neighborhoods, in a metric space, a subset is (relatively) compact if and only if it is sequentially (relatively) compact.

Remark 1.3.2. Clearly, a subset $C \subseteq X$ is precompact if and only if, for each $\varepsilon > 0$ there exists a finite family of closed balls centered in points of C and having radii ε, whose union includes C.

Theorem 1.3.1. *If (X, d) is a complete metric space, then a subset of it is sequentially relatively compact if and only if it is precompact.*

Problem 1.3.1. *Let X be a Banach space. If $C \subseteq X$ is compact and $\{S(t) : C \to X;\ t \geq 0\}$ is a family of possibly nonlinear operators satisfying:*

(i) *there exists a continuous function $\ell : \mathbb{R}_+ \to \mathbb{R}_+$ such that*

$$\|S(t)\xi - S(t)\eta\| \leq \ell(t)\|\xi - \eta\|$$

for each $t \in \mathbb{R}_+$ and $\xi, \eta \in C$ and

(ii) $\lim_{h \downarrow 0} S(h)\xi = \xi$ *for each $\xi \in C$,*

then

$$\lim_{h \downarrow 0} S(h)\xi = \xi$$

uniformly for $\xi \in C$.

Theorem 1.3.2. *The closed convex hull of a (weakly) compact subset in a Banach space is (weakly) compact.*

See Dunford–Schwartz [**90**], Theorem 6, p. 416 and Theorem 4, p. 434. Let us also recall the Schauder Fixed Point Theorem, i.e.,

Theorem 1.3.3. *Let X be a Banach space. If $K \subseteq X$ is nonempty, closed and convex, the mapping $P : K \to K$ is continuous and $P(K)$ is relatively compact, then P has at least one fixed point $\xi \in K$, i.e. there exists $\xi \in K$ such that $P(\xi) = \xi$.*

See Dunford–Schwartz [**90**], Theorem 5, p. 456.

Theorem 1.3.4. *A subset in a Banach space is weakly compact if and only if it is weakly sequentially compact.*

See Edwards [**91**],Theorem 8.12.1, p. 549 and Theorem 8.12.7, p. 551.

Theorem 1.3.5. *Let X be reflexive. A subset in X is weakly relatively sequentially compact if and only if it is norm-bounded.*

See Hille–Phillips [**107**], Theorem 2.10.3, p. 38.

Problem 1.3.2. *If X is reflexive and B, C are two nonempty, closed, bounded and convex subsets in X, then $B + C$ is nonempty, closed, bounded and convex.*

Lemma 1.3.1. *Let X be a Banach space, K a compact subset in X and let $\mathcal{F}$ be a family of Bochner integrable functions from $[\tau, T]$ to K. Then*

$$\left\{ \int_\tau^T f(t)\,dt;\ f \in \mathcal{F} \right\}$$

is relatively compact in X.

The compactness Lemma 1.3.1 is an extension, from continuous to Lebesgue integrable functions, of Vrabie [**176**], Lemma A.1.3, p. 295. For a further extension and proof, see also Lemma 1.5.1 in Section 1.5.

The next result is an infinite dimensional version of the well-known Arzelà–Ascoli Theorem.

Theorem 1.3.6. *Let X be a Banach space. A subset $\mathfrak{F}$ in $C([\tau,T];X)$ is relatively compact if and only if:*

(i) *$\mathfrak{F}$ is equicontinuous on $[\tau,T]$;*

(ii) *there exists a dense subset D in $[\tau,T]$ such that, for each $t\in D$, $\mathfrak{F}(t)=\{f(t);\ f\in\mathfrak{F}\}$ is relatively compact in X.*

See Vrabie [**175**], Theorem A.2.1, p. 296.

Problem 1.3.3. *Let X be a Banach space and let $(u_n)_n$ be a sequence of continuous functions from $[\tau,T]$ to X such that there exist a sequence $\varepsilon_n\downarrow 0$ and a function $m:[0,T-\tau]\to\mathbb{R}_+$, with $\lim_{h\downarrow 0}m(h)=m(0)=0$ such that*

$$\|u_n(t)-u_n(\widetilde{t})\|\leq m(|t-\widetilde{t}|)+\varepsilon_n$$

for each $n\in\mathbb{N}$ and each $t,\widetilde{t}\in[\tau,T]$. Then $\{u_n;\ n\in\mathbb{N}\}$ is equi-uniformly continuous on $[\tau,T]$.

Definition 1.3.2. A subset $\mathfrak{F}\subseteq L^1(\Omega,\mu;X)$ is called *uniformly integrable*[3] if for each $\varepsilon>0$ there exists $\delta(\varepsilon)>0$ such that

$$\int_E \|f(t)\|\,d\mu(t)\leq\varepsilon$$

for each $f\in\mathfrak{F}$ and each $E\in\Sigma$ satisfying $\mu(E)\leq\delta(\varepsilon)$.

Remark 1.3.3. It is easy to see that, whenever $\ell\in L^1(\Omega,\mu)$, the set $F_\ell=\{f\in L^1(\Omega,\mu;X);\ \|f(t)\|\leq\ell(t),\ \text{a.e. for } t\in\Omega\}$ is uniformly integrable.

Problem 1.3.4. *Prove that if $\mathfrak{F}\subseteq L^1(\tau,T;X)$ is uniformly integrable, then it is bounded.*

Theorem 1.3.7. *Let (Ω,Σ,μ) a finite measure space and let X be a reflexive Banach space. Then $\mathfrak{F}\subseteq L^1(\Omega,\mu;X)$ is weakly compact if and only if it is bounded and uniformly integrable.*

See Diestel–Uhl [**83**], Theorem 1, p. 101.

---

[3]Some authors, as for instance Roubíček [**150**], use for this notion the name of *equi-absolutely-continuous* which, in fact, is more adequate. However, for tradition reasons, we prefer the most widely circulated term of uniformly integrable. See, for instance Diestel-Uhl [**83**], Definition 10, p. 74.

Theorem 1.3.8. *Let (Ω, Σ, μ) be a finite measure space and let X be a Banach space. Let $\mathfrak{F} \subseteq L^1(\Omega, \mu; X)$ be bounded and uniformly integrable. If for each $\varepsilon > 0$ there exist a weakly compact subset $C_\varepsilon \subseteq X$ and a measurable subset $E_\varepsilon \in \Sigma$ with $\mu(\Omega \setminus E_\varepsilon) \leq \varepsilon$ and $f(E_\varepsilon) \subseteq C_\varepsilon$ for all $f \in \mathfrak{F}$, then $\mathfrak{F}$ is weakly relatively compact in $L^1(\Omega, \mu; X)$.*

See Diestel [**82**], or Diestel–Uhl [**83**], p. 117.

Corollary 1.3.1. *If $C \subseteq X$ is weakly compact, then*

$$\{f \in L^1(\tau, T; X);\ f(t) \in C \text{ a.e. for } t \in [\tau, T]\}$$

is weakly relatively compact in $L^1(\tau, T; X)$.

1.4. C_0-semigroups

Let X be a Banach space and let $\mathcal{L}(X)$ be the Banach space of all linear bounded operators from X to X endowed with the operator norm $\|\cdot\|_{\mathcal{L}(X)}$, defined by $\|U\|_{\mathcal{L}(X)} = \sup_{\|x\|\leq 1} \|Ux\|$ for each $U \in \mathcal{L}(X)$.

Definition 1.4.1. A family of operators, $\{S(t) : X \to X;\ t \geq 0\}$, in $\mathcal{L}(X)$ is a C_0-*semigroup on* X, if:

(i) $S(0) = I$;
(ii) $S(t+s) = S(t)S(s)$ for each $t, s \geq 0$;
(iii) $\lim_{t \downarrow 0} S(t)x = x$, for each $x \in X$.

A family $\{S(t) : X \to X;\ t \in \mathbb{R}\}$ in $\mathcal{L}(X)$, satisfying (i), (ii) for each $t, s \in \mathbb{R}$ and (iii) with $t \to 0$ instead of $t \downarrow 0$, is called a C_0-*group*.

Definition 1.4.2. By definition, *the infinitesimal generator*, or simply *generator*, of the semigroup of linear operators $\{S(t) : X \to X;\ t \geq 0\}$ is the operator $A : D(A) \subseteq X \to X$, defined by

$$D(A) = \left\{x \in X;\ \exists \lim_{t \downarrow 0} \frac{1}{t}\,(S(t)x - x)\right\} \quad \text{and} \quad Ax = \lim_{t \downarrow 0} \frac{1}{t}\,(S(t)x - x)\,.$$

Equivalently, we say that A *generates* $\{S(t) : X \to X;\ t \geq 0\}$.

Theorem 1.4.1. *If $\{S(t) : X \to X;\ t \geq 0\}$ is a C_0-semigroup, then there exist $M \geq 1$, and $a \in \mathbb{R}$ such that*

$$\|S(t)\|_{\mathcal{L}(X)} \leq Me^{at} \tag{1.4.1}$$

for each $t \geq 0$.

See Vrabie [**175**], Theorem 2.3.1, p. 41.

A C_0-semigroup satisfying (1.4.1) is called of type (M, a). If $M = 1$ and $a = 0$ the C_0-semigroup is called *of contractions* or *of nonexpansive mappings*.

Theorem 1.4.2. *Let $\{S(t) : X \to X;\ t \geq 0\}$ be a C_0-semigroup of type (M, a). Then there exists a norm on X, equivalent to the initial one, such that, with respect to this new norm, the C_0-semigroup is of type $(1, a)$.*

See Vrabie [**175**], Lemma 3.3.1, p. 57.

Theorem 1.4.3. *Let $A : D(A) \subseteq X \to X$ be the infinitesimal generator of a C_0-semigroup $\{S(t) : X \to X;\ t \geq 0\}$. Then:*

(i) *for each $x \in X$ and each $t \geq 0$, we have*

$$\lim_{h \downarrow 0} \frac{1}{h} \int_t^{t+h} S(\tau)x\, d\tau = S(t)x\,;$$

(ii) *for each $x \in X$ and each $t > 0$, we have*

$$\int_0^t S(\tau)x\, d\tau \in D(A) \quad \text{and} \quad A\left(\int_0^t S(\tau)x\, d\tau\right) = S(t)x - x\,;$$

(iii) *for each $x \in D(A)$ and each $t \geq 0$, we have $S(t)x \in D(A)$. In addition, the mapping $t \mapsto S(t)x$ is of class C^1 on $[0, +\infty)$ and satisfies*

$$\frac{d}{dt}\,(S(t)x) = AS(t)x = S(t)Ax\,;$$

(iv) *for each $x \in D(A)$ and each $0 \leq s \leq t < +\infty$, we have*

$$\int_s^t AS(\tau)x\, d\tau = \int_s^t S(\tau)Ax\, d\tau = S(t)x - S(s)x.$$

Theorem 1.4.4. *Let $A : D(A) \subseteq X \to X$ be the infinitesimal generator of a C_0-semigroup $\{S(t) : X \to X;\ t \geq 0\}$. Then $D(A)$ is dense in X, and A is a closed operator.*

If $A : D(A) \subseteq X \to X$ is a linear operator, *the resolvent set* $\rho(A)$ is the set of all numbers λ, called *regular values*, for which the range of $\lambda I - A$, i.e., $R(\lambda I - A) = (\lambda I - A)(D(A))$ is dense in X and $R(\lambda; A) = (\lambda I - A)^{-1}$ is continuous from $R(\lambda I - A)$ to X.

Theorem 1.4.5. *A linear operator $A : D(A) \subseteq X \to X$ is the infinitesimal generator of a C_0-semigroup of contractions if and only if:*

(i) *A is densely defined and closed and*

(ii) *$(0, +\infty) \subseteq \rho(A)$ and, for each $\lambda > 0$, we have*

$$\|R(\lambda; A)\|_{\mathcal{L}(X)} \leq \frac{1}{\lambda}.$$

See Vrabie [**175**], Theorem 3.1.1, p. 51.

Theorem 1.4.6. *The linear operator* $A : D(A) \subseteq X \to X$ *is the infinitesimal generator of a* C_0*-semigroup of type* (M, a) *if and only if*:

(i) A *is densely defined and closed, and*
(ii) $(a, +\infty) \subseteq \rho(A)$ *and, for each* $\lambda > a$ *and each* $n \in \mathbb{N}$, *we have*

$$\|R(\lambda; A)^n\|_{\mathcal{L}(X)} \leq \frac{M}{(\lambda - a)^n}.$$

See Vrabie [**175**], Theorem 3.3.1, p. 56.

Let $J : X \rightsquigarrow X^*$ be the duality mapping on X.

Definition 1.4.3. A linear operator $A : D(A) \subseteq X \to X$ is *dissipative* if for each $x \in X$ there exists $x^* \in J(x)$ such that $(Ax, x^*) \leq 0$.

Theorem 1.4.7. *A linear operator* $A : D(A) \subseteq X \to X$ *is dissipative if and only if, for each* $x \in D(A)$ *and* $\lambda > 0$, *we have*

$$\lambda\|x\| \leq \|(\lambda I - A)x\|. \tag{1.4.2}$$

See Vrabie [**175**], Theorem 3.4.1, p. 59.

Theorem 1.4.8. *Let* $A : D(A) \subseteq X \to X$ *be a densely defined linear operator. Then* A *generates a* C_0*-semigroup of contractions on* X *if and only if*

(i) A *is dissipative, and*
(ii) *there exists* $\lambda > 0$ *such that* $\lambda I - A$ *is surjective.*

Moreover, if A *generates a* C_0*-semigroup of contractions, then* $\lambda I - A$ *is surjective for any* $\lambda > 0$, *and we have* $(Ax, x^*) \leq 0$ *for each* $x \in D(A)$ *and each* $x^* \in J(x)$.

See Vrabie [**175**], Theorem 3.4.2, p. 60.

1.5. Mild solutions

In this section we include some facts referring to the relationship between C_0-semigroups and ordinary differential equations in Banach spaces.

First, let us observe that, from (iii) in Theorem 1.4.3, it follows that, if $A : D(A) \subseteq X \to X$ is the infinitesimal generator of a C_0-semigroup $\{S(t) : X \to X ;\ t \geq 0\}$, then, for each $\tau \geq 0$ and $\xi \in D(A)$, the function $u : [\tau, +\infty) \to X$, defined by $u(t) = S(t - \tau)\xi$ for each $t \geq \tau$, is the unique C^1-solution of the homogeneous abstract Cauchy problem

$$\begin{cases} u'(t) = Au(t) \\ u(\tau) = \xi. \end{cases} \tag{1.5.1}$$

For this reason, it is quite natural to consider that, for each $\xi \in X$, the function u, defined as above, is a solution for (1.5.1) in a generalized sense.

The aim of this section is to extend this concept of generalized solution to the nonhomogeneous problem

$$\begin{cases} u'(t) = Au(t) + f(t) \\ u(\tau) = \xi, \end{cases} \tag{1.5.2}$$

where A is as before, $\xi \in X$ and $f \in L^1(\tau, T; X)$.

Definition 1.5.1. The function $u : [\tau, T] \to X$ is called *classical or C^1-solution* of the problem (1.5.2), if u is continuous on $[\tau, T]$, continuously differentiable on $(\tau, T]$, $u(t) \in D(A)$ for each $t \in (\tau, T]$ and it satisfies $u'(t) = Au(t) + f(t)$ for each $t \in (\tau, T]$ and $u(\tau) = \xi$.

Definition 1.5.2. The function $u : [\tau, T] \to X$ is called *absolutely continuous, or strong solution*, of the problem (1.5.2), if u is absolutely continuous on $[\tau, T]$, $u' \in L^1(\tau, T; X)$, $u(t) \in D(A)$ a.e. for $t \in (\tau, T)$, and it satisfies $u'(t) = Au(t) + f(t)$ a.e. for $t \in (\tau, T)$ and $u(\tau) = \xi$.

Remark 1.5.1. Each classical solution of (1.5.2) is a strong solution of the same problem, but not conversely.

The next result, known as the *Duhamel Principle*, is fundamental in understanding how to extend the concept of generalized solution to nonhomogeneous problems of the form (1.5.2).

Theorem 1.5.1. *Each strong solution of* (1.5.2) *is given by so-called variation of constants formula*

$$u(t) = S(t - \tau)\xi + \int_\tau^t S(t - s)f(s)\, ds, \tag{1.5.3}$$

for each $t \in [\tau, T]$. *In particular, each classical solution of the problem* (1.5.2) *is given by* (1.5.3).

See Vrabie [**175**], Theorem 8.1.1, p. 184.

Simple examples show that, when X is infinite-dimensional and A is unbounded, the problem (1.5.2) may fail to have any strong solution, no matter how regular the datum f is. See Vrabie [**175**], Example 8.1.1, p. 185. This observation justifies why, in the case of infinite-dimensional spaces X, the variation of constants formula can be promoted to the rank of definition. Namely, we introduce

Definition 1.5.3. The function $u : [\tau, T] \to X$, defined by (1.5.3), is called *mild solution* of the problem (1.5.2) on $[\tau, T]$.

We will next recall a necessary and sufficient condition in order that a given set of mild solutions be relatively compact in $C([\tau, T]; X)$.

Definition 1.5.4. The operator $Q : X \times L^1(\tau,T;X) \to C([\tau,T];X)$, defined by $Q(\xi,f) = u$, where u is the unique mild solution of the problem (1.5.2), corresponding to $\xi \in X$ and $f \in L^1(\tau,T;X)$, is called *the mild solution operator* attached to the problem (1.5.2).

Remark 1.5.2. The operator Q is Lipschitz with constant $Me^{|a|(T-\tau)}$, where $M \geq 1$ and $a \in \mathbb{R}$ are given by Theorem 1.4.1, and therefore it maps bounded subsets in $X \times L^1(\tau,T;X)$ into bounded subsets in $C([\tau,T];X)$.

Theorem 1.5.2. *Let $A : D(A) \subseteq X \to X$ be the generator of a C_0-semigroup $\{S(t) : X \to X;\ t \geq 0\}$, let $\mathfrak{D}$ be a bounded subset in X, and $\mathfrak{F}$ a uniformly integrable subset in $L^1(\tau,T;X)$. Then $Q(\mathfrak{D},\mathfrak{F})$ is relatively compact in $C([\theta,T];X)$ for each $\theta \in (\tau,T)$, if and only if there exists a dense subset D in $[\tau,T]$ such that, for each $t \in D$, the t-section of $Q(\mathfrak{D},\mathfrak{F})$, i.e., $Q(\mathfrak{D},\mathfrak{F})(t) = \{Q(\xi,f)(t)\,;\ (\xi,f) \in \mathfrak{D} \times \mathfrak{F}\}$, is relatively compact in X. Moreover, if the latter condition is satisfied and $\tau \in D$, then $Q(\mathfrak{D},\mathfrak{F})$ is relatively compact even in $C([\tau,T];X)$.*

See Vrabie [**175**], Theorem 8.4.1, p. 194.

Definition 1.5.5. The C_0-semigroup $\{S(t) : X \to X\,;\ t \geq 0\}$ is called *compact* if for each $t > 0$, $S(t)$ is a compact operator.

A very useful consequence of Theorem 1.5.2 is the following sufficient condition of relative compactness of the set $Q(\mathfrak{D},\mathfrak{F})$ in $C([\tau,T];X)$.

Theorem 1.5.3. *Let $A : D(A) \subseteq X \to X$ be the generator of a compact C_0-semigroup, let $\xi \in X$, $\mathfrak{D} = \{\xi\}$ and let $\mathfrak{F}$ be a uniformly integrable subset in $L^1(\tau,T;X)$. Then $Q(\mathfrak{D},\mathfrak{F})$ is relatively compact in $C([\tau,T];X)$.*

See Vrabie [**175**], Theorem 8.4.2, p. 196.

We conclude with a useful extension of Lemma 1.3.1.

Lemma 1.5.1. *Let $K \subseteq X$ be compact, let $\mathfrak{F}$ be a family of Lebesgue integrable functions from $[\tau,T]$ to K and let $\{S(t) : X \to X;\ t \geq 0\}$ be a C_0-semigroup on X. Then, for every $t \in [\tau,T]$, the set*

$$\left\{\int_\tau^t S(t-s)f(s)\,ds\,;\, f \in \mathfrak{F}\right\}$$

is relatively compact in X.

Proof. Since $(s,x) \mapsto S(s)x$ is continuous from $\mathbb{R}_+ \times X$ to X and $[0,t-\tau] \times K$ is compact, it follows that $\{S(s)x\,;\ (s,x) \in [0,t-\tau] \times K\}$ is compact. The conclusion follows from the simple observation that

$$\left\{\int_\tau^t S(t-s)f(s)ds\,;\, f \in \mathfrak{F}\right\} \subseteq \frac{1}{t-\tau}\overline{\operatorname{conv}}\{S(s)x\,;\ (s,x) \in [0,T-\tau] \times K\}$$

while, by Theorem 1.3.2, the latter set is compact. □

1.6. Evolutions governed by m-dissipative operators

Let X be a Banach space and let $\|\cdot\|$ be the norm on X. If $x, y \in X$, we denote by $[x, y]_+$ the *right directional derivative of the norm* calculated at x in the direction y, i.e.

$$[x, y]_+ = \lim_{h \downarrow 0} \frac{\|x + hy\| - \|x\|}{h}$$

and by $(x, y)_+$ the *right directional derivative of* $\frac{1}{2}\|\cdot\|^2$ calculated at x in the direction y, i.e.

$$(x, y)_+ = \lim_{h \downarrow 0} \frac{\|x + hy\|^2 - \|x\|^2}{2h}.$$

Analogously, we denote by $[x, y]_-$ the *left directional derivative of the norm* calculated at x in the direction y, i.e.

$$[x, y]_- = \lim_{h \uparrow 0} \frac{\|x + hy\| - \|x\|}{h}$$

and by $(x, y)_-$ the *left directional derivative of* $\frac{1}{2}\|\cdot\|^2$ calculated at x in the direction y, i.e.

$$(x, y)_- = \lim_{h \uparrow 0} \frac{\|x + hy\|^2 - \|x\|^2}{2h}.$$

Exercise 1.6.1. *Show that:*

(i) $(x, y)_\pm = \|x\| [x, y]_\pm$;

(ii) $|[x, y]_\pm| \leq \|y\|$;

(iii) $|[x, y]_\pm - [x, z]_\pm| \leq \|y - z\|$;

(iv) $[x, y]_+ = -[-x, y]_- = -[x, -y]_-$;

(v) $[ax, by]_\pm = b[x, y]_\pm$ *for* $a, b > 0$;

(vi) $[x, y+z]_+ \leq [x, y]_+ + [x, z]_+$ *and* $[x, y+z]_- \geq [x, y]_- + [x, z]_-$;

(vii) $[x, y+z]_+ \geq [x, y]_+ + [x, z]_-$ *and* $[x, y+z]_- \leq [x, y]_- + [x, z]_+$;

(viii) $[x, y + ax]_\pm = [x, y]_\pm + a\|x\|$ *for* $a \in \mathbb{R}$;

(ix) *if* $u : [\tau, T] \to X$ *is differentiable from the right at* $t \in [\tau, T)$ *(differentiable from the left at* $t \in (\tau, T]$*), then both* $s \mapsto \|u(s)\|$ *and* $s \mapsto \|u(s)\|^2$ *are differentiable from the right (left) at* t *and*

$$\tfrac{d^\pm}{dt}(\|u(\cdot)\|)(t) = [u(t), u'_\pm(t)]_\pm$$

$$\tfrac{d^\pm}{dt}(\|u(\cdot)\|^2)(t) = 2(u(t), u'_\pm(t))_\pm,$$

where $u'_\pm(t) = \frac{d^\pm}{dt}(u(\cdot))(t)$;

(x) $\|\cdot\|$ *is Gâteaux differentiable at each* $x \in X$, $x \neq 0$, *if and only if, for each* $x \in X \setminus \{0\}$ *and each* $y \in X$, *we have*

$$[x, y]_+ = -[-x, y]_+.$$

We recall that $J : X \rightsquigarrow X^*$ denotes the duality mapping on X.

Proposition 1.6.1. *For each* $x, y \in X$, *we have:*

(i) *there exists* $x_+^* \in J(x)$ *such that*

$$\|x\|[x, y]_+ = \sup\{(y, x^*);\ x^* \in J(x)\} = (y, x_+^*);$$

(ii) *there exists* $x_-^* \in J(x)$ *such that*

$$\|x\|[x, y]_- = \inf\{(y, x^*);\ x^* \in J(x)\} = (y, x_-^*).$$

See Miyadera [**129**], Theorem 2.5, p. 16.

If $A : X \rightsquigarrow X$, we say that A is an *operator* or *multi-function* and we write $A : D(A) \subseteq X \rightsquigarrow X$, to signify that $A(x) \neq \emptyset$ if and only if $x \in D(A)$. For simplicity, for each $x \in D(A)$, we denote $Ax = A(x)$. Whenever A is single-valued on $D(A)$, we shall identify A with a function defined on $D(A)$, i.e., with its unique selection and we shall write $A : D(A) \subseteq X \to X$ and $Ax = y$ instead of $A : D(A) \subseteq X \rightsquigarrow X$ and of $Ax = \{y\}$. Obviously, each function $f : D(f) \subseteq X \to X$ can be identified with a single-valued operator whose domain is $D(f)$. If $A : D(A) \subseteq X \rightsquigarrow X$ is an operator, then $R(A) = \cup_{x \in D(A)} Ax$. If A and B are operators and $\lambda \in \mathbb{R}$, then $R(A)$, A^{-1}, $A + B$, AB and λA are defined in the usual sense of relations in $X \times X$.

We say that the operator $A : D(A) \subseteq X \rightsquigarrow X$ is *dissipative* if

$$(x_1 - x_2, y_1 - y_2)_- \leq 0$$

for each $x_i \in D(A)$ and $y_i \in Ax_i$, $i = 1, 2$, and *m-dissipative* if it is dissipative and, for each $\lambda > 0$, or equivalently for some $\lambda > 0$, $R(I - \lambda A) = X$.

Let $A : D(A) \subseteq X \rightsquigarrow X$ be an m-dissipative operator, let $\xi \in \overline{D(A)}$, $f \in L^1(\tau, T; X)$ and let us consider the differential equation

$$u'(t) \in Au(t) + f(t). \tag{1.6.1}$$

Definition 1.6.1. A function $u : [\tau, T] \to X$ is called a *strong solution* of (1.6.1) on $[\tau, T]$ if:

(S_1) $u(t) \in D(A)$ a.e. for $t \in (\tau, T)$;

(S_2) $u(t) \in W^{1,1}_{\text{loc}}((\tau, T]; X)$ and there exists $g \in L^1_{\text{loc}}((\tau, T]; X)$, such that

$$\begin{cases} g(t) \in Au(t) & \text{a.e. for } t \in (\tau, T) \\ u'(t) = g(t) + f(t) & \text{a.e. for } t \in (\tau, T). \end{cases}$$

A *strong solution* of (1.6.1) on $[\tau, T)$ is a function, $u : [\tau, T) \to X$, whose restriction to each interval of the form $[\tau, T_0]$, with $\tau < T_0 < T$, is a strong solution of (1.6.1) on $[\tau, T_0]$.

Since whenever X is infinite dimensional, the problem (1.6.1) may have no strong solution, another general concept was introduced. Namely:

Definition 1.6.2. A C^0-*solution* of the problem (1.6.1) is a function u in $C([\tau, T]; X)$ satisfying: for each $\tau < c < T$ and $\varepsilon > 0$ there exist

(i) $\tau = t_0 < t_1 < \cdots < c \leq t_n < T$, $t_k - t_{k-1} \leq \varepsilon$ for $k = 1, 2, \ldots, n$;

(ii) $f_1, \ldots, f_n \in X$ with $\displaystyle\sum_{k=1}^{n} \int_{t_{k-1}}^{t_k} \|f(t) - f_k\| \, dt \leq \varepsilon$;

(iii) $v_0, \ldots, v_n \in X$ satisfying:
$\dfrac{v_k - v_{k-1}}{t_k - t_{k-1}} \in Av_k + f_k$ for $k = 1, 2, \ldots, n$ and such that
$\|u(t) - v_k\| \leq \varepsilon$ for $t \in [t_{k-1}, t_k)$, $k = 1, 2, \ldots, n$.

A function $v_\varepsilon : [\tau, t_n] \to D(A)$, defined by $v_\varepsilon(t) = v_k$ for $t \in [t_{k-1}, t_k)$, $k = 1, 2, \ldots, n$, where t_k, v_k and f_k, for $k = 1, 2, \ldots, n$, are as above, is called an ε-*difference scheme-solution*, or briefly, ε-*DS-solution*.

Theorem 1.6.1. *Let X be a Banach space and let $A : D(A) \subseteq X \rightsquigarrow X$ be m-dissipative. Then, for each $\xi \in \overline{D(A)}$ and $f \in L^1(\tau, T; X)$, there exists a unique C^0-solution $u : [\tau, T] \to \overline{D(A)}$, of* (1.6.1), *which satisfies $u(\tau) = \xi$.*

See Lakshmikantham-Leela [**119**], Theorem 3.6.1, p. 116.

Definition 1.6.3. Let C be a nonempty and closed subset in X. A family $\{S(t) : C \to C;\ t \geq 0\}$ is a *semigroup of nonexpansive mappings* or *semigroup of contractions* on C, if:

(i) $S(0) = I$;

(ii) $S(t + s) = S(t)S(s)$ for all $t, s \in [0, \infty)$;

(iii) for each $\xi \in C$ the function $s \mapsto S(s)\xi$ is continuous at $s = 0$;

(iv) $\|S(t)\xi - S(t)\eta\| \leq \|\xi - \eta\|$ for all $t \geq 0$ and $\xi, \eta \in C$.

Exercise 1.6.2. *Prove that if $\{S(t) : C \to C;\ t \geq 0\}$ is a semigroup of nonexpansive mappings, then the mapping $(t, \xi) \mapsto S(t)\xi$ is continuous from $\mathbb{R}_+ \times C$ to C.*

Remark 1.6.1. In the case in which A is single-valued, linear and, of course, m-dissipative, u is a C^0-solution of (1.6.1) on $[\tau, T]$ in the sense of Definition 1.6.2 if and only if u is a mild solution on $[\tau, T]$ in the sense of Definition 1.5.3. This is an easy consequence of the fact that in the linear case, each mild solution can be approximated uniformly with strong solutions of some suitably chosen approximate problems.

See Vrabie [**173**], Theorem 1.8.2, p. 29.

We denote by $u(\cdot,\tau,\xi,f):[\tau,T]\to\overline{D(A)}$ the unique C^0-solution of (1.6.1) satisfying $u(\tau,\tau,\xi,f)=\xi$.

Theorem 1.6.2. *Let X be a Banach space, let $A:D(A)\subseteq X\rightsquigarrow X$ be m-dissipative, let $\xi,\eta\in\overline{D(A)}$, $f,g\in L^1(\tau,T;X)$ and let $\widetilde{u}=u(\cdot,\tau,\xi,f)$ and $\widetilde{v}=u(\cdot,\tau,\eta,g)$. We have*

$$\|\widetilde{u}(t)-\widetilde{v}(t)\|\leq\|\xi-\eta\|+\int_\tau^t[\,\widetilde{u}(s)-\widetilde{v}(s),f(s)-g(s)\,]_+\,ds \tag{1.6.2}$$

and

$$\|\widetilde{u}(t)-\widetilde{v}(t)\|^2\leq\|\xi-\eta\|^2+2\int_\tau^t(\widetilde{u}(s)-\widetilde{v}(s),f(s)-g(s))_+\,ds, \tag{1.6.3}$$

for each $t\in[\,\tau,T\,]$. Moreover, for each $\tau\leq\nu\leq t\leq T$, we have

$$u(t,\tau,\xi,f)=u(t,\nu,u(\nu,\tau,\xi,f),f\,|_{[\nu,T\,]}). \tag{1.6.4}$$

See Vrabie [**173**], Section 1.7.

The relation (1.6.4) is known under the name of *evolution property.*

Exercise 1.6.3. *Prove that $\{S(t):\overline{D(A)}\to\overline{D(A)},\ t\geq0\}$, where, for each $t\geq0$ and $\xi\in\overline{D(A)}$, $S(t)\xi=u(t,0,\xi,0)$, is a semigroup of nonexpansive mappings (called the semigroup of nonexpansive mappings generated by A on $\overline{D(A)}$).*

Exercise 1.6.4. *If $A:D(A)\subseteq X\rightsquigarrow X$ is m-dissipative, $\xi,\eta\in\overline{D(A)}$ and $f,g\in L^1(\tau,T;X)$, then $\widetilde{u}=u(\cdot,\tau,\xi,f)$ and $\widetilde{v}=u(\cdot,\tau,\eta,g)$ satisfy*

$$\|\widetilde{u}(t)-\widetilde{v}(t)\|\leq\|\xi-\eta\|+\int_\tau^t\|f(s)-g(s)\|\,ds, \tag{1.6.5}$$

for each $t\in[\,\tau,T\,]$.

Theorem 1.6.3. *Let X be a Banach space, let $A:D(A)\subseteq X\rightsquigarrow X$ be m-dissipative, $\xi\in\overline{D(A)}$ and $f\in L^1(\tau,T;X)$. Then $\widetilde{u}:[\,\tau,T\,]\to\overline{D(A)}$ coincides with $u(\cdot,\tau,\xi,f)$ if and only if it is continuous and*

$$\|\widetilde{u}(t)-x\|^2\leq\|\widetilde{u}(s)-x\|^2+2\int_s^t(\widetilde{u}(\theta)-x,f(\theta)+y)_+\,d\theta$$

for each $x\in D(A)$, each $y\in Ax$ and each $\tau\leq s\leq t\leq T$.

See, for instance, Lakshmikantham-Leela [**119**], Theorem 3.5.1, p. 104 and Miyadera [**128**], Theorem 5.18, p. 157.

Definition 1.6.4. The semigroup $\{S(t) : \overline{D(A)} \to \overline{D(A)},\ t \geq 0\}$ is *compact* if, for each $t > 0$, $S(t)$ is a compact operator.

Problem 1.6.1. *If X is a Banach space and C is a nonempty subset of X for which there exists a family $\{S(t) : C \to X\,;\, t > 0\}$ of compact operators such that*

$$\lim_{t \downarrow 0} S(t)\xi = \xi$$

for each $\xi \in C$, then C is separable. As a consequence, if the semigroup $\{S(t) : \overline{D(A)} \to \overline{D(A)},\ t \geq 0\}$ is compact then $\overline{D(A)}$ is separable.

We have the following two compactness results.

Theorem 1.6.4. *Let X be a Banach space, let $A : D(A) \subseteq X \rightsquigarrow X$ be an m-dissipative operator, $\xi \in \overline{D(A)}$ and G a uniformly integrable subset in $L^1(\tau, T; X)$. Then the following conditions are equivalent:*

(i) *the set $\{u(\cdot, \tau, \xi, g)\,;\ g \in G\}$ is relatively compact in $C([\,\tau, T\,]; X)$;*
(ii) *there exists a dense subset E in $[\,\tau, T\,]$ such that, for each $t \in E$, $\{u(t, \tau, \xi, g)\,;\ g \in G\}$ is relatively compact in X.*

See Vrabie [**172**] or Vrabie [**173**], Theorem 2.3.1, p. 45.
A very useful consequence of Theorem 1.6.4 is:

Theorem 1.6.5. *Let X be a Banach space, let $A : D(A) \subseteq X \rightsquigarrow X$ be an m-dissipative operator and let us assume that A generates a compact semigroup. Let $\xi \in \overline{D(A)}$ and let G be uniformly integrable in $L^1(\tau, T; X)$. Then the set $\{u(\cdot, \tau, \xi, g)\,;\ g \in G\}$ is relatively compact in $C([\,\tau, T\,]; X)$.*

See Baras [**8**] or Vrabie [**173**], Theorem 2.3.3, p. 47.
An extension of Theorem 1.6.5 is stated below.

Theorem 1.6.6. *Let X be a Banach space, let $A : D(A) \subseteq X \rightsquigarrow X$ be an m-dissipative operator and let us assume that A generates a compact semigroup. Let $B \subseteq \overline{D(A)}$ be bounded and let G be uniformly integrable in $L^1(\tau, T; X)$. Then, for each $\delta \in (\tau, T)$, the set $\{u(\cdot, \tau, \xi, g)\,;\ (\xi, g) \in B \times G\}$ is relatively compact in $C([\,\delta, T\,]; X)$. If, in addition, B is relatively compact, then $\{u(\cdot, \tau, \xi, g)\,;\ (\xi, g) \in B \times G\}$ is relatively compact even in $C([\,\tau, T\,]; X)$.*

See Vrabie [**173**], Theorems 2.3.2 and 2.3.3, pp. 46–47.

1.7. Examples of m-dissipative operators

To fix the idea, let us first recall some notations. If Ω is a nonempty and open subset in $\mathbb{R}^n$ with boundary Γ, we denote by $C_0^\infty(\Omega)$ the space of C^∞- real functions with compact support in Ω. Further, if $1 \leq p < \infty$

and $m \in \mathbb{N}$, $W^{m,p}(\Omega)$ denotes the space of all functions $u : \Omega \to \mathbb{R}$ which, together with their partial derivatives up to the order m, in the sense of distributions over Ω, belong to $L^p(\Omega)$. Endowed with the norm

$$\|u\|_{W^{m,p}(\Omega)} = \left(\sum_{0 \le |\kappa| \le m} \|\mathcal{D}^\kappa u\|^p_{L^p(\Omega)} \right)^{1/p},$$

$W^{m,p}(\Omega)$ is a separable real Banach space, densely and continuously imbedded in $L^p(\Omega)$. Here, as usual, if $\kappa = (\kappa_1, \kappa_2, \ldots, \kappa_n)$ is a multi-index, we denote by

$$\mathcal{D}^\kappa u = \frac{\partial^{\kappa_1 + \kappa_2 + \cdots + \kappa_n} u}{\partial x_1^{\kappa_1} \partial x_2^{\kappa_2} \ldots \partial x_n^{\kappa_n}},$$

where the partial derivatives are in sense of distributions over Ω. We denote by $W_0^{m,p}(\Omega)$ the closure of $C_0^\infty(\Omega)$ in $W^{m,p}(\Omega)$, by $H^1(\Omega) = W^{1,2}(\Omega)$, $H_0^1(\Omega) = W_0^{1,2}(\Omega)$, $H^{-1}(\Omega) = [\,H_0^1(\Omega)\,]^*$ and $H^2(\Omega) = W^{2,2}(\Omega)$.

Finally, we make the conventional notation $W^{0,p}(\Omega) = L^p(\Omega)$.

Theorem 1.7.1. *Let Ω be a nonempty, open and bounded subset in $\mathbb{R}^n$ whose boundary Γ is of class C^1. Let $m \in \mathbb{N}$ and $p, q \in [\,1, \infty)$.*

(i) *If $mp < n$ and $q < \frac{np}{n-mp}$, then $W^{m,p}(\Omega)$ is compactly imbedded in $L^q(\Omega)$.*
(ii) *If $mp = n$ and $q \in [\,1, \infty)$, then $W^{m,p}(\Omega)$ is compactly imbedded in $L^q(\Omega)$.*
(iii) *If $mp > n$, then $W^{m,p}(\Omega)$ is compactly imbedded in $C(\overline{\Omega})$.*

Example 1.7.1. *The Laplace operator in $L^2(\Omega)$.* Let Ω be a nonempty and open subset in $\mathbb{R}^n$, let $X = L^2(\Omega)$, and let us consider the operator A on X, defined by:

$$\begin{cases} D(A) = \{u \in H_0^1(\Omega);\ \Delta u \in L^2(\Omega)\} \\ Au = \Delta u, \quad \text{for each} \quad u \in D(A). \end{cases}$$

Theorem 1.7.2. *The Laplace operator Δ with homogeneous Dirichlet boundary conditions on $L^2(\Omega)$, i.e., the linear operator A, defined above, is the infinitesimal generator of a C_0-semigroup of contractions. If Ω is bounded with C^1 boundary, then the C_0-semigroup generated by A on $L^2(\Omega)$ is compact.*

See Vrabie [**175**], Theorem 4.1.2, p. 79.

Example 1.7.2. *The Laplace operator in $L^1(\Omega)$.* Let Ω be a nonempty, bounded and open subset in $\mathbb{R}^n$ with C^2 boundary Γ, let $X = L^1(\Omega)$, and let us consider the operator A on X, defined by:

$$\begin{cases} D(A) = \{u \in W_0^{1,1}(\Omega);\ \Delta u \in L^1(\Omega)\} \\ Au = \Delta u, \quad \text{for each} \quad u \in D(A). \end{cases}$$

Theorem 1.7.3. *The Laplace operator Δ with homogeneous Dirichlet boundary conditions on $L^1(\Omega)$, i.e., the linear operator A, defined above, is the infinitesimal generator of a compact C_0-semigroup of contractions on $L^1(\Omega)$.*

See Vrabie [**175**], Theorem 7.2.7, p. 160 and Remark 4.1.3, p. 82.

We also mention the following consequence of the *maximum principle for elliptic equations.*

Theorem 1.7.4. *Let Ω be a nonempty, bounded and open subset in $\mathbb{R}^n$ with C^2 boundary Γ, let $f \in L^1(\Omega)$, $\lambda \geq 0$ and let u be the unique solution of the elliptic problem*

$$\begin{cases} \lambda u - \Delta u = f & \text{in } \Omega \\ u = 0 & \text{on } \Gamma. \end{cases}$$

If $f(x) \geq 0$ a.e. for $x \in \Omega$, then $u(x) \geq 0$ a.e. for $x \in \Omega$.

See Protter–Weinberger [**148**], Theorem 6, p. 64.

We also need the following corollary of the *maximum principle for parabolic equations.*

Theorem 1.7.5. *Let Ω be a nonempty, bounded and open subset in $\mathbb{R}^n$ with C^2 boundary Γ and let $\{S(t) : L^1(\Omega) \to L^1(\Omega),\ t \geq 0\}$ be the C_0-semigroup generated by the Laplace operator with homogeneous Dirichlet boundary conditions on $L^1(\Omega)$. If $\eta_1, \eta_2 \in L^1(\Omega)$ satisfy $\eta_1(x) \leq \eta_2(x)$ a.e. for $x \in \Omega$, then, for each $t \geq 0$, we have $[S(t)\eta_1](x) \leq [S(t)\eta_2](x)$ a.e. for $x \in \Omega$.*

This is a consequence of Protter–Weinberger [**148**], Theorem 5, p. 173.

Example 1.7.3. Let $X = L^p(\mathbb{R}^n)$, with $1 \leq p < \infty$, and let $a \in \mathbb{R}^n$. Let us define the operator $A : D(A) \subseteq X \to X$ by

$$\begin{cases} D(A) = \{u \in X;\ a\nabla u \in X\} \\ Au = a\nabla u = \displaystyle\sum_{i=1}^{n} a_i \frac{\partial u}{\partial x_i}, \quad \text{for } u \in D(A), \end{cases}$$

where the partial derivatives are in the sense of distributions over $\mathbb{R}^n$.

Theorem 1.7.6. *The operator A defined as above is the infinitesimal generator of the C_0-group of isometries $\{G(t) : X \to X;\ t \in \mathbb{R}\}$, given by*

$$[G(t)f](x) = f(x + ta)$$

for each $f \in X$, $t \in \mathbb{R}$ and $x \in \mathbb{R}^n$.

See Vrabie [**175**], Theorem 4.4.1, p. 88.

Example 1.7.4. As before, the operator Δ is, in this example, the Laplace operator in the sense of distributions over Ω. If $\varphi : D(\varphi) \subseteq \mathbb{R} \rightsquigarrow \mathbb{R}$, and $u : \Omega \to D(\varphi)$, we denote by

$$\mathcal{S}_\varphi(u) = \{v \in L^1(\Omega);\ v(x) \in \varphi(u(x)),\ \text{a.e. for } x \in \Omega\}.$$

We say that $\varphi : D(\varphi) \subseteq \mathbb{R} \rightsquigarrow \mathbb{R}$ is *maximal monotone*[4] if $-\varphi$ is m-dissipative.

Theorem 1.7.7. *Let Ω be a nonempty, bounded and open subset in $\mathbb{R}^n$ with C^1 boundary Γ and let $\varphi : D(\varphi) \subseteq \mathbb{R} \rightsquigarrow \mathbb{R}$ be maximal monotone with $0 \in \varphi(0)$. Then the operator $\Delta\varphi : D(\Delta\varphi) \subseteq L^1(\Omega) \rightsquigarrow L^1(\Omega)$, defined by*

$$\begin{cases} D(\Delta\varphi) = \{u \in L^1(\Omega);\ \exists v \in \mathcal{S}_\varphi(u) \cap W_0^{1,1}(\Omega),\ \Delta v \in L^1(\Omega)\} \\ \Delta\varphi(u) = \{\Delta v;\ v \in \mathcal{S}_\varphi(u) \cap W_0^{1,1}(\Omega)\} \cap L^1(\Omega) \text{ for } u \in D(\Delta\varphi), \end{cases}$$

is m-dissipative on $L^1(\Omega)$. If, in addition, $\varphi : \mathbb{R} \to \mathbb{R}$ is continuous on $\mathbb{R}$ and C^1 on $\mathbb{R} \setminus \{0\}$ and there exist two constants $C > 0$ and $a > 0$ if $n \le 2$ and $a > (n-2)/n$ if $n \ge 3$ such that

$$\varphi'(r) \ge C|r|^{a-1}$$

for each $r \in \mathbb{R} \setminus \{0\}$, then $\Delta\varphi$ generates a compact semigroup.

See Vrabie [**173**], Theorem 2.7.1, p. 70 and Lemma 2.7.2, p. 71.

Theorem 1.7.8. *In the hypotheses of Theorem 1.7.7, if $q > 1$ is such that $L^q(\Omega) \subseteq H^{-1}(\Omega)$, then, for each arbitrary but fixed $\xi \in L^q(\Omega)$, the mapping $f \mapsto u(\cdot, \tau, \xi, f)$ is weakly-strongly sequentially continuous from $L^1(\tau, T; L^q(\Omega))$ to $C([\tau, T]; L^1(\Omega))$.*

See Diaz–Vrabie [**80**].
From Theorem 1.7.8, we deduce

Theorem 1.7.9. *In the hypotheses of Theorem 1.7.7, for each arbitrary but fixed $\xi \in L^1(\Omega)$, the mapping $f \mapsto u(\cdot, \tau, \xi, f)$ is weakly-strongly sequentially continuous from $L^1(\tau, T; L^1(\Omega))$ to $C([\tau, T]; L^1(\Omega))$.*

---

[4]The name comes from the property that, in the case of a Hilbert space H, an operator $A : D(A) \subseteq H \rightsquigarrow H$, with $-A$ dissipative, is called *monotone* and an operator B is m-dissipative if and only if it is *maximal dissipative*, i.e., if its graph is not strictly contained in the graph of another dissipative operator.

Proof. Let $\xi \in L^1(\Omega)$, let $f \in L^1(\tau, T; L^1(\Omega))$ and let $(f_n)_n$ be a sequence in $L^1(\tau, T; L^1(\Omega))$ such that $\lim_n f_n = f$ weakly in $L^1(\tau, T; L^1(\Omega))$. As, by Fubini Theorem 1.2.5, $L^1([\tau, T] \times \Omega) = L^1(\tau, T; L^1(\Omega))$, we have that $\lim_n f_n = f$ weakly in $L^1([\tau, T] \times \Omega)$. Let $k \in \mathbb{N}$ be arbitrary but fixed and let us define $P_k : L^1([\tau, T] \times \Omega) \to L^1([\tau, T] \times \Omega)$ by

$$P_k(g)(t,x) = \begin{cases} g(t,x) & \text{if } |g(t,x)| \le k \\ 0 & \text{if } |g(t,x)| > k \end{cases}$$

for each $g \in L^1([\tau, T] \times \Omega)$. Clearly, $(f_n)_n$ is bounded in $L^1([\tau, T] \times \Omega)$, say by $M > 0$. Throughout this proof, we denote by $\|\cdot\|_{L^1}$ the norm of $L^1([\tau, T] \times \Omega)$ and by $\|\cdot\|_{L^\infty}$ the norm of $L^\infty([\tau, T] \times \Omega)$. Since

$$k\mu(\{(s,y) \in [\tau, T] \times \Omega;\ |f_n(s,y)| > k\}) \le \int_{|f_n(t,x)|>k} |f_n(s,y)|ds\,dy \le M,$$

we get

$$\mu(\{(s,y) \in [\tau, T] \times \Omega;\ |f_n(s,y)| > k\}) \le \frac{M}{k}$$

for $n, k = 1, 2, \ldots$. Further, since

$$\|P_k f_n - f_n\|_{L^1} = \int_{|f_n(t,x)|>k} |f_n(s,y)|ds\,dy \tag{1.7.1}$$

for each $k, n \in \mathbb{N}$ and, by Theorem 1.3.7, $\{f_n;\ n \in \mathbb{N}\}$ is uniformly integrable, from (1.7.1) we deduce

$$\lim_k P_k f_n = f_n \tag{1.7.2}$$

strongly in $L^1([\tau, T] \times \Omega)$, uniformly for $n = 1, 2, \ldots$. Since $\lim_n f_n = f$, weakly in $L^1([\tau, T] \times \Omega)$, from (1.7.2), it follows that, for each arbitrary but fixed element g in the dual of $L^1([\tau, T] \times \Omega)$, i.e., $g \in L^\infty([\tau, T] \times \Omega)$, we have

$$\lim_{\substack{n\to\infty \\ k\to\infty}} |(P_k f_n - P_k f, g)| = 0. \tag{1.7.3}$$

Indeed, let us observe that

$$\begin{aligned} |(P_k f_n - P_k f, g)| &\le |(P_k f_n - f_n, g)| + |(f_n - f, g)| + |(f - P_k f, g)| \\ &\le [\, \|P_k f_n - f_n\|_{L^1} + \|f - P_k f\|_{L^1}]\|g\|_{L^\infty} + |(f_n - f, g)|, \end{aligned}$$

and thus (1.7.2) and $\lim_n f_n = f$, weakly in $L^1([\tau, T] \times \Omega)$, imply (1.7.3). Next, take $(\xi_p)_p$ in $L^q(\Omega)$ with $\lim_p \xi_p = \xi$ strongly in $L^1(\Omega)$. We have

$$\begin{aligned} &\|u(t,\tau,\xi,f_n) - u(t,\tau,\xi,f)\| \le \|u(t,\tau,\xi,f_n) - u(t,\tau,\xi_p,f_n)\| \\ &+\|u(t,\tau,\xi_p,f_n) - u(t,\tau,\xi_p,P_k f_n)\| + \|u(t,\tau,\xi_p,P_k f_n) - u(t,\tau,\xi_p,P_k f)\| \\ &+\|u(t,\tau,\xi_p,P_k f) - u(t,\tau,\xi_p,f)\| + \|u(t,\tau,\xi_p,f) - u(t,\tau,\xi,f)\| \\ &\le 2\|\xi - \xi_p\| + \|u(t,\tau,\xi_p,f_n) - u(t,\tau,\xi_p,P_k f_n)\| \end{aligned}$$

$$+\|u(t,\tau,\xi_p,P_kf_n)-u(t,\tau,\xi_p,P_kf)\|+\|u(t,\tau,\xi_p,P_kf)-u(t,\tau,\xi_p,f)\|,$$

where $\|\cdot\|$ stands for the norm in $L^1(\Omega)$. Let $\varepsilon>0$. Fix $p=p(\varepsilon)$ such that

$$\|\xi-\xi_p\|\le\varepsilon.$$

In view of (1.7.3) and Theorem 1.7.8, for this fixed p, we can find $n_1(\varepsilon)\in\mathbb{N}$ such that

$$\|u(t,\tau,\xi_p,P_kf_n)-u(t,\tau,\xi_p,P_kf)\|\le\varepsilon$$

for each $n,k\in\mathbb{N}$, $n\ge n_1(\varepsilon)$ and $k\ge n_1(\varepsilon)$. Furthermore, in view of (1.7.2), for the very same $\varepsilon>0$ and $p=p(\varepsilon)$, there exists $n_2(\varepsilon)\in\mathbb{N}$, such that we have both

$$\|u(t,\tau,\xi_p,f_n)-u(t,\tau,\xi_p,P_kf_n)\|\le\|f_n-P_kf_n\|_{L^1}\le\varepsilon,$$

$$\|u(t,\tau,\xi_p,P_kf)-u(t,\tau,\xi_p,f)\|\le\|f-P_kf\|_{L^1}\le\varepsilon$$

for each $k\in\mathbb{N}$, $k\ge n_2(\varepsilon)$ and each $n\in\mathbb{N}$. Set $n(\varepsilon)=\max\{n_1(\varepsilon),n_2(\varepsilon)\}$. We have

$$\|u(t,\tau,\xi,f_n)-u(t,\tau,\xi,f)\|\le5\varepsilon$$

for each $n\in\mathbb{N}$, $n\ge n(\varepsilon)$, and this completes the proof. □

A nonlinear variant of Theorem 1.7.4 is stated below.

Theorem 1.7.10. *Let Ω be a nonempty, bounded and open subset in $\mathbb{R}^n$ with C^1 boundary Γ and let $\varphi: D(\varphi)\subseteq\mathbb{R}\rightsquigarrow\mathbb{R}$ be maximal monotone with $0\in\varphi(0)$. Let*

$$\begin{cases} u_i-\Delta v_i=f_i & \text{in } \Omega\\ v_i\in\varphi(u_i) & \text{in } \Omega\\ u_i=0 & \text{on } \Gamma,\end{cases}$$

for $i=1,2$. If $f_1,f_2\in L^1(\Omega)$ satisfy $f_1(x)\le f_2(x)$ a.e. for $x\in\Omega$, then $u_1(x)\le u_2(x)$ a.e. for $x\in\Omega$.

See Benilan [**17**].

1.8. Differential and integral inequalities

Let us first introduce

Definition 1.8.1. A function $\omega:\mathbb{R}_+\to\mathbb{R}_+$ which is continuous, nondecreasing and the only C^1-solution of the Cauchy problem

$$\begin{cases} x'(t)=\omega(x(t))\\ x(0)=0\end{cases}$$

is $x\equiv0$ is called a *uniqueness function*.

Remark 1.8.1. If $\omega : \mathbb{R}_+ \to \mathbb{R}_+$ is a uniqueness function, then, for each $m > 0$, $m\omega$ is a uniqueness function too. Clearly, $m\omega$ is continuous and nondecreasing. To conclude, we have merely to observe that $x = x(t)$ is a solution of $x'(t) = \omega(x(t))$ on $[0, T)$ if and only if $y = y(s) = x(ms)$ is a solution of the equation $y'(s) = m\omega(y(s))$ on $[0, \frac{T}{m})$.

Similarly, if $m > 0$, $x \mapsto \omega(mx)$ is a uniqueness function too.

Lemma 1.8.1. *Let $\omega : \mathbb{R}_+ \to \mathbb{R}_+$ be a uniqueness function and let $(b_k)_k$ be strictly decreasing to 0 and $(a_k)_k$ decreasing to 0. Then there exists $T > 0$ such that, for $k = 1, 2, \dots$, each noncontinuable solution $z_k : [0, T_k) \to \mathbb{R}_+$, of the Cauchy problem*

$$\begin{cases} z'(t) = \omega(z(t)) + a_k \\ z(0) = b_k, \end{cases}$$

is defined at least on $[0, T]$, i.e., $T < T_k$, and, for any such sequence $(z_k)_k$, we have

$$z_{k+1}(t) < z_k(t)$$

for $k = 1, 2, \dots$, and each $t \in [0, T]$.

Proof. We observe that, for $k = 1, 2, \dots$ and $t \in [0, \min\{T_k, T_{k+1}\})$,

$$z_{k+1}(t) < z_k(t). \tag{1.8.1}$$

Indeed, if this is not the case, there would exist $t_0 \in (0, \min\{T_k, T_{k+1}\})$ such that

$$\begin{cases} z_k(t) > z_{k+1}(t) & \text{for each } t \in [0, t_0) \\ z_k(t_0) = z_{k+1}(t_0). \end{cases}$$

Hence

$$\begin{aligned} 0 = z_k(t_0) - z_{k+1}(t_0) &= b_k - b_{k+1} \\ + \int_0^{t_0} [\omega(z_k(s)) &- \omega(z_{k+1}(s))]\, ds + t_0(a_k - a_{k+1}) \\ &\geq b_k - b_{k+1} + t_0(a_k - a_{k+1}) > 0, \end{aligned}$$

i.e., $0 > 0$. This contradiction can be eliminated only if $z_{k+1}(t) < z_k(t)$ for each $t \in [0, \min\{T_k, T_{k+1}\})$. In order to complete the proof, it suffices to show that, for $k = 1, 2, \dots$, we have $T_k \leq T_{k+1}$ and so, we can take T any number in $(0, T_1)$. To this aim, let us assume by contradiction that there exists $k = 1, 2, \dots$, such that $T_k > T_{k+1}$. Since T_{k+1} is finite and z_{k+1} is nondecreasing on $[0, T_{k+1})$, it follows that $\lim_{t \uparrow T_{k+1}} z_{k+1}(t) = +\infty$. See for instance Vrabie [**176**], Theorem 2.4.3, p. 69. From (1.8.1), we deduce

$$+\infty = \lim_{t \uparrow T_{k+1}} z_{k+1}(t) \leq z_k(T_{k+1}) < +\infty$$

which is absurd. Hence the supposition that there exists $k = 1, 2, \dots$ satisfying $T_k > T_{k+1}$ is false and the proof is complete. $\square$

Lemma 1.8.2. *Let* $\omega : \mathbb{R}_+ \to \mathbb{R}_+$ *be a uniqueness function and let* $(\gamma_k)_k$ *be strictly decreasing to* 0. *Let* $(x_k)_k$ *be a bounded sequence of measurable functions, from* $[0, \widetilde{T}]$ *to* $\mathbb{R}_+$, *such that*

$$x_k(t) \leq \gamma_k + \int_0^t \omega(x_k(s))\, ds$$

for $k = 1, 2, \dots$ *and for each* $t \in [0, \widetilde{T}]$. *Then there exists* $T \in (0, \widetilde{T}]$ *such that* $\lim_k x_k(t) = 0$ *uniformly for* $t \in [0, T]$.

Proof. Let

$$y_k(t) = \gamma_k + \int_0^t \omega(x_k(s))\, ds$$

for $k = 1, 2, \dots$ and $t \in [0, \widetilde{T}]$. Clearly y_k is absolutely continuous and, since $x_k(t) \leq y_k(t)$ and ω is nondecreasing, we have

$$\begin{cases} y_k'(t) \leq \omega(y_k(t)) \\ y_k(0) = \gamma_k. \end{cases}$$

By Lemma 1.8.1, there exists $T \in (0, \widetilde{T}]$ such that each noncontinuable solution $z_k : [0, T_k) \to \mathbb{R}_+$ of the Cauchy problem

$$\begin{cases} z'(t) = \omega(z(t)) \\ z(0) = b_k, \end{cases}$$

where $b_k = 2\gamma_k$, is defined at least on $[0, T]$ and, for any such sequence $(z_k)_k$, we have $z_{k+1}(t) < z_k(t)$ for each $k \in \mathbb{N}$ and each $t \in [0, T]$. In addition, we have

$$0 \leq x_k(t) \leq y_k(t) \leq z_k(t) \tag{1.8.2}$$

for $k = 1, 2, \dots$ and $t \in [0, T]$. The first inequality is ensured by hypothesis, the second one by the definition of y_k, while the third one follows by contradiction. Indeed, if we assume that for some $t_1 \in (0, T]$, $y_k(t_1) > z_k(t_1)$ (we notice that t_1 cannot be 0 because $y_k(0) = \gamma_k < 2\gamma_k = b_k = z_k(0)$), then there would exist $t_0 \in (0, t_1)$ such that $y_k(t) < z_k(t)$ for $t \in [0, t_0)$ and $y_k(t_0) = z_k(t_0)$. But, in this case, since ω is nondecreasing, we deduce

$$y_k(t_0) \leq \gamma_k + \int_0^{t_0} \omega(y_k(s))\, ds$$

$$< 2\gamma_k + \int_0^{t_0} \omega(z_k(s))\, ds = b_k + \int_0^{t_0} \omega(z_k(s))\, ds = z_k(t_0) = y_k(t_0),$$

i.e., $y_k(t_0) < y_k(t_0)$, which is impossible.

Since $(z_k)_k$ is nonincreasing, it is uniformly bounded on $[0,T]$. So, $(\omega(z_k))_k$, i.e., $(z'_k)_k$, is uniformly bounded on $[0,T]$ and thus $(z_k)_k$ is equicontinuous on $[0,T]$. As it is nonincreasing, it is uniformly convergent to a solution z of the Cauchy problem $z'(t) = \omega(z(t))$, $z(0) = 0$. But ω is a uniqueness function, and therefore we get $z \equiv 0$. Passing to the limit in (1.8.2) for $k \to \infty$, we get $\lim_k x_k(t) = \lim_k y_k(t) = 0$ uniformly for $t \in [0,T]$ and this completes the proof. □

Problem 1.8.1. *Let $\omega : \mathbb{R}_+ \to \mathbb{R}_+$ be a uniqueness function. Prove that if $x : [0,T] \to \mathbb{R}_+$ is absolutely continuous and*

$$\begin{cases} x'(t) \le \omega(x(t)) \\ x(0) = 0 \end{cases}$$

a.e. for $t \in [0,T]$, then $x \equiv 0$ on $[0,T]$.

Definition 1.8.2. By a Carathéodory uniqueness function we mean a function $\omega : I \times \mathbb{R}_+ \to \mathbb{R}_+$ satisfying:

(i) for each $x \in \mathbb{R}_+$, $t \mapsto \omega(t,x)$ is measurable;

(ii) a.e. for $t \in I$, $x \mapsto \omega(t,x)$ is continuous and nondecreasing;

(iii) there exist $\ell \in L^1(I)$ and $\varphi \in C(\mathbb{R}_+)$ such that $\omega(t,x) \le \ell(t)\varphi(x)$ a.e. for $t \in I$ and for each $x \in \mathbb{R}_+$;

(iv) for each $\tau \in I$, the only absolutely continuous solution of the Cauchy problem,

$$\begin{cases} x'(t) = \omega(t, x(t)) \\ x(\tau) = 0, \end{cases}$$

is $x \equiv 0$.

Using similar arguments as those in the proof of Lemma 1.8.2, we deduce

Lemma 1.8.3. *Let $\omega : [\tau, \widetilde{T}] \times \mathbb{R}_+ \to \mathbb{R}_+$ be a Carathéodory uniqueness function and let $(\gamma_k)_k$ be strictly decreasing to 0. Let $(x_k)_k$ be a bounded sequence of measurable functions from $[\tau, \widetilde{T}]$ to $\mathbb{R}_+$ such that*

$$x_k(t) \le \gamma_k + \int_\tau^t \omega(s, x_k(s))\, ds$$

for $k = 1, 2, \ldots$ and for each $t \in [\tau, \widetilde{T}]$. Then there exists $T \in (\tau, \widetilde{T}]$ such that $\lim_k x_k(t) = 0$ uniformly for $t \in [\tau, T]$.

Problem 1.8.2. *Let $\omega : [\tau, T] \times \mathbb{R}_+ \to \mathbb{R}_+$ be a Carathéodory uniqueness function. Prove that if $x : [\tau, T] \to \mathbb{R}_+$ is absolutely continuous and*

$$\begin{cases} x'(t) \le \omega(t, x(t)) \\ x(\tau) = 0 \end{cases}$$

a.e. for $t \in [\tau, T]$, *then* $x \equiv 0$ *on* $[\tau, T]$.

If $x : [\tau, T] \to \mathbb{R}$, we denote by $[D_+ x](t)$ the *right lower Dini derivative* of the function x at t, i.e.

$$[D_+ x](t) = \liminf_{h \downarrow 0} \frac{x(t+h) - x(t)}{h}. \tag{1.8.3}$$

Proposition 1.8.1. *If* $x : [\tau, T] \to \mathbb{R}$ *is continuous and* $[D_+ x](t) \leq 0$ *for each* $t \in [\tau, T)$, *then* x *is nonincreasing on* $[\tau, T]$.

See Hobson [**109**], p. 365.

We conclude this section with a slight extension of the Gronwall Lemma to measurable functions.

Lemma 1.8.4. *Let* $m \in \mathbb{R}$ *and let* $x, k : [\tau, T) \to \mathbb{R}$ *be measurable with* $k \in L^1(\tau, T)$ *and* $k(s) \geq 0$ *a.e. for* $s \in [\tau, T)$. *Let us assume that* $s \mapsto k(s)x(s)$ *is locally integrable on* $[\tau, T)$ *and*

$$x(t) \leq m + \int_\tau^t k(s)x(s)\, ds$$

for every $t \in [\tau, T)$. *Then*

$$x(t) \leq m e^{\int_\tau^t k(s)\, ds}$$

for every $t \in [\tau, T)$.

Problem 1.8.3. *Prove Lemma* 1.8.4.

CHAPTER 2

Specific preliminary results

Unlike Chapter 1, which was mainly concerned with a general background, our aim here is to gather some concepts and results which, although general, are essentially focused on the specific topic of the book. After proving the Brezis-Browder Ordering Principle, we discuss a basic lemma ensuring the existence of projections. Then, we introduce and study the concept of tangent set at a point to a given set and continue with an excursion to various types of tangent cones: Bouligand, Federer, Clarke and Bony. Further on, we state and prove some fundamental results on l.s.c. and u.s.c. multi-functions and add several technical results referring to measures of noncompactness. Finally, we prove some infinite variants and consequences of Scorza Dragoni type theorems.

2.1. Brezis-Browder Ordering Principle

The goal of this section is to prove a general and very simple principle concerning preorder relations. This principle, similar to Zorn's Lemma, unifies a number of various results in nonlinear functional analysis, and is based on

The Axiom of Dependent Choice. *Let $\mathcal{S}$ be a nonempty set and let $\mathcal{R} \subseteq \mathcal{S} \times \mathcal{S}$ be a binary relation with the property that, for each $\xi \in \mathcal{S}$, the set $\{\eta \in \mathcal{S};\ \xi \mathcal{R} \eta\}$ is nonempty. Then, for each $\xi \in \mathcal{S}$, there exists a sequence $(\xi_k)_k$ in $\mathcal{S}$ such that $\xi_0 = \xi$ and $\xi_k \mathcal{R} \xi_{k+1}$ for each $k \in \mathbb{N}$.*

To begin with, let us recall some definitions and notations. Let $\mathcal{S}$ be a nonempty set. A binary relation $\preceq \subseteq \mathcal{S} \times \mathcal{S}$ is a *preorder on* $\mathcal{S}$ if it is reflexive, i.e., $\xi \preceq \xi$ for each $\xi \in \mathcal{S}$, and transitive, i.e., $\xi \preceq \eta$ and $\eta \preceq \zeta$ imply $\xi \preceq \zeta$.

Definition 2.1.1. Let $\mathcal{S}$ be a nonempty set, $\preceq \subseteq \mathcal{S} \times \mathcal{S}$ a preorder on $\mathcal{S}$, and let $\mathcal{N} : \mathcal{S} \to \mathbb{R} \cup \{+\infty\}$ be an increasing function. An $\mathcal{N}$-*maximal element* is an element $\overline{\xi} \in \mathcal{S}$ satisfying $\mathcal{N}(\xi) = \mathcal{N}(\overline{\xi})$, for every $\xi \in \mathcal{S}$ with $\overline{\xi} \preceq \xi$.

We may now proceed to the statement of the main result in this section, i.e., Brezis-Browder Ordering Principle.

Theorem 2.1.1. *Let $\mathcal{S}$ be a nonempty set, $\preceq \subseteq \mathcal{S} \times \mathcal{S}$ a preorder on $\mathcal{S}$ and let $\mathcal{N} : \mathcal{S} \to \mathbb{R} \cup \{+\infty\}$ be a function. Suppose that:*

(i) *each increasing sequence in $\mathcal{S}$ is bounded from above;*
(ii) *the function $\mathcal{N}$ is increasing.*

Then, for each $\xi_0 \in \mathcal{S}$, there exists an $\mathcal{N}$-maximal element $\overline{\xi} \in \mathcal{S}$ satisfying $\xi_0 \preceq \overline{\xi}$.

Proof. Suppose first that the function $\mathcal{N}$ is bounded from above. For each $\xi \in \mathcal{S}$, let us denote

$$\mathcal{S}(\xi) = \{\eta \in \mathcal{S}\,;\ \xi \preceq \eta\}$$

and

$$\beta(\xi) = \sup\{\mathcal{N}(\eta)\,;\ \eta \in \mathcal{S}(\xi)\}.$$

Let us consider a fixed element $\xi_0 \in \mathcal{S}$. On $\mathcal{S}(\xi_0)$, we introduce the binary relation, $\mathcal{R}$, as follows: $\xi\mathcal{R}\eta$ if $\eta \in \mathcal{S}(\xi)$ and

$$\mathcal{N}(\eta) > \frac{1}{2}\left(\mathcal{N}(\xi) + \beta(\xi)\right).$$

Suppose by contradiction that the conclusion of the theorem would be false for ξ_0, i.e., no point of $\mathcal{S}(\xi_0)$ is $\mathcal{N}$-maximal. It follows that, for each $\xi \in \mathcal{S}(\xi_0)$, we have $\beta(\xi) > \mathcal{N}(\xi)$; therefore, there exists $\eta \in \mathcal{S}(\xi)$ satisfying

$$\mathcal{N}(\eta) > \beta(\xi) - \frac{\beta(\xi) - \mathcal{N}(\xi)}{2},$$

that is, $\xi\mathcal{R}\eta$. We can apply the Axiom of Dependent Choice to deduce the existence of a sequence $(\xi_k)_k$ in $\mathcal{S}(\xi_0)$ such that $\xi_k \preceq \xi_{k+1}$ and

$$\mathcal{N}(\xi_{k+1}) > \frac{1}{2}\left(\mathcal{N}(\xi_k) + \beta(\xi_k)\right) \tag{2.1.1}$$

(and therefore $\mathcal{N}(\xi_{k+1}) > \mathcal{N}(\xi_k)$), for $k = 1, 2, \ldots$.

We have thus constructed an increasing sequence $(\xi_k)_k$ with the property that the sequence $(\mathcal{N}(\xi_k))_k$ is strictly increasing. By the assumption (i), $(\xi_k)_k$ is bounded from above in $\mathcal{S}$, i.e., there exists $\overline{\xi} \in \mathcal{S}$ such that $\xi_k \preceq \overline{\xi}$ for every $k \in \mathbb{N}$. We show that $\overline{\xi}$ is $\mathcal{N}$-maximal. Indeed, let $\eta \in \mathcal{S}(\overline{\xi})$ be arbitrary fixed, hence $\eta \in \mathcal{S}(\xi_k)$ for each $k \in \mathbb{N}$. Since $\mathcal{N}$ is bounded from above, the sequence $(\mathcal{N}(\xi_k))_k$ is bounded from above, and thus convergent. Moreover, by the assumption (ii),

$$\lim_k \mathcal{N}(\xi_k) \leq \mathcal{N}(\overline{\xi}) \leq \mathcal{N}(\eta). \tag{2.1.2}$$

On the other hand, from (2.1.1) we have

$$2\mathcal{N}(\xi_{k+1}) - \mathcal{N}(\xi_k) \geq \beta_k \geq \mathcal{N}(\eta),$$

for $k = 1, 2, \ldots$. Passing to the limit as $k \to \infty$, we obtain

$$\mathcal{N}(\overline{\xi}) \geq \lim_k \mathcal{N}(\xi_k) \geq \mathcal{N}(\eta),$$

which combined with (2.1.2) shows that $\overline{\xi}$ is $\mathcal{N}$-maximal. This contradicts our initial hypothesis about the elements of the subset $\mathcal{S}(\xi_0)$, and completes the proof of theorem under the extra-condition that $\mathcal{N}$ is bounded from above.

Consider now the general case, and let us define the auxiliary function $\mathcal{N}_1 : \mathcal{S} \to (-\frac{\pi}{2}, \frac{\pi}{2}]$ by

$$\mathcal{N}_1(\xi) = \begin{cases} \arctan(\mathcal{N}(\xi)) & \text{if } \mathcal{N}(\xi) < +\infty \\ \dfrac{\pi}{2} & \text{if } \mathcal{N}(\xi) = +\infty. \end{cases}$$

The function $\mathcal{N}_1$ is increasing and bounded from above. Therefore there exists an element $\overline{\xi} \in \mathcal{S}$ which verifies the conclusion with $\mathcal{N}_1$ instead of $\mathcal{N}$. But $\mathcal{N}_1(\overline{\xi}) = \mathcal{N}_1(\xi)$ if and only if $\mathcal{N}(\overline{\xi}) = \mathcal{N}(\xi)$, which completes the proof in the general case. □

Here and thereafter, $\operatorname{dist}(x; D)$ is the distance from the point $x \in X$ to the set $D \subseteq X$, i.e., $\operatorname{dist}(x; D) = \inf_{y \in D} \|x - y\|$.

Problem 2.1.1. *Let X be a Banach space and M a closed subset in X. Let $\{S(t) : M \to M;\ t \geq 0\}$ be a semigroup of nonexpansive mappings on M. Let K be a nonempty and closed subset in M. Suppose that*

$$\liminf_{t \downarrow 0} \frac{1}{t} \operatorname{dist}(S(t)x; K) = 0, \tag{2.1.3}$$

for all $x \in K$. Then, $S(t)x \in K$ for all $x \in K$ and $t \geq 0$.

2.2. Projections

Let X be a real Banach space with norm $\|\cdot\|$. We say that $\xi \in X$ *has projection on* K if there exists $\eta \in K$ such that $\|\xi - \eta\| = \operatorname{dist}(\xi; K)$. Any $\eta \in K$ enjoying the above property is called *a projection* of ξ on K, and the set of all projections of ξ on K is denoted by $\Pi_K(\xi)$.

Definition 2.2.1. A subset $K \subseteq X$ is *locally closed* if for every $\xi \in K$ there exists $\rho > 0$ such that $K \cap D(\xi, \rho)$ is closed.

Definition 2.2.2. A subset $K \subseteq X$ is *locally compact* if for every $\xi \in K$ there exists $\rho > 0$ such that $K \cap D(\xi, \rho)$ is compact.

Remark 2.2.1. Obviously every closed set is locally closed. Furthermore, every open set D is locally closed. There exist however locally closed sets which are neither open, nor closed, as for example an open half plane in $\mathbb{R}^3$.

Problem 2.2.1. *Show that if K is closed relative to an open set D, it is locally closed. Conversely, if K is locally closed, there exists an open set D such that $K \subseteq D$ and K is closed relative to D.*

The next lemma will prove useful in the sequel.

Lemma 2.2.1. *Let X be a Banach space and let $K \subseteq X$ be locally compact. Then the set of all $\xi \in X$ for which $\Pi_K(\xi)$ is nonempty is a neighborhood of K.*

Proof. Let $\xi \in K$. Since K is locally compact, there exists $\rho > 0$ such that $K \cap D(\xi, \rho)$ is compact. To complete the proof, it suffices to show that, for each $\eta \in X$ satisfying $\|\eta - \xi\| < \rho/2$, $\Pi_K(\eta)$ is nonempty. Let η as above. There exists a sequence $(\zeta_k)_k$ in K such that $(\|\zeta_k - \eta\|)_k$ converges to $\operatorname{dist}(\eta; K)$. Since $\operatorname{dist}(\eta; K) \le \|\eta - \xi\| < \rho/2$, we have $\|\zeta_k - \eta\| < \rho/2$ for all $k \in \mathbb{N}$ sufficiently large. Therefore, $\|\zeta_k - \xi\| < \rho$ for k sufficiently large. Since $K \cap D(\xi, \rho)$ is compact, we can suppose, by taking a subsequence if necessary, that $(\zeta_k)_k$ converges to a point $\zeta \in K \cap D(\xi, \rho)$. Thus $\zeta \in \Pi_K(\eta)$, and this completes the proof. □

If X is finite dimensional, each locally closed set is locally compact and so we have

Corollary 2.2.1. *Let X be a finite dimensional Banach space and let $K \subseteq X$ be locally closed. Then the set of all $\xi \in X$ for which $\Pi_K(\xi)$ is nonempty is a neighborhood of K.*

Definition 2.2.3. A set $K \subseteq X$ is called *proximal* if there exists a neighborhood V (called *proximal neighborhood*) of K such that $\Pi_K(\xi) \neq \emptyset$ for each $\xi \in V$. If V is a proximal neighborhood of K, any single-valued selection π_K of Π_K is called a *projection subordinated to V*.

In this terminology, Lemma 2.2.1 is actually saying that every locally compact set in a Banach space is proximal. Other examples of proximal sets are weakly closed sets in Hilbert spaces. In this case, V is the whole space. In particular, each nonempty, strongly closed and convex set K, in a Hilbert space, is proximal. In addition, in this case, i.e., when K is convex, then $\xi \mapsto \Pi_K(\xi)$ is single-valued from V to K and thus it can be identified with a function.

2.3. Tangent sets

We begin by introducing a tangency concept which will prove useful in the study of existence properties for differential inclusions. Here and thereafter, if C and D are subsets in X, dist $(C; D)$ denotes the usual distance between C and D, i.e.

$$\text{dist}\,(C; D) = \inf_{x\in C, y\in D} \|x - y\|.$$

Also, if $x \in X$ and $C \subseteq X$, we denote by

$$x + C = \{y \in X;\ \text{there exists } z \in C \text{ such that } y = x + z\}.$$

Definition 2.3.1. Let $K \subseteq X$ and $\xi \in K$. The set $E \subseteq X$ is *tangent* to the set K at the point ξ if, for each $\rho > 0$, we have

$$\liminf_{h\downarrow 0} \frac{1}{h}\,\text{dist}\,(\xi + hE; K \cap D(\xi, \rho)) = 0. \tag{2.3.1}$$

We denote by $\mathcal{TS}_K(\xi)$ the class of all sets which are tangent to K at the point ξ.

The next problem will prove very useful later.

Problem 2.3.1. *Let $K \subseteq X$, $\xi \in K$ and $E \subseteq X$. Prove that the following conditions are equivalent:*

(i) $E \in \mathcal{TS}_K(\xi)$;
(ii) *there exist two sequences, $(h_n)_n$ in $\mathbb{R}_+$ with $h_n \downarrow 0$ and $(\eta_n)_n$ in E such that $\lim_n h_n\eta_n = 0$, and $\liminf_n \frac{1}{h_n}\text{dist}\,(\xi + h_n\eta_n; K) = 0$;*
(iii) *for each $\varepsilon > 0$, $\rho > 0$ and $\delta > 0$ there exist $h \in (0, \delta)$, $p \in D(0, \varepsilon)$ and $\eta \in E$ such that $\xi + h\eta + hp \in K \cap D(\xi, \rho)$;*
(iv) *there exist three sequences, $(h_n)_n$ in $\mathbb{R}_+$ with $h_n \downarrow 0$, $(\eta_n)_n$ in E with $\lim_n h_n\eta_n = 0$ and $(p_n)_n$ in X with $\lim_n p_n = 0$, such that $\xi + h_n\eta_n + h_np_n \in K$ for $n = 1, 2, \ldots$.*

Let us denote by $\mathcal{B}(X)$ the class of all bounded subsets in X.

Problem 2.3.2. *Let $K \subseteq X$, $\xi \in K$ and $E \in \mathcal{B}(X)$. Prove that the following conditions are equivalent:*

(i) $E \in \mathcal{TS}_K(\xi)$;
(ii) $\liminf_{h\downarrow 0} \frac{1}{h}\,\text{dist}\,(\xi + hE; K) = 0$;
(iii) *there exist two sequences, $(h_n)_n$ in $\mathbb{R}_+$ with $h_n \downarrow 0$ and $(\eta_n)_n$ in E, such that $\liminf_n \frac{1}{h_n}\text{dist}\,(\xi + h_n\eta_n; K) = 0$;*
(iv) *for each $\varepsilon > 0$ there exist $\eta \in E$, $\delta \in (0, \varepsilon)$ and $p \in X$ with $\|p\| \le \varepsilon$, such that $\xi + \delta\eta + \delta p \in K$;*
(v) *there exist three sequences, $(h_n)_n$ in $\mathbb{R}_+$ with $h_n \downarrow 0$, $(\eta_n)_n$ in E and $(p_n)_n$ in X with $\lim_n p_n = 0$, such that $\xi + h_n\eta_n + h_np_n \in K$ for $n = 1, 2, \ldots$.*

Remark 2.3.1. One may ask why we are not simply defining the tangent set E by merely imposing (ii) in Problem 2.3.2 instead of (2.3.1). See Definition 2.3.1. To answer this question we have to observe that, whenever E is unbounded, (ii) in Problem 2.3.2 is not enough to keep the local character of the tangency concept, i.e. $\mathcal{TS}_K(\xi) = \mathcal{TS}_{K\cap D(\xi,\rho)}(\xi)$ for each $\rho > 0$. To justify this observation, let us analyze the example below.

Example 2.3.1. Let $X = \mathbb{R}^2$, $E = \{(x,\lambda) \in \mathbb{R}\times\mathbb{R};\ f(x) \le \lambda\}$ and $K = \{(x,\mu) \in \mathbb{R}\times\mathbb{R};\ g(x) \ge \mu\}$, where

$$f(x) = \begin{cases} 0, & |x| \ge 2 \\ 2 - |x|, & |x| < 2 \end{cases}, \quad g(x) = \begin{cases} 0, & |x| \ge 1 \\ -|x|, & |x| < 1, \end{cases}$$

and $\xi = (0,0) \in K$. See Figure 2.3.1. Then, although (ii) in Problem 2.3.2 is satisfied, (2.3.1) is not, unless $\rho \ge 1$.

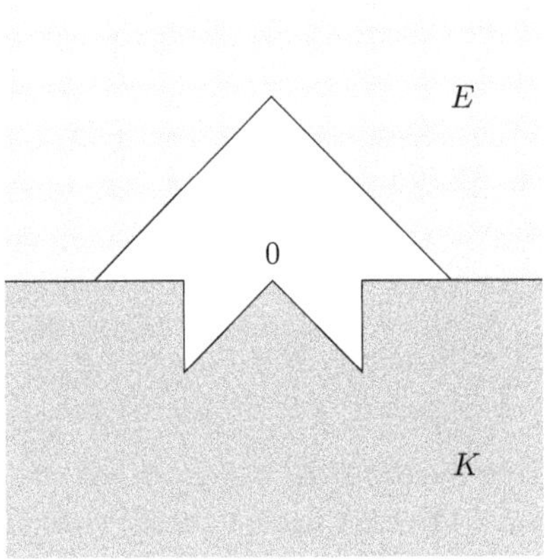

Figure 2.3.1

Definition 2.3.2. The *Hausdorff-Pompeiu distance* between the sets $B, C \in \mathcal{B}(X)$ is defined by $\text{dist}_{HP}(B;C) = \max\{e(B;C), e(C;B)\}$, where, for each $B, C \in \mathcal{B}(X)$, $e(B;C)$ is the *excess* of B over C, defined by $e(B;C) = \sup_{x\in B} \text{dist}\,(x;C)$.

Problem 2.3.3. *Prove that, for each* $B, C \in \mathcal{B}(X)$, *we have*

$$\text{dist}_{HP}(B;C) = \inf\{\varepsilon > 0; B \subseteq C + D(0,\varepsilon),\ C \subseteq B + D(0,\varepsilon)\}.$$

Problem 2.3.4. *Prove that, for each* $B, C \in \mathcal{B}(X)$, *we have*

$$\text{dist}_{HP}(B;C) = \sup_{x\in X} |\text{dist}\,(x;B) - \text{dist}\,(x;C)|.$$

See Beer [**13**], Section 3.2.

Problem 2.3.5. *Let $K \subseteq X$, $\xi \in K$ and $E \in \mathcal{B}(X)$. Prove that:*

(i) *if $0 \in E$, then $E \in \mathcal{TS}_K(\xi)$;*
(ii) *$\{0\} \in \mathcal{TS}_K(\xi)$;*
(iii) *if $E \in \mathcal{TS}_K(\xi)$ and $E \subseteq D$ then $D \in \mathcal{TS}_K(\xi)$;*
(iv) *$E \in \mathcal{TS}_K(\xi)$ if and only if $\overline{E} \in \mathcal{TS}_K(\xi)$;*
(v) *if $E \in \mathcal{TS}_K(\xi)$, then for each $\lambda > 0$, we have $\lambda E \in \mathcal{TS}_K(\xi)$;*
(vi) *for each $\xi \in K$, the set $\mathcal{TS}_K(\xi)$ is closed from the left with respect to the excess e, i.e., if $E \in \mathcal{B}(X)$ and $(E_n)_n$ is a sequence in $\mathcal{TS}_K(\xi)$ such that $\lim_n e(E_n; E) = 0$, then $E \in \mathcal{TS}_K(\xi)$. In particular, for each $\xi \in K$, the set $\mathcal{TS}_K(\xi)$ is closed with respect to the Hausdorff-Pompeiu distance.*

Remark 2.3.2. One may ask whether or not, there are tangent sets E which do not contain "tangent vectors", i.e. $E \in \mathcal{TS}_K(\xi)$ but, for each $\eta \in E$, $\{\eta\} \notin \mathcal{TS}_K(\xi)$. The answer to this question is in the affirmative, even though E is bounded and closed. See Examples 2.4.1 and 2.4.2 and Problem 2.4.3 in the next section.

2.4. Bouligand–Severi tangent vectors

In this section we introduce several concepts of tangent vectors to a set K at a point $\xi \in K$. We begin with the most natural one.

Definition 2.4.1. Let $K \subseteq X$ and $\xi \in K$. The vector $\eta \in X$ is *tangent in the sense of Bouligand–Severi* to the set K at the point ξ if

$$\liminf_{h \downarrow 0} \frac{1}{h} \operatorname{dist}(\xi + h\eta; K) = 0. \tag{2.4.1}$$

We denote by $\mathcal{T}_K(\xi)$ the set of all vectors which are tangent in the sense of Bouligand–Severi to the set K at the point ξ. See Figure 2.4.1.

Remark 2.4.1. By the natural injection $\eta \mapsto \{\eta\}$, $\mathcal{T}_K(\xi)$ is identified with a subclass of $\mathcal{TS}_K(\xi)$. Therefore, in the sequel, by $\mathcal{T}_K(\xi) \subseteq \mathcal{TS}_K(\xi)$ we mean the natural inclusion induced by the injection above.

Problem 2.4.1. *Prove that whenever K is a closed cone, $\mathcal{T}_K(0) = K$.*

Problem 2.4.2. *Prove that whenever E is compact, $E \in \mathcal{TS}_K(\xi)$ if and only if $E \cap \mathcal{T}_K(\xi) \neq \emptyset$.*

We may ask whether the result above remains true if E is merely weakly compact. The answer to this question is in the negative, even though K is compact. More precisely, we give below an example of a compact set K,

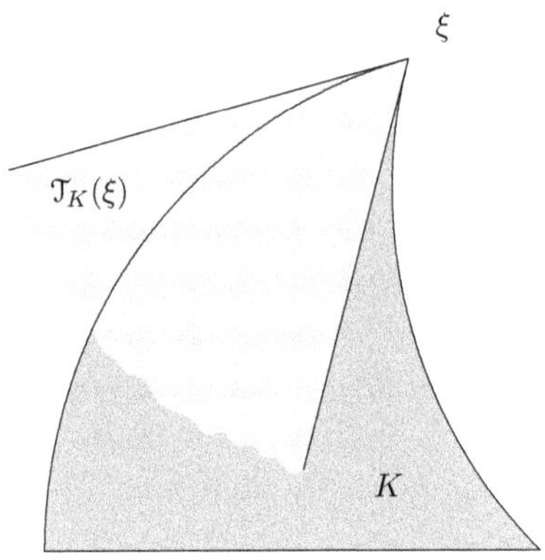

Figure 2.4.1

a point $\xi \in K$ and a weakly compact set E, in a reflexive Banach space (in fact even a Hilbert space), such that $E \in \mathcal{TS}_K(0)$ but, nevertheless, $E \cap \mathcal{T}_K(0) = \emptyset$.

Example 2.4.1. Let $X = \ell_2$ be the space of all real sequences $(x_k)_k$, with $\sum_{k=1}^{\infty} x_k^2 < \infty$, endowed with its usual norm $\|\cdot\|$, defined by

$$\|(x_k)_k\| = \left(\sum_{k=1}^{\infty} x_k^2\right)^{1/2}$$

for each $(x_k)_k \in \ell_2$. Let $\{e_1, e_2, \dots\}$ be the standard orthonormal basis in ℓ_2, $\xi \in \ell_2$ with $\|\xi\| = 2$, let $(h_k)_k$ be a sequence in $(0,1]$, $h_k \downarrow 0$, $f_n = \xi + e_n$, for $n = 1, 2, \dots$, let $E = \xi + D(0,1)$ and $K = \{h_n f_n\,; n = 1, 2, \dots\} \cup \{0\}$. Clearly K is compact, $f_n \in E$, for $n = 1, 2, \dots$, and

$$\operatorname{dist}(h_n E; K) = 0,$$

for $n = 1, 2, \dots$. Thus $E \in \mathcal{TS}_K(0)$. However, $\mathcal{T}_K(0) = \{0\}$. Indeed, the inclusion $\{0\} \subseteq \mathcal{T}_K(0)$ is obvious. Now, if we assume by contradiction that there exists $\eta \in \ell_2$, $\eta \neq 0$, with $\eta \in \mathcal{T}_K(0)$, then there would exist $(t_n)_n$ in $(0,1)$ with $t_n \downarrow 0$ and a sequence of natural numbers $(k_n)_n$, such that

$$\lim_n \frac{1}{t_n}\|t_n\eta - h_{k_n} f_{k_n}\| = 0,$$

or equivalently

$$\lim_n \left\|\eta - \frac{h_{k_n}}{t_n} f_{k_n}\right\| = 0. \tag{2.4.2}$$

Let us observe that $(k_n)_n$ cannot have constant subsequences. Indeed, if we assume that $(k_n)_n$ has a constant subsequence, denoted for simplicity again by $(k_n)_n = (\widetilde{k})_n$, then, since $\|f_{\widetilde{k}}\| = \|\xi + e_{\widetilde{k}}\| \geq \|\xi\| - \|e_{\widetilde{k}}\| \geq 1$, we deduce

$$\lim_n \left\| \eta - \frac{h_{\widetilde{k}}}{t_n} f_{\widetilde{k}} \right\| \geq \lim_n \left(\frac{h_{\widetilde{k}}}{t_n} \|f_{\widetilde{k}}\| - \|\eta\| \right) = \infty,$$

which contradicts (2.4.2). Therefore we have $k_n \to \infty$ as $n \to \infty$.

A similar argument shows that $(\frac{h_{k_n}}{t_n})_n$ is necessarily bounded. Hence, we may assume, without loss of generality, that it is convergent. In addition,

$$\lim_n \frac{h_{k_n}}{t_n} = m > 0,$$

because otherwise, in view of (2.4.2), we would get a contradiction, i.e., $\|\eta\| = 0$. Therefore, again by (2.4.2), we get $\lim_n \|\eta - m f_{k_n}\| = 0$, which shows that $f_{k_n} \to \frac{1}{m}\eta$. But this is impossible because $(f_n)_n$ cannot have strongly convergent subsequences. To justify the last assertion, it suffices to observe that $\|f_i - f_j\| = \sqrt{2}$, for $i, j = 1, 2, \ldots$, $i \neq j$. The contradiction $f_{k_n} \to \frac{1}{m}\eta$ can be eliminated only if $\eta \notin \mathcal{T}_K(0)$. Thus $\mathcal{T}_K(0) = \{0\}$ and since each $f \in E$ satisfies $\|f\| \geq 1$, $E \cap \mathcal{T}_K(0) = \emptyset$, as claimed.

In the example below, we allow K to be a noncompact, closed cone. More precisely, we have

Example 2.4.2. Let $X = \ell_2$ and $\{e_1, e_2, \ldots\}$ be as in Example 2.4.1, let

$$H = \{(x_k)_k \in \ell_2;\ 0 \leq x_k \leq 1,\ k = 1, 2, \ldots\}$$

and

$$G = \overline{\text{conv}}\{e_n;\ n = 1, 2, \ldots\}.$$

Obviously, $G \subseteq H$. Let $(y_k)_k \in \ell_2$ be such that $y_k > 0$ for $k = 1, 2, \ldots$ and let

$$E = y + G \subseteq y + H = \{(x_k)_k \in \ell_2;\ y_k \leq x_k \leq y_k + 1,\ k = 1, 2, \ldots\}.$$

Let $f_n = y + e_n + \frac{1}{n} e_n$, $n = 1, 2, \ldots$, and let us define

$$K = \{\lambda f_n;\ \lambda \geq 0,\ n = 1, 2, \ldots\}.$$

Clearly K is a cone. In addition, if $\lambda_k f_{n_k} \to \mu$, then, either $\lambda_k \to 0$, or $\lambda_k \to \lambda \neq 0$, case in which $(f_{n_k})_k$ must be almost constant, i.e., there exists $k_0 = 1, 2, \ldots$ such that, for each $p, m \geq k_0$, we have $f_{n_p} = f_{n_m}$. In both cases, we have $\mu \in K$ which shows that K is a closed cone. Accordingly, $\mathcal{T}_K(0) = K$. See Problem 2.4.1.

We next prove that $E \in \mathcal{TS}_K(0)$. To this aim, let $h_n \downarrow 0$ and let us choose $\eta_n = y + e_n$ and $p_n = \frac{1}{n} e_n$. We have $h_n \eta_n + h_n p_n = h_n f_n \in K$. In view of the equivalence between (i) and (v) in Problem 2.3.2, it follows that $E \in \mathcal{TS}_K(0)$[1].

Finally, we will show that $E \cap \mathcal{T}_K(0) = \emptyset$. To this aim, we will prove that $\mathcal{T}_K(0) \subseteq X \setminus E$. So, let $\lambda f_n \in \mathcal{T}_K(0) = K$. We have

$$\lambda f_n = \lambda y + \lambda \left(1 + \frac{1}{n}\right) e_n$$

$$= \lambda y_1 e_1 + \lambda y_2 e_2 + \cdots + \lambda \left(y_n + 1 + \frac{1}{n}\right) e_n + \ldots.$$

We distinguish between two cases: $\lambda \in [0,1)$ and $\lambda \in [1,\infty)$. If $\lambda \in [0,1)$, then $\lambda y_1 < y_1$ and therefore $\lambda f_n \notin y + H$. Since $E \subseteq y + H$, this shows that $\lambda f_n \notin E$. If $\lambda \in [1,\infty)$, then $\lambda\left(y_n + 1 + \frac{1}{n}\right) > y_n + 1$. Accordingly, $\lambda f_n \notin y + H$, which implies that $\lambda f_n \notin E$. Hence $E \cap \mathcal{T}_K(0) = \emptyset$ and this completes the proof.

Problem 2.4.3. *Let $X = C([0,1])$ which, endowed with the usual sup-norm, defined by $\|f\| = \sup_{t\in[0,1]} |f(t)|$ for $f \in C([0,1])$, is a nonreflexive Banach space. Let*

$$K = \{f \in C([0,1]) \,; \text{there exists } t \in [0,1] \text{ with } f(t) \le 0\},$$

$$E = \{f \in C([0,1]) \,;\ t \le f(t) \le 1 \text{ for all } t \in [0,1] \text{ and } f(0) = f(1) = 1\}.$$

Let $\xi = 0 \in K$. Show that $E \in \mathcal{TS}_K(\xi)$ but, nevertheless, $E \cap \mathcal{T}_K(\xi) = \emptyset$.

Proposition 2.4.1. *For each $\xi \in K$, the set $\mathcal{T}_K(\xi)$ is a closed cone.*

Proof. Let $\xi \in K$. According to Definition 2.4.1, $\eta \in \mathcal{T}_K(\xi)$ if (2.4.1) holds true. So, let $s > 0$ and let us observe that

$$\liminf_{t\downarrow 0} \frac{1}{t} \operatorname{dist}(\xi + ts\eta; K) = s \liminf_{t\downarrow 0} \frac{1}{ts} \operatorname{dist}(\xi + ts\eta; K)$$

$$= s \liminf_{\tau\downarrow 0} \frac{1}{\tau} \operatorname{dist}(\xi + \tau\eta; K) = 0.$$

Hence $s\eta \in \mathcal{T}_K(\xi)$. In order to complete the proof, it remains to be shown that $\mathcal{T}_K(\xi)$ is a closed set. To this aim let $(\eta_k)_k$ be a sequence of elements in $\mathcal{T}_K(\xi)$, convergent to η. We have

$$\frac{1}{t} \operatorname{dist}(\xi + t\eta; K) \le \frac{1}{t}\|t(\eta - \eta_k)\| + \frac{1}{t} \operatorname{dist}(\xi + t\eta_k; K)$$

---

[1] In fact, we have proved a stronger result, i.e., that

$$\lim_{h\downarrow 0} \frac{1}{h} \operatorname{dist}(\xi + hE; K) = 0.$$

$$= \|\eta - \eta_k\| + \frac{1}{t}\,\text{dist}\,(\xi + t\eta_k; K)$$

for $k = 1, 2, \ldots$. So $\liminf_{t\downarrow 0} \frac{1}{t}\,\text{dist}\,(\xi + t\eta; K) \leq \|\eta - \eta_k\|$ for $k = 1, 2, \ldots$. Since $\lim_k \|\eta - \eta_k\| = 0$, it follows that (2.4.1) holds true, and this completes the proof. □

Problem 2.4.4. *Give another proof to Proposition* 2.4.1 *by using* (vi) *in Problem* 2.3.5.

The cone $\mathcal{T}_K(\xi)$ is called *the contingent cone* to K at ξ.

Proposition 2.4.2. *A vector $\eta \in X$ belongs to the cone $\mathcal{T}_K(\xi)$ if and only if for every $\varepsilon > 0$ there exist $h \in (0, \varepsilon)$ and $p \in D(0, \varepsilon)$ with the property*

$$\xi + h(\eta + p) \in K.$$

Proof. Obviously $\eta \in \mathcal{T}_K(\xi)$ if and only if, for every $\varepsilon > 0$, there exist $h \in (0, \varepsilon)$ and $z \in K$ such that $\frac{1}{h}\|\xi + h\eta - z\| \leq \varepsilon$. Now, let us define $p = \frac{1}{h}(z - \xi - h\eta)$, and let us observe that we have both $\|p\| \leq \varepsilon$, and $\xi + h(\eta + p) = z \in K$, thereby completing the proof. □

A simple but useful consequence is

Corollary 2.4.1. *A vector $\eta \in X$ belongs to the cone $\mathcal{T}_K(\xi)$ if and only if there exist two sequences $(h_m)_m$ in $\mathbb{R}_+$ and $(p_m)_m$ in X with $h_m \downarrow 0$, $\lim_m p_m = 0$ and such that $\xi + h_m(\eta + p_m) \in K$ for each $m \in \mathbb{N}$.*

Remark 2.4.2. We notice that, if ξ is an interior point of the set K, then $\mathcal{T}_K(\xi) = X$. Indeed, in this case there exists $\rho > 0$ with $D(\xi, \rho) \subset K$ and, therefore, for $t > 0$ sufficiently small, $\xi + t\eta \in D(\xi, \rho) \subseteq K$. Obviously, for such numbers $t > 0$, we have $\text{dist}\,(\xi + t\eta; K) = 0$, from where it follows the condition in Definition 2.4.1.

Problem 2.4.5. *Let $r > 0$, $\Sigma = \{x \in X;\ \|x\| = r\}$ and $\xi \in \Sigma$. Prove that $\eta \in \mathcal{T}_\Sigma(\xi)$ if and only if $(\xi, \eta)_+ = 0$. Further, prove that $\eta \in \mathcal{T}_{D(0,r)}(\xi)$ if and only if $(\xi, \eta)_+ \leq 0$.*

Proposition 2.4.3. *If $\eta \in \mathcal{T}_K(\xi)$ then, for every function $h \mapsto \eta_h$ from $(0, 1)$ to X satisfying $\lim_{h\downarrow 0} \eta_h = \eta$, we have*

$$\liminf_{h\downarrow 0} \frac{1}{h}\text{dist}\,(\xi + h\eta_h; K) = 0. \tag{2.4.3}$$

If there exists a function $h \mapsto \eta_h$ from $(0, 1)$ to X satisfying both $\lim_{h\downarrow 0} \eta_h = \eta$ and (2.4.3), *then $\eta \in \mathcal{T}_K(\xi)$.*

Proof. Let $\eta \in \mathcal{T}_K(\xi)$ and let $h \mapsto \eta_h$ be any function satisfying $\lim_{h\downarrow 0} \eta_h = \eta$. Since

$$\text{dist}\,(\xi + h\eta_h; K) \le h\|\eta_h - \eta\| + \text{dist}\,(\xi + h\eta; K)$$

for each $h \in (0,1)$, (2.4.1) implies (2.4.3) and this completes the first assertion.

Next, let us assume that there exists $h \mapsto \eta_h$ from $(0,1)$ to X such that $\lim_{h\downarrow 0} \eta_h = \eta$ and satisfying (2.4.3). Since

$$\text{dist}\,(\xi + h\eta; K) \le h\|\eta - \eta_h\| + \text{dist}\,(\xi + h\eta_h; K)$$

for each $h \in (0,1)$, it follows that (2.4.1) holds also true and this complete the proof. □

An immediate consequence of Proposition 2.4.3 is

Proposition 2.4.4. *A necessary and sufficient condition in order that a vector $\eta \in X$ to belong to the cone $\mathcal{T}_K(\xi)$ is to exist a function $h \mapsto \eta_h$ from $(0,1)$ to X satisfying both $\lim_{h\downarrow 0} \eta_h = \eta$ and (2.4.3).*

Theorem 2.4.1. *Let $K_1, K_2 \subseteq X$ be locally closed. If $\xi \in K_1 \cap K_2$ is an interior point of $K_1 \cup K_2$, then we have*

$$\mathcal{T}_{K_1\cap K_2}(\xi) = \mathcal{T}_{K_1}(\xi) \cap \mathcal{T}_{K_2}(\xi).$$

Proof. Obviously, for each $\xi \in K_1 \cap K_2$, $\mathcal{T}_{K_1\cap K_2}(\xi) \subseteq \mathcal{T}_{K_1}(\xi) \cap \mathcal{T}_{K_2}(\xi)$. To prove that, whenever, in addition, ξ is in the interior of $K_1 \cup K_2$, the converse inclusion holds true, let $\xi \in K_1 \cap K_2$ and let $\eta \in \mathcal{T}_{K_1}(\xi) \cap \mathcal{T}_{K_2}(\xi)$. By Corollary 2.4.1, there exist four sequences $(h_m)_m$, $(\widetilde{h}_m)_m$ in $\mathbb{R}_+$, $(p_m)_m$ and $(\widetilde{p}_m)_m$ in X with $h_m \downarrow 0$, $\widetilde{h}_m \downarrow 0$, $\lim_m p_m = 0$, $\lim_m \widetilde{p}_m = 0$ and such that $\xi + h_m(\eta + p_m) \in K_1$ and $\xi + \widetilde{h}_m(\eta + \widetilde{p}_m) \in K_2$ for each $m \in \mathbb{N}$. Now, since ξ is in the interior of $K_1 \cup K_2$, there exists $\rho > 0$ such that $D(\xi, \rho) \subseteq K_1 \cup K_2$. Since K_1 and K_2 are locally closed, diminishing $\rho > 0$ if necessary, we may assume that both $K_1 \cap D(\xi, \rho)$ and $K_2 \cap D(\xi, \rho)$ are closed. Let $m_0 \in \mathbb{N}$ be such that, for each $m \ge m_0$, we have both $\xi_m = \xi + h_m(\eta + p_m) \in D(\xi, \rho)$ and $\widetilde{\xi}_m = \xi + \widetilde{h}_m(\eta + \widetilde{p}_m) \in D(\xi, \rho)$. As a consequence, if $m \ge m_0$, the line segment $[\,\xi_m, \widetilde{\xi}_m\,] \subseteq D(\xi, \rho) \subseteq K_1 \cup K_2$. Since $D(\xi, \rho)$ is connected, while $D(\xi, \rho) \cap K_1$ and $D(\xi, \rho) \cap K_2$ are closed, there exists $\eta_m \in [\,\xi_m, \widetilde{\xi}_m\,] \cap K_1 \cap K_2$. Since $\eta_m \in [\,\xi_m, \widetilde{\xi}_m\,]$, there exists $\theta_m \in [\,0,1\,]$ such that $\eta_m = (1-\theta_m)\xi_m + \theta_m\widetilde{\xi}_m$. So, denoting

$$t_m = (1-\theta_m)h_m + \theta_m\widetilde{h}_m \quad \text{and} \quad q_m = \frac{(1-\theta_m)h_m}{t_m}p_m + \frac{\theta_m\widetilde{h}_m}{t_m}\widetilde{p}_m,$$

we have

$$\eta_m = \xi + t_m(\eta + q_m) \in K_1 \cap K_2.$$

Finally, observing that $t_m \downarrow 0$ and $\lim_m q_m = 0$, and using Corollary 2.4.1, we get the conclusion. □

Definition 2.4.2. Let $K \subseteq X$ and $\xi \in K$. The vector $\eta \in X$ is *tangent in the sense of Federer* to the set K at the point ξ if

$$\lim_{h \downarrow 0} \frac{1}{h} \operatorname{dist}(\xi + h\eta; K) = 0.$$

We denote by $\mathcal{F}_K(\xi)$ the set of all points $\eta \in X$ which are tangent in the sense of Federer to K at ξ.

Remark 2.4.3. One may easily see that $\mathcal{F}_K(\xi)$ is a cone which is included in $\mathcal{T}_K(\xi)$.

As we will next show, if K is proximal, $f : K \to X$ is continuous and the norm of X is Gâteaux differentiable at each $\xi \in X$, $\xi \neq 0$, the following surprising equivalence holds true: $f(\xi) \in \mathcal{T}_K(\xi)$ for each $\xi \in K$ if and only if $f(\xi) \in \mathcal{F}_K(\xi)$ for each $\xi \in K$, and this in spite of the fact that $\mathcal{F}_K(\xi) \neq \mathcal{T}_K(\xi)$.

2.5. Other types of tangent vectors

The proximal normal cone. The next concept depends on the norm considered, in the sense that it is not preserved by equivalent norms.

Definition 2.5.1. Let $\xi \in K$. We say that $\nu \in X$ is *metric normal to K at ξ* if there exist $\eta \in X$ and $\rho > 0$ such that $D(\eta, \rho)$ contains ξ on its boundary, its interior has empty intersection with K, and $\nu = \eta - \xi$.

Remark 2.5.1. In the case in which V is a proximal neighborhood of K, $\nu \in X$ is metric normal to K at ξ if and only if there exist $\eta \in V$ and $\lambda > 0$ such that $\xi \in \Pi_K(\eta)$ and $\nu = \lambda(\eta - \xi)$.

We emphasize that this concept is essentially dependent of the norm considered. More precisely, it may happen that a set K has a normal vector with respect to a given norm, but doesn't have normal vectors with respect to another norm, even though the two norms are equivalent. Figure 2.5.1 illustrates a point ξ at which there is one metric normal vector with respect to the ℓ_∞ norm on $\mathbb{R}^2$, i.e. $\|(x, y)\|_\infty = \max\{|x|, |y|\}$, but there is no metric normal vector with respect to the usual Euclidian norm.

Definition 2.5.2. The *proximal normal cone to K at $\xi \in K$* is the set of all $\zeta \in X$ of the form $\zeta = \lambda\nu$, where ν is metric normal to K at ξ and

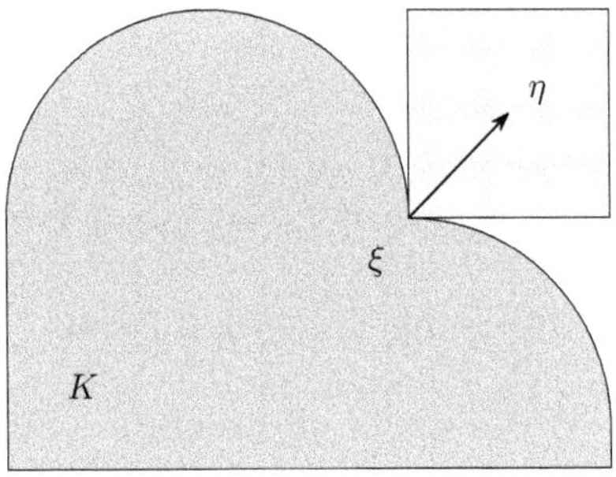

Figure 2.5.1

$\lambda \geq 0$, whenever such a metric normal ν exists, and $\{0\}$ if there is no metric normal to K at ξ. We denote this cone by $\mathcal{N}_K(\xi)$.

Remark 2.5.2. We have

$$\mathcal{N}_K(\xi) = \{\nu;\ \exists \lambda > 0,\ \operatorname{dist}(\xi + \lambda\nu; K) = \lambda\|\nu\|\}\,.$$

The use of the term "cone" in Definition 2.5.2 is justified by the simple observation that, for each $\xi \in K$, $\mathcal{N}_K(\xi)$ is a cone in the usual sense, i.e., for each $\zeta \in \mathcal{N}_K(\xi)$ and $\lambda > 0$, we have $\lambda\zeta \in \mathcal{N}_K(\xi)$.

Let now $\mathbb{S}_K(\xi)$ be the set of all $\eta \in X$ such that $\eta - \xi$ is metric normal to K at ξ whenever such a metric normal exists, and $\mathbb{S}_K(\xi) = \{\xi\}$ otherwise. Let $\eta \in \mathbb{S}_K(\xi)$, and let $E(\xi, \eta) = \{\zeta \in X;\ \|\eta - \zeta\| \geq \|\eta - \xi\|\}$. Since $K \subseteq E(\xi, \eta)$, for each $\eta \in \mathbb{S}_K(\xi)$, we have

$$\mathcal{T}_K(\xi) \subseteq \mathcal{B}_K(\xi), \tag{2.5.1}$$

where

$$\mathcal{B}_K(\xi) = \bigcap_{\eta \in \mathbb{S}_K(\xi)} \mathcal{T}_{E(\xi,\eta)}(\xi).$$

One may easily see that $\mathcal{B}_K(\xi)$ is a cone in X.

Definition 2.5.3. The set $\mathcal{B}_K(\xi)$, defined as above, is called the *Bony tangent cone to K at $\xi \in K$*, and its elements are *tangents in the sense of Bony* to K at ξ.

Remark 2.5.3. Taking into account the definitions of both $E(\xi, \eta)$ and $[\cdot, \cdot]_+$, we easily deduce that $\zeta \in \mathcal{T}_{E(\xi,\eta)}(\xi)$ if and only if $[\xi - \eta, \zeta]_+ \geq 0$. Therefore, $\zeta \in \mathcal{B}_K(\xi)$ if and only if $[-\nu, \zeta]_+ \geq 0$ for each ν which is metric normal to K at ξ. In particular, when X is a Hilbert space with inner product $\langle\cdot,\cdot\rangle$, taking into account that

$$[x,y]_+ = \begin{cases} \dfrac{\langle x,y\rangle}{\|x\|} & \text{if } x \neq 0 \\ \|y\| & \text{if } x = 0, \end{cases}$$

we easily deduce that, for each $\xi \in K$ and each $\eta \in \mathcal{S}_K(\xi)$, $\mathcal{T}_{E(\xi,\eta)}(\xi)$ is a closed half-space having the exterior normal $\eta - \xi$. Therefore, in this case, we have

$$\mathcal{B}_K(\xi) = (\mathcal{N}_K(\xi))^*,$$

where $(\mathcal{N}_K(\xi))^*$ is the so-called *conjugate cone* of $\mathcal{N}_K(\xi)$ i.e.,

$$(\mathcal{N}_K(\xi))^* = \{\eta \in X\,;\ \langle\nu,\eta\rangle \leq 0,\ \text{for each } \nu \in \mathcal{N}_K(\xi)\}.$$

Remark 2.5.4. If there is no metric normal vector to K at ξ, we may easily see that $\mathcal{B}_K(\xi) = X$. See Figure 2.5.2.

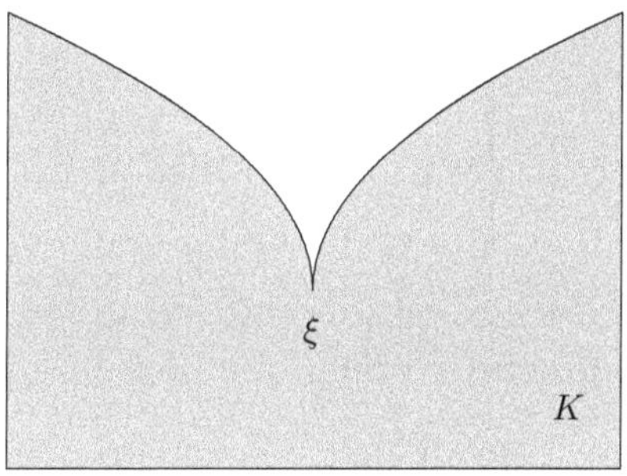

Figure 2.5.2

The Clarke's tangent cone. We are now ready to study another useful tangency concept.

Definition 2.5.4. Let $K \subseteq X$ and $\xi \in K$. The vector $\eta \in X$ is *tangent in the sense of Clarke* to the set K at the point ξ if

$$\lim_{\substack{h\downarrow 0\\ \mu\to\xi\\ \mu\in K}} \frac{1}{h}\,\mathrm{dist}\,(\mu + h\eta; K) = 0.$$

We denote by $\mathcal{C}_K(\xi)$ the set of all vectors $\eta \in X$ which are tangent to $\xi \in K$ in the sense of Clarke. We can easily verify that $\mathcal{C}_K(\xi)$ is a closed convex cone.

Remark 2.5.5. One may easily see that, for each K and each $\xi \in K$, we have

$$\mathcal{C}_K(\xi) \subseteq \mathcal{F}_K(\xi) \subseteq \mathcal{T}_K(\xi) \subseteq \mathcal{B}_K(\xi).$$

The inclusions above may be strict as the following example shows.

Example 2.5.1. Let $K \subseteq \mathbb{R}^2$ be defined as $K = K_1 \cup K_2$, where

$$K_1 = \{(x, y)\,;\ (x, y) \in \mathbb{R}^2,\ y \leq |x|\}$$

and

$$K_2 = \{(0, 1/2^m)\,;\ m \in \mathbb{N}\}$$

and let $\xi = (0, 0) \in K$. Then, one may easily verify that

$$\begin{cases} \mathcal{C}_K(\xi) = \{0\} \\ \mathcal{F}_K(\xi) = K_1 \\ \mathcal{T}_K(\xi) = K_1 \cup \{(0, y)\,;\ y \geq 0\} \\ \mathcal{B}_K(\xi) = \mathbb{R}^2. \end{cases}$$

For a multi-function $F : K \rightsquigarrow X$, we define

$$\liminf_{\substack{\xi \to \xi_0 \\ \xi \in K}} F(\xi) = \left\{ \eta \in X\,;\ \lim_{\substack{\xi \to \xi_0 \\ \xi \in K}} \operatorname{dist}\,(\eta; F(\xi)) = 0 \right\}.$$

Lemma 2.5.1. *If the norm $\|\cdot\|$ is Gâteaux differentiable at each $x \in X$, $x \neq 0$, and $K \subseteq X$ is proximal*[2]*, then, for each $\xi_0 \in K$, we have*

$$\liminf_{\substack{\xi \to \xi_0 \\ \xi \in K}} \mathcal{B}_K(\xi) \subseteq \mathcal{C}_K(\xi_0). \tag{2.5.2}$$

Proof. Let $\eta \neq 0$, $\eta \in \liminf_{\substack{\xi \to \xi_0 \\ \xi \in K}} \mathcal{B}_K(\xi)$. It follows that, for each $\varepsilon > 0$, there exists $\theta > 0$ such that, for each $\phi \in K \cap D(\xi_0, \theta)$, we have

$$D(\eta, \varepsilon) \cap \mathcal{B}_K(\phi) \neq \emptyset. \tag{2.5.3}$$

Take a sufficiently small θ so that, for all $\xi \in K \cap D(\xi_0, \frac{\theta}{4})$ and $t \in [\,0, \frac{\theta}{4\|\eta\|}\,]$, we have $\Pi_K(\xi + t\eta) \neq \emptyset$. Since K is proximal, this is always possible. With ξ and t as above, let us define $g(t) = \operatorname{dist}\,(\xi + t\eta; K)$. In order to prove that $\eta \in \mathcal{C}_K(\xi_0)$, it suffices to show that $g(t) \leq \varepsilon t$ for each $t \in [\,0, \frac{\theta}{4\|\eta\|}\,]$. Further,

[2]See Definition 2.2.3.

since $g(0) = 0$, it suffices to show that $g'(t) \leq \varepsilon$, whenever $g'(t)$ exists and $g(t) \neq 0$. Indeed, since g is Lipschitz and $g(0) = 0$, we have

$$g(t) = \int_{[0,t]\setminus E} g'(s)\, ds,$$

where $E = \{s \in [0,t];\ g(s) = 0\}$, because $\int_E g'(s)\, ds = 0$.

So, let $t \in [0, \frac{\theta}{4\|\eta\|}]$ with $g(t) \neq 0$. Then there exists $\phi \in \Pi_K(\xi + t\eta)$, with $\phi \neq \xi + t\eta$. Let us observe that we have $\phi \in K \cap D(\xi_0, \theta)$. Indeed,

$$\|\phi - \xi_0\| \leq \|\xi + t\eta - \phi\| + \|\xi + t\eta - \xi_0\| \leq 2\|\xi + t\eta - \xi_0\| \leq \theta,$$

as claimed. Now, for a sufficiently small $h > 0$, we obtain

$$g(t+h) - g(t) \leq \|\xi + t\eta + h\eta - \phi\| - \|\xi + t\eta - \phi\|.$$

Dividing by h and letting $h \downarrow 0$, we get

$$g'(t) \leq [\xi + t\eta - \phi, \eta]_+. \tag{2.5.4}$$

Taking into account that $\phi \in \Pi_K(\xi + t\eta)$, from Definition 2.5.1, we deduce that $\xi + t\eta - \phi$ is metric normal to K at ϕ. In view of (2.5.3), there exists $w \in \mathcal{B}_K(\phi)$ with $\|\eta - w\| \leq \varepsilon$.

Since $\|\cdot\|$ is Gâteaux differentiable at each $x \in X$, $x \neq 0$, by (x) in Exercise 1.6.1, we know that, for each $x \in X \setminus \{0\}$ and each $y \in X$, we have $[x, y]_+ = -[-x, y]_+$. Since $\xi + t\eta - \phi \neq 0$, from the observation above and Remark 2.5.3, we conclude that

$$[\xi + t\eta - \phi, w]_+ \leq 0.$$

Using (ii) in Exercise 1.6.1 and the fact that $\|\eta - w\| \leq \varepsilon$, we deduce

$$[\xi + t\eta - \phi, \eta]_+ \leq [\xi + t\eta - \phi, \eta - w]_+ + [\xi + t\eta - \phi, w]_+ \leq \varepsilon.$$

From this inequality and (2.5.4), we get $g'(t) \leq \varepsilon$, as claimed. Thus (2.5.2) holds and this completes the proof. □

Problem 2.5.1. *Show that if X is finite dimensional then*

$$\mathcal{C}_K(\xi_0) \subseteq \liminf_{\substack{\xi \to \xi_0 \\ \xi \in K}} \mathcal{T}_K(\xi).$$

If, in addition, K is locally closed and $\|\cdot\|$ is Gâteaux differentiable, we have also

$$\liminf_{\substack{\xi \to \xi_0 \\ \xi \in K}} \mathcal{B}_K(\xi) = \mathcal{C}_K(\xi_0).$$

Proposition 2.5.1. *Let $K \subseteq X$ be proximal and let $f : K \to X$ be continuous. Let us assume that the norm $\|\cdot\|$ is Gâteaux differentiable at each $x \in X$, $x \neq 0$. Then, the following conditions are equivalent:*

(i) *for each* $\xi \in K$, $f(\xi) \in \mathcal{C}_K(\xi)$;
(ii) *for each* $\xi \in K$, $f(\xi) \in \mathcal{T}_K(\xi)$;
(iii *for each* $\xi \in K$, $f(\xi) \in \mathcal{B}_K(\xi)$.

In general, if $\mathcal{G} : K \rightsquigarrow X$ *is such that* $\mathcal{C}_K(\xi) \subseteq \mathcal{G}(\xi) \subseteq \mathcal{B}_K(\xi)$ *for each* $\xi \in K$, *then each one of the conditions above is equivalent to*

(iv) *for each* $\xi \in K$, $f(\xi) \in \mathcal{G}(\xi)$.

Proof. In view of Remark 2.5.5, it suffices to show that (iii) implies (i). But this easily follows from Lemma 2.5.1 and this completes the proof. □

2.6. Multi-functions

In this section we include several basic notions and results referring to multi-functions, i.e., to functions whose values are sets. Let K and X be topological spaces and let $F : K \rightsquigarrow X$ be a given multi-function, i.e., a function $F : K \to 2^X$.

Definition 2.6.1. The multi-function $F : K \rightsquigarrow X$ is *lower semicontinuous (l.s.c.) at* $\xi \in K$ if for every open set V in X with $F(\xi) \cap V \neq \emptyset$ there exists an open neighborhood U of ξ such that $F(\eta) \cap V \neq \emptyset$ for each $\eta \in U \cap K$. We say that F is *lower semicontinuous (l.s.c.) on* K if it is l.s.c. at each $\xi \in K$.

By a *selection* of the multi-function $F : K \rightsquigarrow X$ we mean a function $f : K \to X$ satisfying $f(x) \in F(x)$ for each $x \in K$.

Theorem 2.6.1. *Let* K *be a metric space,* X *a Banach space and let* $F : K \rightsquigarrow X$ *be a l.s.c. multi-function with nonempty, closed and convex values. Then, for each* $\xi \in K$ *and each* $\eta \in F(\xi)$, *there exists a continuous selection* $f : K \to X$ *of* F *such that* $f(\xi) = \eta$.

For the proof of this result, known as Michael Continuous Selection Theorem, see Deimling [**77**], Theorem 24.1, p. 303.

Definition 2.6.2. The multi-function $F : K \rightsquigarrow X$ is *upper semicontinuous (u.s.c.) at* $\xi \in K$ if for every open neighborhood V of $F(\xi)$ there exists an open neighborhood U of ξ such that $F(\eta) \subseteq V$ for each $\eta \in U \cap K$. We say that F is *upper semicontinuous (u.s.c.) on* K if it is u.s.c. at each $\xi \in K$.

Problem 2.6.1. *Let* $f_i : \mathbb{R} \to \mathbb{R}$, $i = 1, 2$, *be two bounded functions with* f_1 *l.s.c. and* f_2 *u.s.c. in the usual sense, i.e., for each* $y \in \mathbb{R}$ *we have* $\liminf_{x \to y} f_1(x) = f_1(y)$ *and* $\limsup_{x \to y} f_2(x) = f_2(y)$. *Let* Ω *be a nonempty, open and bounded subset in* $\mathbb{R}^n$. *Prove that the multi-function* $F : L^1(\Omega) \rightsquigarrow L^1(\Omega)$, *defined by taking* $F(u)$ *as the set of all functions*

$f \in L^1(\Omega)$ satisfying $f_1(u(x)) \leq f(x) \leq f_2(u(x))$ a.e. for $x \in \Omega$, has nonempty, convex and weakly compact values and is strongly-weakly u.s.c. on $L^1(\Omega)$, in the sense of Definition 2.6.2.

The next two lemmas will prove useful later.

Lemma 2.6.1. *If $F : K \rightsquigarrow X$ is a nonempty and (weakly) compact valued, (strongly-weakly) u.s.c. multi-function, then, for each compact subset C of K, $\cup_{\xi \in C} F(\xi)$ is (weakly) compact. In particular, in both cases, for each compact subset C of K, there exists $M > 0$ such that $\|\eta\| \leq M$ for each $\xi \in C$ and each $\eta \in F(\xi)$.*

Proof. Let C be a (weakly) compact subset in K and let $\{D_\sigma\,;\ \sigma \in \Gamma\}$ be an arbitrary (weakly) open covering of $\cup_{\xi \in C} F(\xi)$. Since F is (weakly) compact valued, for each $\xi \in C$ there exists $n(\xi) \in \mathbb{N}$ such that

$$F(\xi) \subseteq \bigcup_{1 \leq k \leq n(\xi)} D_{\sigma_k}.$$

But F is (strongly-weakly) u.s.c. and therefore there exists a (weakly) open neighborhood $U(\xi)$ of ξ such that

$$F(U(\xi) \cap K) \subseteq \bigcup_{1 \leq k \leq n(\xi)} D_{\sigma_k}.$$

The family $\{U(\xi)\,;\ \xi \in C\}$ is an (weakly) open covering of C. As C is compact, there exists a finite family $\{\xi_1, \xi_2, \ldots, \xi_p\}$ in C such that

$$F(C) \subseteq \bigcup_{1 \leq j \leq p} F(U(\xi_j) \cap K) \subseteq \bigcup_{1 \leq j \leq p} \bigcup_{1 \leq k \leq n(\xi_j)} D_{\sigma_k},$$

and this completes the proof. □

Lemma 2.6.2. *Let X be a Banach space and K a nonempty subset in X. Let $F : K \rightsquigarrow X$ be a nonempty, closed and convex valued, strongly-weakly u.s.c. multi-function*[3], *and let $u_m : [\,0, T\,] \to K$ and $f_m \in L^1(0, T; X)$ be such that $f_m(t) \in F(u_m(t))$ for each $m \in \mathbb{N}$ and a.e. for $t \in [\,0, T\,]$.*

If $\lim_m u_m(t) = u(t)$ *a.e. for $t \in [\,0, T\,]$ and* $\lim_m f_m = f$ *weakly in $L^1(0, T; X)$, then $f(t) \in F(u(t))$ a.e. for $t \in [\,0, T\,]$.*

Proof. By Corollary 1.1.1, there exists a sequence $(g_m)_m$ of convex combinations of $\{f_k\,;\ k \geq m\}$, i.e., $g_m \in \operatorname{conv}\{f_m, f_{m+1}, \ldots\}$ for each $m \in \mathbb{N}$, which converges strongly in $L^1(0, T; X)$ to f. By a classical result

[3] Of course if F is strongly-strongly u.s.c. it is strongly-weakly u.s.c. too and thus the conclusion of Lemma 2.6.2 holds true also in this case.

due to Lebesgue, we know that there exists a subsequence (g_{m_p}) of (g_m) which converges almost everywhere on $[0, T]$ to f. Denote by $\mathcal{T}$ the set of all $s \in [0, T]$ such that both $\left(g_{m_p}(s)\right)_p$ and $(u_m(s))_m$ are convergent to $f(s)$ and to $u(s)$ respectively and, in addition, $f_m(s) \in F(u_m(s))$ for each $m \in \mathbb{N}$. Clearly $[0, T] \setminus \mathcal{T}$ has null measure. Let $s \in \mathcal{T}$ and let E be an open half-space in X including $F(u(s))$. Since F is strongly-weakly u.s.c. at $u(s)$, $(u_m(s))_m$ converges to $u(s)$ and E is a weak neighborhood of $F(u(s))$, there exists $m(E)$ belonging to $\mathbb{N}$, such that $F(u_m(s)) \subseteq E$ for each $m \geq m(E)$. From the relation above, taking into account that $f_m(s) \in F(u_m(s))$ for each $m \in \mathbb{N}$ and a.e. for $s \in [0, T]$, we easily conclude that

$$g_{m_p}(s) \in \overline{\text{conv}}\left(\bigcup_{m \geq m(E)} F(u_m(s))\right)$$

for each $p \in \mathbb{N}$ with $m_p \geq m(E)$. Passing to the limit for $p \to +\infty$ in the relation above we deduce that $f(s) \in \overline{E}$. Since $F(u(s))$ is closed and convex, it is the intersection of all closed half-spaces which include it. So, inasmuch as E was arbitrary, we finally get $f(s) \in F(u(s))$ for each $s \in \mathcal{T}$ and this completes the proof. □

2.7. Measures of noncompactness

Let X be a Banach space and let $\mathcal{B}(X)$ the family of all bounded subsets of X. Let $\varepsilon > 0$ and let us denote by

$$\mathcal{B}_\varepsilon(X) = \{B \in \mathcal{B}(X);\ \text{diam}\,(B) \leq \varepsilon\},$$

where, as expected, diam (B) is the *diameter* of the set B, i.e.

$$\text{diam}\,(B) = \sup\{\|x - y\|;\ x, y \in B\}.$$

Definition 2.7.1. The function $\alpha : \mathcal{B}(X) \to \mathbb{R}_+$, defined by

$$\alpha(B) = \inf\left\{\varepsilon > 0; \exists B_1, B_2, \ldots, B_{n(\varepsilon)} \in \mathcal{B}_\varepsilon(X),\ \ B \subseteq \bigcup_{i=1}^{n(\varepsilon)} B_i\right\}$$

is called the *Kuratowski measure of noncompactness on* X.

Definition 2.7.2. The function $\beta : \mathcal{B}(X) \to \mathbb{R}_+$, defined by

$$\beta(B) = \inf\left\{\varepsilon > 0; \exists x_1, x_2, \ldots, x_{n(\varepsilon)} \in X,\ B \subseteq \bigcup_{i=1}^{n(\varepsilon)} D(x_i, \varepsilon)\right\}$$

is called the *Hausdorff measure of noncompactness on* X.

Remark 2.7.1. One may easily see that, for each $B \in \mathcal{B}(X)$ and $r > 0$ with $B \subseteq D(0, r)$, we have

$$\beta(B) \leq r.$$

Remark 2.7.2. Clearly $\alpha(B) = 0$ ($\beta(B) = 0$) if and only if B is relatively compact. Therefore, since whenever X is finite dimensional, $\mathcal{B}(X)$ coincides with the class of relatively compact subsets of X, by the definition of both α and β, it follows that, in this case, $\alpha \equiv \beta \equiv 0$. Therefore, in that follows we will assume that X is infinite dimensional, case in which either of the two functions α and β estimates the "magnitude of the lack of compactness".

Problem 2.7.1. *Let* $\gamma : \mathcal{B}(X) \to \mathbb{R}_+$ *be either* α *or* β. *Prove that:*

(i) $\gamma(B) \leq \operatorname{diam}(B)$ *and* $\gamma(B) = \gamma(\overline{B})$;
(ii) $\gamma(B) = 0$ *if and only if* B *is relatively compact*;
(iii) $\gamma(\lambda B) \leq |\lambda|\gamma(B)$ *for each* $\lambda \in \mathbb{R}$ *and* $B \in \mathcal{B}(X)$;
(iv) $\gamma(B + C) \leq \gamma(B) + \gamma(C)$ *for each* $B, C \in \mathcal{B}(X)$;
(v) *if* $B \subseteq C$ *then* $\gamma(B) \leq \gamma(C)$;
(vi) $\gamma(B \cup C) = \max\{\gamma(B), \gamma(C)\}$;
(vii) $\gamma(\operatorname{conv}(B)) = \gamma(B)$;
(viii) γ *is Lipschitz continuous with respect to the Hausdorff-Pompeiu distance. More precisely,* $|\gamma(B) - \gamma(C)| \leq L \operatorname{dist}_{HP}(B; C)$ *for each* $B, C \in \mathcal{B}(X)$, *where* $L = 2$ *if* $\gamma = \alpha$ *and* $L = 1$ *if* $\gamma = \beta$.

See Deimling [**77**], Proposition 7.2, p. 41.

Problem 2.7.2. *Let* Y *be a subspace in* X. *If* $B \in \mathcal{B}(X)$, *we define*

$$\beta_Y(B) = \inf\left\{\varepsilon > 0; \exists x_1, x_2, \ldots, x_{n(\varepsilon)} \in Y,\ B \subseteq \bigcup_{i=1}^{n(\varepsilon)} D(x_i, \varepsilon)\right\}.$$

Show that for each $B \in \mathcal{B}(Y)$ *we have*

$$\beta(B) \leq \beta_Y(B) \leq \alpha(B) \leq 2\beta(B).$$

For details see Akhmerov–Kamenskii–Potapov–Rodkina–Sadovskii [**3**], Theorem 1.1.7, p. 4 and Mönch [**130**].

Lemma 2.7.1. *Let* X *be a separable Banach space. Then for each sequence of finite dimensional subspaces* $(X_n)_n$, *with* $X_n \subseteq X_{n+1}$ *for all* $n \in \mathbb{N}$ *and* $X = \overline{\cup_n X_n}$, *and for every bounded and countable set* $\{x_m;\ m \in \mathbb{N}\}$, *we have*

$$\beta(\{x_m;\ m \in \mathbb{N}\}) = \lim_n \limsup_k \operatorname{dist}(x_k; X_n).$$

Proof. We begin by proving the inequality

$$\beta(\{x_m;\ m \in \mathbb{N}\}) \leq \lim_n \limsup_k \operatorname{dist}(x_k; X_n).$$

To this aim, let $n \in \mathbb{N}$ and $\varepsilon > 0$ be arbitrary, and let

$$r_n = \limsup_k \operatorname{dist}(x_k; X_n).$$

Let us choose $k(\varepsilon, n) \in \mathbb{N}$ such that

$$\operatorname{dist}(x_k; X_n) \leq r_n + \varepsilon$$

for each $k \in \mathbb{N}$, $k \geq k(\varepsilon, n)$. Let us define

$$P = \bigcup_{k \geq k(\varepsilon,n)} \{u \in X_n;\ \operatorname{dist}(x_k; X_n) = \|x_k - u\|\}.$$

Then, the set $Q = \{x_k;\ 1 \leq k \leq k(\varepsilon, n)\} \cup P$ is relatively compact. So, there exists a finite set $\{u_i;\ i = 1, 2, \ldots, n(\varepsilon)\}$ such that

$$Q \subseteq \bigcup_{i=1}^{n(\varepsilon)} D(u_i, \varepsilon).$$

Then

$$\{x_m;\ m \in \mathbb{N}\} \subseteq Q \bigcup \bigcup_{u \in P} D(u, r_n + \varepsilon) \subseteq \bigcup_{u \in Q} D(u, r_n + \varepsilon) \subseteq \bigcup_{i=1}^{n(\varepsilon)} D(u_i, r_n + 2\varepsilon)$$

and consequently

$$\beta(\{x_m;\ m \in \mathbb{N}\}) \leq r_n + 2\varepsilon.$$

Since $\varepsilon > 0$ is arbitrary and $X_n \subseteq X_{n+1}$ for each $n \in \mathbb{N}$, we deduce

$$\beta(\{x_m;\ m \in \mathbb{N}\}) \leq \inf_n \limsup_k \operatorname{dist}(x_k; X_n) = \lim_n \limsup_k \operatorname{dist}(x_k; X_n).$$

Next, we will prove the converse inequality, i.e.,

$$\beta(\{x_m;\ m \in \mathbb{N}\}) \geq \lim_n \limsup_k \operatorname{dist}(x_k; X_n).$$

To this end, let $\varepsilon > 0$ be arbitrary and let $r = \beta(\{x_m;\ m \in \mathbb{N}\})$. Then there exists a finite set $\{u_i;\ i = 1, 2, \ldots, m(\varepsilon)\}$ such that

$$\{x_m;\ m \in \mathbb{N}\} \subseteq \bigcup_{i=1}^{m(\varepsilon)} D(u_i, r + \varepsilon).$$

Since $\cup_n X_n$ is dense in X, and $X_n \subseteq X_{n+1}$ for each $n \in \mathbb{N}$, there exists $n(\varepsilon) \in \mathbb{N}$ such that

$$\sup\{\operatorname{dist}(u_i; X_n);\ i = 1, 2, \ldots, m(\varepsilon)\} \leq \varepsilon$$

for each $n \in \mathbb{N}$, $n \geq n(\varepsilon)$. Accordingly,

$$\begin{aligned}\operatorname{dist}(x_k; X_n) &\leq \inf\{\|x_k - u_i\|;\ i = 1, 2, \ldots, m(\varepsilon)\}\\ &+ \sup\{\operatorname{dist}(u_i; X_n);\ i = 1, 2, \ldots, m(\varepsilon)\} \leq r + \varepsilon + \varepsilon\end{aligned}$$

for each $n \in \mathbb{N}$, $n \geq n(\varepsilon)$. Hence

$$2\varepsilon + \beta(\{x_m;\ m \in \mathbb{N}\}) \geq \limsup_k \operatorname{dist}(x_k; X_n).$$

Since the right hand side is decreasing as a function of n, we can pass to the limit for $n \to \infty$ and, since $\varepsilon > 0$ is arbitrary, we get

$$\beta(\{x_m;\ m \in \mathbb{N}\}) \geq \lim_n \limsup_k \operatorname{dist}(x_k; X_n),$$

as claimed. □

Lemma 2.7.2. *Let X be a separable Banach space and $\{F_m;\ m \in \mathbb{N}\}$ a subset in $L^1(\tau, T; X)$ for which there exists $\ell \in L^1(\tau, T; \mathbb{R}_+)$ such that*

$$\|F_m(s)\| \leq \ell(s)$$

for each $m \in \mathbb{N}$ and a.e. for $s \in [\,\tau, T\,]$. Then the mapping

$$s \mapsto \beta(\{F_m(s);\ m \in \mathbb{N}\})$$

is integrable on $[\,\tau, T\,]$ and, for each $t \in [\,\tau, T\,]$, we have

$$\beta\left(\left\{\int_\tau^t F_m(s)\,ds;\ m \in \mathbb{N}\right\}\right) \leq \int_\tau^t \beta(\{F_m(s);\ m \in \mathbb{N}\})\,ds. \tag{2.7.1}$$

Proof. Let $\{x_n;\ n \in \mathbb{N}\} \subseteq X$ be a countable and dense set in X. For $n \in \mathbb{N}$, let us define X_n as the space spanned by $\{x_1, x_2, \ldots, x_n\}$. Clearly X_n is finite dimensional, and $\overline{\cup_n X_n} = X$. So, we are in the hypotheses of Lemma 2.7.1, and therefore, for a.a. $s \in [\,0, T\,]$, we have

$$\beta(\{F_m(s);\ m \in \mathbb{N}\}) = \lim_n \limsup_k \operatorname{dist}(F_k(s); X_n).$$

Since for each $n, m \in \mathbb{N}$, the distance function $x \mapsto \operatorname{dist}(x; X_n)$ is Lipschitz and F_m is measurable, this shows that $s \mapsto \beta(\{F_m(s);\ m \in \mathbb{N}\})$ is measurable. Moreover, we have $\beta(\{F_m(s);\ m \in \mathbb{N}\}) \leq \ell(s)$ a.e. for $s \in [\,0, T\,]$ and thus, $s \mapsto \beta(\{F_m(s);\ m \in \mathbb{N}\})$ is integrable.

For each $n \in \mathbb{N}$ and $x \in X$ let us fix $x_n \in X_n$ such that

$$\operatorname{dist}(x; X_n) = \|x - x_n\|.$$

Since X_n is a subspace in X, for each $x, y \in X$ and $a \in \mathbb{R}$, we have $x_n + y_n \in X_n$ and $ax_n \in X_n$. Hence

$$\begin{aligned}\operatorname{dist}(x + y; X_n) &\leq \|(x + y) - (x_n + y_n)\|\\ &\leq \|x - x_n\| + \|y - y_n\| = \operatorname{dist}(x; X_n) + \operatorname{dist}(y; X_n)\end{aligned}$$

and

$$\operatorname{dist}(ax; X_n) \le \|ax - ax_n\| = |a| \operatorname{dist}(x; X_n).$$

A simple argument, involving Riemann sums and the two inequalities above, shows that, for each $n, m \in \mathbb{N}$, we have

$$\operatorname{dist}\left(\int_\tau^t F_m(s)\, ds; X_n\right) \le \int_\tau^t \operatorname{dist}(F_m(s); X_n)\, ds.$$

From Lemma 2.7.1 and this inequality, we deduce

$$\beta\left(\left\{\int_\tau^t F_m(s)\, ds;\ m \in \mathbb{N}\right\}\right) = \lim_n \limsup_k \operatorname{dist}\left(\int_\tau^t F_k(s)\, ds; X_n\right)$$

$$\le \lim_n \limsup_k \int_\tau^t \operatorname{dist}(F_k(s); X_n)\, ds.$$

By the Fatou Lemma 1.2.1 and the Lebesgue Dominated Convergence Theorem 1.2.3, we get

$$\lim_n \limsup_k \int_\tau^t \operatorname{dist}(F_k(s); X_n)\, ds \le \lim_n \int_\tau^t \limsup_k \operatorname{dist}(F_k(s); X_n)\, ds$$

$$= \int_\tau^t \lim_n \limsup_k \operatorname{dist}(F_k(s); X_n)\, ds.$$

Using once again Lemma 2.7.1, we obtain

$$\int_\tau^t \lim_n \limsup_k \operatorname{dist}(F_k(s); X_n)\, ds = \int_\tau^t \beta(\{F_m(s);\ m \in \mathbb{N}\})\, ds,$$

from where the conclusion. □

Remark 2.7.3. If X is not separable, there is a simple trick which may very often be useful. Namely, let $\{F_m;\ m \in \mathbb{N}\}$ be a subset in $L^1(\tau, T; X)$ for which there exists $\ell \in L^1(\tau, T; \mathbb{R}_+)$ such that

$$\|F_m(s)\| \le \ell(s)$$

for each $m \in \mathbb{N}$ and a.e. for $s \in [\tau, T]$. In view of Theorem 1.2.1, there exists a separable and closed subspace Y of X such that $F_m \in L^1(\tau, T; Y)$ for $m = 1, 2, \ldots$. Let us observe that the restriction of the mapping β_Y — see Problem 2.7.2 — to $\mathcal{B}(Y)$ coincides with the Hausdorff measure of noncompactness on Y. Then, from Lemma 2.7.2, it follows that the function $s \mapsto \beta_Y(\{F_m(s);\ m \in \mathbb{N}\})$ is integrable on $[\tau, T]$ and, for each $t \in [\tau, T]$, we have

$$\beta_Y\left(\left\{\int_\tau^t F_m(s)\, ds;\ m \in \mathbb{N}\right\}\right) \le \int_\tau^t \beta_Y(\{F_m(s);\ m \in \mathbb{N}\})\, ds.$$

Lemma 2.7.3. *Let $(u_n)_n$ be a bounded sequence in X such that*

$$\lim_k \beta(\{u_n;\ n \geq k\}) = 0.$$

Then $\{u_n;\ n \in \mathbb{N}\}$ is relatively compact.

Proof. In view of (vi) in Problem 2.7.1, we have

$$\beta(\{u_n;\ n \geq k\}) = \beta(\{u_n;\ n \in \mathbb{N}\}) = 0$$

for $k = 0, 1, \ldots$, and this completes the proof. □

Although extremely simple, the result above is a very useful compactness argument and therefore we decided to display it as a lemma.

2.8. Scorza Dragoni type theorems

We denote by λ the Lebesgue measure on $\mathbb{R}$, and by $\mathcal{L}$ the class of all Lebesgue measurable subsets in $\mathbb{R}$. Furthermore, if X is a topological space, we denote by $\mathcal{B}(X)$, or simply $\mathcal{B}$, the class of all Borel measurable subsets in X.

Definition 2.8.1. Let X and Y be Banach spaces, K a nonempty subset in X and I an open interval. A function $f : I \times K \to Y$ is a *Carathéodory function* if:

(C_1) for each $\xi \in K$, $t \mapsto f(t, \xi)$ is measurable on I;
(C_2) for a.a. $t \in I$, $u \mapsto f(t, u)$ is continuous on K;
(C_3) for each $\rho > 0$, there exists $\ell_\rho \in L^1_{\text{loc}}(I)$ such that

$$\|f(t, u)\| \leq \ell_\rho(t)$$

for a.a. $t \in I$ and for all $u \in D(0, \rho) \cap K$.

A function $f : I \times K \to Y$ satisfying (C_1), (C_2) and

(C_4) for each $\xi \in K$ there exist $\rho > 0$ and $\ell \in L^1_{\text{loc}}(I)$ such that

$$\|f(t, u)\| \leq \ell(t)$$

for a.a. $t \in I$ and for all $u \in D(\xi, \rho) \cap K$

is called a *locally Carathéodory function.*

First we recall the following result of Scorza Dragoni [**152**] type.

Theorem 2.8.1. *Let X and Y be two separable metric spaces and let $f : I \times X \to Y$ be a function such that $f(\cdot, u)$ is measurable for every $u \in X$ and $f(t, \cdot)$ is continuous for almost every $t \in I$. Then, for each $\varepsilon > 0$, there exists a closed set $A \subseteq I$ such that $\lambda(I \setminus A) < \varepsilon$ and the restriction of f to $A \times X$ is continuous.*

See Kucia [**117**].

Remark 2.8.1. The conclusion of Theorem 2.8.1 remains true in the case $Y = \mathbb{R}$, the function $f(t, \cdot)$ is lower semicontinuous (or upper semicontinuous) for almost every $t \in I$ and $f(\cdot, \cdot)$ is $\mathcal{L} \otimes \mathcal{B}$ measurable. It also holds if I is replaced by any Lebesgue measurable subset in $\mathbb{R}$.

Theorem 2.8.2. *Let X be a real Banach space, K a nonempty and separable subset in X and $f : I \times K \to X$ a Carathéodory function. Then there exists a negligible subset $\mathcal{Z}$ of I such that, for each $t \in I \setminus \mathcal{Z}$ and each function $u : \mathbb{R}_+ \times I \to K$ which is continuous with respect to the second variable and such that there exists $u(t) = \lim_{(h,s)\downarrow(0,t)} u(h, s)$, we have*

$$\lim_{h\downarrow 0} \frac{1}{h} \int_t^{t+h} \|f(s, u(h, s)) - f(t, u(t))\| \, ds = 0. \tag{2.8.1}$$

Proof. Since I can be represented as an at most countable union of finite length intervals, it suffices to consider the case when I is of finite length. For each $\gamma > 0$, we shall obtain a set $L_\gamma \subset I$, with $\lambda(I \setminus L_\gamma) < \gamma$, and such that (2.8.1) holds for all $t \in L_\gamma$. Finally, since $\lambda(I \setminus L_\gamma) < \gamma$, it will suffice to consider $\mathcal{Z} = \cap_m (I \setminus L_{1/m})$.

Let $\gamma > 0$ and let us observe that, by virtue of Theorem 2.8.1, it follows that there exists a compact set $A_\gamma \subseteq I$ such that $\lambda(I \setminus A_\gamma) < \gamma$, and the restriction of f to $A_\gamma \times K$ is continuous.

We define $L_\gamma \subseteq A_\gamma$ as the set of density points of A_γ which are also Lebesgue points of the functions $\widetilde{\ell}_m : I \to \mathbb{R}$, given by $\widetilde{\ell}_m(t) = \ell_m(t)\chi_{I \setminus A_\gamma}(t)$, where ℓ_m is given by (C_3), $m = 1, 2, \ldots$. It is known that $\lambda(L_\gamma) = \lambda(A_\gamma)$ and so, by the definition of a density point, for $t \in L_\gamma$, we have

$$\lim_{\lambda(J)\to 0} \frac{\lambda(A_\gamma \cap J)}{\lambda(J)} = 1, \quad \lim_{\lambda(J)\to 0} \frac{1}{\lambda(J)} \int_J |\widetilde{\ell}_m(s) - \widetilde{\ell}_m(t)| ds = 0, \tag{2.8.2}$$

where J denotes arbitrary intervals of positive length containing t.

Let $t \in L_\gamma$ and let us consider a function $u : \mathbb{R}_+ \times I \to K$ which is continuous in the second variable and such that there exists the limit $u(t) = \lim_{(h,s)\downarrow(0,t)} u(h, s)$. Let $\varepsilon > 0$ be arbitrary. Then there exists $\delta > 0$ such that, for all $\theta \in A_\gamma \cap [\,t, t + \delta\,]$ and $h \in (0, \delta)$,

$$\|f(\theta, u(h, \theta)) - f(t, u(t))\| \leq \frac{\varepsilon}{3}. \tag{2.8.3}$$

Moreover, there exists $m \geq 1$ such that $\|u(h, \theta)\| < m$ for all $\theta \in [t, t + \delta]$ and $h \in (0, \delta)$. By taking a smaller δ if necessary, in view of (2.8.2), we can also assume that both inequalities

$$\frac{1}{h} \int_{[\,t,t+h\,]\setminus A_\gamma} \ell_m(\theta) d\theta \leq \frac{\varepsilon}{3}, \tag{2.8.4}$$

and

$$\frac{\lambda([t,t+h]\setminus A_\gamma)}{h}\|f(t,u(t))\| \le \frac{\varepsilon}{3} \tag{2.8.5}$$

hold true for every $h \in (0,\delta)$. Then, by (2.8.3), for $h \in (0,\delta)$, we have

$$\frac{1}{h}\int_{[t,t+h]\cap A_\gamma} \|f(\theta,u(h,\theta)) - f(t,u(t))\|d\theta \le \frac{\varepsilon}{3}\frac{\lambda([t,t+h]\cap A_\gamma)}{h} \le \frac{\varepsilon}{3},$$

while by (2.8.4) and (2.8.5) we have

$$\frac{1}{h}\int_{[t,t+h]\setminus A_\gamma} \|f(\theta,u(h,\theta)) - f(t,u(t))\|d\theta$$

$$\le \frac{1}{h}\int_{[t,t+h]\setminus A_\gamma} (\ell_m(\theta) + \|f(t,u(t))\|)d\theta$$

$$\le \frac{1}{h}\int_{[t,t+h]\setminus A_\gamma} \ell_m(\theta)d\theta + \frac{\lambda([t,t+h]\setminus A_\gamma)}{h}\|f(t,u(t))\| \le \frac{2}{3}\varepsilon.$$

Finally, we have

$$\frac{1}{h}\int_t^{t+h} \|f(\theta,u(h,\theta)) - f(t,u(t))\|d\theta \le \varepsilon,$$

for all $h \in (0,\delta)$ and this completes the proof. □

The case of a locally Carathéodory function is analyzed below. First we need the following Carathéodory variant of a Dugundji [**89**] type extension result.

Theorem 2.8.3. *Let X, Y be Banach spaces, let $C \subseteq X$ be a closed set and I a nonempty and open interval and let $f : I \times C \to Y$ be a locally Carathéodory function. Then f has a locally Carathéodory extension, $F : I \times X \to Y$ such that $F(t,u) \in \operatorname{conv}(f(t,C))$ a.e. for $t \in I$ and for each $u \in X$. In particular, F satisfies the inequality (C_3) with the very same function ℓ as f does.*

To prove Theorem 2.8.3 consider the first variable, t, as a parameter and just repeat the arguments in the proof of Theorem 7.2, p. 44 in Deimling [**77**], by observing that the resulting extension is a Carathéodory function.

Theorem 2.8.4. *Let X, Y be Banach spaces, K a nonempty, locally closed and separable subset in X and $f : I \times K \to Y$ a locally Carathéodory function. Then, for each $\xi \in K$ there exist $\rho > 0$ and a negligible subset $\mathcal{Z}$ of I such that, for each $t \in I \setminus \mathcal{Z}$ and each $u : \mathbb{R}_+ \times I \to D(\xi,\rho) \cap K$ which* *is continuous with respect to the second variable and for which there exists* $u(t) = \lim_{(h,s)\downarrow(0,t)} u(h,s)$, *we have*

$$\lim_{h\downarrow 0} \frac{1}{h} \int_t^{t+h} \|f(s,u(h,s)) - f(t,u(t))\|\,ds = 0. \tag{2.8.6}$$

Proof. Since K is locally closed and f is locally Carathéodory it follows that for each $\xi \in K$ there exist $\rho > 0$ and $\ell \in L^1_{\mathrm{loc}}(I)$ such that $D(\xi,\rho) \cap K$ is closed and

$$\|f(t,u)\| \leq \ell(t)$$

a.e. for $t \in I$ and for all $u \in D(\xi,\rho) \cap K$. Now, let $\widetilde{f} : I \times X \to Y$ be an extension of $f_{|I\times[D(\xi,\rho)\cap K]}$ to $I \times X$ which is measurable with respect to the first variable and continuous with respect to the second one. The existence of such an extension is ensured by Theorem 2.8.3. Let $r : X \to D(\xi,\rho)$ be defined by

$$r(u) = \begin{cases} u & \text{if } u \in D(\xi,\rho) \\ \xi + \dfrac{\rho}{\|u-\xi\|}(u-\xi) & \text{if } u \in X \setminus D(\xi,\rho) \end{cases}$$

and let $\hat{f} : I \times X \to Y$ be given by $\hat{f}(t,u) = \widetilde{f}(t,r(u))$ for $t \in I$ and for all $u \in X$. Clearly $\hat{f}$ is Carathéodory and thus Theorem 2.8.2 applies. But $\hat{f}_{|I\times[D(\xi,\rho)\cap K]} = f_{|I\times[D(\xi,\rho)\cap K]}$ and this completes the proof. $\square$

In order to extend Theorem 2.8.4, the following topological result is needed.

Proposition 2.8.1. *Each open covering of a separable metric space has at least one finite or countable subcovering.*[4]

See for instance Engelking [**92**], Corollary, p. 177.

Theorem 2.8.5. *Let X and Y be real Banach spaces, K a nonempty, locally closed and separable subset in X and $f : I \times K \to Y$ a locally Carathéodory function. Then, there exists a negligible subset $\mathcal{Z}$ of I such that, for each $t \in I \setminus \mathcal{Z}$ and each $u : \mathbb{R}_+ \times I \to K$ which is continuous with respect to the second variable and for which there exists $u(t) = \lim_{(h,s)\downarrow(0,t)} u(h,s)$, we have*

$$\lim_{h\downarrow 0} \frac{1}{h} \int_t^{t+h} \|f(s,u(h,s)) - f(t,u(t))\|\,ds = 0. \tag{2.8.7}$$

[4] In fact Proposition 2.8.1 says that each separable metric space is a Lindelöf space.

Proof. Since K is locally closed, for each $\xi \in K$, there exists $\rho > 0$ such that $D(\xi, \rho) \cap K$ is closed. Next, we apply Theorem 2.8.4, and get a negligible subset $\mathcal{Z}_\rho$ in I such that, for each $u : \mathbb{R}_+ \times I \to D(\xi, \rho) \cap K$ which is continuous in the second variable and for which there exists $u(t) = \lim_{(h,s)\downarrow(0,t)} u(h, s)$, we have

$$\lim_{h\downarrow 0} \frac{1}{h} \int_t^{t+h} \|f(s, u(h, s)) - f(t, u(t))\| \, ds = 0.$$

Since K is a metric and separable space, thanks to Proposition 2.8.1, the covering $\cup_{\xi\in K} D(\xi, \rho)$ of K has a countable subcovering $\cup_{n\in\mathbb{N}} D(\xi_n, \rho_n)$. So, the negligible set $\mathcal{Z} = \cup_{n\in\mathbb{N}} \mathcal{Z}_{\rho_n}$ satisfies the conclusion of Theorem 2.8.5. □

Corollary 2.8.1. *Let K be a nonempty separable subset of a Banach space X, $\{S(t) : X \to X;\ t \geq 0\}$ a C_0-semigroup on X and $f : I \times K \to X$ a Carathéodory function. Then there exists a negligible subset $\mathcal{Z}$ of I such that, for each $t \in I \setminus \mathcal{Z}$ and each function $u : \mathbb{R}_+ \times I \to K$ which is continuous with respect to the second variable and such that there exists $u(t) = \lim_{(h,s)\downarrow(0,t)} u(h, s)$, we have*

$$\lim_{h\downarrow 0} \frac{1}{h} \int_t^{t+h} S(t + h - s) f(s, u(h, s)) ds = f(t, u(t)). \tag{2.8.8}$$

Proof. We take the same $\mathcal{Z}$ as in Theorem 2.8.2. Fix $t \in I \setminus \mathcal{Z}$ and observe that, since $s \mapsto S(s)f(t, u(t))$ is continuous, we have

$$\lim_{h\downarrow 0} \frac{1}{h} \int_t^{t+h} S(t + h - s) f(t, u(t)) ds = f(t, u(t)).$$

Next, for h small, with $M \geq 1$ and $a \in \mathbb{R}$ given by Theorem 1.4.1, we have

$$\left\| \frac{1}{h} \int_t^{t+h} S(t + h - s) f(s, u(h, s)) ds - \frac{1}{h} \int_t^{t+h} S(t + h - s) f(t, u(t)) ds \right\|$$

$$\leq M e^{(t+1)|a|} \frac{1}{h} \int_t^{t+h} \|f(s, u(h, s)) - f(t, u(t))\| ds.$$

The conclusion follows from Theorem 2.8.2. □

Corollary 2.8.2. *Let K be a nonempty locally closed and separable subset of a Banach space X, $\{S(t) : X \to X;\ t \geq 0\}$ a C_0-semigroup on X and $f : I \times K \to X$ a locally Carathéodory function. Then, for each $\xi \in K$, there exist $\rho > 0$ and a negligible subset $\mathcal{Z}$ of I such that, for each $t \in I \setminus \mathcal{Z}$ and each function $u : \mathbb{R}_+ \times I \to D(\xi, \rho) \cap K$ which is continuous with respect* *to the second variable and for which there exists* $u(t) = \lim_{(h,s)\downarrow(0,t)} u(h,s)$, *we have*

$$\lim_{h\downarrow 0} \frac{1}{h} \int_t^{t+h} S(t+h-s)f(s,u(h,s))ds = f(t,u(t)). \tag{2.8.9}$$

Proof. Just repeat the proof of Corollary 2.8.1 by using Theorem 2.8.4 instead of Theorem 2.8.2. □

Part 1

Ordinary differential equations and inclusions

CHAPTER 3

Nagumo type viability theorems

The aim of this chapter is to introduce the reader to the fundamentals of the viability theory for ordinary differential equations in general Banach spaces. We notice that here we confine ourselves to consider only ordinary differential equations driven by continuous right-hand sides. After explaining what viability exactly means, we prove a necessary condition for viability expressed in the terms of the Nagumo Tangency Condition. We then pass to the statement of the main sufficient (in fact necessary and sufficient) conditions for viability. Next, we prove a technical lemma ensuring the existence of a sequence of approximate solutions and continue with the complete proofs of the sufficient conditions for viability. We show how to get viability in the nonautonomous case, by using the already established theory in the autonomous one. We conclude with several results concerning noncontinuable and global solutions.

3.1. Necessary conditions for viability

Let X be a real Banach space, K a nonempty subset in X, $f : K \to X$ a given function and let us consider the Cauchy problem

$$\begin{cases} u'(t) = f(u(t)) \\ u(0) = \xi. \end{cases} \tag{3.1.1}$$

Definition 3.1.1. A *solution* of (3.1.1) on $[0, T]$ is an everywhere differentiable function $u : [0, T] \to K$ satisfying $u'(t) = f(u(t))$ for each $t \in [0, T]$ and $u(0) = \xi$. A *solution* of (3.1.1) on the semi-open interval $[0, T)$ is defined by analogy.

Definition 3.1.2. The set K is *viable* with respect to f if for each $\xi \in K$ there exist $T > 0$ and a solution $u : [0, T] \to K$ of (3.1.1).

Remark 3.1.1. In the case in which f is continuous on K, the function u in Definition 3.1.2 is a fortiori of class C^1. So, whenever f is continuous, K is viable with respect to f if and only if for each $\xi \in K$ there exists

$T > 0$ such that the Cauchy problem (3.1.1) has at least one C^1 solution $u : [0, T] \to K$.

We can now proceed to the main result in this section, i.e., a necessary condition for viability.

Theorem 3.1.1. *If K is viable with respect to $f : K \to X$, then, for each $\xi \in K$, we have $f(\xi) \in \mathcal{F}_K(\xi)$.*

Proof. Let $\xi \in K$. We have to prove that

$$\lim_{h \downarrow 0} \frac{1}{h} \text{dist}\,(\xi + hf(\xi); K) = 0.$$

See Definition 2.4.2. Since K is viable with respect to f, there exists a differentiable function $u : [0, T] \to X$ with $u(s) \in K$ for all $s \in [0, T]$ and satisfying both $u'(s) = f(u(s))$ for every $s \in [0, T]$ and $u(0) = \xi$. Consequently, we have

$$\text{dist}\,(\xi + hf(\xi); K) \leq \|\xi + hf(\xi) - u(h)\| = h \left\| f(u(0)) - \frac{u(h) - u(0)}{h} \right\|.$$

Therefore

$$0 \leq \lim_{h \downarrow 0} \frac{1}{h} \text{dist}\,(\xi + hf(\xi); K) \leq \lim_{h \downarrow 0} \left\| f(u(0)) - \frac{u(h) - u(0)}{h} \right\| = 0,$$

i.e. $f(\xi) \in \mathcal{F}_K(\xi)$ and this achieves the proof. □

Corollary 3.1.1. *If K is viable with respect to $f : K \to X$ then we have*

$$f(\xi) \in \mathcal{T}_K(\xi), \tag{3.1.2}$$

for each $\xi \in K$.

Proof. The conclusion follows from Remark 2.4.3. □

From Corollary 3.1.1 and Proposition 2.4.3, we deduce

Theorem 3.1.2. *If K is viable with respect to $f : K \to X$ then, for each family of functions $\{f_h;\ h \in (0, 1)\}$, $f_h : K \to X$, satisfying*

$$\lim_{h \downarrow 0} f_h(x) = f(x),$$

pointwise on K, we have

$$\liminf_{h \downarrow 0} \frac{1}{h} \text{dist}\,(\xi + hf_h(\xi); K) = 0$$

for each $\xi \in K$.

We notice that, in Theorem 3.1.1, we don't need to assume that K has some special topological properties, or that f is continuous. However, as we can see from the example below, if X is infinite dimensional and f is merely continuous, the tangency condition $f(\xi) \in \mathcal{T}_K(\xi)$ (and even the stronger one $f(\xi) \in \mathcal{F}_K(\xi)$), for all $\xi \in K$, is far from being sufficient for the viability of K with respect to f.

Example 3.1.1. Let $X = c_0$ be the space of all real sequences $(x_n)_n$ with $\lim_n x_n = 0$. This space, endowed with the sup-norm defined by $\|(x_n)_n\|_\infty = \sup\{|x_n|\,;\ n = 1, 2, \dots\}$ for each $(x_n)_n$ in X, is a real Banach space. Let $f : X \to X$ be defined by $f = (f_k)_k$, where

$$f_k((x_n)_n) = 2\sqrt{|x_k|} \quad k = 1, 2, \dots$$

for each $(x_n)_n \in X$. Take $K = X$ and let us observe that, inasmuch as K is open, for each $\xi \in K$ we have $\mathcal{T}_K(\xi) = X$. Therefore f satisfies the tangency condition (3.1.2) for each $\xi \in K$. On the other hand, f is continuous on $X = c_0$ and consequently, thanks to Remark 3.1.1, X is viable with respect to f if and only if, for each $\xi \in X$, there exists $\delta > 0$ such that the autonomous Cauchy problem

$$\begin{cases} u' = f(u) \\ u(0) = \xi \end{cases}$$

have at least one C^1 solution $u : [\,0, \delta\,] \to X$. But $u : [\,0, \delta\,] \to X$ is a solution of the problem above if and only if $(u_k)_k : [\,0, \delta\,] \to X$ is a solution of the system of infinitely many differential equations

$$\begin{cases} u'_k = 2\sqrt{|u_k|} \\ u_k(0) = \xi_k \qquad k = 1, 2, \dots. \end{cases}$$

Let us assume that the above Cauchy problem or, equivalently, the above system, corresponding to the specific choice $\xi = (1/k^2)_k$, has at least one solution $(u_k)_k : [\,0, \delta\,] \to X$, with $\delta > 0$. This system contains infinitely many uncoupled differential equations with separate variables whose solutions, if there exist, are necessarily of the form

$$u_k(t) = (t + 1/k)^2$$

for each $k = 1, 2, \dots$ and each $t \in [\,0, \delta\,]$. Hence $(u_k)_k$ is defined by these equalities. But, for all $t > 0$, we have $\lim_k u_k(t) = t^2$, in contradiction with the fact that $(u_k(t))_k$ belongs to c_0, i.e., $\lim_k u_k(t) = 0$. This contradiction can be eliminated only if the Cauchy problem in question has no solution.

So, in spite of the fact that f is continuous and satisfies the tangency condition (3.1.2) in Corollary 3.1.1 with $K = X$, the latter is not viable with respect to f.

3.2. Sufficient conditions for viability

The nonexistence phenomenon in Example 3.1.1 is due to the lack of some extra conditions on f as "compactness" or "locally Lipschitz properties". The aim of the next three sections is to show that, whenever we add one, or a combination, of the two mentioned properties, the continuity of f along with the tangency condition in Corollary 3.1.1 is sufficient for the viability of a locally closed set K with respect to f.

The goal of this section is to state several sufficient conditions of viability of a set K with respect to a function f. It should be noticed that, as these conditions are also necessary, we will formulate all of them as necessary and sufficient conditions, the necessity part of each one following from Theorem 3.1.1. Let X and Y be Banach spaces, let K be a nonempty subset in Y and let $f : K \to X$ be a given function[1].

Definition 3.2.1. A function $f : K \to X$ is called *locally compact* if it is continuous and for each $\eta \in K$ there exists $\rho > 0$ such that $f(D_Y(\eta, \rho) \cap K)$ is relatively compact in X. The function f is called *compact* if it is continuous and carries bounded subsets in K into relatively compact subsets in X.

Remark 3.2.1. Clearly, each compact function is locally compact. If $K = X = Y$ and, in addition, X is finite dimensional, each locally compact function is compact. In the latter case each continuous function is locally compact. However, we notice that, when $K \subseteq X$ and K does not coincide with X, even if the latter is finite dimensional, there exist locally compact functions which are not compact. Furthermore, if K is locally compact and f is continuous, then f is locally compact even though X is infinite dimensional.

Definition 3.2.2. If $K \subseteq Y$, we say that $f : K \to X$ is *locally Lipschitz* if for each $\xi \in K$ there exist $\rho > 0$ and $L > 0$ such that

$$\|f(u) - f(v)\| \leq L\|u - v\| \tag{3.2.1}$$

for each $u, v \in D(\xi, \rho) \cap K$. It is *globally Lipschitz* if there exists $L > 0$ such that (3.2.1) holds for each $u, v \in K$.

Definition 3.2.3. Let Y and X be two Banach spaces and let $D \subseteq Y$. A function $f : D \to X$ is called *locally β-compact* if it is continuous and, for each $y \in D$, there exist $r > 0$ and a uniqueness function $\omega : \mathbb{R}_+ \to \mathbb{R}_+$, such

[1] In fact only two specific situations will be considered: the first one when $Y = X$, which corresponds to the autonomous case, and the second one when $Y = \mathbb{R} \times X$, which corresponds to the nonautonomous one.

that $f(D_Y(y,r)\cap D)$ is bounded and, for each subset C in $D_Y(y,r)\cap D$, we have

$$\beta_X(f(C)) \leq \omega(\beta_Y(C)), \tag{3.2.2}$$

where β_X is the Hausdorff measure of noncompactness on X and β_Y is the Hausdorff measure of noncompactness on Y.

A function $f : D \to X$ is called *β-compact* if it is continuous and, for each bounded subset C in D, (3.2.2) is satisfied.

In order to simplify the notation, in all that follows, whenever any possibility of confusion will be ruled out by the context, we will denote both functions β_X and β_Y with the very same symbol, β.

Remark 3.2.2. If Y is finite dimensional and $D \subseteq Y$ is locally closed, each continuous function $f : D \to X$ is locally β-compact. Also, if Y is finite dimensional and $D \subseteq Y$ is closed, each continuous function $f : D \to X$ is β-compact.

Remark 3.2.3. Each locally compact function is locally β-compact. Moreover, each locally Lipschitz function is locally β-compact. Since the sum of each two locally β-compact functions is locally β-compact, it follows that each function f of the form $f = f_1 + f_2$, with f_1 locally compact and f_2 locally Lipschitz, is locally β-compact. Also each compact function, as well as each globally Lipschitz function is β-compact. Finally, each β-compact function is locally β-compact.

Theorem 3.2.1. *Let X be a Banach space, let $K \subseteq X$ be a nonempty and locally closed set and let $f : K \to X$ be a locally β-compact function. A necessary and sufficient condition in order that K be viable with respect to f is to exist the family of functions $\{f_h;\ h \in (0,1)\}$, $f_h : K \to X$, satisfying $\lim_{h\downarrow 0} f_h(\xi) = f(\xi)$ for each $\xi \in K$, and such that*

$$\liminf_{h\downarrow 0} \frac{1}{h}\operatorname{dist}(\xi + hf_h(\xi); K) = 0$$

for each $\xi \in K$.

Proposition 2.4.4 implies that Theorem 3.2.1 is equivalent to

Theorem 3.2.2. *Let X be a Banach space, let $K \subseteq X$ be a nonempty and locally closed set and let $f : K \to X$ be a locally β-compact function. A necessary and sufficient condition in order that K be viable with respect to f is the tangency condition*

$$f(\xi) \in \mathfrak{T}_K(\xi) \tag{3.2.3}$$

for each $\xi \in K$.

The next two results are immediate corollaries of Theorem 3.2.2.

Theorem 3.2.3. *Let X be a Banach space, let $K \subseteq X$ be a nonempty and locally closed set and let $f : K \to X$ be a locally Lipschitz function. Then, a necessary and sufficient condition in order that K be viable with respect to f is the tangency condition* (3.2.3).

Theorem 3.2.4. *Let X be a Banach space, let $K \subseteq X$ be a nonempty and locally closed set and let $f : K \to X$ be a locally compact function. Then, a necessary and sufficient condition in order that K be viable with respect to f is the tangency condition* (3.2.3).

Theorem 3.2.5. *Let X be a Banach space, let $K \subseteq X$ be nonempty and locally closed and let $f : K \to X$ be continuous. Let us assume that K is proximal and the norm $\|\cdot\|$ is Gâteaux differentiable at each $x \in X$, $x \neq 0$. Then the following conditions are equivalent:*

(i) *for every $\xi \in K$, $f(\xi) \in \mathcal{C}_K(\xi)$;*
(ii) *for every $\xi \in K$, $f(\xi) \in \mathcal{T}_K(\xi)$;*
(iii) *for every $\xi \in K$, $f(\xi) \in \mathcal{B}_K(\xi)$;*
(iv) *the set K is viable with respect to f.*

In general, if $\mathcal{G} : K \rightsquigarrow X$ is such that $\mathcal{C}_K(\xi) \subseteq \mathcal{G}(\xi) \subseteq \mathcal{B}_K(\xi)$ for each $\xi \in K$, then each one of the conditions above is equivalent to

(v) *for every $\xi \in K$, $f(\xi) \in \mathcal{G}(\xi)$.*

Finally, from Theorem 3.2.4, we easily deduce the autonomous version of the celebrated Nagumo Viability Theorem, i.e.,

Theorem 3.2.6. *Let X be finite dimensional, let $K \subseteq X$ be nonempty and locally closed and let $f : K \to X$ be continuous. Then, a necessary and sufficient condition in order that K be viable with respect to f is the tangency condition* (3.2.3).

3.3. Existence of ε-approximate solutions

In this section and the next one, we will prove Theorem 3.2.2 which, as we already have mentioned, is equivalent to Theorem 3.2.1. As the necessity part follows from Theorem 3.1.1, here we will focus our attention only on the sufficiency.

The first step is concerned with the existence of "approximate solutions" to the autonomous Cauchy problem below

$$\begin{cases} u'(t) = f(u(t)) \\ u(0) = \xi, \end{cases} \tag{3.3.1}$$

where $K \subseteq X$ and $f : K \to X$ are as in Theorem 3.2.2 and $\xi \in K$.

Thus, let $\xi \in K$. Since K is locally closed, there exists $\rho > 0$ such that the set $D(\xi, \rho) \cap K$ be closed. Next, diminishing $\rho > 0$ if necessary, we can choose $M > 0$ and $T > 0$ such that

$$\|f(x)\| \leq M \tag{3.3.2}$$

for every $x \in D(\xi, \rho) \cap K$, and

$$T(M+1) \leq \rho. \tag{3.3.3}$$

The possibility of diminishing ρ in order to find $M > 0$ satisfying (3.3.2) is a consequence of the fact that f is continuous and thus locally bounded, i.e., f is bounded on $D(\xi, \rho) \cap K$ provided $\rho > 0$ is small enough. Finally, taking a sufficiently small $T > 0$, we obtain (3.3.3).

Lemma 3.3.1. *Let X be a Banach space, let $K \subseteq X$ be nonempty and locally closed and let $f : K \to X$ be continuous and satisfying $f(\xi) \in \mathcal{T}_K(\xi)$ for each $\xi \in K$. Let $\xi \in K$, $\rho > 0$, $M > 0$ and $T > 0$ be fixed as above.*

Then, for each $\varepsilon \in (0, 1)$, there exist $\sigma : [\,0, T\,] \to [\,0, T\,]$ nondecreasing, $g : [\,0, T\,] \to X$ Riemann integrable and $u : [\,0, T\,] \to X$ continuous, such that:

(i) $t - \varepsilon \leq \sigma(t) \leq t$ *for each* $t \in [\,0, T\,]$;
(ii) $\|g(t)\| \leq \varepsilon$ *for each* $t \in [\,0, T\,]$;
(iii) $u(\sigma(t)) \in D(\xi, \rho) \cap K$ *for all* $t \in [\,0, T\,]$ *and* $u(T) \in D(\xi, \rho) \cap K$;
(iv) $u(t) = \xi + \displaystyle\int_0^t f(u(\sigma(s)))\, ds + \int_0^t g(s)\, ds$ *for each* $t \in [\,0, T\,]$.

Before proceeding to the proof of Lemma 3.3.1 we introduce

Definition 3.3.1. A triple (σ, g, u), satisfying (i), (ii), (iii) and (iv) in Lemma 3.3.1, is called an *ε-approximate solution* to the Cauchy problem (3.3.1) on the interval $[\,0, T\,]$.

We may now pass to the proof of Lemma 3.3.1.

Proof. We begin by showing the existence of an ε-approximate solution on an interval $[\,0, \delta\,]$ with $\delta \in (0, T\,]$. As, for every $\xi \in K$, f satisfies the tangency condition $f(\xi) \in \mathcal{T}_K(\xi)$, from Proposition 2.4.2, it follows that there exist $\delta \in (0, T\,]$, $\delta \leq \varepsilon$ and $p \in X$ with $\|p\| \leq \varepsilon$, such that

$$\xi + \delta f(\xi) + \delta p \in K.$$

Now let us define $\sigma : [\,0, \delta\,] \to [\,0, \delta\,]$, $g : [\,0, \delta\,] \to X$ and $u : [\,0, \delta\,] \to X$ by

$$\begin{cases} \sigma(t) = 0 & \text{for } t \in [\,0, \delta\,] \\ g(t) = p & \text{for } t \in [\,0, \delta\,] \\ u(t) = \xi + t f(\xi) + t p & \text{for } t \in [\,0, \delta\,]. \end{cases}$$

One can readily see that the triple (σ, g, u) is an ε-approximate solution to the Cauchy problem (3.3.1) on the interval $[0, \delta]$. Indeed the conditions (i), (ii) and (iv) are obviously fulfilled, while (iii) follows from (3.3.2), (3.3.3) and (i). To show the latter assertion, we observe that, for every $t \in [0, \delta]$, $u(\sigma(t)) = \xi$ and therefore $u(\sigma(t)) \in D(\xi, \rho) \cap K$. Clearly $u(\delta) \in K$. On the other hand, by (3.3.2) and (3.3.3), we deduce

$$\|u(\delta) - \xi\| \leq \delta\|f(\xi)\| + \delta\|p\| \leq T(M+1) \leq \rho.$$

Thus (iii) is satisfied. Next, we will prove the existence of an ε-approximate solution defined on the whole interval $[0, T]$. To this aim we shall make use of Brezis–Browder Theorem 2.1.1, as follows. Let $\mathcal{S}$ be the set of all ε-approximate solutions to the problem (3.3.1) having the domains of definition of the form $[0, c]$ with $c \in (0, T]$. On $\mathcal{S}$ we define the relation $\preceq$ by

$$(\sigma_1, g_1, u_1) \preceq (\sigma_2, g_2, u_2)$$

if the domain of definition $[0, c_1]$ of the first triple is included in the domain of definition $[0, c_2]$ of the second triple and the two ε-approximate solutions coincide on the common part of the domains. Obviously $\preceq$ is a preorder relation on $\mathcal{S}$. Let us first show that each increasing sequence $((\sigma_m, g_m, u_m))_m$ is bounded from above. Indeed, let $((\sigma_m, g_m, u_m))_m$ be an increasing sequence, and let $c^* = \lim_m c_m$, where $[0, c_m]$ denotes the domain of definition of (σ_m, g_m, u_m). Clearly, $c^* \in (0, T]$. We will show that there exists at least one element, $(\sigma^*, g^*, u^*) \in \mathcal{S}$, defined on $[0, c^*]$ and satisfying $(\sigma_m, g_m, u_m) \preceq (\sigma^*, g^*, u^*)$ for each $m \in \mathbb{N}$. In order to do this, let us prove first that there exists $\lim_m u_m(c_m)$. For each $m, k \in \mathbb{N}$, $m \leq k$, we have $u_m(s) = u_k(s)$ for all $s \in [0, c_m]$. Taking into account of (iii), (iv) and (3.3.2), we deduce

$$\|u_m(c_m) - u_k(c_k)\| \leq \int_{c_m}^{c_k} [\|f(u_k(\sigma_k(\theta)))\| + \|g_k(\theta)\|]\, d\theta \leq (M + \varepsilon)|c_k - c_m|$$

for every $m, k \in \mathbb{N}$, which proves that there exists $\lim_m u_m(c_m)$. Since for every $m \in \mathbb{N}$, $u_m(c_m) \in D(\xi, \rho) \cap K$, and the latter is closed, it readily follows that $\lim_m u_m(c_m) \in D(\xi, \rho) \cap K$. Furthermore, because all the functions in the set $\{\sigma_m;\ m \in \mathbb{N}\}$ are nondecreasing, with values in $[0, c^*]$, and satisfy $\sigma_m(c_m) \leq \sigma_p(c_p)$ for every $m, p \in \mathbb{N}$ with $m \leq p$, there exists $\lim_m \sigma_m(c_m)$ and this limit belongs to $[0, c^*]$. This shows that we can define the triple of functions $(\sigma^*, g^*, u^*) : [0, c^*] \to [0, c^*] \times X \times X$ by

$$\sigma^*(t) = \begin{cases} \sigma_m(t) & \text{for } t \in [0, c_m] \\ \lim_m \sigma_m(c_m) & \text{for } t = c^* \end{cases}$$
$$g^*(t) = \begin{cases} g_m(t) & \text{for } t \in [0, c_m] \\ 0 & \text{for } t = c^* \end{cases}$$
$$u^*(t) = \begin{cases} u_m(t) & \text{for } t \in [0, c_m] \\ \lim_m u_m(c_m) & \text{for } t = c^*. \end{cases}$$

One can easily see that (σ^*, g^*, u^*) is an ε-approximate solution which is an upper bound for $((\sigma_m, g_m, u_m))_m$. Now, let us define the function $\mathcal{N} : \mathcal{S} \to \mathbb{R}$ by $\mathcal{N}((\sigma, g, u)) = c$, where $[0, c]$ is the domain of definition of (σ, g, u). Clearly $\mathcal{N}$ satisfies the hypotheses of Brezis-Browder Theorem 2.1.1. Then, $\mathcal{S}$ contains at least one $\mathcal{N}$-maximal element $(\overline{\sigma}, \overline{g}, \overline{u})$, defined on $[0, \overline{c}]$. In other words, if $(\widetilde{\sigma}, \widetilde{g}, \widetilde{u}) \in \mathcal{S}$, defined on $[0, \widetilde{c}]$, satisfies $(\overline{\sigma}, \overline{g}, \overline{u}) \preceq (\widetilde{\sigma}, \widetilde{g}, \widetilde{u})$, then we necessarily have $\overline{c} = \widetilde{c}$. We will next show that $\overline{c} = T$. Indeed, let us assume by contradiction that $\overline{c} < T$. Since

$$\|\overline{u}(\overline{c}) - \xi\| \leq \int_0^{\overline{c}} \|f(\overline{u}(\overline{\sigma}(s)))\| \, ds + \int_0^{\overline{c}} \|\overline{g}(s)\| \, ds \leq \overline{c}(M + \varepsilon)$$
$$\leq \overline{c}(M + 1) < T(M + 1),$$

we deduce that

$$\|\overline{u}(\overline{c}) - \xi\| < \rho. \tag{3.3.4}$$

Then, as $\overline{u}(\overline{c}) \in K$ and $f(\overline{u}(\overline{c})) \in \mathcal{T}_K(\overline{u}(\overline{c}))$, there exist $\delta \in (0, T - \overline{c})$, $\delta \leq \varepsilon$ and $p \in X$, $\|p\| \leq \varepsilon$, such that $\overline{u}(\overline{c}) + \delta f(\overline{u}(\overline{c})) + \delta p \in K$. From (3.3.4), it follows that we can diminish δ if necessary, in order to have

$$\|\overline{u}(\overline{c}) + \delta f(\overline{u}(\overline{c})) + \delta p - \xi\| \leq \rho. \tag{3.3.5}$$

Let us define the functions $\sigma : [0, \overline{c} + \delta] \to [0, \overline{c} + \delta]$ and $g : [0, \overline{c} + \delta] \to X$ by

$$\sigma(t) = \begin{cases} \overline{\sigma}(t) & \text{for } t \in [0, \overline{c}] \\ \overline{c} & \text{for } t \in (\overline{c}, \overline{c} + \delta] \end{cases}, \quad g(t) = \begin{cases} \overline{g}(t) & \text{for } t \in [0, \overline{c}] \\ p & \text{for } t \in (\overline{c}, \overline{c} + \delta]. \end{cases}$$

Clearly, g is Riemann integrable on $[0, \overline{c} + \delta]$ and $\|g(t)\| \leq \varepsilon$ for every $t \in [0, \overline{c} + \delta]$. We define $u : [0, \overline{c} + \delta] \to X$ by

$$u(t) = \begin{cases} \overline{u}(t) & \text{for } t \in [0, \overline{c}] \\ \overline{u}(\overline{c}) + (t - \overline{c}) f(\overline{u}(\overline{c})) + (t - \overline{c}) p & \text{for } t \in (\overline{c}, \overline{c} + \delta]. \end{cases}$$

Obviously

$$u(t) = \xi + \int_0^t f(\overline{u}(\sigma(\theta))) \, d\theta + \int_0^t g(\theta) \, d\theta$$

for every $t \in [0, \overline{c} + \delta]$ and thus σ, g and u satisfy the conditions (i), (ii) and (iv). Let us observe that

$$u(\sigma(t)) = \begin{cases} \overline{u}(\overline{\sigma}(t)) & \text{for } t \in [0, \overline{c}] \\ \overline{u}(\overline{c}) & \text{for } t \in [\overline{c}, \overline{c} + \delta], \end{cases}$$

and so $u(\sigma(t)) \in D(\xi, \rho) \cap K$ which proves (iii). Furthermore, from the choice of δ and p, we have $u(\overline{c} + \delta) = \overline{u}(\overline{c}) + \delta f(\overline{u}(\overline{c})) + \delta p \in K$. Moreover, from (3.3.5), we conclude $\|u(\overline{c} + \delta) - \xi\| = \|\overline{u}(\overline{c}) + \delta f(\overline{u}(\overline{c})) + \delta p - \xi\| \leq \rho$ and consequently u satisfies (iii). Thus $(\sigma, g, u) \in \mathcal{S}$.

Finally, inasmuch as $(\overline{\sigma}, \overline{g}, \overline{u}) \preceq (\sigma, g, u)$ and $\overline{c} < \overline{c} + \delta$, it follows that $(\overline{\sigma}, \overline{g}, \overline{u})$ is not an $\mathcal{N}$-maximal element. But this is absurd. This contradiction can be eliminated only if each $\mathcal{N}$-maximal element in the set $\mathcal{S}$ is defined on $[0, T]$. □

3.4. Convergence of ε-approximate solutions

The goal of this section is to prove the convergence of a suitably chosen sequence of ε-approximate solutions. We may now proceed to the proof of Theorem 3.2.2.

Proof. Let us consider a sequence $(\varepsilon_n)_n$ in $(0, 1)$, decreasing to 0, and let $((\sigma_n, g_n, u_n))_n$ be a sequence of ε_n-approximate solutions of (3.3.1) on $[0, T]$. Diminishing $\rho > 0$ and $T > 0$ if necessary, we may assume that the conclusion of Lemma 3.3.1 holds true and, in addition, f is β-compact on $D(\xi, \rho) \cap K$.

We will show first that $(u_n)_n$ has at least one convergent subsequence in the sup-norm. Let $M > 0$ as in (3.3.2) and let us observe that, by (i), (ii) and (iv)[2], we have

$$\|u_n(t) - u_n(\sigma_n(t))\| \leq (M + 1)\varepsilon_n \tag{3.4.1}$$

for each $t \in [0, T]$.

We consider first the case when X is separable. From (3.4.1), (iv), the fact that f is β-compact on $D(\xi, \rho) \cap K$, Lemma 2.7.2 and Remark 2.7.1, it follows

$$\begin{aligned} &\beta(\{u_n(t);\ n \geq k\}) \\ &\leq \beta\left(\left\{\int_0^t f(u_n(\sigma_n(s)))\,ds;\ n \geq k\right\}\right) + \beta\left(\left\{\int_0^t g_n(s)\,ds;\ n \geq k\right\}\right) \\ &\leq \int_0^t \beta\left(\{f(u_n(\sigma_n(s)));\ n \geq k\}\right) ds + \int_0^t \beta\left(\{g_n(s);\ n \geq k\}\right) ds \end{aligned}$$

[2]Within this proof, all the quotations to items like (i)∼(iv) refer to the corresponding items in Lemma 3.3.1.

$$\leq \int_0^t \omega(\beta(\{u_n(\sigma_n(s));\ n \geq k\}))\, ds + T\varepsilon_k$$

$$\leq \int_0^t \omega(\beta(\{u_n(s);\ n \geq k\} + \{u_n(\sigma_n(s)) - u_n(s);\ n \geq k\}))\, ds + T\varepsilon_k$$

$$\leq \int_0^t \omega(\beta\{u_n(s);\ n \geq k\} + (M+1)\varepsilon_k)\, ds + T\varepsilon_k.$$

For $k = 1, 2, \ldots$, let us denote by $x_k(t) = \beta(\{u_n(t);\ n \geq k\}) + (M+1)\varepsilon_k$, and by $\gamma_k = (M + T + 1)\varepsilon_k$. The inequality above rewrites as

$$x_k(t) \leq \gamma_k + \int_0^t \omega(x_k(s))\, ds$$

for each $t \in [0, T]$. By Lemma 1.8.2, diminishing $T > 0$ if necessary, we may assume that $\lim_k x_k(t) = 0$ uniformly for $t \in [0, T]$. But this shows that $\lim_k \beta(\{u_n(t);\ n \geq k\}) = 0$ and thus we are in the hypotheses of Lemma 2.7.3. So, for each $t \in [0, T]$, $\{u_n(t);\ n = 1, 2, \ldots\}$ is relatively compact. From (ii), (iv) and (3.3.2), we conclude that $\{u_n;\ n = 1, 2, \ldots\}$ is equicontinuous. By Arzelà–Ascoli Theorem 1.3.6, we deduce that there exists $u \in C([0, T]; X)$ such that, on a subsequence at least,

$$\lim_n u_n(t) = u(t)$$

uniformly for $t \in [0, T]$. In view of (3.4.1), we have also

$$\lim_n u_n(\sigma_n(t)) = u(t)$$

uniformly for $t \in [0, T]$. From (iii) and the fact that $D(\xi, \rho) \cap K$ is closed, we conclude that $u(t) \in D(\xi, \rho) \cap K$ for each $t \in [0, T]$. Passing to the limit for $n \to \infty$ in (iv) with u, σ and g substituted by u_n, σ_n and g_n respectively, we deduce that

$$u(t) = \xi + \int_0^t f(u(s))\, ds$$

for each $t \in [0, T]$. Thus u is a solution of (3.3.1). This completes the proof of Theorem 3.2.2 in the case when X is separable.

If X is not separable, there exists a separable and closed subspace, Y, of X such that $u_n(t), f(u_n(\sigma_n(t))), g_n(t) \in Y$ for $n = 1, 2, \ldots$ and $t \in [0, T]$. From Problem 2.7.2 and from the monotonicity of ω, we have

$$\beta_Y(f(C)) \leq 2\beta(f(C)) \leq 2\omega(\beta(C)) \leq 2\omega(\beta_Y(C)),$$

for each set $C \subseteq D(\xi, \rho) \cap K \cap Y$. In view of Remark 1.8.1, 2ω is a uniqueness function too. Repeating the routine above, with β replaced by β_Y and ω replaced by 2ω, using Remark 2.7.3 instead of Lemma 2.7.2 and the fact that the restriction of β_Y — as defined in Problem 2.7.2 — to $\mathcal{B}(Y)$ is the Hausdorff measure of noncompactness on Y, we conclude that, for each $t \in [0,T]$, $\{u_n(t);\ n = 1,2,\dots\}$ is relatively compact in Y. But Y is a subspace of X and, from now on, we have to repeat the same arguments as in the case when X is separable. This completes the proof. $\square$

Problem 3.4.1. *Give a direct proof of Nagumo's Theorem* 3.2.6 *avoiding the use of the measure of noncompactness.*

Problem 3.4.2. *Give a direct proof of Theorem* 3.2.3 *avoiding the use of the measure of noncompactness.*

Problem 3.4.3. *Give a direct proof of Theorem* 3.2.4 *avoiding the use of the measure of noncompactness.*

3.5. Extension to the nonautonomous case

In this section we will show how the results established before for the autonomous equation $u'(t) = f(u(t))$ extend to the nonautonomous one $u'(t) = f(t, u(t))$. So, let X be a real Banach space, $\mathcal{C}$ a nonempty subset in $\mathbb{R} \times X$, $f : \mathcal{C} \to X$ a given function and let us consider the Cauchy problem for the nonautonomous differential equation

$$\begin{cases} u'(t) = f(t, u(t)) \\ u(\tau) = \xi. \end{cases} \tag{3.5.1}$$

Definition 3.5.1. A *solution* of (3.5.1) on $[\tau, T]$ is an everywhere differentiable function $u : [\tau, T] \to X$ satisfying:

(i) $(t, u(t)) \in \mathcal{C}$ for each $t \in [\tau, T]$ and
(ii) $u'(t) = f(t, u(t))$ for each $t \in [\tau, T]$ and $u(\tau) = \xi$.

Definition 3.5.2. The set $\mathcal{C}$ is *viable* with respect to f if for each $(\tau, \xi) \in \mathcal{C}$ there exist $T \in \mathbb{R}$, $T > \tau$, and a solution $u : [\tau, T] \to X$ of (3.5.1).

Remark 3.5.1. If f is continuous, $\mathcal{C}$ is viable with respect to f if and only if for each $(\tau, \xi) \in \mathcal{C}$ there exists $T \in \mathbb{R}$, $T > \tau$ such that the Cauchy problem (3.5.1) have at least one C^1 solution $u : [\tau, T] \to X$.

We will rewrite the nonautonomous problem above as an autonomous one in the space $\mathfrak{X} = \mathbb{R} \times X$, endowed with the norm

$$\|(t, u)\|_{\mathfrak{X}} = \sqrt{|t|^2 + \|u\|^2},$$

for each $(t, u) \in \mathfrak{X}$.[3] Namely, set $z(s) = (t(s), u(s))$ and $F(z) = (1, f(z))$, for $s \in [0, T - \tau]$. Then, the Cauchy problem above is equivalent to

$$\begin{cases} z'(s) = F(z(s)) \\ z(0) = (\tau, \xi). \end{cases} \tag{3.5.2}$$

So, all the viability results proved before extend in an obvious way to the nonautonomous case via the transformations above. For instance, we have

Theorem 3.5.1. *If $\mathcal{C}$ is viable with respect to $f : \mathcal{C} \to X$ then, for each $(\tau, \xi) \in \mathcal{C}$, we have $(1, f(\tau, \xi)) \in \mathcal{F}_{\mathcal{C}}(\tau, \xi)$.*

Corollary 3.5.1. *If $\mathcal{C}$ is viable with respect to $f : \mathcal{C} \to X$ then, for each $(\tau, \xi) \in \mathcal{C}$, we have $(1, f(\tau, \xi)) \in \mathcal{T}_{\mathcal{C}}(\tau, \xi)$.*

Remark 3.5.2. If $\mathcal{C}$ is a cylindrical domain, i.e. $\mathcal{C} = I \times K$ with I an interval open to the right and K a subset in X, then the tangency conditions:

(i) $(1, f(\tau, \xi)) \in \mathcal{T}_{\mathcal{C}}(\tau, \xi)$ and
(ii) $f(\tau, \xi) \in \mathcal{T}_K(\xi)$

are equivalent.

Problem 3.5.1. *Let $f : I \times K \to X$. Assume that, for some $\xi \in K$, $t \mapsto f(t, \xi)$ is continuous from the right at $\tau \in I$. Then $f(\tau, \xi) \in \mathcal{T}_K(\xi)$ if and only if*

$$\liminf_{h \downarrow 0} \frac{1}{h} \operatorname{dist}\left(\xi + \int_{\tau}^{\tau+h} f(\theta, \xi)\, d\theta; K\right) = 0. \tag{3.5.3}$$

Theorem 3.5.2. *Let X be a Banach space, $\mathcal{C} \subseteq \mathbb{R} \times X$ a nonempty and locally closed set and let $f : \mathcal{C} \to X$ be a locally β-compact function. A necessary and sufficient condition in order that $\mathcal{C}$ be viable with respect to f is that*

$$(1, f(\tau, \xi)) \in \mathcal{T}_{\mathcal{C}}(\tau, \xi) \tag{3.5.4}$$

for each $(\tau, \xi) \in \mathcal{C}$.

Proof. Let us observe that $u : [\tau, T] \to X$ is a solution of (3.5.1) if and only if $z : [0, T - \tau] \to \mathcal{C}$, $z(s) = (s + \tau, u(s + \tau))$ is a solution of the autonomous Cauchy problem (3.5.2). Since F is continuous and satisfies both $F(z) \in \mathcal{T}_{\mathcal{C}}(z)$ for each $z \in \mathcal{C}$ and

$$\beta_{\mathfrak{X}}(F(B)) = \beta_{\mathfrak{X}}(\{1\} \times f(B)) = \beta_X(f(B)) \leq \omega(\beta_{\mathfrak{X}}(B))$$

for each bounded subset B in $\mathcal{C}$, the conclusion follows from Theorem 3.2.2. $\square$

[3] In fact, all the results which will follow, except for Theorem 3.5.6, remain unchanged if, on $\mathfrak{X}$, we consider any other equivalent norm.

The next two results are immediate corollaries of Theorem 3.5.2.

Theorem 3.5.3. *Let X be a Banach space, $\mathcal{C} \subseteq \mathbb{R} \times X$ a nonempty and locally closed set and let $f : \mathcal{C} \to X$ be a locally compact function. Then, a necessary and sufficient condition in order that $\mathcal{C}$ be viable with respect to f is the tangency condition* (3.5.4).

Theorem 3.5.4. *Let X be a Banach space, $\mathcal{C} \subseteq \mathbb{R} \times X$ a nonempty and locally closed set and let $f : \mathcal{C} \to X$ be a locally Lipschitz function. Then, a necessary and sufficient condition in order that $\mathcal{C}$ be viable with respect to f is the tangency condition* (3.5.4).

From Theorem 3.5.3, we deduce the Nagumo Viability Theorem, i.e.,

Theorem 3.5.5. *Let X be finite dimensional, let $\mathcal{C} \subseteq \mathbb{R} \times X$ be a nonempty and locally closed set and let $f : \mathcal{C} \to X$ be a continuous function. Then, a necessary and sufficient condition in order that $\mathcal{C}$ be viable with respect to f is the tangency condition* (3.5.4).

We notice that, since the results above concern the forward existence property, all extend to the slightly more general case of forward locally closed sets defined below.

Definition 3.5.3. A set $\mathcal{C} \subseteq \mathbb{R} \times X$ is called *forward locally closed* if for each $(\tau, \xi) \in \mathcal{C}$ there exist $T > \tau$ and $\rho > 0$ such that $([\tau, T] \times D(\xi, \rho)) \cap \mathcal{C}$ is closed.

Using Proposition 2.5.1 and Remark 3.5.2, we deduce

Theorem 3.5.6. *Let X be a Banach space, $\mathcal{X} = \mathbb{R} \times X$, $\mathcal{K} \subseteq \mathcal{X}$ a nonempty and locally closed set and let $f : \mathcal{K} \to X$ be continuous. Let us assume that $\mathcal{K}$ is proximal and the norm, $\|\cdot\|$, on X is Gâteaux differentiable at each $x \in X$, $x \neq 0$, and $\mathcal{X}$ is endowed with the norm $\|(t,u)\|_{\mathcal{X}} = \sqrt{|t|^2 + \|u\|^2}$, for each $(t,u) \in \mathcal{X}$. Then the following conditions are equivalent*:

(i) *for every $(\tau, \xi) \in \mathcal{K}$, $(1, f(\tau, \xi)) \in \mathcal{C}_{\mathcal{K}}(\tau, \xi)$;*
(ii) *for every $(\tau, \xi) \in \mathcal{K}$, $(1, f(\tau, \xi)) \in \mathcal{T}_{\mathcal{K}}(\tau, \xi)$;*
(iii) *for every $(\tau, \xi) \in \mathcal{K}$, $(1, f(\tau, \xi)) \in \mathcal{B}_{\mathcal{K}}(\tau, \xi)$;*
(iv) *the set $\mathcal{K}$ is viable with respect to f.*

In general, if $\mathcal{G} : \mathcal{K} \rightsquigarrow \mathbb{R} \times X$ is such that $\mathcal{C}_{\mathcal{K}}(\tau, \xi) \subseteq \mathcal{G}(\tau, \xi) \subseteq \mathcal{B}_{\mathcal{K}}(\tau, \xi)$ for each $(\tau, \xi) \in \mathcal{K}$, then each one of the conditions above is equivalent to

(v) *for every $(\tau, \xi) \in \mathcal{K}$, $(1, f(\tau, \xi)) \in \mathcal{G}(\tau, \xi)$.*

Proof. Let us observe that the norm

$$\|(t,x)\|_{\mathcal{X}} = \sqrt{|t|^2 + \|x\|^2}$$

is Gâteaux differentiable at each $(t,x) \in \mathcal{X}$, $(t,x) \neq (0,0)$. $\square$

Remark 3.5.3. Since in the construction of the ε-approximate solutions only a countable set of values of both t and ξ is needed, one may ask whether or not we can relax the tangency condition (3.5.4) to hold merely for (τ,ξ) in a certain dense, possibly countable, subset in $\mathcal{C}$. The next example shows that the answer to this natural question is in the negative.

Example 3.5.1. Let $g : [0,1) \to \mathbb{R}$ be a function which is continuous, strictly increasing and $g'(t) = 0$ a.e. for $t \in [0,1)$. An example of such a function was given by Zaanen-Luxemburg [**186**]. See also Gelbaum-Olmsted [**100**], Example 30, p. 105. Let $\mathcal{C} \subseteq \mathbb{R}\times\mathbb{R}$ the graph of g. Then the function $f \equiv 0$ satisfies the tangency condition $(1,0) \in \mathcal{T}_{\mathcal{C}}((\tau, g(\tau)))$ for a.a. $\tau \in [0,1)$, but the Cauchy problem $u'(t) = 0$ and $u(0) = 0$ has no solution in the sense of Definition 3.1.1. So, we cannot replace the "everywhere" tangency condition (3.5.4) with an "almost everywhere" one.

The following "cylindrical" nonautonomous variant of Lemma 3.3.1 will prove useful in the sequel.

Lemma 3.5.1. *Let X be a real Banach space, K a nonempty, locally closed subset in X, I a nonempty and open interval and $f : I \times K \to X$ a continuous function satisfying the tangency condition $f(\tau,\xi) \in \mathcal{T}_K(\xi)$ for each $(\tau,\xi) \in I \times K$. Then, for each $(\tau,\xi) \in I \times K$ and each continuous extension $F : I \times X \to X$ of f, there exist $\rho > 0$, $T \in I$, $T > \tau$ and $M > 0$ and such that $D(\xi,\rho) \cap K$ is closed and, for each $\varepsilon \in (0,1)$, there exist one family of nonempty and pairwise disjoint intervals: $\mathcal{P}_T = \{[t_m, s_m);\ m \in \Gamma\}$, with Γ finite or countable, and two functions, $r : [\tau, T] \to X$ Riemann integrable, and $u : [\tau, T] \to X$ continuous, satisfying:*

(i) $\bigcup_{m\in\Gamma}[t_m, s_m) = [\tau, T)$ *and* $s_m - t_m \leq \varepsilon$ *for each* $m \in \Gamma$;
(ii) $u(t_m) \in D(\xi,\rho) \cap K$ *for each* $m \in \Gamma$ *and* $u(T) \in D(\xi,\rho) \cap K$;
(iii) $\|F(t,v)\| \leq M$ *for each* $(t,v) \in [\tau, T] \times D(\xi,\rho)$;
(iv) $\|r(t)\| \leq \varepsilon$ *for each* $t \in [\tau, T]$;
(v) *for each* $m \in \Gamma$ *and each* $t \in [t_m, s_m)$, *u satisfies*

$$u(t) = u(t_m) + \int_{t_m}^{t} f(\theta, u(t_m))\, d\theta + \int_{t_m}^{t} r(\theta)\, d\theta\,;$$

(vi) *for each* $m \in \Gamma$, *we have*

$$\sup_{s\in[t_m,s_m)} \|f(s,u(t_m)) - F(s,u(s))\| \leq \varepsilon.$$

Problem 3.5.2. *Use Problem* 3.5.1 *to prove Lemma* 3.5.1.

Another class of functions which play a crucial role in ordinary differential equations is defined below.

Definition 3.5.4. A function $f : I \times K \to X$ is called *locally almost-dissipative* if for each $(\tau, \xi) \in I \times K$ there exist $T > \tau$ and $\rho > 0$ such that $[\tau, T] \subseteq I$, $D(\xi, \rho) \cap K$ is closed, and there exist a continuous extension $F : [\tau, \sup I) \times X \to X$ of f and a Carathéodory uniqueness function $\omega : I \times \mathbb{R}_+ \to \mathbb{R}_+$ such that

$$[\, u - v, F(t, u) - F(t, v) \,]_+ \leq \omega(t, \|u - v\|)$$

for each $(t, u), (t, v) \in [\, \tau, T\,] \times D(\xi, \rho)$.

The next result refers to the viability of a cylindrical set with respect to an almost dissipative function.

Theorem 3.5.7. *Let I be a nonempty and open interval, $K \subseteq X$ a nonempty and locally closed set and let $f : I \times K \to X$ be continuous and locally almost dissipative. Then, a necessary and sufficient condition in order that $I \times K$ be viable with respect to f is that, for each $(\tau, \xi) \in I \times K$, the tangency condition $f(\tau, \xi) \in \mathcal{T}_K(\xi)$ be satisfied.*

Problem 3.5.3. *Use Lemma* 3.5.1 *to prove Theorem* 3.5.7.

We conclude this section with another "cylindrical" nonautonomous variant of Lemma 3.3.1 whose proof, based on Problem 3.5.1, being very similar with that one of Lemma 3.3.1 is left to the reader. We notice that this variant will prove useful in Chapter 4 when dealing with sufficient conditions for invariance. See for instance the proof of Theorem 4.1.3.

Lemma 3.5.2. *Let X be a real Banach space, K a nonempty, locally closed subset in X, I a nonempty and open interval and $f : I \times K \to X$ a continuous function satisfying the tangency condition $f(\tau, \xi) \in \mathcal{T}_K(\xi)$ for each $(\tau, \xi) \in I \times K$. Then, for each $(\tau, \xi) \in I \times K$, there exist $\rho > 0$, $T \in I$, $T > \tau$ and $M > 0$ and such that $D(\xi, \rho) \cap K$ is closed and, for each $\varepsilon \in (0, 1)$, there exist $\sigma : [\,\tau, T\,] \to [\,\tau, T\,]$ nondecreasing, $g : [\,\tau, T\,] \to X$ Riemann integrable and $u : [\,\tau, T\,] \to X$ continuous, such that:*

(i) $t - \varepsilon \leq \sigma(t) \leq t$ *for each* $t \in [\,\tau, T\,]$;
(ii) $\|g(t)\| \leq \varepsilon$ *for each* $t \in [\,\tau, T\,]$;
(iii) $\|f(t, u)\| \leq M$ *for each* $t \in [\,\tau, T\,]$ *and* $u \in D(\xi, \rho) \cap K$;
(iv) $u(\sigma(t)) \in D(\xi, \rho) \cap K$ *for all* $t \in [\,\tau, T\,]$ *and* $u(T) \in D(\xi, \rho) \cap K$;
(v) $u(t) = \xi + \int_\tau^t f(s, u(\sigma(s)))\, ds + \int_\tau^t g(s)\, ds$ *for each* $t \in [\,\tau, T\,]$;
(vi) $\|u(\sigma(t)) - u(t)\| \leq (M + 1)\varepsilon$ *for each* $t \in [\,\tau, T\,]$.

Problem 3.5.4. *Give a direct proof of Theorem* 3.5.5 *avoiding the use of the measure of noncompactness.*

Problem 3.5.5. *Give a direct proof of Theorem* 3.5.4 *avoiding the use of the measure of noncompactness.*

3.6. Global solutions

Let $\mathcal{C} \subseteq \mathbb{R} \times X$ be nonempty and let $f : \mathcal{C} \to X$. In this section we will prove some results concerning the existence of noncontinuable, or even global solutions to the Cauchy problem

$$\begin{cases} u'(t) = f(t, u(t)) \\ u(\tau) = \xi. \end{cases} \tag{3.6.1}$$

We recall that a solution $u : [\tau, T) \to X$ to (3.6.1) is called *noncontinuable*, if there is no other solution $v : [\tau, \widetilde{T}) \to X$ of the same equation, with $T < \widetilde{T}$ and satisfying $u(t) = v(t)$ for all $t \in [\tau, T)$. The solution u is called *global* if $T = T_{\mathcal{C}}$, where

$$T_{\mathcal{C}} = \sup\{t \in \mathbb{R};\ \text{there exists } \eta \in X,\ \text{with } (t, \eta) \in \mathcal{C}\}. \tag{3.6.2}$$

The next theorem follows from Brezis–Browder Theorem 2.1.1.

Theorem 3.6.1. *Let X be a Banach space, let $\mathcal{C} \subseteq \mathbb{R} \times X$ be nonempty and let $f : \mathcal{C} \to X$. The following conditions are equivalent*:

(i) *$\mathcal{C}$ is viable with respect to f*;

(ii) *for each $(\tau, \xi) \in \mathcal{C}$ there exists at least one noncontinuable solution $u : [\tau, T) \to X$ of* (3.6.1).

Proof. Clearly (ii) implies (i). To prove that (i) implies (ii) it suffices to show that every solution u can be continued up to a noncontinuable one. To this aim, we will make use of Brezis–Browder Theorem 2.1.1. Let $\mathcal{S}$ be the set of all solutions to (3.6.1), defined at least on $[\tau, T)$, and coinciding with u on that interval. On the set $\mathcal{S}$ which, by virtue of (i), is nonempty, we define the binary relation $\preceq$ by $u \preceq v$ if the domain $[\tau, T_v)$ of v is larger that the domain $[\tau, T_u)$ of u, i.e., $T_u \leq T_v$, and $u(t) = v(t)$ for all $t \in [\tau, T_u)$. Clearly $\preceq$ is a preorder on $\mathcal{S}$. Next, let $(u_m)_m$ be an increasing sequence in $\mathcal{S}$, and let us denote by $[\tau, T_m)$ the domain of definition of u_m. Let $T^* = \lim_m T_m$, which is finite, or not, and let us define $u^* : [\tau, T^*) \to X$ by $u^*(t) = u_m(t)$ for each $t \in [\tau, T_m)$. Since $(T_m)_m$ is increasing and $u_m(t) = u_k(t)$ for each $m \leq k$ and each $t \in [\tau, T_m)$, u^* is well-defined and belongs to $\mathcal{S}$. Moreover, u^* is an upper bound of $(u_m)_m$. Thus each increasing sequence in $\mathcal{S}$ is bounded from above. Moreover, the function $\mathcal{N} : \mathcal{S} \to \mathbb{R}$, defined by $\mathcal{N}(v) = T_v$, for each $v \in \mathcal{S}$, is increasing, and therefore we are in the hypotheses of Theorem 2.1.1. Accordingly, for $u \in \mathcal{S}$, there exists at least one element $\overline{u} \in \mathcal{S}$ with $u \preceq \overline{u}$ and, in addition, $\overline{u} \preceq \widetilde{u}$ implies $T_{\widetilde{u}} = T_{\overline{u}}$. But this means that $\overline{u}$ is noncontinuable, and, of course, that it extends u. The proof is complete. □

Remark 3.6.1. Notice that in Theorem 3.6.1 we don't need to assume $\mathcal{C}$ to be locally closed or f to be continuous.

Theorem 3.6.2. *Let X be a Banach space, let $\mathcal{K}$ be a nonempty and locally closed subset in $\mathbb{R} \times X$ and let $f : \mathcal{K} \to X$ be a continuous function. Let us assume that $\mathcal{K}$ is proximal and the norm $\|\cdot\|$ is Gâteaux differentiable at each $x \in X$, $x \neq 0$. Then, a necessary and sufficient condition in order that for each $(\tau, \xi) \in \mathcal{K}$ there exists at least one noncontinuable solution to* (3.6.1) *is anyone of the five equivalent conditions in Theorem* 3.5.6.

Proof. In view of Theorem 3.5.6, each one of the conditions (i)$\sim$ (v) is equivalent to the viability of $\mathcal{K}$ with respect to f. The conclusion follows from Theorem 3.6.1, and the proof is complete. □

We conclude this section with a result concerning the existence of global solutions.

Definition 3.6.1. A function $f : \mathcal{C} \to X$ is called *positively sublinear* if there exist three continuous functions $a : \mathbb{R} \to \mathbb{R}_+$, $b : \mathbb{R} \to \mathbb{R}_+$, and $c : \mathbb{R} \to \mathbb{R}_+$ such that

$$\|f(t,\xi)\| \leq a(t)\|\xi\| + b(t)$$

for each $(t, \xi) \in K_+^c(f)$, where

$$K_+^c(f) = \{(t,\xi) \in \mathcal{C};\ \|\xi\| > c(t) \ \text{ and } [\xi, f(t,\xi)]_+ > 0\}.$$

We recall that here $[\xi, \eta]_+$ is the right directional derivative of the norm $\|\cdot\|$ calculated at ξ in the direction η.

Remark 3.6.2. There are three important specific cases in which f is positively sublinear:

(i) when f is bounded on $\mathcal{C}$;
(ii) when f is globally Lipschitz;
(iii) when f satisfies the "sign condition" $[\xi, f(t,\xi)]_+ \leq 0$ for each $(t,\xi) \in \mathcal{C}$.

Definition 3.6.2. The set $\mathcal{C}$ is X-closed if for each sequence $((t_n, \xi_n))_n$ in $\mathcal{C}$ with $\lim_n (t_n, \xi_n) = (t, \xi)$, with $t < T_{\mathcal{C}}$, where $T_{\mathcal{C}}$ is given by (3.6.2), it follows that $(t, \xi) \in \mathcal{C}$.

A typical example of X-closed set is $\mathcal{C} = I \times K$ with I a nonempty and open to the right interval and $K \subseteq X$ nonempty and closed.

Theorem 3.6.3. *Let X be a Banach space, let $\mathcal{C} \subseteq \mathbb{R} \times X$ be nonempty and let $f : \mathcal{C} \to X$ be a given function. If $\mathcal{C}$ is X-closed, f maps bounded subsets in $\mathcal{C}$ into bounded subsets in X, is positively sublinear, and $\mathcal{C}$ is viable with respect to f, then each solution of (3.6.1) can be continued up to a global one, i.e. defined on $[\tau, T_{\mathcal{C}})$, where $T_{\mathcal{C}}$ is given by (3.6.2).*

Proof. Since $\mathcal{C}$ is viable with respect to f, by Theorem 3.6.1, for each $(\tau, \xi) \in \mathcal{C}$, there exists at least one noncontinuable solution $u : [\tau, T) \to X$ to (3.6.1). We will show that $T = T_{\mathcal{C}}$. To this aim, let us assume the contrary, i.e., that $T < T_{\mathcal{C}}$. In particular this means that $T < +\infty$. Integrating from τ to t the equality $[u(s), u'(s)]_+ = \frac{d^+}{ds}(\|u(\cdot)\|)(s)$ for $s \in [\tau, T)$, we get

$$\|u(t)\| = \|\xi\| + \int_\tau^t [u(s), f(s, u(s))]_+ \, ds$$

for each $t \in [\tau, T)$. Let us denote by

$$\begin{aligned} E_t &= \{s \in [\tau, t];\ [u(s), f(s, u(s))]_+ > 0 \text{ and } \|u(s)\| > c(s)\}, \\ G_t &= \{s \in [\tau, t];\ [u(s), f(s, u(s))]_+ \leq 0\}, \\ H_t &= \{s \in [\tau, t];\ \|u(s)\| \leq c(s)\}. \end{aligned}$$

Using (ii) in Exercise 1.6.1, we get

$$[u, v]_+ \leq \|v\|$$

for each $u, v \in X$. From this inequality, taking into account that, for each $t \in [\tau, T]$, $H_t \subseteq H_T$, we get

$$\begin{aligned} \|u(t)\| &\leq \|\xi\| + \int_{E_t} [a(s)\|u(s)\| + b(s)]\, ds + \int_{H_t \setminus G_t} [u(s), f(s, u(s))]_+ \, ds \\ &\leq \|\xi\| + \int_{E_t} [a(s)\|u(s)\| + b(s)]\, ds + \int_{H_T} \|f(s, u(s))\| \, ds. \end{aligned}$$

But f maps bounded subsets in $\mathcal{C}$ into bounded subsets in X and therefore there exists $m > 0$ such that

$$\sup\{\|f(s, u(s))\|; s \in H_T\} \leq m.$$

Hence,

$$\|u(t)\| \leq \|\xi\| + (T - \tau)m + \int_\tau^T b(s)\, ds + \int_\tau^T a(s)\|u(s)\|\, ds.$$

By the Gronwall Lemma 1.8.4, it follows that u is bounded on $[\tau, T)$. Accordingly, $f(\cdot, u(\cdot))$ is bounded on $[\tau, T)$ and so there exists $\lim_{t \uparrow T} u(t) = u^*$. Since $\mathcal{C}$ is X-closed and $T < T_{\mathcal{C}}$, it follows that $(T, u^*) \in \mathcal{C}$. Using this observation and recalling that $\mathcal{C}$ is viable with respect to f, we conclude that

u can be continued to the right of T. But this is absurd, because u is noncontinuable. This contradiction can be eliminated only if $T = T_{\mathcal{C}}$, and this achieves the proof. □

As, in view of Lemma 2.2.1, each compact set K is proximal and, by (i) in Remark 3.6.2, each continuous function $f : I \times K \to X$ (with K compact) is positively sublinear, from Theorem 3.5.6, we get:

Corollary 3.6.1. *Let X be a Banach space, $K \subseteq X$ a nonempty and compact subset of X, I an open interval and let $f : I \times K \to X$ be a continuous function. Let us assume that the norm $\|\cdot\|$ is Gâteaux differentiable at each $x \in X$, $x \neq 0$. A necessary and sufficient condition in order that for each $(\tau, \xi) \in I \times K$ to exist at least one solution, $u : [\tau, \sup I) \to K$, of* (3.6.1) *is any one of the five equivalent conditions in Theorem* 3.5.6.

CHAPTER 4

Problems of invariance

Here we introduce the reader to the basic problems on (local) invariance referring mainly to ordinary differential equations governed by continuous functions. After some simple preliminary results, we focus our attention on two sufficient conditions for local invariance expressed in terms of certain comparison inequalities coupled either with the viability of the set in question or with the Nagumo Tangency Condition. We continue with the main general sufficient condition for local invariance based on the so-called Exterior Tangency Condition. In the specific case of proximal sets, we show that viability combined with a very general comparison condition implies invariance. Next, we answer the question: "when does tangency imply exterior tangency?" and we conclude with some results on the relationship between local invariance and monotonicity.

4.1. Preliminary facts

Let X be a real Banach space, D an open subset in X, K a nonempty and locally closed subset of D, $f : I \times D \to X$ a given function and let us consider the Cauchy problem for the nonautonomous differential equation

$$\begin{cases} u'(t) = f(t, u(t)) \\ u(\tau) = \xi. \end{cases} \tag{4.1.1}$$

Definition 4.1.1. The subset K is *locally invariant* with respect to f if for each $(\tau, \xi) \in I \times K$ and each solution $u : [\tau, c] \to D$, $c \in I$, $c > \tau$, of (4.1.1), there exists $T \in (\tau, c]$ such that $u(t) \in K$ for each $t \in [\tau, T]$. It is *invariant* if it satisfies the above condition of local invariance with $T = c$.

Problem 4.1.1. Prove that if K is closed and locally invariant with respect to f, then it is invariant with respect to f.

The relationship between viability and local invariance is clarified in

Proposition 4.1.1. *Let X be a Banach space, let $D \subseteq X$ be open, $K \subseteq D$ nonempty and locally closed and $f : D \to X$ locally β-compact. If K is locally invariant with respect to f, then K is viable with respect to f.*

Proof. By Theorem 3.5.2 combined with Remarks 2.4.2 and 3.5.2, for each $(\tau,\xi)\in I\times K$ there exists $u:[\tau,c]\to D$, solution of (4.1.1). Since K is locally invariant, there exists $T\le c$ such that $u(t)\in K$ for each $t\in[0,T]$. The proof is complete. □

The converse of this assertion is no longer true, as we can see from the following example.

Example 4.1.1. Let $D=\mathbb{R}$, $K=\{0\}$ and let $f:\mathbb{R}\to\mathbb{R}$ be defined by $f(u)=3\sqrt[3]{u^2}$ for every $u\in\mathbb{R}$. Then K is viable with respect to f but K is not locally invariant with respect to f, because the differential equation $u'(t)=f(u(t))$ has at least two solutions which satisfy $u(0)=0$, i.e., $u\equiv 0$ and $v(t)=t^3$.

A simple necessary and sufficient condition of invariance is stated below.

Theorem 4.1.1. *Let X be a Banach space, let $D\subseteq X$ be open, $K\subseteq D$ nonempty and locally closed and $f:D\to X$ locally β-compact. Assume further that the associated Cauchy problem has the uniqueness property. Then, a necessary and sufficient condition in order that K be locally invariant with respect to f is that, for every $(t,\xi)\in I\times K$,*

$$f(t,\xi)\in\mathcal{T}_K(\xi). \tag{4.1.2}$$

Proof. The necessity part follows from Proposition 4.1.1, Corollary 3.5.1 and Remark 3.5.2 while the sufficiency follows from Theorem 3.5.2 combined with the uniqueness property. □

Whenever $K\subseteq D$ and $f:I\times D\to X$, we agree to say that $I\times K$ is viable with respect to f, if $I\times K$ is viable with respect to $f_{|_{I\times K}}$. So, Theorem 4.1.1 says that, in general, if $I\times K$ is viable with respect to f and (4.1.1) has the uniqueness property, then $I\times K$ is locally invariant with respect to f. The preceding example shows that this is no longer true if we assume that K is viable with respect to f and merely $u'(t)=f_{|_{I\times K}}(t,u(t))$ has the uniqueness property. The next example reveals another interesting fact about local invariance. It shows that the local invariance of K with respect to f can take place even if $u'(t)=f_{|_{I\times K}}(t,u(t))$ has not the uniqueness property.

Example 4.1.2. Let $K=\{(x,y)\in\mathbb{R}^2;\ y\le 0\}$ and let $f:\mathbb{R}^2\to\mathbb{R}^2$ be defined by

$$f((x,y))=\begin{cases}(1,0) & \text{if } (x,y)\in\mathbb{R}^2\setminus K\\ (1,-3\sqrt[3]{y^2}) & \text{if } (x,y)\in K.\end{cases}$$

Obviously K is locally invariant with respect to $f_{|K}$ but $u'(t) = f_{|K}(u(t))$ doesn't not have the uniqueness property. The latter assertion follows from the remark that, from each point, $(x, 0)$ (on the boundary of K), we have at least two solutions to $u'(t) = f(u(t))$, $u(t) = (t+x, 0)$ and $v(t) = (t+x, -t^3)$ satisfying $u(0) = v(0) = (x, 0)$. See Figure 4.1.1.

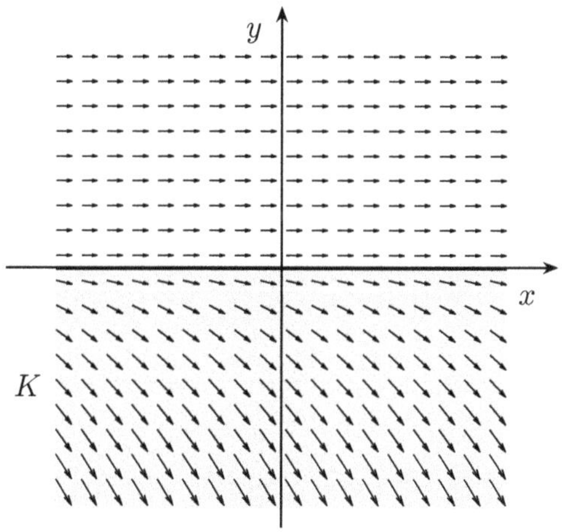

Figure 4.1.1

The result below shows that from the viability of $I \times K$ with respect to f and an appropriate comparison property of f we get local invariance of K with respect to f. First we introduce

Definition 4.1.2. A function $\omega : I \times \mathbb{R}_+ \to \mathbb{R}_+$ is a *comparison function* if $\omega(t, 0) = 0$ for each $t \in I$, and for each $[\tau, T) \subseteq I$, the only continuous function $x : [\tau, T) \to \mathbb{R}_+$, satisfying

$$\begin{cases} [D_+x](t) \leq \omega(t, x(t)) & \text{for all } t \in [\tau, T) \\ x(\tau) = 0, \end{cases} \tag{4.1.3}$$

is the null function. We recall that $[D_+x](t)$ is given by (1.8.3).

Theorem 4.1.2. *Let X be a Banach space, let I be an open interval, $D \subseteq X$ an open set and let $K \subseteq D$ be closed. Let us assume that $I \times K$ is viable with respect to $f : I \times D \to X$ and there exists an open neighborhood $V \subseteq D$ of K and one comparison function $\omega : I \times \mathbb{R}_+ \to \mathbb{R}_+$ such that*

$$[\xi_1 - \xi_2, f(t, \xi_1) - f(t, \xi_2)]_+ \leq \omega(t, \|\xi_1 - \xi_2\|) \tag{4.1.4}$$

for each $t \in I$, $\xi_1 \in V \setminus K$ and $\xi_2 \in K$. Then K is locally invariant with respect to f.

Proof. Let $V \subseteq D$ be the open neighborhood of K such that f satisfies (4.1.4), and let $\omega : I \times \mathbb{R}_+ \to \mathbb{R}_+$ be the corresponding comparison function. Let $(\tau, \xi) \in I \times K$, let $u : [\tau, c] \to D$ be any solution to (4.1.1). Diminishing c if necessary, we may assume that $u(t) \in V$ for each $t \in [\tau, c]$. We prove that $u(t) \in K$ for each $t \in [\tau, c]$. To this end, let us assume by contradiction that there exists $t_1 \in [\tau, c]$ such that $u(t_1) \in V \setminus K$. Let $\tau \leq t_0 < t_1$ be such that $u(t) \in V \setminus K$ for every $t \in (t_0, t_1]$ and $u(t_0) \in K$. This is possible because K is closed and u is continuous. Let $v : [t_0, d] \to K$ be any solution of $v'(t) = f(t, v(t))$ which satisfies $v(t_0) = u(t_0)$. Such a solution exists because K is viable with respect to f. Let $t_2 = \min\{d, t_1\}$. Let $g : [t_0, t_2] \to \mathbb{R}_+$ be defined by $g(t) = \|u(t) - v(t)\|$ for each $t \in [t_0, t_2]$. Let $t \in [t_0, t_2)$ and $h > 0$ with $t + h \in [t_0, t_2]$. We have

$$\frac{g(t+h) - g(t)}{h} \leq \alpha(h) + \frac{\|u(t) - v(t) + h(u'(t) - v'(t))\| - \|u(t) - v(t)\|}{h},$$

where

$$\alpha(h) = \left\| \frac{u(t+h) - u(t)}{h} - u'(t) \right\| + \left\| \frac{v(t+h) - v(t)}{h} - v'(t) \right\|.$$

Since $u'(t) = f(t, u(t))$, $v'(t) = f(t, v(t))$ and $\lim_{h \downarrow 0} \alpha(h) = 0$, passing to the inf-limit in the inequality above for $h \downarrow 0$ and taking into account that V, K, and f satisfy (4.1.4), we get

$$[D_+ g](t) \leq \omega(t, g(t))$$

for each $t \in [t_0, t_2)$. So, $g(t) \equiv 0$ which implies that $u(t) = v(t)$ for all $t \in [t_0, t_2]$. Since $v(t) \in K$, we arrived at a contradiction. This contradiction can be eliminated only if $u(t) \in K$ for each $t \in [\tau, c]$. The proof is complete. □

Exercise 4.1.1. *Show that if* (4.1.4) *is satisfied for each $t \in I$ and $\xi_1, \xi_2 \in D$, then the Cauchy problem* (4.1.1) *has the uniqueness property.*

If f is continuous, instead of the viability assumption, we may suppose that the tangency condition (4.1.2) holds true. The price one has to pay is that, in (4.1.4), ω must be not only a comparison function but also a uniqueness function.

Theorem 4.1.3. *Let X be a Banach space, let I be an open interval, $D \subseteq X$ an open set and let $K \subseteq D$ be closed. Let $f : I \times D \to X$ be a continuous function such that the tangency condition* (4.1.2) *holds for every $(t, \xi) \in I \times K$. Assume further that there exists an open neighborhood*

$V \subseteq D$ of K and one Carathéodory uniqueness function[1] $\omega : I \times \mathbb{R}_+ \to \mathbb{R}_+$, *such that* (4.1.4) *is satisfied for each* $t \in I$, $\xi_1 \in V \setminus K$ *and* $\xi_2 \in K$. *Then* K *is locally invariant with respect to* f.

Proof. Let $V \subseteq D$ be the open neighborhood of K such that f satisfies (4.1.4), let $(\tau, \xi) \in I \times K$ and let $u : [\,\tau, c\,] \to D$ be any solution to (4.1.1). Diminishing c if necessary, we may assume that $u(t) \in V$ for each $t \in [\,\tau, c\,]$. We prove that $u(t) \in K$ for each $t \in [\,\tau, c\,]$. To this end, let us assume by contradiction that there exists $t_1 \in [\,\tau, c\,]$ such that $u(t_1) \in V \setminus K$. Let $\tau \le t_0 < t_1$ be such that $u(t) \in V \setminus K$ for every $t \in (t_0, t_1\,]$ and $u(t_0) \in K$. This is possible because K is closed and u is continuous. Let $((\sigma_n, g_n, u_n))_n$ a sequence of ε_n-approximate solutions of the problem

$$\begin{cases} v'(t) = f(t, v(t)) \\ v(t_0) = u(t_0) \end{cases}$$

on $[\,t_0, T\,]$, with $T \le t_1$, whose existence is ensured by Lemma 3.5.2. Since $s \mapsto f(s, u_n(\sigma_n(s)))$ is integrable (being measurable and bounded by (iii) in Lemma 3.5.2), by (v) in the same lemma, we have

$$u_n'(t) = f(t, u_n(\sigma_n(t))) + g_n(t)$$

a.e. for $t \in [\,t_0, T\,]$. Thus we have

$$\|u_n(\sigma_n(t)) - u(t)\| \le \widetilde{M}\varepsilon_n$$

$$+ \int_{t_0}^{t} [\,u_n(\sigma_n(s)) - u(s), f(s, u_n(\sigma_n(s))) - f(s, u(s))\,]_+ \, ds,$$

and thus

$$\|u_n(\sigma_n(t)) - u(t)\| \le \widetilde{M}\varepsilon_n + \int_{t_0}^{t} \omega(s, \|u_n(\sigma_n(s)) - u(s)\|) \, ds$$

for $n = 1, 2, \ldots$ and $t \in [\,t_0, T\,]$. Here $\widetilde{M} > 0$ does not depend on n and t. See (ii) and (vi) in Lemma 3.5.2. Applying Lemma 1.8.3, we conclude that there exists $T_0 \in (t_0, T\,]$ such that $\lim_n u_n(\sigma_n(t)) = u(t)$ uniformly on $[\,t_0, T_0\,]$. From (iv) in Lemma 3.5.2, recalling that $D(\xi, \rho) \cap K$ is closed, we deduce that $u(t) \in K$ for each $t \in [t_0, T_0]$ thereby contradicting the definition of t_0. The contradiction can be eliminated only of $u(t) \in K$ for each $t \in [\,\tau, c\,]$ and this completes the proof. □

[1]See Definition 1.8.2.

4.2. Sufficient conditions for local invariance

Our first sufficient condition for local invariance says that, whenever there exists an open neighborhood $V \subseteq D$ of K such that f satisfies the "exterior tangency" condition

$$\liminf_{h \downarrow 0} \frac{1}{h} \left[\operatorname{dist}(\xi + hf(t,\xi); K) - \operatorname{dist}(\xi; K)\right] \leq \omega(t, \operatorname{dist}(\xi; K)) \tag{4.2.1}$$

for each $(t,\xi) \in I \times V$, where ω is a comparison function in the sense of Definition 4.1.2, then K is locally invariant with respect to f. More precisely, we have

Theorem 4.2.1. *Let X be a Banach space, $K \subseteq D \subseteq X$, with K locally closed and D open, and let $f : I \times D \to X$. If there exist an open neighborhood $V \subseteq D$ of K and a comparison function $\omega : I \times \mathbb{R}_+ \to \mathbb{R}_+$ such that f satisfies* (4.2.1), *then K is locally invariant with respect to f.*

Proof. Let $V \subseteq D$ be the open neighborhood of K such that f satisfies (4.2.1), and let $\omega : I \times \mathbb{R}_+ \to \mathbb{R}_+$ be the corresponding comparison function. Let $\xi \in K$ and let $u : [\,\tau, c\,] \to V$ be any solution to (4.1.1). Diminishing c if necessary, we may assume that there exists $\rho > 0$ such that $D(\xi,\rho) \cap K$ is closed and $u(t) \in D(\xi, \rho/2)$ for each $t \in [\,\tau, c\,]$. Let $g : [\,\tau, c\,] \to \mathbb{R}_+$ be defined by $g(t) = \operatorname{dist}(u(t); K)$ for each $t \in [\,\tau, c\,]$. Let $t \in [\,\tau, c)$ and $h > 0$ with $t + h \in [\,\tau, c\,]$. We have

$$\begin{aligned} g(t+h) &= \operatorname{dist}(u(t+h); K) \\ &\leq h \left\| \frac{u(t+h) - u(t)}{h} - u'(t) \right\| + \operatorname{dist}(u(t) + hu'(t); K). \end{aligned}$$

Therefore

$$\frac{g(t+h) - g(t)}{h} \leq \alpha(h) + \frac{\operatorname{dist}(u(t) + hu'(t); K) - \operatorname{dist}(u(t); K)}{h},$$

where

$$\alpha(h) = \left\| \frac{u(t+h) - u(t)}{h} - u'(t) \right\|.$$

Since $u'(t) = f(t, u(t))$ and $\lim_{h \downarrow 0} \alpha(h) = 0$, passing to the inf-limit in the inequality above for $h \downarrow 0$, and taking into account that V, K, and f satisfy (4.2.1), we get

$$[D_+ g](t) \leq \omega(t, g(t))$$

for each $t \in [\,\tau, c)$. So, $g(t) \equiv 0$ which means that $u(t) \in \overline{K} \cap D(\xi, \rho/2)$. But $\overline{K} \cap D(\xi, \rho/2) \subseteq K \cap D(\xi, \rho)$, and this achieves the proof. □

4.3. Viability and comparison imply invariance

Definition 4.3.1. Let $K \subseteq D \subseteq X$ be proximal. See Definition 2.2.3. We say that a function $f : I \times D \to X$ has *the comparison property with respect to* (D, K) if there exist a proximal neighborhood $V \subseteq D$ of K, one projection $\pi_K : V \to K$ subordinated to V, and one comparison function $\omega : I \times \mathbb{R}_+ \to \mathbb{R}_+$, such that

$$[\xi - \pi_K(\xi), f(t,\xi) - f(t, \pi_K(\xi))]_+ \leq \omega(t, \|\xi - \pi_K(\xi)\|) \tag{4.3.1}$$

for each $(t, \xi) \in I \times V$.

Since (4.3.1) is always satisfied for $(t, \xi) \in I \times K$, in Definition 4.3.1, we have merely to assume that (4.3.1) holds for each $(t, \xi) \in I \times (V \setminus K)$.

Definition 4.3.2. Let $K \subseteq D \subseteq X$ be proximal. We say that the function $f : I \times D \to X$ is:

(i) (D, K)-*Lipschitz* if there exist a proximal neighborhood $V \subseteq D$ of K, a projection $\pi_K : V \to K$ subordinated to V, and $L > 0$, such that

$$\|f(t, \xi) - f(t, \pi_K(\xi))\| \leq L\|\xi - \pi_K(\xi)\|$$

for each $(t, \xi) \in I \times (V \setminus K)$;

(ii) (D, K)-*dissipative* if there exist a proximal neighborhood $V \subseteq D$ of K and a projection $\pi_K : V \to K$ subordinated to V, such that

$$[\xi - \pi_K(\xi), f(t, \xi) - f(t, \pi_K(\xi))]_+ \leq 0$$

for each $(t, \xi) \in I \times (V \setminus K)$.

Remark 4.3.1. We notice that, if we assume that (4.3.1), or either of the conditions (i), or (ii) in Definition 4.3.2 is satisfied for ξ replaced by ξ_1 and $\pi_K(\xi)$ replaced by ξ_2 with $\xi_1, \xi_2 \in V$ then, for each $[\tau, T] \subseteq I$ and $\xi \in K$, there exists at most one solution $u : [\tau, T] \to K$ to (4.1.1) satisfying $u(\tau) = \xi$. See Exercise 4.1.1. Of contrary, in this frame, it may happen that, for certain (or for all) $[\tau, T] \subseteq I$ and $\xi \in K$, (4.1.1) has at least two solutions $u, v : [\tau, T] \to K$ satisfying $u(\tau) = v(\tau) = \xi$.

Let V be a proximal neighborhood of K, and let $\pi_K : V \to K$ be a projection subordinated to V. If $f : I \times V \to K$ is a function with the property that, for each $t \in I$ and $\eta \in K$, the restriction of $f(t, \cdot)$ to the "segment"

$$V_\eta = \{\xi \in V \setminus K\, ;\ \pi_K(\xi) = \eta\}$$

is dissipative, then f is (D, K)-dissipative.

Clearly, if f is either (D, K)-Lipschitz, or (D, K)-dissipative, then it has the comparison property with respect to (D, K). We notice that there are examples showing that there exist functions f which, although neither (D,K)-Lipschitz, nor (D,K)-dissipative, do have the comparison property. Moreover, there exist functions which, although (D,K)-Lipschitz, are not Lipschitz on D, as well as, functions which although (D,K)-dissipative, are not dissipative on D. In fact, these two properties describe merely the local behavior of f at the interface between K and $D \setminus K$. We include below two examples: the first one of a (D,K)-Lipschitz function which is not locally Lipschitz, and the second one of a function which, although non-dissipative, is (D,K)-dissipative. We notice that both examples refer to the autonomous case.

Example 4.3.1. The graph of a (D,K)-Lipschitz function $f : \mathbb{R} \to \mathbb{R}$ which is not Lipschitz is illustrated in Figure 4.3.1. Here $f(x) = x \sin \frac{1}{x}$ for

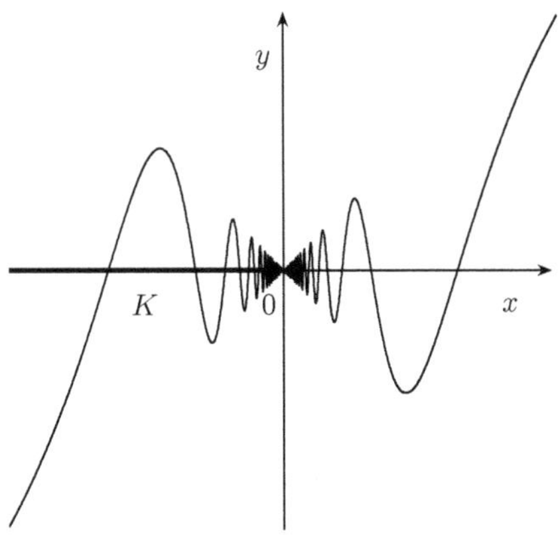

Figure 4.3.1

$x \neq 0$ and $f(0) = 0$, $K = (-\infty, 0]$ and D is an arbitrary open subset in $\mathbb{R}$ including K.

Example 4.3.2. The graph of a function $f : \mathbb{R} \to \mathbb{R}$ which is (D,K)-dissipative but not dissipative is illustrated in Figure 4.3.2. This time, $f(x) = x(\alpha^2 - x^2)$, K is $[-\alpha, \alpha]$ and D is an arbitrary open subset in $\mathbb{R}$ including K.

Theorem 4.3.1. *Let X be a Banach space, $K \subseteq D \subseteq X$, with K proximal and locally closed and D open, and let $f : I \times D \to X$. If f has the comparison property with respect to (D,K), and for every $(t,\xi) \in I \times K$, $f(t,\xi) \in \mathcal{T}_K(\xi)$, then f satisfies the exterior tangency condition* (4.2.1). *Therefore, K is locally invariant with respect to f.*

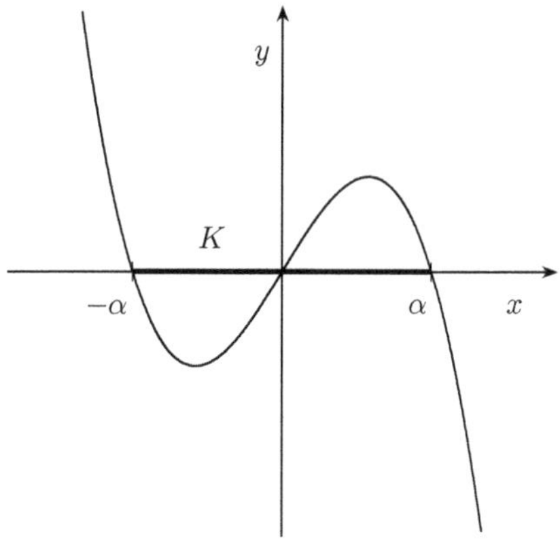

Figure 4.3.2

Proof. Let $V \subseteq D$ be a proximal neighborhood of K and π_K one projection subordinated to V as in Definition 4.3.1. Let $\xi \in V$ and $h > 0$. Taking into account that $\|\xi - \pi_K(\xi)\| = \text{dist}\,(\xi; K)$, we have

$$\text{dist}\,(\xi + hf(t,\xi); K) - \text{dist}\,(\xi; K) \leq \|\xi - \pi_K(\xi) + h[f(t,\xi) - f(t, \pi_K(\xi))]\|$$
$$-\|\xi - \pi_K(\xi)\| + \text{dist}\,(\pi_K(\xi) + hf(t, \pi_K(\xi)); K).$$

Dividing by h, passing to the inf-limit for $h \downarrow 0$, and using (2.4.1), we get

$$\liminf_{h \downarrow 0} \frac{1}{h}[\text{dist}\,(\xi + hf(t,\xi); K) - \text{dist}\,(\xi; K)]$$
$$\leq [\xi - \pi_K(\xi), f(t,\xi) - f(t, \pi_K(\xi))]_+ \leq \omega(t, \|\xi - \pi_K(\xi)\|).$$

But this inequality shows that (4.2.1) holds. To complete the proof we apply Theorem 4.2.1. □

In the specific case in which f is continuous, we have

Theorem 4.3.2. *Let X be a Banach space, $K \subseteq D \subseteq X$, with K locally compact and D open and let $f : I \times D \to X$ be continuous. Let us assume that the norm $\|\cdot\|$ is Gâteaux differentiable at each $x \in X$, $x \neq 0$ and f has the comparison property with respect to (D, K). If one of the four conditions below is satisfied:*

(i) *for every $(t,\xi) \in I \times K$, $f(t,\xi) \in \mathcal{C}_K(\xi)$;*
(ii) *for every $(t,\xi) \in I \times K$, $f(t,\xi) \in \mathcal{T}_K(\xi)$;*
(iii) *for every $(t,\xi) \in I \times K$, $f(t,\xi) \in \mathcal{B}_K(\xi)$;*
(iv) *the set K is viable with respect to f,*

then f satisfies the exterior tangency condition (4.2.1). *Therefore, K is invariant with respect to f.*

In general, if $\mathcal{G} : K \rightsquigarrow X$ satisfies $\mathcal{C}_K(\xi) \subseteq \mathcal{G}(\xi) \subseteq \mathcal{B}_K(\xi)$ for each $\xi \in K$ and

(v) *for every $(t, \xi) \in I \times K$, $f(t, \xi) \in \mathcal{G}(\xi)$,*

then (4.2.1) *is satisfied too.*

Proof. The conclusion follows from Theorems 3.2.5 and 4.3.1. □

4.4. When tangency does imply exterior tangency?

Next, we will prove that, in special circumstances, the tangency condition (4.1.2) for a function $f : I \times K \to X$ comes from the exterior tangency condition (4.2.1) for a suitably defined extension $\widetilde{f} : I \times D \to X$ of f. More precisely, we have

Theorem 4.4.1. *Let X be a Banach space, let $K \subseteq X$ be proximal and let $f : I \times K \to X$ be a given function satisfying* (4.1.2). *If $V \subseteq X$ is a proximal neighborhood of K and $r : V \to K$ is a projection subordinated to V, then $\widetilde{f} : I \times V \to X$, defined by $\widetilde{f}(t, \cdot) = f(t, r(\cdot))$, satisfies* (4.2.1).

Proof. Let $\xi \in V$ and $h > 0$. We have

$$\begin{aligned} &\operatorname{dist}(\xi + h\widetilde{f}(t, \xi); K) - \operatorname{dist}(\xi; K) \\ &\leq \|\xi - r(\xi)\| + \operatorname{dist}(r(\xi) + hf(t, r(\xi)); K) - \|\xi - r(\xi)\| \\ &= \operatorname{dist}(r(\xi) + hf(t, r(\xi)); K). \end{aligned}$$

Dividing by $h > 0$ and passing to lim inf for $h \downarrow 0$, we get

$$\liminf_{h \downarrow 0} \frac{1}{h}[\operatorname{dist}(\xi + h\widetilde{f}(t, \xi); K) - \operatorname{dist}(\xi; K)] \leq 0.$$

So, (4.2.1) holds true with $\omega \equiv 0$, and the proof is complete. □

It should be noticed that, the conclusion of Theorem 4.4.1 is no longer true if we are looking for a continuous extension $\widetilde{f}$ of a continuous function f satisfying (3.2.3), as the next example shows.

Example 4.4.1. Let $K = K_1 \cup K_2$, where $K_1 = \{(x, 3\sqrt[3]{x^2})\,;\ x \in \mathbb{R}_+\}$, $K_2 = \{(x, 3\sqrt[3]{x^2})\,;\ x \in \mathbb{R}_-\}$. If $\xi \in K_1$, we define $f(\xi)$ as the unit clockwise oriented tangent vector to K_1 at ξ, and if $\xi \in K_2$, we define $f(\xi)$ as the unit counterclockwise oriented tangent vector to K_1 at ξ. Let us observe that $f((0,0)) = (0,1)$. Thus $f : K \to \mathbb{R}^2$ is continuous and $f(\xi) \in \mathcal{T}_K(\xi)$ for each $\xi \in K$. By virtue of Theorem 3.2.6, K is viable with respect to f.

Let $\widetilde{f}$ be any continuous extension of f to an open neighborhood V of the origin. We may assume that, for each $v \in V$, $\widetilde{f}_2(v) \geq \frac{1}{2}$.

In fact, the equation $u'(t) = f(u(t))$ subjected to $u(0) = (0,0)$ has two local solutions $u, v : [0, \delta] \to K$, with $u([0,\delta]) \subseteq K_1$ and $v([0,\delta]) \subseteq K_2$. Diminishing $\delta > 0$, we may assume that no solution to $u'(t) = \widetilde{f}(u(t))$, $u(0) = (0,0)$, can escape from V. Now, if we assume that K is invariant with respect to $\widetilde{f}$, we have

$$F_{0,(0,0)}(\delta) = \{u(\delta)\,;\ u'(t) = \widetilde{f}(u(t)),\ \text{for all } t \in [0,\delta],\ u(0) = (0,0)\} \subseteq K,$$

and by virtue of a classical result due to Kneser (see Theorem 7.7.1), we know that $F_{0,(0,0)}(\delta)$ is connected, and therefore, we conclude that there exists at least one solution $w : [0,\delta] \to K$ of $u'(t) = \widetilde{f}(u(t))$, $u(0) = (0,0)$ with $w(\delta) = (0,0)$. But this is impossible, because $w_2(\delta) \geq \frac{1}{2}\delta$.

However, if f is continuous and K is smooth enough, by the very same proof we deduce

Theorem 4.4.2. *Let X be a Banach space, $K \subseteq X$, let $f : I \times K \to X$ be a continuous function satisfying* (4.1.2). *If there exist a proximal neighborhood $V \subseteq X$ of K and a continuous projection $r : V \to K$ subordinated to V, then f can be extended to a continuous function $\widetilde{f} : I \times V \to X$ satisfying* (4.2.1).

4.5. Local invariance and monotonicity

Let X be a Banach space, let K be a nonempty subset in X and $\preceq$ a preorder on X. The preorder $\preceq$ is *locally invariant with respect to* f if for each $(\tau,\xi) \in I \times K$ and each solution $u : [\tau, c] \to D$, $c \in I$, $\tau < c$, of (4.1.1), there exists $T \in (\tau, c]$ such that $u(t) \in K$ for each $t \in [\tau, T]$ and u is $\preceq$-monotone on $[\tau, T]$, i.e., for each $\tau \leq s \leq t \leq T$, we have $u(s) \preceq u(t)$. Here and thereafter, if $\xi \in K$, we denote by $\mathcal{P}(\xi) = \{\eta \in K;\ \xi \preceq \eta\}$.

Remark 4.5.1. The preorder $\preceq$ is locally invariant with respect to f if and only if for each $(\tau,\xi) \in I \times K$ and each solution $u : [\tau, c] \to D$, $c \in I$, $\tau < c$, of (4.1.1), there exists $T \in (\tau, c]$ such that, for each $s \in [\tau, T]$ and $t \in [s, T]$, we have $u(t) \in \mathcal{P}(u(s))$.

Proposition 4.5.1. *The preorder $\preceq$ is locally invariant with respect to f if and only if for each $\xi \in K$, the set $\mathcal{P}(\xi)$ is locally invariant with respect to f.*

Corollary 4.5.1. *If for each $\xi \in K$ there exists an open neighborhood $V \subseteq D$ of $\mathcal{P}(\xi)$ and a comparison function, $\omega : I \times \mathbb{R}_+ \to \mathbb{R}_+$, such that*

$$\liminf_{h \downarrow 0} \frac{1}{h}[\operatorname{dist}(\eta + hf(t,\eta); \mathcal{P}(\xi)) - \operatorname{dist}(\eta; \mathcal{P}(\xi))] \leq \omega(t, \operatorname{dist}(\eta; \mathcal{P}(\xi)))$$

for each $(t,\eta) \in I \times V$, then $\preceq$ is locally invariant with respect to f.

Corollary 4.5.2. *If, for each $\xi \in K$, f has the comparison property with respect to $(D, \mathcal{P}(\xi))$ and, for each $(t,\eta) \in I \times \mathcal{P}(\xi)$, we have*

$$\liminf_{h \downarrow 0} \frac{1}{h}\operatorname{dist}(\eta + hf(t,\eta); \mathcal{P}(\xi)) = 0,$$

then $\preceq$ is locally invariant with respect to f.

CHAPTER 5

Viability under Carathéodory conditions

In this chapter we extend to the case of Carathéodory solutions the viability results in Chapter 3, referring to classical solutions. We notice that in this, a fortiori nonautonomous, case, due to some reasons explained below, we will confine ourselves to consider only cylindrical sets. After showing that an a.e. Nagumo-type tangency condition is necessary for Carathéodory viability, we state and prove that the very same a.e. Nagumo-type tangency condition is also sufficient under some natural Carathéodory-type extra-conditions combined with appropriate either compactness or Lipschitz conditions on the right hand-side. Finally, we focus our attention on the existence of noncontinuable or even global Carathéodory solutions.

5.1. Necessary conditions for Carathéodory viability

Let X be a real Banach space, K a nonempty subset in X, $f : I \times K \to X$ a given function and let us consider the Cauchy problem

$$\begin{cases} u'(t) = f(t, u(t)) \\ u(\tau) = \xi. \end{cases} \tag{5.1.1}$$

Definition 5.1.1. By a *Carathéodory solution* of (5.1.1) on $[\tau, T]$ we mean an absolutely continuous function $u : [\tau, T] \to K$ which is a.e. differentiable on $[\tau, T]$, with $u' \in L^1(\tau, T; X)$ and satisfying the differential equation in (5.1.1) for a.a. $t \in [\tau, T]$ and $u(\tau) = \xi$. A *Carathéodory solution* of (5.1.1) on the semi-open interval $[\tau, T)$ is defined similarly, with the mention that here, we have to impose that $u' \in L^1_{\text{loc}}([\tau, T); X)$.

Remark 5.1.1. Since the function u in Definition 5.1.1 is a fortiori in $W^{1,1}(\tau, T; X)$ for each $[\tau, T] \subseteq I$, if u is a Carathéodory solution of (5.1.1) on $[\tau, T]$, then

$$u(t) = u(s) + \int_s^t u'(\theta)\, d\theta$$

for each $\tau \leq s \leq t \leq T$.

Definition 5.1.2. The set $I \times K$ is *Carathéodory viable* with respect to f if for each $(\tau, \xi) \in I \times K$ there exist $T \in I$, $T > \tau$, and a Carathéodory solution $u : [\tau, T] \to K$ of (5.1.1)

We can now proceed to the main result in this section, i.e., a necessary condition for viability.

Theorem 5.1.1. *Let X be a Banach space. If $K \subseteq X$ is separable and the set $I \times K$ is Carathéodory viable with respect to the Carathéodory function $f : I \times K \to X$, then there exists a negligible subset $\mathcal{Z}$ of I such that, for each (τ, ξ) in $(I \setminus \mathcal{Z}) \times K$, we have $f(\tau, \xi) \in \mathfrak{F}_K(\xi)$.*

Proof. Since K is separable and f is a Carathéodory function, we are in the hypotheses of Theorem 2.8.2. Therefore, there exists a negligible subset $\mathcal{Z}$ of I such that for each $\tau \in I \setminus \mathcal{Z}$ and each continuous function $u : [\tau, T) \to K$, we have

$$\lim_{h \downarrow 0} \frac{1}{h} \int_{\tau}^{\tau+h} \|f(s, u(s)) - f(\tau, u(\tau))\| \, ds = 0. \tag{5.1.2}$$

Let $(\tau, \xi) \in (I \setminus \mathcal{Z}) \times K$. In view of Definition 2.4.2, we have to prove that $\lim_{h \downarrow 0} \frac{1}{h} \operatorname{dist}(\xi + hf(\tau, \xi); K) = 0$. Since $I \times K$ is Carathéodory viable with respect to f, there exist $T \in I$, $T > \tau$, and an almost everywhere differentiable function $u : [\tau, T] \to K$ satisfying $u'(s) = f(s, u(s))$ for a.a. $s \in [\tau, T]$ and $u(\tau) = \xi$. In view of Remark 5.1.1, we have

$$\frac{u(\tau + h) - u(\tau)}{h} = \frac{1}{h} \int_{\tau}^{\tau+h} f(s, u(s)) \, ds.$$

Thus

$$\lim_{h \downarrow 0} \frac{1}{h} \operatorname{dist}(\xi + hf(\tau, \xi); K) \leq \lim_{h \downarrow 0} \frac{1}{h} \|u(\tau) + hf(\tau, \xi) - u(\tau + h)\|$$

$$\leq \lim_{h \downarrow 0} \left\| f(\tau, \xi) - \frac{u(\tau + h) - u(\tau)}{h} \right\| = \lim_{h \downarrow 0} \left\| f(\tau, \xi) - \frac{1}{h} \int_{\tau}^{\tau+h} f(s, u(s)) \, ds \right\|$$

$$\leq \lim_{h \downarrow 0} \frac{1}{h} \int_{\tau}^{\tau+h} \|f(\tau, u(\tau)) - f(s, u(s))\| \, ds.$$

Thanks to (5.1.2), $f(\tau, \xi) \in \mathfrak{F}_K(\xi)$, which completes the proof. □

Corollary 5.1.1. *Let X be a Banach space. If $K \subseteq X$ is separable and $I \times K$ is Carathéodory viable with respect to the Carathéodory function $f : I \times K \to X$, then there exists a negligible subset $\mathcal{Z}$ of I such that, for each (τ, ξ) in $(I \setminus \mathcal{Z}) \times K$, we have $f(\tau, \xi) \in \mathfrak{T}_K(\xi)$.*

Proof. The conclusion is a simple consequence of Theorem 5.1.1 and Remark 2.4.3. □

Theorem 5.1.2. *Let X be a Banach space. If $K \subseteq X$ is separable and locally closed and $I \times K$ is Carathéodory viable with respect to the locally Carathéodory function $f : I \times K \to X$, then there exists a negligible subset $\mathcal{Z}$ of I such that, for each (τ, ξ) in $(I \setminus \mathcal{Z}) \times K$, we have $f(\tau, \xi) \in \mathcal{F}_K(\xi)$.*

Proof. The proof is similar with the one of Theorem 5.1.1, with the mention that here, instead of Theorem 2.8.2, we have to use Theorem 2.8.5. □

The next example shows why, in this context, we cannot consider the fully general case of a noncylindrical domain $\mathcal{C}$ as in the case when we were looking for classical, i.e., C^1 solutions. At a first glace, it seems that we cannot do this because the usual reduction of the nonautonomous case $u'(t) = f(t, u(t))$ to the autonomous one $z'(t) = F(z(t))$, with $z = (t, u)$ and $F(z) = (1, f(z))$, cannot work, and this, due to the fact that whenever f is a Carathéodory function, F may fail to be continuous. Surprisingly, this is not the only reason why we cannot extend the classical viability theory to noncylindrical domains in the Carathéodory case, as we can see from the next "autonomous" example below.

Example 5.1.1. Let $\mathcal{C} \subseteq \mathbb{R} \times \mathbb{R}$ be the graph of the Cantor function $g : [0, 1) \to \mathbb{R}$ which is continuous, strictly increasing and $g'(t) = 0$ a.e. for $t \in [0, 1)$. See Gelbaum–Olmsted [**100**], **30**, p. 105. Then the "autonomous" function $f \equiv 0$ satisfies the tangency condition $(1, 0) \in \mathcal{T}_{\mathcal{C}}(\tau, g(\tau))$ for a.a. $\tau \in [0, 1)$, but the Cauchy problem $u'(t) = 0$ and $u(0) = 0$ has no Carathéodory solution. So, the situation is even worse than the one observed in Example 3.5.1, i.e., when under similar circumstances, we have concluded that the same Cauchy problem has no solution in the sense of Definition 3.1.1. This shows that, in the case of a noncylindrical set $\mathcal{C}$, no matter how regular is the right hand side $f : \mathcal{C} \to X$, we cannot replace the "everywhere" tangency condition with an "almost everywhere" one in order to get Carathéodory viability.

5.2. Sufficient conditions for Carathéodory viability

The next class of functions will play a crucial role in the sequel.

Definition 5.2.1. A *compact-Carathéodory function* $f : I \times K \to X$ is a locally Carathéodory function satisfying

(C_5) for each fixed $\xi \in K$ and each $\tau \in I$ there exist $T > \tau$ and $\rho > 0$ such that $[\tau, T] \subseteq I$ and, for each $\varepsilon > 0$, there exists a subset H_ε

in $[\tau, T]$ with $\lambda(H_\varepsilon) \leq \varepsilon$ and such that the set

$$\{f(t,u)\,;\ (t,u) \in ([\tau,T] \setminus H_\varepsilon) \times (D(\xi,\rho) \cap K)\}$$

is relatively compact in X.

A *Lipschitz-Carathéodory function* is a function satisfying (C_1), (C_4) in Definition 2.8.1 and

(C_6) for each $\xi \in K$, there exist $\rho > 0$ and $\mathcal{L} \in L^1_{\rm loc}(I)$, such that

$$\|f(t,u) - f(t,v)\| \leq \mathcal{L}(t)\|u - v\|$$

for a.a. $t \in I$ and for all $u, v \in D(\xi,\rho) \cap K$.

A *locally β-compact-Carathéodory function* is a function satisfying (C_1), (C_2), (C_4) in Definition 2.8.1 and

(C_7) for each $\xi \in K$, there exist $\rho > 0$ and a uniqueness Carathéodory function $\omega : I \times \mathbb{R}_+ \to \mathbb{R}_+$ in the sense of Definition 1.8.2 such that

$$\beta(f(t,C)) \leq \omega(t, \beta(C))$$

a.e. for $t \in I$ and for each $C \subseteq D(\xi,\rho) \cap K$.

An *β-compact-Carathéodory function* is a function satisfying (C_1), (C_2), (C_3) in Definition 2.8.1 and

(C_8) there exists a uniqueness Carathéodory function $\omega : I \times \mathbb{R}_+ \to \mathbb{R}_+$ in the sense of Definition 1.8.2 such that

$$\beta(f(t,C)) \leq \omega(t, \beta(C))$$

a.e. for $t \in I$ and for each bounded subset C in K.

Remark 5.2.1. If X is finite dimensional, each locally Carathéodory function is a β-compact-Carathéodory function. Furthermore, compact-Carathéodory functions as well as Lipschitz-Carathéodory functions are β-compact-Carathéodory functions.

Theorem 5.2.1. *Let X be a separable Banach space, K a locally closed subset in X and $f : I \times K \to X$ a locally β-compact-Carathéodory function. Then $I \times K$ is Carathéodory viable with respect to f if and only if there exists a negligible subset $\mathcal{Z}$ in I such that, for each $(t,\xi) \in (I \setminus \mathcal{Z}) \times K$, we have*

$$f(t,\xi) \in \mathcal{T}_K(\xi). \tag{5.2.1}$$

Theorem 5.2.2. *Let X be a separable Banach space, K a locally closed subset in X and $f : I \times K \to X$ a Lipschitz-Carathéodory function. Then, a necessary and sufficient condition in order that $I \times K$ be Carathéodory viable with respect to f is the tangency condition* (5.2.1).

Theorem 5.2.3. *Let X be a separable Banach space, K a locally closed subset in X and $f : I \times K \to X$ a compact-Carathéodory function. Then, a necessary and sufficient condition in order that $I \times K$ be Carathéodory viable with respect to f is the tangency condition* (5.2.1).

From Theorem 5.2.3, we easily deduce

Theorem 5.2.4. *Let X be finite dimensional, let K be a locally closed subset in X and let $f : I \times K \to X$ be a locally Carathéodory function. Then, a necessary and sufficient condition in order that $I \times K$ be Carathéodory viable with respect to f is the tangency condition* (5.2.1).

Finally, we have

Theorem 5.2.5. *Let X be a separable Banach space, K a nonempty and locally closed subset in X and let $f : I \times K \to X$ be locally Carathéodory. Let us assume that K is proximal and the norm $\|\cdot\|$ is Gâteaux differentiable at each $x \in X$, $x \neq 0$. Then the following conditions are equivalent:*

(i) *there exists a negligible subset $\mathcal{Z}$ of I such that for every $(t, \xi) \in (I \setminus \mathcal{Z}) \times K$, $f(t, \xi) \in \mathcal{C}_K(\xi)$;*
(ii) *there exists a negligible subset $\mathcal{Z}$ of I such that for every $(t, \xi) \in (I \setminus \mathcal{Z}) \times K$, $f(t, \xi) \in \mathcal{T}_K(\xi)$;*
(iii) *there exists a negligible subset $\mathcal{Z}$ of I such that for every $(t, \xi) \in (I \setminus \mathcal{Z}) \times K$, $f(t, \xi) \in \mathcal{B}_K(\xi)$;*
(iv) *the set $I \times K$ is Carathéodory viable with respect to f.*

In general, if $\mathcal{G} : K \rightsquigarrow X$ is such that $\mathcal{C}_K(\xi) \subseteq \mathcal{G}(\xi) \subseteq \mathcal{B}_K(\xi)$ for each $\xi \in K$, then each one of the conditions above is equivalent to

(v) *there exists a negligible subset $\mathcal{Z}$ of I such that for every $(t, \xi) \in (I \setminus \mathcal{Z}) \times K$, $f(t, \xi) \in \mathcal{G}(\xi)$.*

5.3. Existence of $(\varepsilon, \mathcal{L})$-approximate Carathéodory solutions

We begin with the following simple but useful result.

Proposition 5.3.1. *Let X be a real Banach space, K a nonempty and separable subset in X and $f : I \times K \to X$ a Carathéodory function. Then, the tangency condition* (5.2.1) *is equivalent to the condition* (5.3.1) *below: there exists a negligible subset $\mathcal{Z}$ of I such that, for every $(t, \xi) \in (I \setminus \mathcal{Z}) \times K$, we have*

$$\liminf_{h \downarrow 0} \frac{1}{h} \operatorname{dist}\left(\xi + \int_t^{t+h} f(\theta, \xi)\, d\theta; K\right) = 0. \tag{5.3.1}$$

Problem 5.3.1. *Prove Proposition* 5.3.1.

The main ingredient in the proof of the sufficiency of Theorem 5.2.1 is the next lemma which offers an existence result for "ε-approximate Carathéodory solutions" of the Cauchy problem

$$\begin{cases} u'(t) = f(t, u(t)) \\ u(\tau) = \xi. \end{cases} \tag{5.3.2}$$

Lemma 5.3.1. *Let X be a separable Banach space, K a nonempty, locally closed subset in X and $f : I \times K \to X$ a locally Carathéodory function satisfying the tangency condition (5.2.1). Let (τ, ξ) be arbitrarily fixed in $I \times K$, let $r > 0$ be such that $D(\xi, r) \cap K$ is closed. Let $\mathcal{Z} = \mathcal{Z}_1 \cup \mathcal{Z}_2 \cup \mathcal{Z}_3$, where $\mathcal{Z}_1$ is the negligible set in (5.2.1), $\mathcal{Z}_2$ is the negligible set in (5.3.1), while $\mathcal{Z}_3$ is the negligible set in I such that for each $t \in I \setminus \mathcal{Z}_3$, $f(t, \cdot)$ is continuous on K.*

Then, there exist $\rho \in (0, r]$, $T \in (\tau, \sup I]$, $\theta_0 \in I \setminus \mathcal{Z}$ and $\mathcal{M}$ in $L^1(\tau, T; \mathbb{R}_+)$, such that for each $\varepsilon \in (0, 1)$ and each open set $\mathcal{L}$ of $\mathbb{R}$, with $\mathcal{Z} \subseteq \mathcal{L}$ and $\lambda(\mathcal{L}) < \varepsilon$, there exist one family of nonempty and pairwise disjoint intervals: $\mathcal{P}_T = \{[t_m, s_m);\ m \in \Gamma\}$, with Γ finite or countable, and three functions, $g \in L^1(\tau, T; X)$, $r : [\tau, T] \to X$ Borel measurable and $u : [\tau, T] \to X$ continuous, satisfying:

(i) $\bigcup_{m \in \Gamma} [t_m, s_m) = [\tau, T)$ *and* $s_m - t_m \leq \varepsilon$ *for each* $m \in \Gamma$;
(ii) *if* $t_m \in \mathcal{L}$ *then* $[t_m, s_m) \subset \mathcal{L}$;
(iii) $u(t_m) \in D(\xi, \rho) \cap K$ *for each* $m \in \Gamma$, $u(T) \in D(\xi, \rho) \cap K$;
(iv) $g(s) = \begin{cases} f(s, u(t_m)) & \text{a.e. on } [t_m, s_m) \text{ if } t_m \notin \mathcal{L} \\ f(\theta_0, u(t_m)) & \text{a.e. on } [t_m, s_m) \text{ if } t_m \in \mathcal{L}; \end{cases}$
(v) $\|g(t)\| \leq \mathcal{M}(t)$ *a.e. for* $t \in [\tau, T]$;
(vi) $\|r(t)\| \leq \varepsilon$ *a.e. for* $t \in [\tau, T]$;
(vii) $u(\tau) = \xi$ *and, for each* $m \in \Gamma$ *and each* $t \in [t_m, T]$, u *satisfies*

$$u(t) = u(t_m) + \int_{t_m}^{t} g(\theta)\, d\theta + \int_{t_m}^{t} r(\theta)\, d\theta.$$

Before proceeding to the proof of Lemma 5.3.1, we introduce:

Definition 5.3.1. Let $\varepsilon > 0$ and $\mathcal{L}$ an open set including the negligible set $\mathcal{Z}$ in Lemma 5.3.1. A quadruple $(\mathcal{P}_T, g, r, u)$, satisfying (i)$\sim$(vii) in Lemma 5.3.1, is called an $(\varepsilon, \mathcal{L})$-*approximate Carathéodory solution* to the Cauchy problem (5.3.2) on the interval $[\tau, T]$.

We may now pass to the proof of Lemma 5.3.1.

Proof. Let $(\tau, \xi) \in I \times K$ be arbitrary and choose $r > 0$ such that $D(\xi, r) \cap K$ is closed and for which there exists $\ell(\cdot) \in L^1_{\text{loc}}(I)$ such that

$$\|f(t, u)\| \leq \ell(t) \tag{5.3.3}$$

for a.a. $t \in I$ and every $u \in D(\xi, r) \cap K$. This is always possible because K is locally closed and f satisfies (C_4) in Definition 2.8.1.

Fix a $\theta_0 \in I \setminus \mathcal{Z}$. Then $v \mapsto f(\theta_0, v)$ is continuous on K. Taking a sufficiently small $\rho \in (0, r]$, we can find $M > 0$ satisfying

$$\|f(\theta_0, v)\| \leq M \tag{5.3.4}$$

for each $v \in D(\xi, \rho) \cap K$. Next, take $T > \tau$ such that $[\tau, T] \subseteq I$ and let us define

$$\mathcal{M}(t) = \max\{M, \ell(t)\}$$

a.e. for $t \in [\tau, T]$. Clearly $\mathcal{M} \in L^1(\tau, T; \mathbb{R}_+)$, and therefore, diminishing $T > \tau$, if necessary, we may assume that

$$T - \tau + \int_\tau^T \mathcal{M}(\theta) d\theta \leq \rho. \tag{5.3.5}$$

We first prove that the conclusion of Lemma 5.3.1 remains true if we replace T as above with a possibly smaller number $\mu \in (\tau, T]$ which, at this stage, is allowed to depend on $\varepsilon \in (0, 1)$. This being done, by using the Brezis-Browder Ordering Principle, we will prove that we can take $\mu = T$, independent of $\varepsilon \in (0, 1)$.

For $\varepsilon \in (0, 1)$ take an open set $\mathcal{L}$ of $\mathbb{R}$ with $\mathcal{Z} \subseteq \mathcal{L}$ and whose Lebesgue measure $\lambda(\mathcal{L}) < \varepsilon$.

Case 1. If $\tau \in \mathcal{L}$, since f satisfies the tangency condition (5.2.1) at (θ_0, ξ), it follows that there exist $\delta \in (0, \varepsilon)$ with $[\tau, \tau + \delta) \subseteq \mathcal{L}$, and $p \in X$ with $\|p\| \leq \varepsilon$, and such that $\xi + \delta f(\theta_0, \xi) + \delta p \in K$.

Let us define $g : [\tau, \tau+\delta] \to X$, $r : [\tau, \tau+\delta] \to X$ and $u : [\tau, \tau+\delta] \to X$ by $g(t) = f(\theta_0, \xi)$, $r(t) = p$, and respectively by

$$u(t) = \xi + \int_\tau^t g(\theta)\, d\theta + \int_\tau^t r(\theta)\, d\theta \tag{5.3.6}$$

for each $t \in [\tau, \tau + \delta]$. By (5.3.4) and the definition of $\mathcal{M}$, we deduce that the family $\mathcal{P}_{\tau+\delta} = \{[\tau, \tau + \delta)\}$ and the functions $\mathcal{M}$, g, r and u satisfy (i)-(vii) with T substituted by $\tau + \delta$.

Case 2. If $\tau \notin \mathcal{L}$, we have $\tau \notin \mathcal{Z}$ and, in view of Proposition 5.3.1, there exist $\delta \in (0, \varepsilon)$ and $p \in X$, with $\|p\| \leq \varepsilon$ and $\xi + \int_\tau^{\tau+\delta} f(\theta, \xi)\, d\theta + \delta p \in K$. Setting $g(\theta) = f(\theta, \xi)$ and $r(\theta) = p$, for $\theta \in [\tau, \tau + \delta]$, and defining u by (5.3.6), we can easily see that, again, the family $\mathcal{P}_{\tau+\delta} = \{[\tau, \tau + \delta)\}$ and the functions $\mathcal{M}$, g, r and u satisfy (i)-(vii) with T substituted by $\tau + \delta$.

Next, we will show that there exists at least one quadruple $(\mathcal{P}_T, g, r, u)$ satisfying (i)∼(vii) on $[\tau, T]$. To this aim we shall use the Brezis-Browder Ordering Principle, as follows. Let $\mathcal{U}$ be the set of all quadruples $(\mathcal{P}_\mu, g, r, u)$ with $\mathcal{L}$ fixed as above and $\mu \leq T$ and satisfying (i)∼(vii) with μ instead of

T. As we have already proved, this set is nonempty. On $\mathcal{U}$ we introduce a partial order as follows. We say that

$$(\mathcal{P}_{\mu_1}, g_1, r_1, u_1) \preceq (\mathcal{P}_{\mu_2}, g_2, r_2, u_2),$$

where $\mathcal{P}_{\mu_k} = \{[t^k_m, s^k_m); m \in \Gamma_k\}$, $k = 1, 2$, if

(O_1) $\mu_1 \leq \mu_2$ and, if $\mu_1 < \mu_2$, there exists $i \in \Gamma_2$ such that $\mu_1 = t^2_i$;

(O_2) for each $m_1 \in \Gamma_1$ there exists $m_2 \in \Gamma_2$ such that:

$$t^1_{m_1} = t^2_{m_2} \quad \text{and} \quad s^1_{m_1} = s^2_{m_2} \, ;$$

(O_3) $g_1(\theta) = g_2(\theta)$, $r_1(\theta) = r_2(\theta)$ and $u_1(\theta) = u_2(\theta)$ for $\theta \in [\,\tau, \mu_1\,]$.

Let us define the function $\mathcal{N} : \mathcal{U} \to \mathbb{R}$ by

$$\mathcal{N}((\mathcal{P}_\mu, g, r, u)) = \mu.$$

It is clear that $\mathcal{N}$ is increasing on $\mathcal{U}$. Let us take now an increasing sequence

$$((\mathcal{P}_{\mu_j}, g_j, r_j, u_j))_{j\in\mathbb{N}}$$

in $\mathcal{U}$ and let us show that it is bounded from above in $\mathcal{U}$. We define an upper bound as follows. First, set

$$\mu^* = \sup\{\mu_j;\ j \in \mathbb{N}\}.$$

If $\mu^* = \mu_j$ for some $j \in \mathbb{N}$, $(\mathcal{P}_{\mu_j}, g_j, r_j, u_j)$ is clearly an upper bound. If $\mu_j < \mu^*$ for each $j \in \mathbb{N}$, we define $\mathcal{P}_{\mu^*} = \{[\,t^j_m, s^j_m);\ j \in \mathbb{N},\ m \in \Gamma_j\}$. In the latter case, $\mathcal{P}_{\mu^*}$ can be written in the form $\mathcal{P}_{\mu^*} = \{[\,t_m, s_m); m \in \mathbb{N}\}$. We define

$$g(t) = g_j(t), \quad r(t) = r_j(t), \quad u(t) = u_j(t)$$

for $j \in \mathbb{N}$ and every $t \in [\,\tau, \mu_j\,]$. Let us observe that $(\mathcal{P}_{\mu^*}, g, r, u)$, where $\mathcal{P}_{\mu^*}$, g, r and u are defined as above, satisfies (i), (ii), the first condition in (iii), (iv), (v) and (vi) with T replaced with μ^*. Notice that (vii) is also satisfied but only on $[\,\tau, \mu^*)$. Obviously we have $u(t_m) \in D(\xi, \rho) \cap K$ for each $m \in \mathbb{N}$. Since g and r are a.e. defined on $[\,\tau, \mu^*\,]$, it remains to prove that u can be extended to $[\,\tau, \mu^*\,]$ and satisfies the second condition in (iii), i.e., $u(\mu^*) \in D(\xi, \rho) \cap K$ and (vii) for $t = \mu^*$. To this aim, let us observe that, in view of (vii) on $[\,\tau, \mu*)$, and of the fact that, by (v), we have $\|g(t)\| \leq \mathcal{M}(t)$, for a.a. $t \in [\,\tau, T\,]$, with $\mathcal{M}$ integrable on $[\,\tau, T\,]$, it follows that u satisfies the Cauchy condition for the existence of the limit for $t \uparrow \mu^*$. Consequently, there exists $\lim_{t\uparrow\mu^*} u(t)$. Accordingly, u can be continuously extended at μ^* by $u(\mu^*) = \lim_{t\uparrow\mu^*} u(t)$. Since $u(t_m) \in D(\xi, \rho) \cap K$ for $m \in \mathbb{N}$, and $D(\xi, \rho) \cap K$ is closed, we easily see that $u(\mu^*) \in D(\xi, \rho) \cap K$ and thus the last condition in (iii) is also satisfied. So, with $u : [\,\tau, \mu^*\,] \to X$, defined as above, we obviously have that $(\mathcal{P}_{\mu^*}, g, r, u)$ satisfies (i)$\sim$(vi). It is also easy to see that (vii) holds for each $m \in \mathbb{N}$ and each $t \in [\,t_m, \mu^*)$. To check (vii) for $t = \mu^*$, we have to fix any $m \in \mathbb{N}$, to take $t = \mu_j$ with $\mu_j > t_m$ in (vii) and to pass to the limit for j tending to ∞ both sides in (vii).

Thus $(\mathcal{P}_{\mu^*}, g, r, u)$ is an upper bound for $((\mathcal{P}_{\mu_j}, g_j, r_j, u_j))_{j\in\mathbb{N}}$. So, the set $\mathcal{U}$, endowed with the partial order $\preceq$, and the function $\mathcal{N}$ satisfy the hypotheses of Brezis-Browder Ordering Principle. Accordingly, there exists at least one $\mathcal{N}$-maximal element $(\mathcal{P}_\nu, g_\nu, r_\nu, u_\nu)$ in $\mathcal{U}$, which means that if $(\mathcal{P}_\nu, g_\nu, r_\nu, u_\nu) \preceq (\mathcal{P}_\sigma, g_\sigma, r_\sigma, u_\sigma)$ then $\nu = \sigma$.

Next, we show that $\nu = T$, where T satisfies (5.3.5). To this aim let us assume by contradiction that $\nu < T$ and let $\xi_\nu = u_\nu(\nu)$ which belongs to $D(\xi, \rho) \cap K$. Throughout the function $\mathcal{M}$ is defined as in the beginning of the proof. In view of (i)$\sim$ (vii), we have

$$\|\xi_\nu - \xi\| \leq \int_\tau^\nu \|g_\nu(\theta)\| \, d\theta + \int_\tau^\nu \|r_\nu(\theta)\| \, d\theta \leq (\nu - \tau)\varepsilon + \int_\tau^\nu \mathcal{M}(\theta) \, d\theta.$$

Recalling that $\nu < T$ and $\varepsilon < 1$, from (5.3.5) we get

$$\|\xi_\nu - \xi\| < \rho. \tag{5.3.7}$$

There are two possibilities: either $\nu \in \mathcal{L}$ or $\nu \notin \mathcal{L}$.

If $\nu \in \mathcal{L}$, we act as in *Case* 1 above with ν instead of τ and with ξ_ν instead of ξ. So, from the tangency condition (5.2.1) combined with (5.3.7), we infer that there exist $\delta \in (0, \varepsilon]$, with $\nu + \delta \leq T$, $[\nu, \nu + \delta) \subseteq \mathcal{L}$ and $p \in X$, satisfying $\|p\| \leq \varepsilon$, such that $\xi_\nu + \delta f(\theta_0, \xi_\nu) + \delta p \in D(\xi, \rho) \cap K$.

If $\nu \notin \mathcal{L}$, we act as in *Case* 2 above with ν instead of τ and with ξ_ν instead of ξ. From Proposition 5.3.1 combined with (5.3.7), we infer that there exist $\delta \in (0, \varepsilon]$, with $\nu + \delta \leq T$, and $q \in X$, satisfying $\|q\| \leq \varepsilon$ and

$$\xi_\nu + \int_\nu^{\nu+\delta} f(\theta, \xi_\nu) \, d\theta + \delta q \in D(\xi, \rho) \cap K.$$

We define $\mathcal{P}_{\nu+\delta} = \mathcal{P}_\nu \cup \{[\nu, \nu + \delta)\}$ and both $g_{\nu+\delta} : [\tau, \nu + \delta] \to X$ and $r_{\nu+\delta} : [\tau, \nu + \delta] \to X$ by

$$g_{\nu+\delta}(t) = \begin{cases} g_\nu(t) & \text{if } t \in [\tau, \nu] \\ f(\theta_0, \xi_\nu) & \text{if } t \in (\nu, \nu + \delta] \end{cases}$$

$$r_{\nu+\delta}(t) = \begin{cases} r_\nu(t) & \text{if } t \in [\tau, \nu] \\ p & \text{if } t \in (\nu, \nu + \delta] \end{cases}$$

if $\nu \in \mathcal{L}$, and

$$g_{\nu+\delta}(t) = \begin{cases} g_\nu(t) & \text{if } t \in [\,\tau, \nu\,] \\ f(t, \xi_\nu) & \text{if } t \in (\,\nu, \nu + \delta\,] \end{cases}$$

$$r_{\nu+\delta}(t) = \begin{cases} r_\nu(t) & \text{if } t \in [\,\tau, \nu\,] \\ q & \text{if } t \in (\,\nu, \nu + \delta\,] \end{cases}$$

if $\nu \notin \mathcal{L}$. Finally, we define $u_{\nu+\delta} : [\,\tau, \nu + \delta\,] \to X$ by

$$u_{\nu+\delta}(t) = \begin{cases} u_\nu(t) & \text{if } t \in [\,\tau, \nu\,] \\ \xi_\nu + \displaystyle\int_\nu^t g_{\nu+\delta}(\theta)\, d\theta + \int_\nu^t r_{\nu+\delta}(\theta)\, d\theta & \text{if } t \in (\nu, \nu + \delta\,]. \end{cases}$$

Since $u_{\nu+\delta}(\nu + \delta) \in K \cap D(\xi, \rho)$, it follows that $(\mathcal{P}_{\nu+\delta}, g_{\nu+\delta}, r_{\nu+\delta}, u_{\nu+\delta})$ satisfies (i)$\sim$(vii) with $\nu + \delta$ instead of T. So, $(\mathcal{P}_{\nu+\delta}, g_{\nu+\delta}, r_{\nu+\delta}, u_{\nu+\delta}) \in \mathfrak{U}$ and

$$(\mathcal{P}_\nu, g_\nu, r_\nu, u_\nu) \preceq (\mathcal{P}_{\nu+\delta}, g_{\nu+\delta}, r_{\nu+\delta}, u_{\nu+\delta}) \text{ with } \nu < \nu + \delta.$$

This contradiction can be avoided only if $\nu = T$. The proof is therefore complete. □

5.4. Convergence of $(\varepsilon, \mathcal{L})$-approximate Carathéodory solutions

The proof of the sufficiency of Theorem 5.2.1 consists in showing the convergence of a suitably chosen sequence of $(\varepsilon, \mathcal{L})$-approximate Carathéodory solutions.

Proof. Let $(\tau, \xi) \in I \times K$ and let $r > 0$ be such that $D(\xi, r) \cap K$ is closed. Let $\rho \in (0, r\,]$, $T \in (\tau, \sup I\,]$, $\theta_0 \in I \setminus \mathcal{Z}$ and $\mathcal{M} \in L^1(\tau, T; \mathbb{R}_+)$ given by Lemma 5.3.1, let $(\varepsilon_n)_n$ be a sequence in $(0, 1)$ strictly decreasing to 0, let $(\mathcal{L}_n)$ be a decreasing sequence of open subsets in $\mathbb{R}$ such that the negligible set $\mathcal{Z}$, defined in Lemma 5.3.1, satisfy $\mathcal{Z} \subseteq \mathcal{L}_n$ and $\lambda(\mathcal{L}_n) < \varepsilon_n$ for every $n \in \mathbb{N}$. Take $\mathcal{L} = \cap_{n \geq 1} \mathcal{L}_n$ and a sequence of $(\varepsilon_n, \mathcal{L}_n)$-approximate Carathéodory solutions $((\mathcal{P}_T^n, g_n, r_n, u_n))_n$ of (5.3.2) whose existence is ensured also by Lemma 5.3.1.

Let us define $\sigma_n : [\,\tau, T\,] \to [\,\tau, T\,]$ by $\sigma_n(t) = t_m^n$ for $t \in [\,t_m^n, s_m^n)$, $\sigma_n(T) = T$, and

$$\mathcal{H}_n = \bigcup_{t_m^n \in \mathcal{L}_n} [\,t_m^n, s_m^n).$$

In view of (ii)[1] we have $\mathcal{H}_n \subset \mathcal{L}_n$ and therefore

$$\lambda(\mathcal{H}_n) < \varepsilon_n. \tag{5.4.1}$$

[1]Except otherwise specified, all references to (i), (ii),..., (vii) are to the corresponding items in Lemma 5.3.1.

Set $\mathcal{E}_n = [\tau, T] \setminus \mathcal{H}_n$, let $t \in [\tau, T]$, and let us define $\mathcal{H}_n^t = [\tau, t] \cap \mathcal{H}_n$ and $\mathcal{E}_n^t = [\tau, t] \cap \mathcal{E}_n$. Since by (vii) we have

$$u_n(t) = \xi + \int_\tau^t g_n(s)\, ds + \int_\tau^t r_n(s)\, ds, \tag{5.4.2}$$

from Lemma 2.7.2, Remark 2.7.1 and (vi), we deduce

$$\begin{aligned}
&\beta(\{u_n(t);\ n \geq k\}) \\
&\leq \beta\left(\left\{\int_\tau^t g_n(s)\, ds;\ n \geq k\right\}\right) + \beta\left(\left\{\int_\tau^t r_n(s)\, ds;\ n \geq k\right\}\right) \\
&\leq \int_\tau^t \beta(\{g_n(s);\ n \geq k\})\, ds + (T-\tau)\varepsilon_k \\
&\leq \int_{\mathcal{E}_k^t} \beta(\{f(s, u_n(\sigma_n(s)));\ n \geq k\})\, ds + \int_{\mathcal{H}_k^t} \beta(\{g_n(s);\ n \geq k\})\, ds + (T-\tau)\varepsilon_k \\
&\leq \int_{\mathcal{E}_k^t} \omega(s, \beta(\{u_n(\sigma_n(s));\ n \geq k\}))ds + \int_{\mathcal{H}_k} \mathcal{M}(s)\, ds + (T-\tau)\varepsilon_k \\
&\leq \int_{\mathcal{E}_k^t} \omega(s, \beta(\{u_n(s);\ n \geq k\} + \{u_n(\sigma_n(s)) - u_n(s);\ n \geq k\}))ds \\
&\quad + \int_{\mathcal{H}_k} \mathcal{M}(s)\, ds + (T-\tau)\varepsilon_k.
\end{aligned}$$

Let $k = 1, 2, \ldots$ and $n \geq k$. From (5.4.2), we have

$$\|u_n(\sigma_n(t)) - u_n(t)\| \leq \sup\left\{\int_{\sigma_n(t)}^t \mathcal{M}(\theta)d\theta + \int_{\sigma_n(t)}^t \|r_n(\theta)\|d\theta\, ;\ t \in [\tau, T]\right\}.$$

Let us denote by

$$\delta_k = \sup\left\{\int_{\sigma_n(t)}^t \mathcal{M}(\theta)\, d\theta + \int_{\sigma_n(t)}^t \|r_n(\theta)\|\, d\theta\, ;\ t \in [\tau, T],\ n \geq k\right\}.$$

By (vii), (v), (vi) and (i), we have

$$\lim_n \|u_n(t) - u_n(\sigma_n(t))\| \leq \lim_n \delta_n = 0. \tag{5.4.3}$$

Thus,

$$\begin{aligned}
&\beta(\{u_n(t);\ n \geq k\}) \\
&\leq \int_{\mathcal{E}_k^t} \omega(s, \beta(\{u_n(s);\ n \geq k\}) + \beta(\{u_n(\sigma_n(s)) - u_n(s);\ n \geq k\}))ds \\
&\quad + \int_{\mathcal{H}_k} \mathcal{M}(s)\, ds + (T-\tau)\varepsilon_k
\end{aligned}$$

$$\leq \int_\tau^t \omega(s, \beta(\{u_n(s);\ n \geq k\}) + \delta_k) ds + \int_{\mathcal{H}_k} \mathcal{M}(s)\, ds + (T - \tau)\varepsilon_k.$$

Denoting by $x_k(t) = \beta(\{u_n(t);\ n \geq k\}) + \delta_k$ and

$$\gamma_k = \int_{\mathcal{H}_k} \mathcal{M}(s)\, ds + (T - \tau)\varepsilon_k + \delta_k,$$

we get

$$x_k(t) \leq \gamma_k + \int_\tau^t \omega(s, x_k(s))\, ds$$

for $k = 1, 2, \ldots$ and $t \in [\,\tau, T\,]$. As $\gamma_n \downarrow 0$, we are in the hypotheses of Lemma 1.8.3, wherefrom we conclude that, diminishing $T > \tau$ if necessary, on a subsequence at least, we have $\lim_k x_k(t) = 0$, which means that $\lim_k \beta(\{u_n(t);\ n \geq k\}) = 0$, uniformly for $t \in [\,\tau, T\,]$. Now, Lemma 2.7.3 comes into play and shows that, for each $t \in [\,\tau, T\,]$, $\{u_n(t);\ n = 1, 2, \ldots\}$ is relatively compact. Since by (v), (vi) and (vii) in Lemma 5.3.1, it follows that $\{u_n;\ n = 1, 2, \ldots\}$ is equicontinuous on $[\,\tau, T\,]$, from Theorem 1.3.6, we conclude that there exists $u \in C([\,\tau, T\,]; X)$ such that, on a subsequence at least, we have $\lim_n u_n(t) = u(t)$ uniformly for $t \in [\,\tau, T\,]$, where $u \in C([\,\tau, T\,]; X)$.

To complete the proof, it remains to show that u is a solution of the Cauchy problem (5.3.2). To this aim, let us remark that by (iii), we have $u_n(\sigma_n(t)) \in D(\xi, \rho) \cap K$ and since $D(\xi, \rho) \cap K$ is closed, from (5.4.3), we deduce that $u(t) \in D(\xi, \rho) \cap K$ for every $t \in [\,\tau, T\,]$. Again (5.4.3) yields

$$\lim_n g_n(s) = \lim_n f(s, u_n(\sigma_n(s))) = f(s, u(s)),$$

for each $s \in [\,\tau, T\,] \setminus \mathcal{L}$. Using Lebesgue Dominated Convergence Theorem 1.2.3 in order to pass to the limit for $n \to \infty$ both sides in (5.4.2), we conclude that

$$u(t) = \xi + \int_\tau^t f(s, u(s))\, ds$$

for each $t \in [\,\tau, T\,]$. This completes the proof. $\square$

Problem 5.4.1. *Give a direct proof to Theorem* 5.2.2, *avoiding the measure of noncompactness.*

Problem 5.4.2. *Give a direct proof to Theorem* 5.2.4, *avoiding the measure of noncompactness.*

Problem 5.4.3. *Give a direct proof to Theorem* 5.2.3, *avoiding the measure of noncompactness.*

5.5. Noncontinuable Carathéodory solutions

We next present some results concerning the existence of noncontinuable, or even global Carathéodory solutions to

$$u'(t) = f(t, u(t)). \tag{5.5.1}$$

A Carathéodory solution $u : [\,\tau, T) \to K$ to (5.5.1) is called *noncontinuable*, if there is no other Carathéodory solution $v : [\,\tau, \widetilde{T}) \to K$ of the same equation, with $T < \widetilde{T}$ and satisfying $u(t) = v(t)$ for all $t \in [\,\tau, T)$. The Carathéodory solution u is called *global* if $T = \sup I$. The next theorem follows from Brezis–Browder Theorem 2.1.1.

Theorem 5.5.1. *Let X be a Banach space, $K \subseteq X$ a nonempty set and let $f : I \times K \to X$. Then, the following conditions are equivalent:*

(i) *$I \times K$ is Carathéodory viable with respect to f;*

(ii) *for each $(\tau, \xi) \in I \times K$ there exists at least one noncontinuable Carathéodory solution $u : [\tau, T) \to K$ of* (5.5.1), *satisfying the initial condition $u(\tau) = \xi$.*

Remark 5.5.1. Notice that in Theorem 5.5.1 we do not assume K to be locally closed or f to be locally Carathéodory.

Definition 5.5.1. A function $f : I \times K \to X$ is called *Carathéodory positively sublinear*, if there exist $a, b \in L^1_{\mathrm{loc}}(I)$, $c \in L^\infty_{\mathrm{loc}}(I)$ and a negligible subset $\mathcal{Z}$ of I such that

$$\|f(t, \xi)\| \le a(t)\|\xi\| + b(t)$$

for each $(t, \xi) \in K^c_+(f)$, where

$$K^c_+(f) = \{(t, \xi) \in (I \setminus \mathcal{Z}) \times K;\ \|\xi\| > c(t) \ \text{ and } [\,\xi, f(t, \xi)\,]_+ > 0\}.$$

Theorem 5.5.2. *Let X be a Banach space, $K \subseteq X$ a nonempty and closed set and let $f : I \times K \to X$ be a locally Carathéodory function. If f is Carathéodory positively sublinear and $I \times K$ is Carathéodory viable with respect to f, then each Carathéodory solution of* (5.5.1) *can be continued up to a global one, i.e., defined on $[\,\tau, \sup I)$.*

Proof. Since $I \times K$ is Carathéodory viable with respect to f, by Theorem 5.5.1, it follows that for each $(\tau, \xi) \in I \times K$ there exists at least one noncontinuable Carathéodory solution, $u : [\,\tau, T) \to K$, of (5.5.1), satisfying $u(\tau) = \xi$. We will show that $T = \sup I$. To this aim, let us assume the contrary, i.e., that $T < \sup I$. In particular this means that $T < +\infty$. As $u'(s) = f(s, u(s))$ for a.a. $s \in [\,\tau, T)$, we deduce

$$\frac{d^+}{ds}(\|u(\cdot)\|)(s) = [\,u(s), f(s, u(s))\,]_+$$

for a.a. $s \in [\tau, T)$. Let $t \in [\tau, T)$. Integrating over $[\tau, t] \subseteq [\tau, T)$ this equality, we get

$$\|u(t)\| = \|\xi\| + \int_\tau^t [u(s), f(s, u(s))]_+ \, ds$$

$$\le \|\xi\| + \int_{E_t} [u(s), f(s, u(s))]_+ \, ds + \int_{H_t \setminus G_t} [u(s), f(s, u(s))]_+ \, ds,$$

where

$$\begin{aligned} E_t &= \{s \in [\tau, t];\ [u(s), f(s, u(s))]_+ > 0 \text{ and } \|u(s)\| > c(s)\}, \\ G_t &= \{s \in [\tau, t];\ [u(s), f(s, u(s))]_+ \le 0\}, \\ H_t &= \{s \in [\tau, t];\ \|u(s)\| \le c(s)\}. \end{aligned}$$

Taking into account that $H_t \subseteq H_T$ and that, by (ii) in Exercise 1.6.1, $[u, v]_+ \le \|v\|$ for each $u, v \in X$, we get

$$\|u(t)\| \le \|\xi\| + \int_{E_t} [a(s)\|u(s)\| + b(s)] \, ds + \int_{H_T} \|f(s, u(s))\| \, ds$$

for each $t \in [\tau, T)$. But f is a Carathéodory function and therefore there exists $\ell \in L^1_{\text{loc}}(I)$ such that

$$\|f(s, u(s))\| \le \ell(s)$$

for a.a. $s \in H_T$. See (C_3) in Definition 2.8.1. Hence

$$\|u(t)\| \le \|\xi\| + \int_\tau^T \ell(s) \, ds + \int_\tau^T b(s) \, ds + \int_\tau^T a(s)\|u(s)\| \, ds$$

for each $t \in [\tau, T)$. By Gronwall Lemma 1.8.4, it follows that u is bounded on $[\tau, T)$.

Using once again the fact that f is Carathéodory, we deduce that $f(\cdot, u(\cdot))$ is bounded on $[\tau, T)$ by a function in $L^1(\tau, T)$ and so, there exists $\lim_{t \uparrow T} u(t) = u^*$. As K is closed and $T < \sup I$, we get $(T, u^*) \in I \times K$. Using this observation and recalling that $I \times K$ is Carathéodory viable with respect to f, we conclude that u can be continued to the right of T. But this is absurd, because u is noncontinuable. This contradiction can be eliminated only if $T = \sup I$, and this completes the proof. □

CHAPTER 6

Viability for differential inclusions

The aim of this chapter is to present the main results on viability and invariance in the case of differential inclusions. We first introduce the notions of exact solution and almost exact solution and we study their relationship. We next consider the autonomous case and we prove some necessary and necessary and sufficient conditions for almost exact viability expressed in terms of the set-tangency concept introduced in Chapter 2. The nonautonomous case is reduced to the autonomous one by using the usual trick of introducing a new unknown function. Some problems concerning the continuation of (almost) exact solutions and the existence of global (almost) exact solutions are also considered. Finally, we establish a sufficient condition of invariance and a necessary condition in the specific case of a finite-dimensional problem.

6.1. Necessary conditions for exact viability

Let X be a real Banach space, K a nonempty subset in X, $F : K \rightsquigarrow X$ a given multi-function and let us consider the Cauchy problem for the differential inclusion

$$\begin{cases} u'(t) \in F(u(t)) \\ u(0) = \xi. \end{cases} \tag{6.1.1}$$

Definition 6.1.1. An *exact solution* of (6.1.1) on $[\,0,T\,]$ is an absolutely continuous function $u : [\,0,T\,] \to K$ which is a.e. differentiable on $[\,0,T\,]$ with $u' \in L^1(0,T;X)$ and satisfies

$$\begin{cases} u'(t) \in F(u(t)) & \text{at each point } t \in [\,0,T\,] \text{ at which u is differentiable} \\ u(0) = \xi. \end{cases}$$

An exact solution of (6.1.1) on the semi-open interval $[\,0,T)$ is defined similarly, noticing that, in this case, we have to impose the weaker constraint $u' \in L^1_{\text{loc}}([\,0,T);X)$.

Definition 6.1.2. An *almost exact solution* of (6.1.1) on $[\,0,T\,]$ is an absolutely continuous function $u : [\,0,T\,] \to K$ which is a.e. differentiable on $[0,T]$ with $u' \in L^1(0,T;X)$ and satisfies

$$\begin{cases} u'(t) \in F(u(t)) & \text{a.e. for } t \in [0,T] \\ u(0) = \xi. \end{cases}$$

An *almost exact solution* of (6.1.1) on the semi-open interval $[0,T)$ is defined similarly, noticing that, in this case, we have merely to impose that $u' \in L^1_{\mathrm{loc}}([0,T);X)$.

Remark 6.1.1. Clearly each exact solution is almost exact. It should be emphasized that, although not obvious, under some very natural continuity assumptions on F, the converse statement is also true as we shall later prove. See Corollary 6.1.1 below.

Remark 6.1.2. If u is an almost exact solution of (6.1.1) on $[0,T]$, we have

$$u(t) = u(s) + \int_s^t u'(\theta)\,d\theta$$

for each $0 \leq s \leq t \leq T$. In view of Remark 6.1.1, the equality above holds also true if u is an exact solution.

Definition 6.1.3. The set K is *exact viable* (*almost exact viable*) with respect to F if for each $\xi \in K$ there exists $T > 0$ such that (6.1.1) has at least one exact solution (almost exact solution) $u : [0,T] \to K$.

Let $u : [0,T] \to X$ and let $t \in [0,T)$. We denote by $\mathcal{D}_+u(t)$ the set of all limit points of the mapping $h \mapsto h^{-1}(u(t+h) - u(t))$ for $h \downarrow 0$. Clearly, if u is right differentiable at t and $u'_+(t)$ is the right derivative of u at t, we have $\mathcal{D}_+u(t) = \{u'_+(t)\}$. In particular, if u is differentiable at t, then $\mathcal{D}_+u(t) = \{u'(t)\}$.

Theorem 6.1.1. *Let X be a Banach space, let $F : K \rightsquigarrow X$ be a strongly-weakly u.s.c. multi-function with nonempty, closed and convex values and let $u : [0,T] \to K$ be an almost exact solution of* (6.1.1). *Then*

$$\mathcal{D}_+u(t) \subseteq F(u(t)) \cap \mathcal{T}_K(u(t))$$

for each $t \in [0,T)$.

Proof. If $t \in [0,T)$ is such that the mapping $h \mapsto h^{-1}(u(t+h) - u(t))$ has no limit point for $h \downarrow 0$ then $\mathcal{D}_+u(t) = \emptyset$, and therefore we have nothing to prove. So, let $t \in [0,T)$ be such that $\mathcal{D}_+u(t)$ is nonempty and let $\eta \in \mathcal{D}_+u(t)$ be arbitrary. Then, there exists $h_k \downarrow 0$ such that

$$\lim_k (h_k)^{-1}(u(t+h_k) - u(t)) = \eta.$$

Since F is strongly-weakly u.s.c. at $u(t)$ and $s \mapsto u(s)$ is continuous, it follows that for each open half space V with $F(u(t)) \subseteq V$, there exists a positive integer $k(V)$ such that, for $k = k(V), k(V)+1, \ldots,$ we have

$$\frac{1}{h_k}(u(t+h_k) - u(t)) = \frac{1}{h_k}\int_t^{t+h_k} u'(s)\,ds \in \overline{V}.$$

Thus $\eta \in \overline{V}$. Since $F(u(t))$ is closed and convex, it is the intersection of all closed half spaces which contain it. Hence $\eta \in F(u(t))$. To complete the proof, we have merely to observe that

$$\lim_k \frac{1}{h_k}\operatorname{dist}(u(t) + h_k\eta; K) \leq \lim_k \frac{1}{h_k}\|u(t) + h_k\eta - u(t+h_k)\| = 0,$$

which shows that $\eta \in \mathcal{T}_K(u(t))$. Since $\eta \in \mathcal{D}_+u(t)$ was arbitrary, we get $\mathcal{D}_+u(t) \subseteq F(u(t)) \cap \mathcal{T}_K(u(t))$, and this completes the proof. □

Corollary 6.1.1. *Let X be a Banach space, $F : K \rightsquigarrow X$ be a strongly-weakly u.s.c. multi-function with nonempty, closed and convex values. Then each almost exact solution of* (6.1.1) *is an exact solution too.*

In view of Corollary 6.1.1, from now on, whenever F is strongly-weakly u.s.c. with nonempty, closed and convex values, we shall speak only about exact solution and exact viability. However, if F is not as mentioned before, we still shall make a distinction between exact and almost exact concepts.

Corollary 6.1.2. *Let X be a Banach space, $F : K \rightsquigarrow X$ be a strongly-weakly u.s.c. multi-function with nonempty, closed and convex values and let $u : [0,T] \to K$ be an exact solution of* (6.1.1). *If $t \in [0,T)$ is such that*

$$F(u(t)) \cap \mathcal{T}_K(u(t)) = \emptyset,$$

then $h \mapsto h^{-1}(u(t+h) - u(t))$ has no limit point for $h \downarrow 0$.

Problem 6.1.1. *Let X be a Banach space, $F : K \rightsquigarrow X$ be a strongly-weakly u.s.c. multi-function with nonempty, closed and convex values. Prove that, whenever K is exact viable with respect to $F : K \rightsquigarrow X$, then the set $C = \{\xi \in K;\ F(\xi) \cap \mathcal{T}_K(\xi) \neq \emptyset\}$ is dense in K.*

Before proceeding to the main result in this section, a measurability lemma is needed. We first introduce

Definition 6.1.4. A set $C \subseteq X$ is *quasi-weakly* (*relatively*) *compact* if, for each $r > 0$, $C \cap D(0,r)$ is weakly (relatively) compact.

Problem 6.1.2. *Prove that each quasi-weakly compact set is closed.*

Remark 6.1.3. If X is reflexive, from Theorems 1.3.4 and 1.3.5, we conclude that each closed and convex set $C \subseteq X$ is quasi-weakly compact.

Lemma 6.1.1. *Let X be a Banach space, let C be a convex and quasi-weakly compact subset in X, let $\rho > 0$ and let $v \in L^1(0,T;X)$, with $v(t) \in C + D(0,\rho)$ a.e. for $t \in [\,0,T\,]$. Then there exist two measurable functions $f : [\,0,T\,] \to C$ and $g : [\,0,T\,] \to D(0,\rho)$ such that $f \in L^1(0,T;X)$ and $v(s) = f(s) + g(s)$ a.e. for $s \in [\,0,T\,]$.*

Proof. In view of Theorem 1.2.1, there exist $\rho_n \downarrow \rho$ and a sequence of countably-valued functions $(v_n)_n$, $v_n : [\,0,T\,] \to C + D(0,\rho_n)$ such that $\lim_n v_n = v$ a.e. uniformly on $[\,0,T\,]$. See Definition 1.2.1. Since v lies in $L^1(0,T;X)$, we may assume with no loss of generality that, in addition, $(v_n)_n$ converges to v in $L^1(0,T;X)$, too. On the other hand, there exist two sequences of countably-valued functions $(f_n)_n$ and $(g_n)_n$, $f_n : [\,0,T\,] \to C$, $g_n : [\,0,T\,] \to D(0,\rho_n)$ such that $v_n = f_n + g_n$, for $n = 1,2,\ldots$. As $\lim_n v_n = v$ a.e. uniformly on $[\,0,T\,]$, we conclude that

$$\begin{aligned}\|f_n(t)\| \leq \|v_n(t)\| + \|g_n(t)\| &\leq \|v(t)\| + \|v_n(t) - v(t)\| + \rho_n \\ &\leq \|v(t)\| + M, \end{aligned} \tag{6.1.2}$$

a.e. for $t \in [\,0,T\,]$, where $M > 0$ is independent of $n = 1,2,\ldots$. From (6.1.2), recalling that C is quasi-weakly compact and $v \in L^1(0,T;X)$, we deduce that, for each $\varepsilon > 0$, there exist a weakly compact subset C_ε of C and a set $E_\varepsilon \subseteq [\,0,T\,]$ such that, for $n = 1,2,\ldots$, $f_n(E_\varepsilon) \subseteq C_\varepsilon$[1] and $\mu([\,0,T\,] \setminus E_\varepsilon) \leq \varepsilon$. Indeed, let $\varepsilon > 0$ be arbitrary, let $r_\varepsilon > 0$ be such that

$$\frac{\|v\|_{L^1(0,T;X)} + TM}{r_\varepsilon} \leq \varepsilon \tag{6.1.3}$$

and let $C_\varepsilon = C \cap D(0,r_\varepsilon)$. Further, let

$$E_\varepsilon = \{t \in [\,0,T\,];\ \|v(t)\| + M \leq r_\varepsilon\}.$$

Clearly C_ε is weakly compact and, in view of (6.1.2), $f_n(E_\varepsilon) \subseteq C_\varepsilon$ for $n = 1,2,\ldots$. It remains to prove that $\mu([\,0,T\,] \setminus E_\varepsilon) \leq \varepsilon$. To this aim, let us observe that

$$\begin{aligned} r_\varepsilon \mu([\,0,T\,] \setminus E_\varepsilon) &\leq \int_{[\,0,T\,]\setminus E_\varepsilon} (\|v(s)\| + M)\, ds \\ &\leq \int_0^T (\|v(s)\| + M)\, ds = \|v\|_{L^1(0,T;X)} + TM. \end{aligned}$$

From this inequality and (6.1.3) we get $\mu([\,0,T\,] \setminus E_\varepsilon) \leq \varepsilon$, as claimed. So, $\{f_n;\ n = 1,2,\ldots\}$ satisfies the hypotheses of Theorem 1.3.8 and thus, it is weakly relatively compact in $L^1(0,T;X)$. In view of Theorem 1.3.4,

---

[1]In fact we have $f_n(t) \in C_\varepsilon$ a.e. for $t \in E_\varepsilon$, but redefining the functions f_n on a set of null measure we arrive at $f_n(E_\varepsilon) \subseteq C_\varepsilon$ without affecting the rest of the properties of f_n needed in that follows.

it is weakly sequentially compact. Hence, we may assume, with no loss of generality, that $\lim_n f_n = f$ weakly in $L^1(0,T;X)$. Then, in view of Corollary 1.1.1, there exists $(\widetilde{f_n})_n$ with $\widetilde{f_n} \in \operatorname{conv}\{f_k;\ k \geq n\}$, such that $\lim_n \widetilde{f_n} = f$ in $L^1(0,T;X)$. So, on a subsequence at least, $\lim_n \widetilde{f_n}(s) = f(s)$ a.e. for $s \in [\,0,T\,]$. But C is convex and, in view of Problem 6.1.2, is closed too. Thus $f(s) \in C$ a.e. for $s \in [\,0,T\,]$. Since $v_n = f_n + g_n$ and $(f_n)_n$ converges weakly in $L^1(0,T;X)$, while $(v_n)_n$ converges in $L^1(0,T;X)$ to v, we conclude that $(g_n)_n$ converges weakly in $L^1(0,T;X)$ to $g = v - f$. Redefining, if necessary, both f and g on a set of null Lebesgue measure, we may assume that $f(s) \in C$, $g(s) \in D(0,\rho)$ for each $s \in [\,0,T\,]$ and $v(s) = f(s) + g(s)$ a.e. for $s \in [\,0,T\,]$. As f and g are measurable, the proof is complete. □

Theorem 6.1.2. *Let X be a Banach space. If $K \subseteq X$ is almost exact viable with respect to the multi-function $F : K \rightsquigarrow X$, then at each point $\xi \in K$ at which F is u.s.c. and $F(\xi)$ is convex and quasi-weakly compact, we have*

$$F(\xi) \in \mathcal{TS}_K(\xi).^{2}$$

Proof. Let $\xi \in K$ be a point at which $F(\xi)$ is convex and quasi-weakly compact and F is u.s.c. Since K is almost exact viable with respect to F, there exists at least one almost exact solution $u : [\,0,T\,] \to K$ of (6.1.1). As u is continuous at $t = 0$, F is u.s.c. at $u(0) = \xi$ and $u'(s) \in F(u(s))$, a.e. for $s \in [\,0,T\,]$, it follows that for each $\rho > 0$ there exists $\delta(\rho) > 0$ such that

$$u'(s) \in F(\xi) + D(0,\rho)$$

a.e. for $s \in [\,0,\delta(\rho)\,]$. So, if $(\rho_n)_n$ is a sequence in $(0,1)$, $\rho_n \downarrow 0$, there exists $h_n \downarrow 0$ such that

$$u'(s) \in F(\xi) + D(0,\rho_n)$$

for $n = 1,2,\ldots$ and a.e. for $s \in [\,0,h_n\,]$. Hence, for each $n = 1,2,\ldots$, there exist f_n and g_n with $f_n(s) \in F(\xi)$, $g_n(s) \in D(0,\rho_n)$ and

$$u'(s) = f_n(s) + g_n(s)$$

a.e. for $s \in [\,0,h_n\,]$. Since $F(\xi)$ is convex and quasi-weakly compact, in view of Lemma 6.1.1, we may assume without loss of generality that both f_n and g_n are integrable. Let $n = 1,2,\ldots$, and let us define

$$\eta_n = \frac{1}{h_n}\int_0^{h_n} f_n(s)\,ds.$$

[2] See Definition 2.3.1.

As $F(\xi)$ is convex and closed (see Problem 6.1.2), we have

$$\eta_n \in F(\xi).$$

Let us observe that

$$\|u(h_n) - \xi - h_n\eta_n\| = \left\|\xi + \int_0^{h_n} u'(s)\,ds - \xi - h_n\eta_n\right\|$$

$$= \left\|\int_0^{h_n} (u'(s) - f_n(s))\,ds\right\| = \left\|\int_0^{h_n} g_n(s)\,ds\right\| \le h_n\rho_n,$$

for $n = 1, 2, \ldots$. Hence we have

$$\lim_n \frac{1}{h_n}\|u(h_n) - \xi - h_n\eta_n\| = 0.$$

Set

$$p_n = \frac{1}{h_n}\left(u(h_n) - \xi - h_n\eta_n\right),$$

and let us observe that $h_n\eta_n = u(h_n) - \xi - h_np_n$. Clearly this shows that $\lim_n h_n\eta_n = 0$. Since $u(h_n) \in K$, for $n = 1, 2, \ldots$, we get

$$\xi + h_n\eta_n + h_np_n \in K$$

for $n = 1, 2, \ldots$. These considerations, along with $\lim_n p_n = 0$, $\eta_n \in F(\xi)$, for $n = 1, 2, \ldots$, and with the equivalence between (i) and (iv) in Problem 2.3.1, show that $F(\xi) \in \mathcal{TS}_K(\xi)$, and this completes the proof. □

Problem 6.1.3. *Prove that in Theorem* 6.1.2 *we can relax the assumption "quasi-weakly compact" to "quasi-weakly relatively compact" without affecting the conclusion.*

In the case in which, in addition, F is compact valued, we get a necessary condition stronger than the one in Theorem 6.1.2. See Remark 2.4.1.

Theorem 6.1.3. *Let X be a Banach space. If the set K is almost exact viable with respect to the multi-function $F : K \rightsquigarrow X$ then, at each point $\xi \in K$ at which F is u.s.c. and $F(\xi)$ is nonempty, convex and compact, we have $F(\xi) \cap \mathcal{T}_K(\xi) \neq \emptyset$.*

Proof. See Theorem 6.1.2 and Problem 2.4.2. □

A consequence of Theorem 6.1.2 and Remark 6.1.3 is stated below.

Theorem 6.1.4. *If X is reflexive and $K \subseteq X$ is almost exact viable with respect to the multi-function $F : K \rightsquigarrow X$ then, at each point $\xi \in K$ at which F is u.s.c. and $F(\xi)$ is nonempty, convex and closed, we have $F(\xi) \in \mathcal{TS}_K(\xi)$.*

6.2. Sufficient conditions for exact viability

The goal of this section is to state several sufficient conditions of viability of a set K with respect to a multi-function F. It should be noticed that some of these conditions are also necessary and therefore we will formulate them as necessary and sufficient conditions, the necessity part of each one following from Theorem 6.1.2. Let X and Y be Banach spaces, D is a nonempty subset in Y and $F : D \rightsquigarrow X$ is a given multi-function[3].

Definition 6.2.1. A multi-function $F : D \rightsquigarrow X$ is *locally compact* if it is u.s.c. and for each $\eta \in D$ there exists $\rho > 0$ such that $F(D_Y(\eta, \rho) \cap D)$ is relatively compact in X. Further, F is called *compact* if it is u.s.c. and carries bounded subsets in D into relatively compact subsets in X.

Remark 6.2.1. Clearly, each compact multi-function is locally compact. If $D = X = Y$ and, in addition, X is finite dimensional, each locally compact multi-function with bounded values is compact. Also when X is finite dimensional, each multi-function with nonempty and compact values is locally compact. However, we notice that, when $D \subsetneq X$ and D does not coincide with X, even if the latter is finite dimensional, there exist locally compact multi-functions which are not compact. Furthermore, if D is locally compact and F is u.s.c. with nonempty and compact values, F is locally compact even though X is infinite dimensional. See Lemma 2.6.1.

Definition 6.2.2. Let Y and X be two Banach spaces and let $D \subseteq Y$. A multi-function $F : D \rightsquigarrow X$ is called *locally β-compact* if it is u.s.c. and, for each $y \in D$, there exist $r > 0$ and a uniqueness function $\omega : \mathbb{R}_+ \to \mathbb{R}_+$ such that $F(D_Y(y, r) \cap D)$ is bounded and, for each set $C \subseteq D_Y(y, r) \cap D$, we have

$$\beta_X(F(C)) \leq \omega(\beta_Y(C)), \tag{6.2.1}$$

where β_X is the Hausdorff measure of noncompactness on X and β_Y is the Hausdorff measure of noncompactness on Y.

A multi-function $F : D \rightsquigarrow X$ is called *β-compact* if it is u.s.c. and, for each bounded subset C in D, (6.2.1) is satisfied.

In order to simplify the notation, in all that follows, whenever any possibility of confusion will be ruled out by the context, we will denote both functions β_X and β_Y with the very same symbol, β.

Remark 6.2.2. One may easily verify that each locally compact multi-function is locally β-compact. Also, each β-compact multi-function is locally

[3]As in the single-valued case, in fact only two specific situations will be considered, the first one when $Y = X$ which corresponds to the autonomous case and the second one when $Y = \mathbb{R} \times X$ which corresponds to the nonautonomous one.

β-compact. Furthermore, if D is locally compact and $F : D \rightsquigarrow X$ is u.s.c. with nonempty and compact values, then F is locally compact and thus locally β-compact. See Lemma 2.6.1. Therefore, if Y is finite dimensional and $D \subseteq Y$ is locally closed, each u.s.c. multi-function $F : D \rightsquigarrow X$ with compact values is locally β-compact. Moreover, if Y is finite dimensional and $D \subseteq Y$ is closed, each u.s.c. multi-function $F : D \rightsquigarrow X$ with compact values is β-compact.

Remark 6.2.3. Each locally β-compact multi-function has relatively compact values, because $\beta(F(\xi)) \leq \omega(\beta(\{\xi\})) = \omega(0) = 0$ for each $\xi \in D$.

Theorem 6.2.1. *Let X be a Banach space, let $K \subseteq X$ be nonempty and locally closed and let $F : K \rightsquigarrow X$ be a locally β-compact multi-function with nonempty, closed and convex values. A necessary and sufficient condition in order that K be exact viable with respect to F is that*

$$F(\xi) \in \mathcal{TS}_K(\xi) \tag{6.2.2}$$

for each $\xi \in K$.

From Theorem 6.2.1, Remark 6.2.3 and Problem 2.4.2 we get

Theorem 6.2.2. *Let X be a Banach space, let $K \subseteq X$ be nonempty and locally closed and let $F : K \rightsquigarrow X$ be a locally β-compact multi-function with nonempty, closed and convex values. A necessary and sufficient condition in order that K be exact viable with respect to F is that*

$$F(\xi) \cap \mathcal{T}_K(\xi) \neq \emptyset \tag{6.2.3}$$

for each $\xi \in K$.

From Theorem 6.2.2 and Remark 6.2.2, we deduce

Theorem 6.2.3. *Let X be finite dimensional, let $K \subseteq X$ be nonempty and locally closed and let $F : K \rightsquigarrow X$ be an u.s.c. multi-function with nonempty, compact and convex values. A necessary and sufficient condition in order that K be exact viable with respect to F is the tangency condition* (6.2.3).

A result of a different topological nature is

Theorem 6.2.4. *Let X be a Banach space, let $K \subseteq X$ be nonempty and locally compact and let $F : K \rightsquigarrow X$ be a strongly-weakly u.s.c. multi-function with nonempty, weakly compact and convex values. Then, a sufficient condition in order that K be exact viable with respect to F is the tangency condition* (6.2.2).

Since in reflexive Banach spaces the class of weakly relatively compact subsets coincides with the class of bounded subsets, from Theorem 6.2.4, we deduce

Corollary 6.2.1. *Let X be a reflexive Banach space, let $K \subseteq X$ be a nonempty and locally compact set and let $F : K \rightsquigarrow X$ be a strongly-weakly u.s.c. multi-function with nonempty, bounded, closed and convex values. Then, a sufficient condition in order that K be exact viable with respect to F is the tangency condition* (6.2.2).

Remark 6.2.4. In the hypotheses of Theorem 6.2.4, if the function $u : [0,T] \to K$ is an exact solution of (6.1.1), by Corollary 6.1.2, we conclude that whenever $F(u(t)) \in \mathcal{TS}_K(u(t))$ but $F(u(t)) \cap \mathcal{T}_K(u(t)) = \emptyset$, then $h \mapsto h^{-1}(u(t+h) - u(t))$ has no limit points as $h \downarrow 0$. So, if an exact solution reaches a point $x \in K$ with $F(x) \in \mathcal{TS}_K(x)$ and $F(x) \cap \mathcal{T}_K(x) = \emptyset$, it crosses x along a completely nonsmooth (at that point) trajectory.

We conclude this section with an example showing that the convexity condition on the values of F is essential in obtaining the viability of a locally closed set K with respect to an u.s.c. multi-function $F : K \rightsquigarrow X$ by means of the tangency condition $F(\xi) \cap \mathcal{T}_K(\xi) \neq \emptyset$ for all $\xi \in K$.

Example 6.2.1. Let $X = \mathbb{R}^2$, $K = D(0,1)$ and $F : K \to \mathbb{R}^2$, defined by $F(\xi) = \{(-1,0),(1,0)\}$ for each $\xi \in K$. Then, one may easily see that K is locally closed (in fact closed and convex), F is u.s.c., satisfies the tangency condition, but, nevertheless, K is not viable with respect to F.

6.3. Existence of ε-approximate exact solutions

The main goal of the next two sections is to prove Theorems 6.2.1 and 6.2.4. As the necessity part follows from Theorem 6.1.2 combined with Remark 6.2.3, here we will focus our attention only to the sufficiency part.

The first step is concerned with the existence of "approximate solutions" to the autonomous Cauchy problem for the differential inclusion below

$$\begin{cases} u'(t) \in F(u(t)) \\ u(0) = \xi, \end{cases} \tag{6.3.1}$$

where $K \subseteq X$ is locally closed, $\xi \in K$ and $F : K \rightsquigarrow X$ is locally bounded. This happens, for instance, under the hypotheses of both Theorems 6.2.1 and 6.2.4, in the latter case, thanks to Lemma 2.6.1. Since K is locally closed, there exists $\rho > 0$ such that the set $D(\xi,\rho) \cap K$ be closed. Next, diminishing $\rho > 0$ if necessary, we can choose $M > 0$ and $T > 0$ such that

$$\|y\| \leq M \tag{6.3.2}$$

for every $x \in D(\xi, \rho) \cap K$ and $y \in F(x)$, and

$$T(M+1) \leq \rho. \tag{6.3.3}$$

The possibility of diminishing ρ in order to find $M > 0$ satisfying (6.3.2) is a consequence of the fact that F is locally bounded, i.e., F is bounded on $D(\xi, \rho) \cap K$ provided $\rho > 0$ is small enough. Finally, taking a sufficiently small $T > 0$, we obtain (6.3.3).

Lemma 6.3.1. *Let X be a Banach space, let $K \subseteq X$ be nonempty and locally closed and let $F : K \rightsquigarrow X$ be locally bounded and satisfying $F(\xi) \in \mathfrak{TS}_K(\xi)$ for each $\xi \in K$. Let $\xi \in K$, $\rho > 0$, $M > 0$ and $T > 0$ be fixed as above. Then, for each $\varepsilon \in (0,1)$, there exist $\sigma : [0,T] \to [0,T]$ nondecreasing, $f : [0,T] \to X$ and $g : [0,T] \to X$ Riemann integrable and $u : [0,T] \to X$ continuous, such that:*

(i) $t - \varepsilon \leq \sigma(t) \leq t$ *for each* $t \in [0,T]$;
(ii) $\|g(t)\| \leq \varepsilon$ *a.e for* $t \in [0,T]$;
(iii) $u(\sigma(t)) \in D(\xi, \rho) \cap K$ *for all* $t \in [0,T]$ *and* $u(T) \in D(\xi, \rho) \cap K$;
(iv) $f(s) \in F(u(\sigma(s)))$ *a.e for* $s \in [0,T]$;
(v) $u(t) = \xi + \int_0^t f(s)\,ds + \int_0^t g(s)\,ds$ *for each* $t \in [0,T]$.

Before proceeding to the proof of Lemma 6.3.1 we introduce:

Definition 6.3.1. A quadruple (σ, f, g, u), satisfying the conditions (i)$\sim$(v) in Lemma 6.3.1, is called an *ε-approximate exact solution* to the Cauchy problem (6.3.1) on the interval $[0,T]$.

We may now pass to the proof of Lemma 6.3.1.

Proof. Let $\varepsilon > 0$ be arbitrary. We begin by showing the existence of an ε-approximate exact solution on an interval $[0,\delta]$ with $\delta \in (0,T]$. By hypothesis, $F(\xi) \in \mathfrak{TS}_K(\xi)$. From the equivalence between (i) and (iv) in Problem 2.3.2, it follows that there exist $\eta \in F(\xi)$, $\delta \in (0,T]$, $\delta \leq \varepsilon$ and $p \in X$ with $\|p\| \leq \varepsilon$, such that

$$\xi + \delta\eta + \delta p \in K.$$

Now let us define $\sigma : [0,\delta] \to [0,\delta]$, $f : [0,\delta] \to X$, $g : [0,\delta] \to X$ and $u : [0,\delta] \to X$ by

$$\begin{cases} \sigma(t) = 0 & \text{for } t \in [0,\delta] \\ f(t) = \eta & \text{for } t \in [0,\delta] \\ g(t) = p & \text{for } t \in [0,\delta] \\ u(t) = \xi + t\eta + tp & \text{for } t \in [0,\delta]. \end{cases}$$

One can readily see that the quadruple (σ, f, g, u) is an ε-approximate exact solution to the Cauchy problem (6.3.1) on the interval $[0, \delta]$. Indeed, the conditions (i), (ii), (iv) and (v) are obviously fulfilled. To show (iii), let us observe that $u(\sigma(t)) = \xi$ and therefore $u(\sigma(t)) \in D(\xi, \rho) \cap K$ for every $t \in [0, \delta]$. Clearly $u(\delta) \in K$. On the other hand, by (6.3.2) and (6.3.3), we deduce

$$\|u(\delta) - \xi\| \leq \delta\|\eta\| + \delta\|p\| \leq T(M+1) \leq \rho.$$

Thus (iii) is satisfied.

Next, we will prove the existence of an ε-approximate exact solution defined on the whole interval $[0, T]$. To this aim we shall make use of Brezis–Browder Theorem 2.1.1, as follows. Let $\mathcal{S}$ be the set of all ε-approximate exact solutions to the problem (6.3.1) having the domains of definition of the form $[0, c]$ with $c \in (0, T]$. On $\mathcal{S}$ we define the relation $\preceq$ by

$$(\sigma_1, f_1, g_1, u_1) \preceq (\sigma_2, f_2, g_2, u_2)$$

if the domain of definition $[0, c_1]$ of the first quadruple is included in the domain of definition $[0, c_2]$ of the second quadruple and the two ε-approximate exact solutions coincide on the common part of the domains. Obviously $\preceq$ is a preorder relation on $\mathcal{S}$. Let us show first that each increasing sequence $((\sigma_m, f_m, g_m, u_m))_m$ is bounded from above. Indeed, let $((\sigma_m, f_m, g_m, u_m))_m$ be an increasing sequence, and let $c^* = \lim_m c_m$, where $[0, c_m]$ denotes the domain of definition of $(\sigma_m, f_m, g_m, u_m)$. Clearly, $c^* \in (0, T]$. We will show that there exists at least one element, $(\sigma^*, f^*, g^*, u^*) \in \mathcal{S}$, defined on $[0, c^*]$ and satisfying $(\sigma_m, f_m, g_m, u_m) \preceq (\sigma^*, f^*, g^*, u^*)$ for each $m \in \mathbb{N}$. In order to do this, we have to prove first that there exists $\lim_m u_m(c_m)$. For each $m, k \in \mathbb{N}$, $m \leq k$, we have $u_m(s) = u_k(s)$ for all $s \in [0, c_m]$. Taking into account (iii), (iv), (v) and (6.3.2), we deduce

$$\|u_m(c_m) - u_k(c_k)\| \leq \int_{c_m}^{c_k} [\|f_k(\theta)\| + \|g_k(\theta)\|]\, d\theta \leq (M+\varepsilon)|c_k - c_m|$$

for every $m, k \in \mathbb{N}$, which proves that there exists $\lim_m u_m(c_m)$. Since for every $m \in \mathbb{N}$, $u_m(c_m) \in D(\xi, \rho) \cap K$, and the latter is closed, it readily follows that $\lim_m u_m(c_m) \in D(\xi, \rho) \cap K$. Furthermore, because all the functions in the set $\{\sigma_m;\ m \in \mathbb{N}\}$ are nondecreasing, with values in $[0, c^*]$, and satisfy $\sigma_m(c_m) \leq \sigma_p(c_p)$ for every $m, p \in \mathbb{N}$ with $m \leq p$, there exists $\lim_m \sigma_m(c_m)$ and this limit belongs to $[0, c^*]$. This shows that we can define the quadruple $(\sigma^*, f^*, g^*, u^*) : [0, c^*] \to [0, c^*] \times X \times X \times X$ as follows. Let

$$\sigma^*(t) = \begin{cases} \sigma_m(t) & \text{for } t \in [0, c_m] \\ \lim_m \sigma_m(c_m) & \text{for } t = c^* \end{cases}$$
$$u^*(t) = \begin{cases} u_m(t) & \text{for } t \in [0, c_m] \\ \lim_m u_m(c_m) & \text{for } t = c^* \end{cases}$$
$$g^*(t) = \begin{cases} g_m(t) & \text{for } t \in [0, c_m] \\ 0 & \text{for } t = c^* \end{cases}$$
$$f^*(t) = \begin{cases} f_m(t) & \text{for } t \in [0, c_m] \\ \eta^* & \text{for } t = c^*, \end{cases}$$

where η^* is an arbitrary but fixed element in $F(u^*(\sigma^*(c^*)))$. One can easily see that $(\sigma^*, f^*, g^*, u^*)$ is an ε-approximate exact solution which is an upper bound for $((\sigma_m, f_m, g_m, u_m))_m$. Let us define the function $\mathcal{N} : \mathcal{S} \to \mathbb{R}$ by $\mathcal{N}((\sigma, f, g, u)) = c$, where $[0, c]$ is the domain of definition of (σ, f, g, u). Clearly $\mathcal{N}$ satisfies the hypotheses of Brezis-Browder Theorem 2.1.1. Then, $\mathcal{S}$ contains at least one $\mathcal{N}$-maximal element $(\overline{\sigma}, \overline{f}, \overline{g}, \overline{u})$, defined on $[0, \overline{c}]$. In other words, if $(\widetilde{\sigma}, \widetilde{f}, \widetilde{g}, \widetilde{u}) \in \mathcal{S}$, defined on $[0, \widetilde{c}]$, satisfies $(\overline{\sigma}, \overline{f}, \overline{g}, \overline{u}) \preceq (\widetilde{\sigma}, \widetilde{f}, \widetilde{g}, \widetilde{u})$, then we necessarily have $\overline{c} = \widetilde{c}$. We will next show that $\overline{c} = T$. Indeed, let us assume by contradiction that $\overline{c} < T$. Since

$$\|\overline{u}(\overline{c}) - \xi\| \le \int_0^{\overline{c}} \|\overline{f}(s)\|\, ds + \int_0^{\overline{c}} \|\overline{g}(s)\|\, ds \le \overline{c}(M + \varepsilon)$$
$$\le \overline{c}(M + 1) < T(M + 1),$$

we deduce that

$$\|\overline{u}(\overline{c}) - \xi\| < \rho. \tag{6.3.4}$$

Then, as $\overline{u}(\overline{c}) \in K$, we have $F(\overline{u}(\overline{c})) \in \mathcal{TS}_K(\overline{u}(\overline{c}))$ and thus, again by the equivalence between (i) and (iv) in Problem 2.3.2, there exist $\overline{\eta} \in F(\overline{u}(\overline{c}))$, $\delta \in (0, T - \overline{c})$, $\delta \le \varepsilon$ and $p \in X$, $\|p\| \le \varepsilon$, such that $\overline{u}(\overline{c}) + \delta\overline{\eta} + \delta p \in K$. From (6.3.4), it follows that we can diminish δ, if necessary[4], in order to have

$$\|\overline{u}(\overline{c}) + \delta\overline{\eta} + \delta p - \xi\| \le \rho. \tag{6.3.5}$$

Let us define the functions $\sigma : [0, \overline{c} + \delta] \to [0, \overline{c} + \delta]$, $f : [0, \overline{c} + \delta] \to X$ and $g : [0, \overline{c} + \delta] \to X$ by

$$\sigma(t) = \begin{cases} \overline{\sigma}(t) & \text{for } t \in [0, \overline{c}] \\ \overline{c} & \text{for } t \in (\overline{c}, \overline{c} + \delta] \end{cases},$$

$$f(t) = \begin{cases} \overline{f}(t) & \text{for } t \in [0, \overline{c}] \\ \overline{\eta} & \text{for } t \in (\overline{c}, \overline{c} + \delta] \end{cases}, \quad g(t) = \begin{cases} \overline{g}(t) & \text{for } t \in [0, \overline{c}] \\ p & \text{for } t \in (\overline{c}, \overline{c} + \delta]. \end{cases}$$

[4]Of course, both $\overline{\eta}$ and p may change with δ.

Clearly, f and g are Riemann integrable on $[0,\overline{c}+\delta]$ and $\|g(t)\| \leq \varepsilon$ for every $t \in [0,\overline{c}+\delta]$. We define $u : [0,\overline{c}+\delta] \to X$ by

$$u(t) = \begin{cases} \overline{u}(t) & \text{for } t \in [0,\overline{c}] \\ \overline{u}(\overline{c}) + (t-\overline{c})\overline{\eta} + (t-\overline{c})p & \text{for } t \in (\overline{c},\overline{c}+\delta]. \end{cases}$$

Let us notice that

$$u(t) = \xi + \int_0^t f(\theta)\,d\theta + \int_0^t g(\theta)\,d\theta$$

for every $t \in [0,\overline{c}+\delta]$. Thus σ, f, g and u satisfy the conditions (i), (ii), (iv) and (v). Since

$$u(\sigma(t)) = \begin{cases} \overline{u}(\overline{\sigma}(t)) & \text{for } t \in [0,\overline{c}] \\ \overline{u}(\overline{c}) & \text{for } t \in (\overline{c},\overline{c}+\delta], \end{cases}$$

it follows that $u(\sigma(t)) \in D(\xi,\rho) \cap K$ and thus (iii) is also satisfied. Furthermore, from the choice of δ and p, we have $u(\overline{c}+\delta) = \overline{u}(\overline{c})+\delta\overline{\eta}+\delta p \in K$. Moreover, from (6.3.5), we conclude $\|u(\overline{c}+\delta) - \xi\| = \|\overline{u}(\overline{c}) + \delta\overline{\eta} + \delta p - \xi\| \leq \rho$ and consequently u satisfies (iii). Thus $(\sigma, f, g, u) \in \mathcal{S}$.

Finally, inasmuch as $(\overline{\sigma}, \overline{f}, \overline{g}, \overline{u}) \preceq (\sigma, f, g, u)$ and $\overline{c} < \overline{c} + \delta$, it follows that $(\overline{\sigma}, \overline{f}, \overline{g}, \overline{u})$ is not an $\mathcal{N}$-maximal element. But this is absurd. This contradiction can be eliminated only if each maximal element in the set $\mathcal{S}$ is defined on $[0,T]$. □

6.4. Convergence of ε-approximate exact solutions

The goal of this section is to prove both Theorems 6.2.1 and 6.2.4. We will do that by showing the convergence of a suitably chosen sequence of ε-approximate exact solutions. Let us consider a sequence $(\varepsilon_n)_n$ in $(0,1)$, decreasing to 0, and let $((\sigma_n, f_n, g_n, u_n))_n$ be a sequence of ε_n-approximate solutions of (6.3.1) on $[0,T]$. Let us observe that, by (i), (ii), (iv) and (v)[5], we have

$$\|u_n(t) - u_n(\sigma_n(t))\| \leq (M+1)\varepsilon_n \tag{6.4.1}$$

for each $t \in [0,T]$.

We begin with the proof of Theorem 6.2.1.

Proof. Let $M > 0$ as in (6.3.2). Diminishing $\rho > 0$, if necessary, we may assume that F is β-compact on $D(\xi,\rho) \cap K$.

---

[5]Within this section, all the quotations to items like (i)∼(v) refer to the corresponding items in Lemma 6.3.1.

We analyze first the case when X is separable. From (6.4.1), (v), the fact that F is β-compact on $D(\xi,\rho)\cap K$, Lemma 2.7.2, Problem 2.7.1 and Remark 2.7.1, it follows that

$$\beta(\{u_n(t);\ n\geq k\})$$
$$\leq \beta\left(\left\{\int_0^t f_n(s)\,ds;\ n\geq k\right\}\right)+\beta\left(\left\{\int_0^t g_n(s)\,ds;\ n\geq k\right\}\right)$$
$$\leq \int_0^t \beta(\{f_n(s);\ n\geq k\})\,ds+\int_0^t \beta\left(\{g_n(s);\ n\geq k\}\right)\,ds$$
$$\leq \int_0^t \omega(\beta\{u_n(\sigma_n(s));\ n\geq k\})\,ds+T\varepsilon_k$$
$$\leq \int_0^t \omega(\beta(\{u_n(s);\ n\geq k\}+\{u_n(\sigma_n(s))-u_n(s);\ n\geq k\}))\,ds+T\varepsilon_k$$
$$\leq \int_0^t \omega(\beta(\{u_n(s);\ n\geq k\})+\beta(\{u_n(\sigma_n(s))-u_n(s);\ n\geq k\}))\,ds+T\varepsilon_k$$
$$\leq \int_0^t \omega(\beta(\{u_n(s);\ n\geq k\})+(M+1)\varepsilon_k)\,ds+T\varepsilon_k.$$

Set $x_k(t)=\beta(\{u_n(t);\ n\geq k\})+(M+1)\varepsilon_k$ and $\gamma_k=(M+T+1)\varepsilon_k$. The inequality above rewrites as

$$x_k(t)\leq \gamma_k+\int_0^t \omega(x_k(s))\,ds$$

for each $t\in[0,T]$. By Lemma 1.8.2, diminishing $T>0$, if necessary, we may assume that $\lim_k x_k(t)=0$ uniformly for $t\in[0,T]$. But this shows that $\lim_k \beta(\{u_n(s);\ n\geq k\})=0$, and thus we are in the hypotheses of Lemma 2.7.3. It follows that, for each $t\in[0,T]$, $\{u_n(t);\ n=1,2,\dots\}$ is relatively compact. By (v) and (6.3.2) we get that $\{u_n;\ n=1,2,\dots\}$ is equicontinuous, and therefore, thanks to Arzelà–Ascoli Theorem 1.3.6, there exists $u\in C([0,T];X)$ such that, on a subsequence at least,

$$\lim_n u_n(t)=u(t)$$

uniformly for $t\in[0,T]$. In view of (6.4.1), we also have

$$\lim_n u_n(\sigma_n(t))=u(t)$$

uniformly for $t\in[0,T]$. From (iii) and the fact that $D(\xi,\rho)\cap K$ is closed, we conclude that $u(t)\in D(\xi,\rho)\cap K$ for each $t\in[0,T]$.

In view of Remark 6.2.3, F has compact and thus weakly compact values. We shall apply Theorem 1.3.8 to show that u is both absolutely continuous and a.e. differentiable on $[0,T]$, and

$$u(t) = \xi + \int_0^t u'(s)\,ds$$

for each $t \in [0,T]$. Indeed, since $f_n(s) \in F(u_n(\sigma_n(s)))$ for $n = 1,2,\ldots$ and $s \in [0,T]$, and, by (6.3.2), $F(D(\xi,\rho) \cap K)$ is bounded, it follows that $\{f_n;\ n = 1,2,\ldots\}$ is uniformly integrable. Moreover, as F is u.s.c. and has weakly compact values, in view of Lemma 2.6.1,

$$\{F(u_n(\sigma_n(t)));\ n = 1,2,\ldots, t \in [0,T]\}$$

is weakly compact. So, by Theorem 1.3.2, its closed convex hull is weakly compact too and thus we are in the hypotheses of Theorem 1.3.8 which, along with Theorem 1.3.4, shows that $(f_n)_n$ has at least one weakly convergent subsequence in $L^1(0,T;X)$ to some function f. Summarizing, we have $\lim_n u_n = u$ uniformly on $[0,T]$ and $\lim_n u_n' = f$ weakly in $L^1(0,T;X)$. Thus u is absolutely continuous and a.e. differentiable on $[0,T]$, $u'(t) = f(t)$ a.e. for $t \in [0,T]$ and u is a primitive of its derivative u'.

It remains to be shown that, at each differentiability point, $t \in [0,T]$, of u, we have $u'(t) \in F(u(t))$. Let $t \in [0,T)$ be a differentiability point of u and let $h > 0$ be such that $t + h \in [0,T]$. Let $n = 1,2,\ldots$ be arbitrary but fixed. We have

$$\frac{1}{h}(u_n(t+h) - u_n(t)) = \frac{1}{h}\int_t^{t+h} f_n(s)\,ds + \frac{1}{h}\int_t^{t+h} g_n(s)\,ds$$

where $f_n(s) \in F(u_n(\sigma_n(s)))$ for each $s \in [t, t+h]$. Let $\varepsilon > 0$. Since F is u.s.c. at $u(t)$, u is continuous, $\lim_n \sigma_n(s) = s$ and $\lim_n u_n(\sigma_n(s)) = u(s)$, uniformly for $s \in [0,T]$, there exists $h(\varepsilon) > 0$ and $n(\varepsilon) \in \mathbb{N}$ such that, for each $h \in (0, h(\varepsilon)]$ and each $n \geq n(\varepsilon)$, we have

$$F(u_n(\sigma_n(s))) \subseteq F(u(t)) + D(0,\varepsilon)$$

for each $s \in [t, t+h]$ and

$$\left\| \frac{1}{h}\int_t^{t+h} g_n(s)\,ds \right\| \leq \varepsilon.$$

Since

$$\frac{1}{h}\int_t^{t+h} f_n(s)\,ds \in \overline{\text{conv}} \bigcup_{s\in[t,t+h]} F(u_n(\sigma_n(s)))$$

and $F(u(t))+D(0,\varepsilon)$ is convex because both $F(u(t))$ and $D(0,\varepsilon)$ are convex, it follows that

$$\frac{1}{h}(u_n(t+h)-u_n(t)) \in F(u(t))+D(0,2\varepsilon)$$

for each $h \in (0,h(\varepsilon)]$ and each $n \geq n(\varepsilon)$. Keeping h fixed in $(0,h(\varepsilon)]$, passing to the limit for $n \to \infty$ in this relation and taking into account that $F(u(t))+D(0,2\varepsilon)$ is closed (because $F(u(t))$ is compact), we get

$$\frac{1}{h}(u(t+h)-u(t)) \in F(u(t))+D(0,2\varepsilon).$$

Finally, passing to the limit for $h \downarrow 0$ in this relation we get $u'(t) \in F(u(t))$. Since the case $t=T$ can be treated similarly by computing the left derivative of u at T, the proof of Theorem 6.2.1 is complete in the case when X is separable.

If X is not separable, there exists a separable and closed subspace, Y, of X such that $u_n(t), f_n(t), g_n(t) \in Y$ for $n = 1,2,\ldots$ and a.e. for $t \in [0,T]$. On the other hand, from Problem 2.7.2, Definition 6.2.2 and the monotonicity of ω, we deduce

$$\beta_Y(F(C)) \leq 2\beta(F(C)) \leq 2\omega(\beta(C)) \leq 2\omega(\beta_Y(C)),$$

for each set $C \subseteq D(\xi,\rho)\cap K \cap Y$. From now on we have to argue as in the last part of the proof of Theorem 3.2.2. The proof is complete. □

We can now proceed to the proof of Theorem 6.2.4.

Proof. Since K is locally compact and F is strongly-weakly u.s.c. and has nonempty and weakly compact values, by Lemma 2.6.1, it follows that F is locally bounded. Then, we can find $M>0$ satisfying (6.3.2). Diminishing $\rho>0$ and $T>0$ if necessary, we may assume that the conclusion of Lemma 6.3.1 holds true and, in addition, $D(\xi,\rho)\cap K$ is compact. Taking into account that, by (iii), $u_n(\sigma_n(t)) \in D(\xi,\rho)\cap K$ for $n = 1,2,\ldots$ and $t \in [0,T]$, it follows that, for each $t \in [0,T]$, $\{u_n(\sigma_n(t));\ n=1,2,\ldots\}$ is relatively compact in X. In view of (6.4.1), we conclude that, for each $t \in [0,T]$, $\{u_n(t);\ n=1,2,\ldots\}$ is relatively compact in X, too.

From (6.3.2) and (ii) and (v), we deduce that $\{u_n;\ n=1,2,\ldots\}$ is equicontinuous on $[0,T]$. By Arzelà–Ascoli Theorem 1.3.6, we conclude that there exists $u \in C([0,T];X)$ such that, on a subsequence at least, we have $\lim_n u_n(t) = u(t)$ uniformly for $t \in [0,T]$. In view of this relation, of (i) and (iii), we deduce that $\lim_n u_n(\sigma_n(t)) = u(t)$ uniformly for $t \in [0,T]$ and $u(t) \in D(\xi,\rho)\cap K$.

As $D(\xi,\rho)\cap K$ is compact and F is strongly-weakly u.s.c. and has convex and weakly compact values, by Lemma 2.6.1, it follows that the set $F(D(\xi,\rho)\cap K)$ is bounded being weakly relatively compact. Then, as $f_n(s)\in F(u_n(\sigma_n(s)))$ for $n=1,2,\ldots$ and $s\in[0,T]$, it follows that $\{f_n;\ n=1,2,\ldots\}$ is uniformly integrable. By Theorem 1.3.8 combined with Theorem 1.3.4, we deduce that $(f_n)_n$ has at least one weakly convergent subsequence in $L^1(0,T;X)$ to some function f. Summarizing, we have $\lim_n u_n = u$ uniformly on $[0,T]$ and $\lim_n u_n' = f$ weakly in $L^1(0,T;X)$. Thus u is absolutely continuous, a.e. differentiable on $[0,T]$, $u'(t)=f(t)$ a.e. for $t\in[0,T]$ and u is a primitive of its derivative u'. From (iv) and Lemma 2.6.2, we conclude that $f(t)\in F(u(t))$ a.e. for $t\in[0,T]$. Thus u is an almost exact solution of (6.1.1) on $[0,T]$. As $\xi\in K$ is arbitrary, this shows that K is almost exact viable with respect to F. To complete the proof, it remains to show that u is even an exact solution of (6.1.1) on $[0,T]$. To this aim, let $t\in[0,T)$ be a differentiability point of u and let E be an arbitrary open half-space with $F(u(t))\subseteq E$. Since E is weakly open too, u is continuous and F is strongly-weakly u.s.c at $u(t)$, there exists $\delta(E)>0$ such that, for each $h\in(0,\delta(E)\,]$, with $t+h\le T$, we have

$$F(u(s))\subseteq E$$

for each $s\in[\,t,t+h\,]$. Consequently, for h as above, we have

$$\bigcup_{s\in[\,t,t+h\,]} F(u(s))\subseteq E.$$

On the other hand, for $n=1,2,\ldots,$

$$\frac{1}{h}(u_n(t+h)-u_n(t))=\frac{1}{h}\int_t^{t+h} f_n(s)\,ds+\frac{1}{h}\int_t^{t+h} g_n(s)\,ds.$$

Since $\lim_n u_n(\sigma_n(s))=u(s)$ uniformly on $[0,T]$ and $f_n(s)\in F(u_n(\sigma_n(s)))$ for each $s\in[0,T]$, there exists $n(E)=1,2,\ldots$ such that, for all $n\ge n(E)$,

$$\frac{1}{h}\int_t^{t+h} f_n(s)\,ds\in\overline{\mathrm{conv}}\bigcup_{s\in[\,t,t+h\,]} F(u_n(\sigma_n(s)))\subseteq\overline{E}.$$

Therefore

$$\frac{1}{h}(u_n(t+h)-u_n(t))-\frac{1}{h}\int_t^{t+h} g_n(s)\,ds\in\overline{E}.$$

Recalling that $\lim_n\|g_n(s)\|=0$ uniformly for $s\in[0,T]$ and passing to the limit successively for $n\to\infty$ and $h\downarrow 0$ in the last relation, we get $u'(t)\in\overline{E}$. Since E is an arbitrary open half-space including $F(u(t))$ and the latter, being convex and closed, is the intersection of all closed half-spaces including it, we conclude that $u'(t)\in F(u(t))$. Since the case $t=T$ can be handled similarly, this achieves the conclusion. □

6.5. The nonautonomous u.s.c. case

In this section we will show how all the results established before for the autonomous differential inclusion $u'(t) \in F(u(t))$ extend to the nonautonomous one $u'(t) \in F(t, u(t))$. So, let X be a real Banach space, $\mathcal{C}$ a nonempty subset in $\mathbb{R} \times X$, $F : \mathcal{C} \rightsquigarrow X$ a given multi-function, $(\tau, \xi) \in \mathcal{C}$ and let us consider the Cauchy problem for the nonautonomous differential inclusion

$$\begin{cases} u'(t) \in F(t, u(t)) \\ u(\tau) = \xi. \end{cases} \tag{6.5.1}$$

Definition 6.5.1. By an *exact solution* of (6.5.1) on $[\tau, T]$ we mean an absolutely continuous function $u : [\tau, T] \to X$, a.e. differentiable on $[\tau, T]$, with $u' \in L^1(\tau, T; X)$, and satisfying:

(i) $(t, u(t)) \in \mathcal{C}$ for each $t \in [\tau, T]$ and

(ii) $u'(t) \in F(t, u(t))$ at every point $t \in [\tau, T]$ at which u is differentiable, and $u(\tau) = \xi$.

An exact solution of (6.5.1) on the semi-open interval $[\tau, T)$ is defined similarly.

Definition 6.5.2. By an *almost exact solution* of (6.5.1) on $[\tau, T]$ we mean an absolutely continuous function $u : [\tau, T] \to X$, a.e. differentiable on $[\tau, T]$, with $u' \in L^1(\tau, T; X)$, and satisfying:

(i) $(t, u(t)) \in \mathcal{C}$ for each $t \in [\tau, T]$ and

(ii) $u'(t) \in F(t, u(t))$ a.e. for $t \in [\tau, T]$, and $u(\tau) = \xi$.

An almost exact solution of (6.5.1) on the semi-open interval $[\tau, T)$ is defined similarly, noticing that, in this case, we have to impose the weaker constraint $u' \in L^1_{\text{loc}}([\tau, T); X)$.

Definition 6.5.3. The set $\mathcal{C}$ is *exact viable* (*almost exact viable*) with respect to F if for each $(\tau, \xi) \in \mathcal{C}$ there exist $T \in \mathbb{R}$, $T > \tau$, and an exact solution (almost exact solution) $u : [\tau, T] \to X$ of (6.5.1).

We will rewrite the nonautonomous problem above as an autonomous one in the space $\mathcal{X} = \mathbb{R} \times X$, endowed with the norm $\|(t, u)\| = |t| + \|u\|$, for each $(t, u) \in \mathcal{X}$. Namely, set $z(s) = (t(s), u(s))$ and $\mathcal{F}(z) = (1, F(z))$, for $s \in [0, T - \tau]$, where $(1, F(z)) = \{(1, y);\ y \in F(z)\}$. Then, the Cauchy problem above is equivalent to

$$\begin{cases} z'(s) \in \mathcal{F}(z(s)) \\ z(0) = (\tau, \xi). \end{cases} \tag{6.5.2}$$

So, all the viability results proved before extend in an obvious way to the nonautonomous case via the transformations above. Namely, we have

Theorem 6.5.1. *Let X be a Banach space. If $\mathcal{C} \subseteq \mathbb{R} \times X$ is almost exact viable with respect to $F : \mathcal{C} \leadsto X$ then, at each point $(\tau, \xi) \in \mathcal{C}$ at which F is u.s.c. and $F(\tau, \xi)$ is convex and quasi-weakly compact, we have $(1, F(\tau, \xi)) \in \mathcal{TS}_{\mathcal{C}}(\tau, \xi)$.*

Theorem 6.5.2. *If X is reflexive and $\mathcal{C} \subseteq \mathbb{R} \times X$ is almost exact viable with respect to $F : \mathcal{C} \leadsto X$, then, at each point $(\tau, \xi) \in \mathcal{C}$ at which F is u.s.c. and $F(\tau, \xi)$ is convex and closed, we have $(1, F(\tau, \xi)) \in \mathcal{TS}_{\mathcal{C}}(\tau, \xi)$.*

Remark 6.5.1. If $\mathcal{C}$ is a cylindrical domain, i.e. $\mathcal{C} = I \times K$ with I an open to the right interval and K a subset in X then, for each $(\tau, \xi) \in \mathcal{C}$, the two tangency conditions below are equivalent.

(i) $(1, F(\tau, \xi)) \in \mathcal{TS}_{\mathcal{C}}(\tau, \xi)$;
(ii) $F(\tau, \xi) \in \mathcal{TS}_K(\xi)$.

Theorem 6.5.3. *Let X be a Banach space, $\mathcal{C} \subseteq \mathbb{R} \times X$ a nonempty and locally closed set and let $F : \mathcal{C} \leadsto X$ be a locally β-compact multi-function with nonempty, closed and convex values. A necessary and sufficient condition in order that $\mathcal{C}$ be exact viable with respect to F is the tangency condition*

$$(1, F(\tau, \xi)) \in \mathcal{TS}_{\mathcal{C}}(\tau, \xi) \tag{6.5.3}$$

for each $(\tau, \xi) \in \mathcal{C}$.

Proof. Let us observe that $u : [\tau, T] \to X$ is an exact solution of (6.5.1) if and only if $z : [0, T - \tau] \to \mathcal{C}$, $z(s) = (s + \tau, u(s + \tau))$ is an exact solution of the autonomous Cauchy problem (6.5.2). Since F is u.s.c. and satisfies both $F(z) \in \mathcal{TS}_{\mathcal{C}}(z)$ for each $z \in \mathcal{C}$ and

$$\beta_{\mathcal{X}}(\mathcal{F}(B)) = \beta_{\mathcal{X}}(\{1\} \times F(B)) = \beta_X(F(B)) \leq \omega(\beta_{\mathcal{X}}(B))$$

for each bounded subset B in $\mathcal{C}$, the conclusion follows from Theorem 6.2.2. □

A corollary of Theorem 6.5.3, Remark 6.2.2 and Problem 2.4.2 is

Theorem 6.5.4. *Let X be a Banach space, $\mathcal{C} \subseteq \mathbb{R} \times X$ a nonempty and locally closed set and let $F : \mathcal{C} \leadsto X$ be a locally β-compact multi-function with nonempty, closed and convex values. Then, a necessary and sufficient condition in order that $\mathcal{C}$ be exact viable with respect to F is the tangency condition*

$$(1, F(\tau, \xi)) \cap \mathcal{T}_{\mathcal{C}}(\tau, \xi) \neq \emptyset \tag{6.5.4}$$

for each $(\tau, \xi) \in \mathcal{C}$.

Problem 6.5.1. *Prove that $(1, y) \in \mathcal{T}_{\mathcal{C}}(\tau, \xi)$ if and only if there exists a sequence $((\tau_n, \xi_n))_n \in \mathcal{C}$ such that $\tau_n \downarrow \tau$, $\lim_n \xi_n = \xi$ and $\lim_n \frac{\xi_n - \xi}{\tau_n - \tau} = y$.*

From Theorem 6.5.4, we easily deduce

Theorem 6.5.5. *Let X be finite dimensional, let $\mathcal{C} \subseteq \mathbb{R} \times X$ be a nonempty and locally closed set and let $F : \mathcal{C} \rightsquigarrow X$ be an u.s.c. multi-function with nonempty, convex and compact values. Then, a necessary and sufficient condition in order that $\mathcal{C}$ be exact viable with respect to F is the tangency condition* (6.5.4).

A nonautonomous version of Theorem 6.2.4 is

Theorem 6.5.6. *Let X be a Banach space, $\mathcal{C} \subseteq \mathbb{R} \times X$ a nonempty and locally compact set and let $F : \mathcal{C} \rightsquigarrow X$ be a strongly-weakly u.s.c. multi-function with nonempty, weakly compact and convex values. Then, a sufficient condition in order that $\mathcal{C}$ be exact viable with respect to F is the tangency condition* (6.5.3).

6.6. Global (almost) exact solutions

Let $\mathcal{C} \subseteq \mathbb{R} \times X$ be nonempty and let $F : \mathcal{C} \rightsquigarrow X$. In this section we will prove some results concerning the existence of noncontinuable, or even global solutions to the Cauchy problem

$$\begin{cases} u'(t) \in F(t, u(t)) \\ u(\tau) = \xi. \end{cases} \tag{6.6.1}$$

An (almost) exact solution $u : [\tau, T) \to X$ of (6.6.1) is called *noncontinuable*, if there is no other (almost) exact solution $v : [\tau, \widetilde{T}) \to X$ of (6.6.1), with $T < \widetilde{T}$ and satisfying $u(t) = v(t)$ for all $t \in [\tau, T)$. The (almost) exact solution u is called *global* if $T = T_{\mathcal{C}}$, with $T_{\mathcal{C}}$ given by (3.6.2). The next theorem follows from Brezis–Browder Theorem 2.1.1.

Theorem 6.6.1. *Let X be a Banach space, $\mathcal{C} \subseteq \mathbb{R} \times X$ be nonempty and let $F : \mathcal{C} \rightsquigarrow X$. The following conditions are equivalent*:

(i) *$\mathcal{C}$ is (almost) exact viable with respect to F*;
(ii) *for each $(\tau, \xi) \in \mathcal{C}$ there exists at least one noncontinuable (almost) exact solution $u : [\tau, T) \to X$ of* (6.6.1).

Since the proof of Theorem 6.6.1 is completely similar with that one of Theorem 3.6.1, we do not enter into details.

Remark 6.6.1. Notice that in Theorem 6.6.1 we do not assume $\mathcal{C}$ to be locally closed or F to be u.s.c.

We conclude this section with a result concerning the existence of global solutions.

Definition 6.6.1. A multi-function $F : \mathcal{C} \rightsquigarrow X$ is called *positively sublinear* if there exist three continuous functions $a : \mathbb{R} \to \mathbb{R}_+$, $b : \mathbb{R} \to \mathbb{R}_+$, and $c : \mathbb{R} \to \mathbb{R}_+$ such that

$$\|f\| \leq a(t)\|\xi\| + b(t)$$

for each $(t, \xi, f) \in K_+^c(F)$, where

$$K_+^c(F) = \{(t, \xi, f) \in \mathcal{C} \times X;\ \|\xi\| > c(t),\ f \in F(t, \xi),\ [\xi, f]_+ > 0\}.$$

Remark 6.6.2. There are three notable specific cases in which F is positively sublinear:

(i) when F is bounded on $\mathcal{C}$;
(ii) when F has sublinear growth with respect to its last argument[6];
(iii) when f satisfies the "sign condition" $[\xi, f]_+ \leq 0$ for each $(t, \xi) \in \mathcal{C}$ and $f \in F(t, \xi)$.

Theorem 6.6.2. *Let X be a Banach space, let $\mathcal{C} \subseteq \mathbb{R} \times X$ be nonempty and let $F : \mathcal{C} \rightsquigarrow X$ be a given multi-function. If $\mathcal{C}$ is X-closed*[7]*, F maps bounded subsets in $\mathcal{C}$ into bounded subsets in X, is positively sublinear, and $\mathcal{C}$ is (almost) exact viable with respect to F, then each (almost) exact solution of* (6.6.1) *can be continued up to a global one, i.e., defined on $[\tau, T_{\mathcal{C}})$, where $T_{\mathcal{C}}$ is given by* (3.6.2).

The proof of Theorem 6.6.2 repeats the same routine as that of the proof of Theorem 3.6.3, with the special mention that $f(s, u(s))$ in that proof should be replaced here by $f(s)$, where $f(s) \in F(s, u(s))$ for $s \in [\tau, T)$.

6.7. Sufficient conditions for invariance

Let X be a real Banach space, D an open subset in X, K a nonempty subset of D, and let us consider the Cauchy problem for the differential inclusion

$$\begin{cases} u'(t) \in F(u(t)) \\ u(0) = \xi, \end{cases} \tag{6.7.1}$$

where $F : D \rightsquigarrow X$ is a given multi-function.

Definition 6.7.1. The subset K is *locally invariant* with respect to F if for each $\xi \in K$ and each almost exact solution $u : [0, c] \to D$, $c > 0$, of (6.7.1), there exists $T \in (0, c]$ such that we have $u(t) \in K$ for each $t \in [0, T]$. It is *invariant* if it satisfies the local invariance condition above with $T = c$.

---

[6]This means that there exist two continuous functions $a : \mathbb{R} \to \mathbb{R}_+$ and $b : \mathbb{R} \to \mathbb{R}_+$ such that $\|f\| \leq a(t)\|\xi\| + b(t)$ for each $(t, \xi) \in \mathcal{C}$ and each $f \in F(t, \xi)$.

[7]See Definition 3.6.2.

Problem 6.7.1. *Show that whenever K is closed and locally invariant with respect to F, then it is invariant with respect to F.*

Our first sufficient condition for local invariance is expressed in terms of the exterior tangency condition: there exists an open neighborhood V of K, with $V \subseteq D$, such that

$$\liminf_{h \downarrow 0} \frac{1}{h} [\text{dist}\,(\xi + h\eta; K) - \text{dist}\,(\xi; K)] \leq \omega(\text{dist}\,(\xi; K)) \tag{6.7.2}$$

for each $\xi \in V$ and each $\eta \in F(\xi)$, where ω is a uniqueness function. See Definition 1.8.1. This tangency condition can be viewed as a multi-valued counterpart of (4.2.1). The main result in this section is

Theorem 6.7.1. *Let X be a Banach space, let $K \subseteq D \subseteq X$, with K locally closed and D open, and let $F : D \rightsquigarrow X$. If (6.7.2) is satisfied, then K is locally invariant with respect to F.*

Proof. Let $V \subseteq D$ be the open neighborhood of K whose existence is ensured by (6.7.2) and let $\omega : \mathbb{R}_+ \to \mathbb{R}_+$ be the corresponding uniqueness function. Let $\xi \in K$ and let $u : [0, c] \to V$ be any almost exact solution to (6.7.1). Diminishing c if necessary, we may assume that there exists $\rho > 0$ such that $D(\xi, \rho) \cap K$ is closed and $u(t) \in D(\xi, \rho/2)$ for each $t \in [0, c]$. Let $g : [0, c] \to \mathbb{R}_+$ be defined by $g(t) = \text{dist}\,(u(t); K)$ for each $t \in [0, c]$. Let us observe that g is absolutely continuous on $[0, c]$. Let $t \in [0, c)$ be such that both $u'(t)$ and $g'(t)$ exist and $u'(t) \in F(u(t))$, and let $h > 0$ with $t + h \in [0, c]$. We have

$$\begin{aligned} g(t+h) &= \text{dist}\,(u(t+h); K) \\ &\leq h \left\| \frac{u(t+h) - u(t)}{h} - u'(t) \right\| + \text{dist}\,(u(t) + hu'(t); K). \end{aligned}$$

Therefore

$$\frac{g(t+h) - g(t)}{h} \leq \gamma(h) + \frac{\text{dist}\,(u(t) + hu'(t); K) - \text{dist}\,(u(t); K)}{h},$$

where

$$\gamma(h) = \left\| \frac{u(t+h) - u(t)}{h} - u'(t) \right\|.$$

Since $\lim_{h \downarrow 0} \gamma(h) = 0$, passing to the inf-limit for $h \downarrow 0$ and taking into account that V, K and F satisfy (6.7.2), we get

$$g'(t) \leq \omega(g(t))$$

a.e. for $t \in [0, c)$. So, in view of Problem 1.8.1, $g(t) \equiv 0$ which means that $u(t) \in \overline{K} \cap D(\xi, \rho/2)$ for all $t \in [0, c)$. But $\overline{K} \cap D(\xi, \rho/2) \subseteq K \cap D(\xi, \rho)$ for each $t \in [0, c)$, and this completes the proof. □

Next, we rephrase some comparison properties we introduced in the single-valued case.

Definition 6.7.2. Let $K \subseteq D \subseteq X$ be proximal. A multi-function $F : D \rightsquigarrow X$ has *the comparison property with respect to* (D, K) if there exist a proximal neighborhood $V \subseteq D$ of K, one projection $\pi_K : V \to K$ subordinated to V, and one uniqueness function $\omega : \mathbb{R}_+ \to \mathbb{R}_+$, such that

$$\sup_{\eta \in F(\xi)} \inf_{\eta_\pi \in F(\pi_K(\xi))} [\xi - \pi_K(\xi), \eta - \eta_\pi]_+ \leq \omega(\|\xi - \pi_K(\xi)\|) \tag{6.7.3}$$

for each $\xi \in V \setminus K$.

Some notable specific cases of multi-functions obeying comparison properties with respect to (D, K) are mentioned below.

Definition 6.7.3. Let $K \subseteq D \subseteq X$ be proximal. The multi-function $F : D \rightsquigarrow X$ is called:

(i) (D, K)*-Lipschitz* if there exist a proximal neighborhood $V \subseteq D$ of K, a projection $\pi_K : V \to K$ subordinated to V, and $L > 0$, such that

$$\sup_{\eta \in F(\xi)} \inf_{\eta_\pi \in F(\pi_K(\xi))} \|\eta - \eta_\pi\| \leq L\|\xi - \pi_K(\xi)\|$$

for each $\xi \in V \setminus K$;

(ii) (D, K)*-dissipative* if there exist a proximal neighborhood $V \subseteq D$ of K, and a projection, $\pi_K : V \to K$, subordinated to V, such that

$$\sup_{\eta \in F(\xi)} \inf_{\eta_\pi \in F(\pi_K(\xi))} [\xi - \pi_K(\xi), \eta - \eta_\pi]_+ \leq 0$$

for each $\xi \in V \setminus K$.

Theorem 6.7.2. *Let X be a Banach space, let $K \subseteq D \subseteq X$, with K proximal and D open, and let $F : D \rightsquigarrow X$. If F has the comparison property with respect to (D, K), and*

$$F(\xi) \subseteq \mathfrak{T}_K(\xi) \tag{6.7.4}$$

for each $\xi \in K$, then (6.7.2) holds true.

Proof. Let $V \subseteq D$ be the open neighborhood of K given by Definition 6.7.2, let $\xi \in V$ and $\eta \in F(\xi)$. Let $\pi_K : V \to K$ be the projection subordinated to V given also by Definition 6.7.2. Let $h \in (0, T]$. Since $\|\xi - \pi_K(\xi)\| = \operatorname{dist}(\xi; K)$, we have

$$\operatorname{dist}(\xi + h\eta; K) - \operatorname{dist}(\xi; K) \leq \|\xi - \pi_K(\xi) + h(\eta - \zeta)\|$$

$$-\|\xi - \pi_K(\xi)\| + \text{dist}\,(\pi_K(\xi) + h\zeta; K),$$

for each $\zeta \in F(\pi_K(\xi))$.

Dividing by h, passing to the lim inf for $h \downarrow 0$, and using (6.7.4), we get

$$\liminf_{h\downarrow 0} \frac{1}{h}[\text{dist}\,(\xi + h\eta; K) - \text{dist}\,(\xi; K)] \leq [\xi - \pi_K(\xi), \eta - \zeta]_+.$$

Since $\zeta \in F(\pi_K(\xi))$ is arbitrary, we have

$$\liminf_{h\downarrow 0} \frac{1}{h}[\text{dist}\,(\xi + h\eta; K) - \text{dist}\,(\xi; K)] \leq \inf_{\zeta\in F(\pi_K(\xi))} [\,\xi - \pi_K(\xi), \eta - \zeta\,]_+.$$

Therefore

$$\liminf_{h\downarrow 0} \frac{1}{h}[\text{dist}\,(\xi + h\eta; K) - \text{dist}\,(\xi; K)] \leq \omega(\|\xi - \pi_K(\xi)\|).$$

But this inequality shows that (6.7.2) holds, and this completes the proof. □

By Theorems 6.7.1 and 6.7.2 we get

Corollary 6.7.1. *Let X be a Banach space, let $K \subseteq D \subseteq X$, with K proximal and locally closed and D open, and let $F : D \rightsquigarrow X$. If F has the comparison property with respect to (D, K) and, for each $\xi \in K$, $F(\xi) \subseteq \mathfrak{T}_K(\xi)$, then K is local invariant with respect to F.*

We conclude this section by showing that, in some circumstances, (6.7.4) is also necessary for the local invariance of K with respect to F. Although the result presented below can be extended to infinite dimensional Banach spaces X as well, we state and prove it only for $X = \mathbb{R}^n$ simply because this is the only case we need in the sequel.

Theorem 6.7.3. *Let $K \subseteq D \subseteq \mathbb{R}^n$, with K nonempty and D open, and let $F : D \rightsquigarrow \mathbb{R}^n$ be l.s.c. with nonempty, closed and convex values. If K is locally invariant with respect to F, then, for each $\xi \in K$, $F(\xi) \subseteq \mathfrak{T}_K(\xi)$.*

Proof. Let $\xi \in K$ and $\eta \in F(\xi)$. Since F is l.s.c. with nonempty, closed and convex values, by the Michael Continuous Selection Theorem 2.6.1, there exists a continuous function $f : D \to \mathbb{R}^n$ such that $f(\xi) = \eta$ and $f(x) \in F(x)$ for each $x \in D$. As D is open, by Peano's Local Existence Theorem, there exists at least one C^1-solution $u : [\,0, T\,] \to D$ of the equation $u'(t) = f(u(t))$ satisfying $u(0) = \xi$. Clearly $u'(0) = f(u(0)) = \eta$ and, in addition, u is an almost exact solution of the differential inclusion (6.7.1). Since $\xi \in K$ and the latter is locally invariant with respect to F, there exists $0 < a \leq T$ such that $u(t) \in K$ for all $t \in [\,0, a\,]$. Now, repeating the same arguments as in the proof of Theorem 3.1.1, we conclude that $\eta = u'(0) \in \mathfrak{F}_K(\xi) \subseteq \mathfrak{T}_K(\xi)$. The proof is complete. □

CHAPTER 7

Applications

In order to illustrate the effectiveness of the abstract developed theory, here we gather several applications. We first show that the viability of a set with respect to a function, defined on a larger open set, implies the viability of the relative closure of that set with respect to that function. We next deal with the viability of an epigraph and we prove a necessary condition in order that a given function be a comparison function. We study the existence problem of monotone solutions for both ordinary differential equations and inclusions. Next, by using viability and invariance arguments, we prove a variant of the well-known Banach Fixed Point Theorem. Further, taking advantage of the infinite-dimensional version of the Nagumo Viability Theorem, we deduce the existence of positive solutions for a pseudo-parabolic semilinear partial differential equation. We continue with the proofs of two well-known results in the classical theory of ordinary differential equations, i.e., Hukuhara and Kneser Theorems. We conclude with an application to the characteristics method for a class of first-order partial differential equations.

7.1. Viability of the relative closure

Let X be a Banach space, I a nonempty and open interval, K a nonempty subset in X, $f : I \times K \to X$ and let us consider the differential equation

$$u'(t) = f(t, u(t)). \tag{7.1.1}$$

Since each solution $u : [\tau, T) \to K$ of (7.1.1) is in fact a solution of a Cauchy problem with the initial data $u(\tau)$, whenever we speak about non-continuable or global solutions of (7.1.1) we mean noncontinuable or global solutions of the corresponding Cauchy problem in which the initial datum ξ is determined by value of the solution at τ. We begin with a simple but useful lemma.

Lemma 7.1.1. *Let X be a Banach space, let $K \subseteq X$ be nonempty and locally closed and let $f : I \times K \to X$ be continuous. If $I \times K$ is viable with respect to f, then for each $(\tau, \xi) \in I \times K$ there exist $\rho > 0$ and $T > \tau$ such* *that* $[\tau, T] \subseteq I$ *and for each* $\eta \in D(\xi, \rho) \cap K$ *each noncontinuable solution* u *of* (7.1.1), *satisfying* $u(\tau) = \eta$, *is defined at least on* $[\tau, T]$.

Proof. Let $(\tau, \xi) \in I \times K$, let $\rho > 0$ be such that $D(\xi, 3\rho) \cap K$ is closed and let $\eta \in D(\xi, \rho) \cap K$. Diminishing $\rho > 0$ if necessary, we can find $M > 0$ and $T > \tau$ such that $[\tau, T] \subseteq I$ and both inequalities

$$\|f(t, x)\| \leq M \tag{7.1.2}$$

for each $(t, x) \in [\tau, T] \times (D(\xi, 3\rho) \cap K)$ and

$$(T - \tau)M \leq \rho \tag{7.1.3}$$

are satisfied. Let O be the interior of $D(\xi, 3\rho)$. Obviously $\xi \in O \cap K$ and therefore the latter is nonempty. Since $I \times K$ is viable with respect to f and O is open, the set $I \times (O \cap K)$ enjoys the same property. So, by virtue of Theorem 3.6.1, we conclude that, for each $\eta \in O \cap K$, there exists a noncontinuable solution $u : [\tau, T_\eta) \to O \cap K$ of (7.1.1), with $u(\tau) = \eta$. We will show that, whenever $\eta \in D(\xi, \rho)$, we have $T_\eta > T$. To this aim, let us observe that, in view of (7.1.2), there exists $\lim_{t \uparrow T_\eta} u(t) = u^*$. Clearly we have $u^* \in K \cap D(\xi, 3\rho)$ and, in addition, $\|u^* - \xi\| = 3\rho$. Indeed, if we assume by contradiction that $\|u^* - \xi\| < 3\rho$, then $u^* \in O \cap K$ and, since $I \times (O \cap K)$ is viable with respect to f, u can be continued to the right of T_η which is absurd as long as u is noncontinuable.

At this point, let us assume by contradiction that $T_\eta \leq T$. Then, in view of (7.1.2) and (7.1.3), we have

$$\|u(t) - \xi\| \leq \|u(t) - \eta\| + \|\eta - \xi\| \leq (T - \tau)M + \rho \leq 2\rho$$

for each $t \in [\tau, T_\eta)$ and therefore

$$\|u^* - \xi\| \leq 2\rho < 3\rho$$

which contradicts $\|\xi - u^*\| = 3\rho$. This contradiction can be eliminated only if $T_\eta > T$ and this completes the proof. □

The "multi-valued" variant of Lemma 7.1.1 below can be proved using similar arguments.

Lemma 7.1.2. *Let* X *be a Banach space, let* $K \subseteq X$ *be nonempty and locally closed and let* $F : K \rightsquigarrow X$ *be locally bounded. If* K *is almost exact viable with respect to* F, *then for each* $\xi \in K$ *there exist* $\rho > 0$ *and* $T > 0$ *such that, for each* $\eta \in D(\xi, \rho) \cap K$, *each noncontinuable almost exact solution* u *of the differential inclusion* $u'(t) \in F(u(t))$, *satisfying* $u(0) = \eta$, *is defined at least on* $[0, T]$.

Proposition 7.1.1. *Let X be a Banach space, $D \subseteq X$ be open, let $K \subseteq D$ be locally compact and let $f : I \times D \to X$ be continuous. If $I \times K$ is viable with respect to f, and $\overline{K}^D$ is the closure of K relative to D, then $I \times \overline{K}^D$ is also viable with respect to f.*

Proof. Let $\tau \in I$ and let $(\xi_k)_k$ be an arbitrary sequence in K which is convergent to some $\xi \in \overline{K}^D$. Since $I \times K$ is viable with respect to f, there exists a sequence $(u_k)_k$ of K-valued noncontinuable solutions to (7.1.1) satisfying $u_k(\tau) = \xi_k$, for $k = 1, 2, \ldots$. In view of Lemma 7.1.1, we know that the intersection of the domains of this sequence contains a nontrivial interval $[\tau, T]$. Diminishing T if necessary, we may assume that there exists $\rho > 0$ such that $u_k(t) \in D(\xi, \rho) \subseteq D$, for all $k \in \mathbb{N}$ and $t \in [\tau, T]$. By a compactness argument involving Arzelà–Ascoli Theorem 1.3.6, we conclude that, on a subsequence at least, we have $\lim_k u_k(t) = u(t)$ uniformly on $[\tau, T]$, where u is a solution of (7.1.1) satisfying $u(\tau) = \xi$. But $u(t) \in \overline{K}^D$ for all $t \in [\tau, T]$, and this completes the proof. □

7.2. Viability of the epigraph

We will next prove a characterization of the viability of the epigraph of a certain function in terms of a differential inequality. We recall that $[D_+x](t)$ denotes the *right lower Dini derivative* of the function x at t, i.e.

$$[D_+x](t) = \liminf_{h \downarrow 0} \frac{x(t+h) - x(t)}{h}.$$

Theorem 7.2.1. *Let $\omega : I \times \mathbb{R}_+ \to \mathbb{R}$ and $v : [\tau, T) \to \mathbb{R}_+$ be continuous, with $[\tau, T) \subseteq I$. Then*

$$\mathrm{epi}(v) = \{(t, \eta);\ v(t) \leq \eta,\ t \in [\tau, T)\}$$

is viable with respect to $(t, y) \mapsto (1, \omega(t, y))$ if and only if v satisfies

$$[D_+v](t) \leq \omega(t, v(t)) \tag{7.2.1}$$

for each $t \in [\tau, T)$.

Proof. *Sufficiency.* We show that, for each $t \in [\tau, T)$, the point $(t, v(t))$, which lies on the boundary $\partial\,\mathrm{epi}(v)$ of $\mathrm{epi}(v)$[1], satisfies the Nagumo's tangency condition (3.2.3). From (7.2.1) it follows that

$$\left[D_+\left(v(\cdot) - \int_\tau^\cdot \omega(s, v(s))\,ds\right)\right](t) \leq 0$$

[1]We notice that, in general, $\{(t, v(t));\ t \in [\tau, T)\}$ does not coincide with $\partial\,\mathrm{epi}(v)$, because whenever there exists $\lim_{t \uparrow T} v(t) = v_T$, then $(T, v_T) \in \partial\,\mathrm{epi}(v)$.

for each $t \in [\tau, T)$. Thus, in view of Proposition 1.8.1, we necessarily have that $t \mapsto v(t) - \int_\tau^t \omega(s, v(s))\, ds$ is nonincreasing on $[\tau, T]$. So, for each $t \in [\tau, T)$ and $h > 0$ such that $t + h < T$, we have

$$\left(t + h, v(t) + \int_t^{t+h} \omega(s, v(s))\, ds\right) \in \operatorname{epi}(v),$$

and therefore

$$\begin{aligned}
&\operatorname{dist}\left((t, v(t)) + h(1, \omega(t, v(t))); \operatorname{epi}(v)\right) \\
&\leq \left\| (t, v(t)) + h(1, \omega(t, v(t))) - \left(t + h, v(t) + \int_t^{t+h} \omega(s, v(s))\, ds\right) \right\| \\
&= \left| h\omega(t, v(t)) - \int_t^{t+h} \omega(s, v(s))\, ds \right|.
\end{aligned}$$

Dividing by $h > 0$ and passing to lim inf for $h \downarrow 0$ we get (3.2.3) and this completes the proof of the sufficiency.

Necessity. Let us assume that $\operatorname{epi}(v)$ is viable with respect to the function $(t, y) \mapsto (1, \omega(t, y))$, let $t \in [\tau, T)$, and let (s, x) be a solution to $s' = 1$, $x' = \omega(s, x)$, satisfying the initial conditions $s(0) = t$ and $x(0) = v(t)$, and which remains in $\operatorname{epi}(v)$. We have

$$\frac{v(t + h) - v(t)}{h} \leq \frac{x(h) - x(0)}{h}.$$

Accordingly

$$[D_+ v](t) \leq \omega(s(0), x(0)) = \omega(t, v(t)),$$

and this completes the proof of the necessity. □

Remark 7.2.1. If ω is increasing with respect to the second variable, then the function v in Theorem 7.2.1 could be assumed to be merely lower semicontinuous. Indeed, it is sufficient to verify the tangency condition for points of the form $(t, v(t))$. In fact, if the tangency condition is satisfied for such points, then it is satisfied for points of the form $(t, v(t) + \lambda)$, with $\lambda > 0$. See Problem 7.2.1.

Corollary 7.2.1. *Let $\omega : I \times \mathbb{R}_+ \to \mathbb{R}_+$ be continuous with $\omega(t, 0) = 0$ for each $t \in I$, and such that, for each $\tau \in I$, the Cauchy problem*

$$\begin{cases} y'(t) = \omega(t, y(t)), \\ y(\tau) = 0 \end{cases} \tag{7.2.2}$$

has only the null solution. Then ω is a comparison function.

Proof. Let $x : [\tau, T) \to \mathbb{R}_+$ be any continuous solution of the problem (7.2.1). By Theorem 7.2.1, epi(x) is viable with respect to the function $(t, y) \mapsto (1, \omega(t, y))$. So, the unique solution $y : [\tau, \sup I) \to \mathbb{R}_+$ of the Cauchy problem (7.2.2) satisfies $0 \leq x(t) \leq y(t) = 0$ for each $t \in [\tau, T)$. $\square$

Problem 7.2.1. *Let the function $\omega : I \times \mathbb{R}_+ \to \mathbb{R}$, be increasing in the second variable. Let $v : [\tau, T) \to \mathbb{R}$, with $[\tau, T) \subseteq I$ and let $\lambda > 0$. Let us assume that for some $t \in [\tau, T)$ we have*

$$(1, \omega(t, v(t))) \in \mathcal{T}_{\mathrm{epi}(v)}(t, v(t)).$$

Show that

$$(1, \omega(t, v(t) + \lambda)) \in \mathcal{T}_{\mathrm{epi}(v)}(t, v(t) + \lambda).$$

Problem 7.2.2. *Let $f : [0, T] \to \mathbb{R}$ be lower semicontinuous and such that the function $t \mapsto t + f(t)$ is increasing. Suppose further that there exists $M \in \mathbb{R}_+$ such that for each $t \in [0, T)$, $[D_+ f](t) \leq M$. Show that f is Lipschitz.*

7.3. Monotone solutions

Let X be a Banach space, $K \subseteq X$ nonempty, and let $\preceq \subseteq K \times K$ be a *preorder* on K, i.e., a reflexive and transitive binary relation. For our later purposes, it is convenient to identify such a relation with the multi-function $\mathcal{P} : K \rightsquigarrow K$, defined by

$$\mathcal{P}(\xi) = \{\eta \in K;\ \xi \preceq \eta\}$$

for each $\xi \in K$, and called also a preorder. The preorder $\preceq \subseteq K \times K$, or $\mathcal{P}$ is *closed* if $\preceq \subseteq K \times K$ is a closed subset in $X \times X$. Let $f : I \times K \to X$. We say that $\preceq \subseteq K \times K$, or $I \times \mathcal{P}$, is *viable with respect to* f if, for each $(\tau, \xi) \in I \times K$, there exist $[\tau, T] \subseteq I$ and a solution $u : [\tau, T] \to X$ of (7.1.1) satisfying $u(\tau) = \xi$, $u(t) \in K$ for each $t \in [\tau, T]$ and u is $\preceq$-monotone on $[\tau, T]$, i.e., for each $\tau \leq s \leq t \leq T$, we have $u(s) \preceq u(t)$. The next lemma is the main tool in our forthcoming analysis.

Lemma 7.3.1. *Let X be a Banach space, K be locally compact in X, let $f : I \times K \to X$ be continuous and let $\mathcal{P}$ be a preorder on K. If $I \times \mathcal{P}$ is viable with respect to f then, for each $\xi \in K$, $I \times \mathcal{P}(\xi)$ is viable with respect to f. If $\mathcal{P}$ is closed in $X \times X$ and, for each $\xi \in K$, $I \times \mathcal{P}(\xi)$ is viable with respect to f, then $I \times \mathcal{P}$ is viable with respect to f.*

Proof. Clearly, if $I \times \mathcal{P}$ is viable with respect to f, then, for all $\xi \in K$, $I \times \mathcal{P}(\xi)$ is viable with respect to f.

Now, if $\mathcal{P}$ is closed, then, for each $\xi \in K$, $\mathcal{P}(\xi)$ is a fortiori closed. Let us assume that, for each $\xi \in K$, $I \times \mathcal{P}(\xi)$ is viable with respect to

f. Let $(\tau, \xi) \in I \times K$. We shall show that there exist $[\tau, T] \subseteq I$ and at least one solution $u : [\tau, T] \to K$ of (7.1.1), with $u(\tau) = \xi$ and such that $u([s, T]) \subseteq \mathcal{P}(u(s))$ for each $s \in [\tau, T]$. To this aim, we proceed in several steps.

In the first step, we note that, by Lemma 7.1.1, there exists $T > \tau$, $T \in I$, such that for every noncontinuable solution $u : [\tau, \widetilde{T}) \to K$ to (7.1.1) with $u(\tau) = \xi$ we have $T < \widetilde{T}$. Since $\mathcal{P}(\xi)$ is viable with respect to f, there exists a solution $u : [\tau, T] \to K$ of (7.1.1) with $u(\tau) = \xi$ and $u([\tau, T]) \subseteq \mathcal{P}(\xi)$.

In the second step, we remark that, for every solution $v : [\tau, T] \to K$ to (7.1.1), with $v(\tau) = \xi$ and $v([\tau, T]) \subseteq \mathcal{P}(\xi)$, and for every $\nu \in [\tau, T)$, there exists a solution $w : [\tau, T] \to K$ to (7.1.1) such that w equals v on $[\tau, \nu]$ and $w([\nu, T]) \subseteq \mathcal{P}(w(\nu))$.

In the third step, we observe that, thanks to the first two steps, for every nonempty and finite subset S of $[\tau, T)$, with $\tau \in S$, there exists a solution $u : [\tau, T] \to K$ of (7.1.1) satisfying both $u(\tau) = \xi$ and $u([s, T]) \subseteq \mathcal{P}(u(s))$ for all $s \in S$.

In the fourth step, we consider a sequence $(S_k)_{k \in \mathbb{N}}$ of nonempty finite subsets of $[\tau, T)$ such that $\tau \in S_k$, $S_k \subseteq S_{k+1}$ for each $k \in \mathbb{N}$, and the set $S = \cup_{k \in \mathbb{N}} S_k$ is dense in $[\tau, T]$. For example, we can take

$$S_k = \{\tau + (i/2^k)(T - \tau);\ i = 0, 1, ..., 2^k - 1\}.$$

Further, we shall make use of the third step to get a sequence of solutions $(u_k : [\tau, T] \to K)_k$ to (7.1.1), satisfying $u_k(\tau) = \xi$ and such that $u_k([s, T]) \subseteq \mathcal{P}(u_k(s))$ for each $k \in \mathbb{N}$ and each $s \in S_k$. Now, by virtue of the Arzelà–Ascoli Theorem 1.3.6, we can assume, extracting a subsequence if necessary, that the sequence $(u_k)_k$ converges uniformly on $[\tau, T]$ to a solution $u : [\tau, T] \to K$ of (7.1.1). Clearly $u(\tau) = \xi$.

In the fifth step, we show that $u([s, T]) \subseteq \mathcal{P}(u(s))$ for all $s \in S$. Indeed, given s as above, there exists $k \in \mathbb{N}$ such that $s \in S_k$. Then $s \in S_m$ and $u_m([s, T]) \subseteq \mathcal{P}(u_m(s))$ for all $m \in \mathbb{N}$ with $k \leq m$. At this point, the closedness of the graph of $\mathcal{P}$ shows that $u([s, T]) \subseteq \mathcal{P}(u(s))$.

In the sixth and final step, taking into account that S is dense in $[\tau, T]$, u is continuous on $[\tau, T]$ and the graph of $\mathcal{P}$ is closed, we conclude that the preceding relation holds for every $s \in [\tau, T]$ and this completes the proof. □

Theorem 7.3.1. *Let X be a Banach space, K be locally compact and let $\mathcal{P}$ be a closed preorder on K. Let $f : I \times K \to X$ be continuous. Then, a necessary and sufficient condition in order that $I \times \mathcal{P}$ be viable with respect to f is that $f(t, \xi) \in \mathcal{T}_{\mathcal{P}(\xi)}(\xi)$ for each $(t, \xi) \in I \times K$.*

Proof. Just apply Lemma 7.3.1, Theorem 3.2.4 and Remark 3.2.1. □

Problem 7.3.1. *Let $w : K \to \mathbb{R}$ be continuous. Suppose that for every $(\tau, \xi) \in I \times K$ there exists a solution $u : [\tau, T] \to K$ of (7.1.1) satisfying $u(\tau) = \xi$ such that $w(u(t)) \leq w(\xi)$ for every $t \in [\tau, T]$. Show that, for every $(\tau, \xi) \in I \times K$ there exists a solution $u : [\tau, T] \to K$ of (7.1.1) satisfying $u(\tau) = \xi$ and such that the function $t \mapsto w(u(t))$ is nondecreasing.*

We conclude this section with some remarks referring to the multi-valued case. Let $F : K \rightsquigarrow X$, and let us consider the differential inclusion (6.1.1). The preorder $\preceq \subseteq K \times K$, or $\mathcal{P}$, is *almost exact viable with respect to F* if for each $\xi \in K$, there exist $T > 0$ and a solution $u : [0, T] \to K$ of (6.1.1) such that u is $\preceq$-monotone on $[0, T]$, i.e., for each $0 \leq s \leq t \leq T$, we have $u(s) \preceq u(t)$.

Using Lemma 7.1.2 and similar arguments as those in the proof of Lemma 7.3.1, we get the following variant of the latter.

Lemma 7.3.2. *Let K be locally compact in X, let $F : K \rightsquigarrow X$ be u.s.c. with convex closed and bounded values. Let $\mathcal{P}$ be a preorder on K. If $\mathcal{P}$ is almost exact viable with respect to F then, for each $\xi \in K$, $\mathcal{P}(\xi)$ is almost exact viable with respect to F. If $\mathcal{P}$ is closed in $X \times X$ and, for each $\xi \in K$, $\mathcal{P}(\xi)$ is almost exact viable with respect to F, then $\mathcal{P}$ is almost exact viable with respect to F.*

7.4. A Banach-type fixed point theorem

We present here a simple extension of Banach fixed point theorem in the frame of closed subsets of a Banach space.

Theorem 7.4.1. *Let X be a Banach space, K a nonempty and closed subset in X and $g : K \to X$ a Lipschitz function with Lipschitz constant $L < 1$. If*

$$\liminf_{h \downarrow 0} \frac{1}{h} \operatorname{dist}\left(\xi + h(g(\xi) - \xi); K\right) = 0$$

for each $\xi \in K$, then g has a unique fixed point.

Proof. Let $f : K \to X$ be defined by $f(x) = g(x) - x$ for each $x \in K$. In view of Theorem 3.2.3, K is viable with respect to f. Therefore, by (ii) in Remark 3.6.2 and Theorem 3.6.3, for each $\xi \in K$, there exists a unique global solution $u(\cdot, \xi) : \mathbb{R}_+ \to K$ of the Cauchy problem

$$\begin{cases} u'(t) = f(u(t)) \\ u(0) = \xi. \end{cases}$$

Fix $T > 0$ and let us define $Q : K \to K$, by $Q\xi = u(T, \xi)$. Multiplying both sides the equality

$$u'(t,\xi) - u'(t,\eta) = g(u(t,\xi)) - g(u(t,\eta)) - u(t,\xi) + u(t,\eta)$$

by e^t, we get successively

$$\frac{d}{dt}\left\{e^t\left[u(t,\xi) - u(t,\eta)\right]\right\} = e^t[g(u(t,\xi)) - g(u(t,\eta))]$$

$$e^t\|u(t,\xi) - u(t,\eta)\| \leq \|\xi - \eta\| + \int_0^t e^s L\|u(s,\xi) - u(s,\eta)\|\, ds$$

for each $t \in \mathbb{R}_+$. From Gronwall Lemma 1.8.4 we deduce

$$\|Q\xi - Q\eta\| \leq e^{(L-1)T}\|\xi - \eta\|$$

and thus Q is contraction. By Banach Fixed Point Theorem, it follows that there exists $\xi \in K$ such that $Q\xi = \xi$. This means that $u(T,\xi) = u(0,\xi)$. Since the equation is autonomous, it follows that $u(\cdot,\xi)$ is a T-periodic solution of $u'(t) = g(u(t)) - u(t)$. So, $u(t,\xi) = u(T + t,\xi)$ for each $t \in \mathbb{R}_+$, or equivalently, $u(t,\xi)$ is a fixed point of Q. As Q has exactly one fixed point, we conclude that $u(t,\xi) = \xi$ each $t \in \mathbb{R}_+$. Consequently $0 = g(\xi) - \xi$ and this completes the proof. □

7.5. Positive solutions to pseudoparabolic PDEs

Let Ω be a nonempty, bounded and open subset in $\mathbb{R}^3$, with smooth boundary Γ and let us consider the following semilinear pseudoparabolic initial-boundary-value problem

$$\begin{cases} u_t = \Delta u_t + \Delta u + g(u) & (t,x) \in Q_T \\ u = 0 & (t,x) \in \Sigma_T \\ u(0,x) = \eta(x) & x \in \Omega. \end{cases} \tag{7.5.1}$$

Here and thereafter $Q_T = (0,T) \times \Omega$ and $\Sigma_T = (0,T) \times \Gamma$.

Theorem 7.5.1. *Let $g : \mathbb{R} \to \mathbb{R}$ be continuous and let us assume that*

$$u + g(u) \geq 0 \tag{7.5.2}$$

for each $u \in \mathbb{R}_+$. Then, for each $\eta \in H^2(\Omega) \cap H_0^1(\Omega)$ with $\eta - \Delta\eta \geq 0$ a.e. on Ω, there exists $T > 0$ such that the problem (7.5.1) has at least one solution $u \in C^1([0,T]; H^2(\Omega) \cap H_0^1(\Omega))$ satisfying both $u(t) - \Delta u(t) \geq 0$ and $u(t) \geq 0$ for each $t \in [0,T]$ and a.e. on Ω. If, in addition, g is positively sublinear, then each solution of (7.5.1) can be continued up to a global one.

Proof. Clearly (7.5.1) is equivalent to

$$\begin{cases} (u-\Delta u)_t = \Delta u + g(u) & (t,x) \in Q_T \\ u = 0 & (t,x) \in \Sigma_T \\ u(0,x) = \eta(x) & x \in \Omega. \end{cases} \tag{7.5.3}$$

Let $A : D(A) \subseteq L^2(\Omega) \to L^2(\Omega)$ be defined by

$$\begin{cases} D(A) = H_0^1(\Omega) \cap H^2(\Omega) \\ Au = \Delta u \text{ for each } u \in D(A), \end{cases}$$

let $J = (I - A)^{-1}$ and let us denote by $u = Jv$. Since $AJ = J - I$, (7.5.3) can be rewritten as an abstract differential equation in the space $L^2(\Omega)$, i.e.

$$\begin{cases} v' = f_1(v) + f_2(v) \\ v(0) = \xi, \end{cases} \tag{7.5.4}$$

where $f_1 : L^2(\Omega) \to L^2(\Omega)$ is defined by

$$f_1(v)(x) = g((Jv)(x))$$

for each $v \in L^2(\Omega)$ and a.e. for $x \in \Omega$, $f_2 = J - I$ and $\xi = (I - A)\eta$.

Since the operator J is continuous from $L^2(\Omega)$ to $H^2(\Omega) \cap H_0^1(\Omega)$, and, in our specific case, i.e. $n = 3$, thanks to (iii) in Theorem 1.7.1, $H^2(\Omega)$ is compactly imbedded in $C(\overline{\Omega})$, it follows that f_1 is well-defined and locally compact (in fact compact). Moreover, f_2 is obviously Lipschitz, being linear continuous.

At this point, let us recall that $\xi = (I - A)\eta \geq 0$ a.e. on Ω. Let

$$K = \{v \in L^2(\Omega)\,;\ v \geq 0 \text{ a.e. on } \Omega\}.$$

In view of (7.5.2), we deduce that $f = f_1 + f_2$ and K satisfy the tangency condition (3.2.3). Indeed, to prove that for each $\xi \in K$

$$\liminf_{s \downarrow 0} \frac{1}{s} \text{dist}\,(\xi + sf_1(\xi) + sf_2(\xi); K) = 0,$$

it suffices to show that, for each $s \in (0,1)$, we have

$$\xi + s(J\xi - \xi) + sg(J\xi) \geq 0 \tag{7.5.5}$$

a.e. on Ω. But this is certainly the case, because $\xi \geq 0$ a.e. on Ω, along with Theorem 1.7.4, implies both $(1-s)\xi \geq 0$ and $J\xi \geq 0$ a.e. on Ω. By virtue of (7.5.2), it follows that $J\xi + g(J\xi) \geq 0$ and thus (7.5.5) holds. In addition, thanks to Remark 3.2.3, f is locally β-compact. So, we are in the hypotheses of Theorem 3.2.2. Therefore K is viable with respect to $f_1 + f_2$. Hence there exists at least one solution, $v : [0,T] \to L^2(\Omega)$, of (7.5.4), with $v(t) \in K$ for each $t \in [0,T]$. But $u(t) = Jv(t)$, and consequently $u(t) - \Delta u(t) \geq 0$ for each $t \in [0,T]$ and a.e. on Ω. Using once again

Theorem 1.7.4, we conclude that $u(t) \geq 0$ for each $t \in [0, T]$ and a.e. on Ω. Since $v \in C^1([0, T]; L^2(\Omega))$, and $u = Jv$, with J linear continuous from $L^2(\Omega)$ to $H^2(\Omega) \cap H_0^1(\Omega)$, we conclude that $u \in C^1([0, T]; H^2(\Omega) \cap H_0^1(\Omega))$, and the proof is complete. □

7.6. Hukuhara Theorem

In this section, by using viability and invariance techniques, we will prove a famous theorem, in ordinary differential equations, due to Hukuhara. To fix the ideas, we first notice that, until now, we focused our attention merely on concepts of viability and invariance which refer to solutions starting from a point τ and defined only on right intervals of the form $[\tau, T]$. Therefore, within this section, we refer to these concepts as to *right viability* and *right invariance*. Similarly, if we are interested in working with solutions starting from a point τ and defined only on left intervals of the form $[T, \tau]$, we will speak about *left viability* and *left invariance*.

Throughout this section $X = \mathbb{R}^n$, $D \subseteq \mathbb{R}^n$ is a nonempty and open subset and I is a nonempty and open interval. Let $f : I \times D \to \mathbb{R}^n$ be a continuous function, and let us consider the nonautonomous differential equation

$$u'(t) = f(t, u(t)). \tag{7.6.1}$$

We begin with some simple but useful propositions. In order to do this, we need to consider the autonomous differential equation, i.e.

$$u'(t) = g(u(t)). \tag{7.6.2}$$

Proposition 7.6.1. *Let $D \subseteq \mathbb{R}^n$ be a nonempty and open set and let $g : D \to \mathbb{R}^n$. Then $K \subseteq D$ is right (left) viable with respect to g if and only if it is the union of a certain family of right (left) trajectories of (7.6.2).*

Proposition 7.6.2. *Let $D \subseteq \mathbb{R}^n$ be a nonempty and open set and let $g : D \to \mathbb{R}^n$. If K is the union of all right (left) trajectories of (7.6.2) in D, issuing from a given subset $C \subseteq D$, then K is locally right (left) invariant with respect to g. In particular, the subset $K \subseteq D$ is locally right (left) invariant with respect to the continuous function g if and only if it is the union of all right (left) trajectories of (7.6.2) issuing from K.*

Proposition 7.6.3. *Let I be a nonempty and open interval, $D \subseteq \mathbb{R}^n$ a nonempty and open set and let $f : I \times D \to \mathbb{R}^n$. The subset $K \subseteq D$ is locally right invariant with respect to f if and only if $D \setminus K$ is locally left invariant with respect to f.*

Let $(\tau, \xi) \in I \times D$, and let us denote by $\mathcal{S}(\tau, \xi)$ the set of all noncontinuable solutions u of (7.6.1) satisfying $u(\tau) = \xi$.

Definition 7.6.1. The *right solution funnel* through $(\tau, \xi) \in I \times D$, $F_{\tau,\xi}$, is defined by

$$F_{\tau,\xi} = \{(s, u(s))\,;\ s \geq \tau,\ u \in \mathcal{S}(\tau, \xi)\}.$$

If $t \geq \tau$, we define the *t-cross section of* $F_{\tau,\xi}$ by

$$F_{\tau,\xi}(t) = \{u(t)\,;\ u \in \mathcal{S}(\tau, \xi)\}.$$

The next compactness result will prove useful in all that follows.

Proposition 7.6.4. *Let I be a nonempty and open interval, $D \subseteq \mathbb{R}^n$ a nonempty and open set and let $f : I \times D \to \mathbb{R}^n$ be a continuous function. Let $(\tau, \xi) \in I \times D$ and let $t > \tau$ be such that, for each $u \in \mathcal{S}(\tau, \xi)$, $u(t)$ is defined. Then*

$$F_{\tau,\xi}([\,\tau, t\,]) = \{(s, u(s))\,;\ s \in [\,\tau, t\,],\ u \in \mathcal{S}(\tau, \xi)\}$$

is compact.

Proof. By Arzelà–Ascoli Theorem 1.3.6, it follows that the restriction of $\mathcal{S}(\tau, \xi)$ to $[\,\tau, t\,]$ is relatively compact in $C([\,\tau, t\,]; \mathbb{R}^n)$. Let $((t_m, u_m(t_m)))_m$ be an arbitrary sequence in $F_{\tau,\xi}([\,\tau, t\,])$, with $u_m : (a_m, b_m) \to D$ for each $m \in \mathbb{N}$. We may assume with no loss of generality that $\lim_m t_m = s$. Obviously, $s \in [\,\tau, t\,] \subseteq (a_m, b_m)$ for every $m \in \mathbb{N}$, and therefore, by the remark above, there exists at least one subsequence of $(u_m)_m$, denoted for simplicity again by $(u_m)_m$, and $u \in \mathcal{S}(\tau, \xi)$, with $\lim_m u_m = u$ uniformly on $[\,\tau, t\,]$. But this shows that $\lim_m (t_m, u_m(t_m)) = (s, u(s))$. To complete the proof we have merely to observe that $(s, u(s)) \in F_{\tau,\xi}([\,\tau, t\,])$. □

From Proposition 7.6.4 we deduce

Corollary 7.6.1. *Let I be a nonempty and open interval, $D \subseteq \mathbb{R}^n$ a nonempty and open set and let $f : I \times D \to \mathbb{R}^n$ be a continuous function. Then, for each $(\tau, \xi) \in I \times K$, the set $F_{\tau,\xi}$ is locally closed.*

Throughout $\partial F_{\tau,\xi}(t)$ denotes the boundary of $F_{\tau,\xi}(t)$. We are now ready to state Hukuhara Theorem, i.e.,

Theorem 7.6.1. *Let I be a nonempty and open interval, D a nonempty and open subset in $\mathbb{R}^n$, $(\tau, \xi) \in I \times D$, $f : I \times D \to \mathbb{R}^n$ a continuous function and let $t > \tau$ be such that, for each $u \in \mathcal{S}(\tau, \xi)$, $u(t)$ is defined. Then, for each $\eta \in \partial F_{\tau,\xi}(t)$ there exists a solution v of* (7.6.1) *with $v(\tau) = \eta$ and such that $v(s) \in \partial F_{\tau,\xi}(s)$ for all $s \in [\,\tau, t\,]$.*

In order to prove Theorem 7.6.1 we need

Theorem 7.6.2. *If I is a nonempty and open interval, D a nonempty and open subset in $\mathbb{R}^n$, $f : I \times D \to \mathbb{R}^n$ a continuous function, $K_1, K_2 \subseteq D$ are locally closed and viable with respect to f, and if $K_1 \cup K_2 = D$, then $K_1 \cap K_2$ is viable with respect to$f$.*

Proof. The conclusion is a consequence of Theorem 2.4.1 combined with Theorem 3.5.5. □

Remark 7.6.1. A result similar to Theorem 7.6.2 holds true trivially in the case of local invariance. More precisely, if $D \subseteq \mathbb{R}^n$ is open, $K_1, K_2 \subseteq D$ are locally closed and locally invariant with respect to f, then $K_1 \cap K_2$ is locally invariant with respect to f.

Let $\tau \in I$ be fixed, let us denote by $\mathcal{D} = \{s \in I;\ s > \tau\} \times D$, and let us define $\mathcal{F} : \mathcal{D} \to \mathbb{R} \times \mathbb{R}^n$, by $\mathcal{F}(t, \xi) = (1, f(t, \xi))$ for each $(t, \xi) \in \mathcal{D}$. Throughout, we denote by $\partial_{\mathcal{D}} F_{\tau,\xi}$ the boundary of $F_{\tau,\xi}$ relative to $\mathcal{D}$, i.e., $\partial_{\mathcal{D}} F_{\tau,\xi} = \overline{(\mathcal{D} \setminus F_{\tau,\xi})}^{\mathcal{D}} \cap \overline{\mathcal{D} \cap F_{\tau,\xi}}^{\mathcal{D}}$. We will deduce Theorem 7.6.1 from a slightly more general result, i.e., Theorem 7.6.3 below.

Theorem 7.6.3. *Let I be a nonempty and open interval, D a nonempty and open subset in $\mathbb{R}^n$ and let $f : I \times D \to \mathbb{R}^n$ be a continuous function. Then, for each $(\tau, \xi) \in I \times D$, the set $\partial_{\mathcal{D}} F_{\tau,\xi}$ is left viable with respect to $\mathcal{F}$.*

Proof. Let us observe that (7.6.1) can be equivalently written as

$$w'(t) = \mathcal{F}(w(t)),$$

where $\mathcal{F}$ is defined as above, and $w = (s, u)$. By the definition of $F_{\tau,\xi}$, we easily deduce that $\mathcal{D} \cap F_{\tau,\xi}$ is right viable and right locally invariant with respect to $\mathcal{F}$, and hence, by Propositions 7.6.1, 7.6.2, 7.6.3, it follows that $\mathcal{D} \setminus F_{\tau,\xi}$ is both left viable and left locally invariant with respect to $\mathcal{F}$. So, thanks to Proposition 7.1.1, we conclude that $\mathcal{K}_1 = \overline{\mathcal{D} \setminus F_{\tau,\xi}}^{\mathcal{D}}$ is left viable with respect to $\mathcal{F}$. Further, also by definition, $\mathcal{D} \cap F_{\tau,\xi}$ is left viable with respect to $\mathcal{F}$, and again by Proposition 7.1.1, it follows that $\mathcal{K}_2 = \overline{\mathcal{D} \cap F_{\tau,\xi}}^{\mathcal{D}}$ is left viable with respect to $\mathcal{F}$. Since $\mathcal{K}_1 \cup \mathcal{K}_2 = \mathcal{D}$, by Problem 2.2.1 and Theorem 7.6.2, we conclude that $\mathcal{K}_1 \cap \mathcal{K}_2 = \partial_{\mathcal{D}} F_{\tau,\xi}$ is left viable with respect to $\mathcal{F}$, and this completes the proof. □

We may now proceed to the proof of Theorem 7.6.1.

Proof. First, let us observe that thanks to Proposition 7.6.4, it follows that $\partial_{\mathcal{D}} F_{\tau,\xi} \subseteq F_{\tau,\xi}$. Hence, in view of Theorem 7.6.3, we know that, for each $(t, u(t)) \in \partial_{\mathcal{D}} F_{\tau,\xi}$, there exists at least one solution $v : [\theta, t] \to D$, with $\tau \leq \theta < t$, $v(t) = u(t)$ and such that $(s, v(s)) \in \partial_{\mathcal{D}} F_{\tau,\xi}(s)$ for each $s \in [\theta, t]$. But, by virtue of Proposition 7.6.4, $F_{\tau,\xi}([\tau, t])$ is compact, and therefore a simple maximality argument shows that we can always extend such a solution to $[\tau, t]$, and this completes the proof. $\square$

Remark 7.6.2. Theorem 7.6.3 implies that, for each $(\widetilde{t}, \widetilde{y}) \in \partial_{\mathcal{D}} F_{\tau,\xi}$, there exists at least one noncontinuable solution, $v(\cdot) : (\sigma, \widetilde{t}\,] \to \mathbb{R}^n$, of (7.6.1), such that $(s, v(s)) \in \partial_{\mathcal{D}} F_{\tau,\xi}$ for all $s \in (\sigma, \widetilde{t}\,] \cap [\,\tau, \widetilde{t}\,]$. If $t > \tau$ is chosen as in Theorem 7.6.1 and $\widetilde{t} \in [\,\tau, t\,]$, then by virtue of both Theorem 7.6.3 and Proposition 7.6.4, it follows that $[\,\tau, \widetilde{t}\,] \subset (\sigma, \widetilde{t}\,]$.

Remark 7.6.3. We notice that

$$\partial F_{\tau,\xi}(t) \subseteq (\partial_{\mathcal{D}} F_{\tau,\xi})(t) = \{v \in X\,;\; (t, v) \in \partial_{\mathcal{D}} F_{\tau,\xi}\}$$

and the inclusion can be strict. So, Theorem 7.6.3 is more general than Theorem 7.6.1 because it considers all of $\partial_{\mathcal{D}} F_{\tau,\xi}$.

7.7. Kneser Theorem

The goal of this section is to give a proof, based on viability techniques, to the celebrated Theorem of Kneser. Namely, with the notations in the preceding section, we have

Theorem 7.7.1. *Let I be a nonempty and open interval, D a nonempty and open subset in $\mathbb{R}^n$, $f : I \times D \to \mathbb{R}^n$ a continuous function, $(\tau, \xi) \in I \times D$, and let $t > \tau$ be such that, for each $u \in \mathcal{S}(\tau, \xi)$, $u(t)$ is defined. Then $F_{\tau,\xi}(t)$ is connected.*

Proof. Let us assume by contradiction that $F_{\tau,\xi}(t)$ is not connected. Then there exist two nonempty subsets C_1, C_2 with $F_{\tau,\xi}(t) = C_1 \cup C_2$ but $C_1 \cap \overline{C}_2 = \overline{C}_1 \cap C_2 = \emptyset$. Let K_1 be the union of the graphs of all right noncontinuable solutions v which, either are not defined at t or, if defined, satisfy

$$\operatorname{dist}(v(t); C_1) \le \operatorname{dist}(v(t); C_2). \tag{7.7.1}$$

Similarly, we define K_2 by reversing the inequality (7.7.1). By virtue of Proposition 7.6.1, both K_1 and K_2 are right viable with respect to $\mathcal{F}$. In addition, K_1 and K_2 are locally closed in $I \times D$ and $K_1 \cup K_2 = I \times D$. So, by Proposition 7.1.1, it follows that both $\overline{K}_1^{\mathcal{D}}$ and $\overline{K}_2^{\mathcal{D}}$ are right viable with respect to $\mathcal{F}$ and, of course, locally closed. In view of Theorem 7.6.2, we conclude that $K = \overline{K}_1^{\mathcal{D}} \cap \overline{K}_2^{\mathcal{D}}$ is right viable with respect to $\mathcal{F}$. As $(\tau, \xi) \in K$, there exists a noncontinuable solution v such that $(s, v(s)) \in K$ for each s in the domain of v. By the choice of t, $v(t)$ is defined and belongs to $C_1 \cup C_2$. Since $(t, v(t)) \in K$, $\operatorname{dist}(v(t); C_1) = \operatorname{dist}(v(t); C_2)$ which must be 0. But this is absurd because $v(t)$ would be either in $C_1 \cap \overline{C}_2$, or in

$\overline{C}_1 \cap C_2$. This contradiction can be eliminated only if $F_{\tau,\xi}(t)$ is connected as claimed. □

7.8. Lyapunov functions

In this section we will show how viability can be used in order to obtain sufficient conditions for asymptotic stability via Lyapunov functions. Let us consider the autonomous differential equation

$$u'(t) = g(u(t)), \tag{7.8.1}$$

where $g : \mathbb{R}^n \to \mathbb{R}^n$ is continuous. Let $V : \mathbb{R}^n \to \mathbb{R}$, and let us define

$$V^*(\xi) = \limsup_{h \downarrow 0} \frac{1}{h}[V(\xi + hg(\xi)) - V(\xi)]$$

and

$$V_*(\xi) = \liminf_{h \downarrow 0} \frac{1}{h}[V(\xi + hg(\xi)) - V(\xi)].$$

If $V^*(\xi) = V_*(\xi)$, we denote this common value by $\dot{V}(\xi)$ and we note that, if V is differentiable, we have $\dot{V}(\xi) = \langle \operatorname{grad} V(\xi), g(\xi) \rangle$. The next result will prove useful in the sequel.

Theorem 7.8.1. *If $V : \mathbb{R}^n \to \mathbb{R}$ is locally Lipschitz, $u : [0, T) \to \mathbb{R}^n$ is any solution to (7.8.1) and $t \in [0, T)$, then*

$$V^*(u(t)) = \limsup_{h \downarrow 0} \frac{1}{h}[V(u(t + h)) - V(u(t))].$$

Proof. We have

$$V^*(u(t)) = \limsup_{h \downarrow 0} \frac{1}{h}[V(u(t) + hg(u(t))) - V(u(t))]$$

$$\leq \limsup_{h \downarrow 0} \frac{1}{h}\left[V\left(u(t) + \int_t^{t+h} g(u(s))\,ds\right) - V(u(t))\right]$$

$$+ \limsup_{h \downarrow 0} \frac{1}{h}\left[-V\left(u(t) + \int_t^{t+h} g(u(s))\,ds\right) + V(u(t) + hg(u(t)))\right].$$

Since

$$\left|\limsup_{h \downarrow 0} \frac{1}{h}\left[-V\left(u(t) + \int_t^{t+h} g(u(s))\,ds\right) + V(u(t) + hg(u(t)))\right]\right|$$

$$\leq L \limsup_{h \downarrow 0} \frac{1}{h}\left\|\int_t^{t+h} g(u(s))\,ds - hg(u(t))\right\| = 0,$$

where $L > 0$ is the Lipschitz constant of V on a suitably chosen neighborhood of $u(t)$, we deduce

$$V^*(u(t)) \leq \limsup_{h\downarrow 0} \frac{1}{h}[V(u(t+h)) - V(u(t))].$$

Similarly, we get

$$V^*(u(t)) \geq \limsup_{h\downarrow 0} \frac{1}{h}[V(u(t+h)) - V(u(t))]$$

and this completes the proof. □

Definition 7.8.1. We say that 0 is *stable* for (7.8.1) if for each $\varepsilon > 0$ there exists $\delta(\varepsilon) \in (0, \varepsilon)$ such that for each $\xi \in \mathbb{R}^n$ with $\|\xi\| \leq \delta(\varepsilon)$, each solution u of (7.8.1), satisfying $u(0) = \xi$, is defined on $[0, +\infty)$, and $\|u(t)\| \leq \varepsilon$ for all $t \geq 0$.

Clearly if 0 is stable for (7.8.1), $\{0\}$ is both viable and locally invariant with respect to g, and thus $g(0) = 0$. In other words, if 0 is stable for (7.8.1), then $u \equiv 0$ is necessarily a solution to (7.8.1), and this is the only one issuing from 0.

Definition 7.8.2. We say that V has positive gradient at $v \in \mathbb{R}$ if

$$\liminf_{\substack{\mathrm{dist}\,(\xi;K)\downarrow 0\\ \xi\notin K}} \frac{V(\xi) - v}{\mathrm{dist}\,(\xi; K)} > 0$$

where $K = V^{-1}((-\infty, v])$.

If V is of class C^1 and $\|\mathrm{grad}\, V\|$ is bounded from below on $V^{-1}(\{v\})$ by a constant $c > 0$, then V has positive gradient at v.

Proposition 7.8.1. *Let $g : \mathbb{R}^n \to \mathbb{R}^n$, let $V : \mathbb{R}^n \to \mathbb{R}$ be continuous, $v \in \mathbb{R}$ and let $K = V^{-1}((-\infty, v])$. If V has positive gradient at v and $V^*(\xi) \leq 0$ whenever $V(\xi) = v$, then K is viable with respect to g.*

Proof. Let us assume by contradiction that K is not viable with respect to g. In view of Theorem 3.2.5, this means that there exists $\xi \in \partial K$ such that $g(\xi)$ is not tangent in the sense of Federer to K at ξ. So, there exist $\gamma > 0$ and a sequence $h_m \downarrow 0$ such that

$$\gamma < \frac{1}{h_m}\mathrm{dist}\,(\xi + h_m g(\xi); K)$$

for all $m \in \mathbb{N}$. As $\xi \in \partial K$, it follows that $V(\xi) = v$. Furthermore, since V has positive gradient at v, there exists $\nu > 0$ such that

$$\nu < \frac{V(\xi + h_m g(\xi)) - v}{\mathrm{dist}\,(\xi + h_m g(\xi); K)} < \frac{1}{h_m \gamma}[V(\xi + h_m g(\xi)) - v]$$

for each $m \in \mathbb{N}$. Hence

$$0 < \nu \leq \frac{1}{\gamma} \limsup_{h \downarrow 0} \frac{V(\xi + hg(\xi)) - V(\xi)}{h} = \frac{1}{\gamma} V^*(\xi),$$

thereby contradicting the hypothesis that $V^*(\xi) \leq 0$. This contradiction can be eliminated only if K is viable with respect to g, and this completes the proof. □

Theorem 7.8.2. *Let us assume that* (7.8.1) *has the uniqueness property. Then* 0 *is stable for* (7.8.1) *if and only if there exist a continuous function* $V : \mathbb{R}^n \to [0, +\infty)$ *and a sequence* $v_m \downarrow 0$ *such that* $V(x) = 0$ *if and only if* $x = 0$ *and* $V^*(x) \leq 0$ *for those* $x \in \mathbb{R}^n$ *for which* $V(x) = v_m$ *for some* $m \in \mathbb{N}$, *and* V *has positive gradient at each* v_m.

Proof. We denote by $u(\cdot, \xi) : [0, +\infty) \to \mathbb{R}^n$ the unique solution of (7.8.1) satisfying $u(0, \xi) = \xi$.

Sufficiency. Let us assume that such a sequence $(v_m)_m$ and function V exist. Let $K_m = V^{-1}([0, v_m])$. Since (7.8.1) has the uniqueness property, by Proposition 7.8.1, it follows that K_m is both viable and locally invariant with respect to g. Let $\varepsilon > 0$ and let us define both $\mu_*(\varepsilon) = \inf_{\|\xi\| = \varepsilon} V(\xi)$ and $\mu^*(\varepsilon) = \sup_{\|\xi\| = \varepsilon} V(\xi)$. Then, $\mu_*(\varepsilon) > 0$. For any $\varepsilon > 0$, choose $m = m(\varepsilon)$ such that $v_m < \mu_*(\varepsilon)$, and choose $\delta(\varepsilon) \in (0, \varepsilon)$ such that $\mu^*(\delta) < v_m$ for each $\delta \in (0, \delta(\varepsilon))$. Let $\xi \in \mathbb{R}^n$ with $\|\xi\| \leq \delta(\varepsilon)$. Clearly, $\xi \in K_m$. If there exists $t > 0$ such that $\|u(t, \xi)\| > \varepsilon$, then there exists $t_0 \in (0, t)$ such that $\|u(t_0, \xi)\| = \varepsilon$. On the other hand, by the choice of m, we have $u(t_0, \xi) \notin K_m$, thereby contradicting the local invariance of K_m with respect to g. This contradiction can be eliminated only if $\|u(t, \xi)\| < \varepsilon$ for all $t \geq 0$, and this completes the proof of the sufficiency.

Necessity. Suppose that 0 is stable for (7.8.1). Let $\delta(\cdot)$ the function in Definition 7.8.1, choose $\varepsilon_1 > 0$, and let us define inductively $\varepsilon_{m+1} = \delta(\varepsilon_m)/2$. Clearly, $\varepsilon_m \downarrow 0$. Let us denote by

$$K_m^0 = \{u(t, \xi) ;\ \|\xi\| \leq \delta(\varepsilon_m) \text{ and } t \geq 0\}.$$

Since K_m^0 contains only points reached at positive time by solutions starting in $D(0, \delta(\varepsilon_m))$, it follows that K_m^0 is viable with respect to g. By Proposition 7.1.1, we deduce that $K_m = \overline{K_m^0}$ is viable with respect to g. Since $u'(t) = g(u(t))$ has the uniqueness property, it follows that K_m is in fact invariant with respect to g. For each $m = 1, 2, \ldots$, let

$$v_m = \sum_{i=m+1}^{\infty} \varepsilon_i.$$

Clearly, $v_m = \frac{1}{2}\sum_{i=m+1}^{\infty} \delta(\varepsilon_{i-1}) \leq \varepsilon_m$ for $m = 1, 2, \ldots$. Let $K_0 = \mathbb{R}^n$ and $v_0 = +\infty$. We have $\ldots \subseteq K_2 \subseteq K_1 \subseteq K_0$ and therefore, we may define $N(\xi) = \max\{m;\ \xi \in K_m\}$. Notice that we want to have $V(\xi) = v_m$ for all $\xi \in \partial K_m$. Define $V(0) = 0$ and

$$V(\xi) = \min\{v_{N(\xi)}, v_{N(\xi)+1} + \text{dist}\,(\xi; K_{N(\xi)+1})\}$$

for $\xi \neq 0$. We begin by proving that V is continuous. First, let us observe that, for each $m \in \mathbb{N}$, the function N is constant on $K_m \setminus K_{m+1}$. Accordingly, the restriction of V to $K_m \setminus K_{m+1}$ is continuous, and so the function V itself is continuous at each interior point of $K_m \setminus K_{m+1}$. Next, we show that for each $m \in \mathbb{N}$, V is continuous on the boundary, ∂K_m, of K_m. Let ξ be arbitrary in ∂K_m. We have both $\|\xi\| \in [\,\delta(\varepsilon_m), \varepsilon_m)$ and $\sup_{\eta \in K_{m+1}} \|\eta\| \leq \varepsilon_{m+1}$. Therefore, $N(\xi) = m$ and $\text{dist}\,(\xi; K_{m+1}) > \delta(\varepsilon_m) - \varepsilon_{m+1} = \delta(\varepsilon_m)/2$. Hence, we have $v_{N(\xi)+1} + \text{dist}\,(\xi; K_{N(\xi)+1}) > v_{N(\xi)}$ and $V(\xi) = v_m$. Choose $\xi_i \notin K_m$ for $i = 1, 2, \ldots$, with $\xi_i \to \xi$. Then, $\lim_{i\to\infty} \text{dist}\,(\xi_i; K_m) = 0$, and

$$\lim_{i\to\infty} V(\xi_i) = v_m = V(\xi),$$

which shows that V is continuous at ξ. It remains to prove that V is continuous at 0. Take $\xi_i \to 0$. Then $N(\xi_i) \to \infty$ and $V(\xi_i) \leq v_{N(\xi_i)}$. But $v_m \to 0$ as $m \to \infty$, so that

$$\lim_{\xi\to 0} V(\xi) = 0 = V(0).$$

Thus V is continuous on $\mathbb{R}^n$.

Finally, we show that $V(\xi) = v_m$ implies $V^*(\xi) \leq 0$. Indeed, if for some $m \in \mathbb{N}$ we have $V(\xi) = v_m$, then $\xi \in K_m$, $\xi + hg(\xi) \in K_{m-1}$ for $h > 0$ sufficiently small, and

$$V(\xi + hg(\xi)) \leq v_m + \text{dist}\,(\xi + hg(\xi); K_m).$$

Since K_m is viable with respect to g, by virtue of Theorem 3.1.1, we have $g(\xi) \in \mathcal{F}_{K_m}(\xi)$, and therefore

$$V^*(\xi) \leq \limsup_{h\downarrow 0} \frac{1}{h}\text{dist}\,(\xi + hg(\xi); K_m) = 0.$$

If $0 < \text{dist}\,(\xi; K_m) \leq v_m - v_{m-1} = \varepsilon_m$, then $V(\xi) = v_m + \text{dist}\,(\xi; K_m)$, so that V has positive gradient at each v_m, thereby completing the proof. □

7.9. The characteristics method for a first order PDE

Let Ω be a nonempty open subset of $\mathbb{R}^n$. A function $w : \Omega \to \mathbb{R}$ is *Severi differentiable* at a point $x \in \Omega$ if, for every $u \in \mathbb{R}^n$, there exists the finite limit

$$Dw(x)(u) = \lim_{\substack{s\downarrow 0 \\ p\to 0}} (1/s)(w(x+s(u+p)) - w(x)),$$

called the *Severi differential of* w *at* x *in the direction* u. If w is Severi differentiable at x, the function $u \mapsto Dw(x)(u)$ from $\mathbb{R}^n$ to $\mathbb{R}$ is the *Severi differential of* w *at* x.

Let $H : \Omega \times \mathbb{R} \rightsquigarrow \mathbb{R}^n \times \mathbb{R}$ be a multi-function with nonempty values, and consider the first order partial differential equation

$$\inf_{(\eta,\theta)\in H(\xi,w(\xi))} (Dw(\xi)(\eta) - \theta) = 0. \tag{7.9.1}$$

By a *solution* to the equation (7.9.1) we mean a *Severi differentiable* function $w : \Omega \to \mathbb{R}$ which satisfies the equality (7.9.1) for all $\xi \in \Omega$.

In order to construct an existence theory for (7.9.1), we take as a model the classical characteristics method and characterize the solutions w of (7.9.1) by means of the behavior of the functions w along the almost exact solutions (u, v) of the differential inclusion

$$(u'(t), v'(t)) \in H(u(t), v(t)), \tag{7.9.2}$$

called *characteristics system*. Since the differential inclusion (7.9.2) is autonomous, we consider only almost exact solutions $(u, v) : [0, T) \to \mathbb{R}^n \times \mathbb{R}$ where $0 < T \leq \infty$. The characterization consists of the conditions:

(C_1) *for every* $\xi \in \Omega$, *there exists an almost exact solution* $(u, v) : [0, T) \to \mathbb{R}^n \times \mathbb{R}$ *of* (7.9.2), *with* $(u(0), v(0)) = (\xi, w(\xi))$, *and such that, for every* $s \in (0, T)$, *we have* $w(u(s)) \leq v(s)$;

(C_2) *for every* $\xi \in \Omega$, *for every almost exact solution* $(u, v) : [0, T) \to \mathbb{R}^n \times \mathbb{R}$ *of* (7.9.2), *with* $(u(0), v(0)) = (\xi, w(\xi))$, *and for every* $s \in (0, T)$, *we have* $v(s) \leq w(u(s))$.

Now we are ready to state a first result concerning (7.9.1).

Theorem 7.9.1. *Let* $H : \Omega \times \mathbb{R} \rightsquigarrow \mathbb{R}^n \times \mathbb{R}$ *be both u.s.c. and l.s.c. with nonempty, compact and convex values, and let* $w : \Omega \to \mathbb{R}$ *be Severi differentiable and such that* H *has the comparison property with respect to* $(\Omega \times \mathbb{R}, \mathrm{hyp}(w))$. *Then* w *is a solution to the equation* (7.9.1) *if and only if it satisfies conditions* (C_1) *and* (C_2).

In fact, in the absence of the differentiability conditions, the function w satisfying (C_1) and (C_2) are just the solutions w of the two differential inequations

$$\inf_{(\eta,\theta)\in H(\xi,w(\xi))} (\underline{D}w(\xi)(\eta) - \theta) \leq 0, \tag{7.9.3}$$

$$0 \leq \inf_{(\eta,\theta)\in H(\xi,w(\xi))} (\overline{D}w(\xi)(\eta) - \theta), \tag{7.9.4}$$

respectively. Here

$$\underline{D}w(\xi)(\eta) = \liminf_{\substack{s \downarrow 0 \\ p \to 0}} (1/s)(w(\xi + s(\eta + p)) - w(\xi)),$$

$$\overline{D}w(\xi)(\eta) = \limsup_{\substack{s \downarrow 0 \\ p \to 0}} (1/s)(w(\xi + s(\eta + p)) - w(\xi)).$$

If $g : D \to \mathbb{R}$, we recall that $\mathrm{epi}\,(g) = \{(\xi, t);\ g(\xi) \leq t,\ \xi \in D\}$ is the *epigraph* of the function g and $\mathrm{hyp}\,(g) = \{(\xi, t);\ g(\xi) \geq t,\ \xi \in D\}$ denotes the *hypograph* of the function g.

Exercise 7.9.1. *Prove that:*

$$\mathrm{epi}\ (\underline{D}w(\xi)) = \mathcal{T}_{\mathrm{epi}(w)}(\xi, w(\xi))$$

$$\mathrm{hyp}\ (\overline{D}w(\xi)) = \mathcal{T}_{\mathrm{hyp}(w)}(\xi, w(\xi)).$$

It is clear that w is differentiable at ξ if and only if $\underline{D}w(\xi)(\eta)$ and $\overline{D}w(\xi)(\eta)$ are finite and equal to each other for all $\eta \in \mathbb{R}^n$.

The epigraph and hypograph equalities above show that $\eta \mapsto \underline{D}w(\xi)(\eta)$ and $\eta \mapsto \overline{D}w(\xi)(\eta)$ are l.s.c. and u.s.c. respectively.

By a *solution of the differential inequation* (7.9.3) (or (7.9.4)) we mean a function $w : \Omega \to \mathbb{R}$ which satisfies the inequality (7.9.3) (or (7.9.4)) for all $\xi \in \Omega$. Clearly a function w is a solution to (7.9.1) if and only if it is a differentiable solution to the couple (7.9.3) and (7.9.4). Since differentiability at a point implies continuity at that point, we conclude that Theorem 7.9.1 above is a natural corollary of Theorem 7.9.2 below.

Theorem 7.9.2. *Let $H : \Omega \times \mathbb{R} \rightsquigarrow \mathbb{R}^n \times \mathbb{R}$ be both u.s.c. and l.s.c. with nonempty, compact and convex values, and let $w : \Omega \to \mathbb{R}$ be continuous such that H has the comparison property with respect to $(\Omega \times \mathbb{R}, \mathrm{hyp}(w))$. Then w is a solution of* (7.9.3) *and* (7.9.4) *if and only if it satisfies conditions* (C_1) *and* (C_2).

We mention that all solutions to every variational problem satisfy both (C_1) and (C_2) (the Bellman optimality principle) with a suitable chosen H. Hence, under rather common hypotheses upon the components of a variational problem, its continuous solution satisfies the inequations (7.9.3) and (7.9.4) (the generalized Bellman "equation"). A typical example is included below.

Example 7.9.1. We consider the time optimal control problem associated to a control system and a target. More precisely, let $F : \mathbb{R}^n \rightsquigarrow \mathbb{R}^n$ be a multi-function and let us consider the differential inclusion

$$u'(t) \in F(u(t)), \tag{7.9.5}$$

and fix a *target set* $\mathcal{T}$ (nonempty and closed). Let $\mathcal{R}$ be the *reachable set*, that is, the set of all initial points which can be transferred to $\mathcal{T}$ by almost exact solutions of (7.9.5). For $\xi \in \mathcal{R}$ define $T(\xi)$ as the minimum (assuming that it exists) of the transition times. An almost exact solution of (7.9.5) that transfers ξ to $\mathcal{T}$ in $T(\xi)$ is called *optimal.* The Bellman optimality principle for the time optimal control problem states that

(C_3) *for every $\xi \in \mathcal{R}$, there exists at least one almost exact solution (an optimal one) $u : [0, T) \to \mathbb{R}^n$ of (7.9.5), with $u(0) = \xi$, and such that, for every $s \in (0, T)$, $w(u(s)) + s \leq w(\xi)$;*

(C_4) *for every $\xi \in \mathcal{R}$, for every almost exact solution $u : [0, T) \to \mathbb{R}^n$ of (7.9.5), with $u(0) = \xi$, and for every $s \in (0, T)$, we have $w(\xi) \leq w(u(s)) + s$.*

It is easy to see that conditions (C_3) and (C_4) are particular cases of (C_1) and (C_2), respectively, for the choice $H(x, y) = F(x) \times \{-1\}$. Therefore, the corresponding partial differential equation (called the Bellman equation) is

$$1 + \inf_{\eta \in F(\xi)} DT(\xi)(\eta) = 0, \quad \xi \in \mathcal{R} \setminus \mathcal{T}.$$

Returning to the general equation (7.9.1), we point out that, in its turn, Theorem 7.9.2 follows from the following anatomized variant of itself.

Theorem 7.9.3. (i) *Let $H : \Omega \times \mathbb{R} \rightsquigarrow \mathbb{R}^n \times \mathbb{R}$ be u.s.c. with nonempty, compact and convex values. A continuous function $w : \Omega \to \mathbb{R}$ is a solution of (7.9.3) if and only if it satisfies (C_1).*

(ii) *Let $H : \Omega \times \mathbb{R} \rightsquigarrow \mathbb{R}^n \times \mathbb{R}$ be l.s.c. with closed and convex values. Let $w : \Omega \to \mathbb{R}$ be a continuous function such that H has the comparison property with respect to $(\Omega \times \mathbb{R}, \mathrm{hyp}(w))$. Then w is a solution of (7.9.4) if and only if it satisfies (C_2).*

Proof. (i) Since $H(\xi, w(\xi))$ is compact and since $\underline{D}w(\xi)$ is l.s.c. on $\mathbb{R}^n$, it follows that "inf" can be replaced by "min" in (7.9.3). By the first equality in Exercise 7.9.1, (7.9.3) states that

$$\phi \neq H(\xi, w(\xi)) \cap \mathcal{T}_{\mathrm{epi}(w)}(\xi, w(\xi)). \tag{7.9.6}$$

Since w is continuous, (7.9.6) is equivalent to (7.9.7) below

$$\phi \neq H(\xi, t) \cap \mathcal{T}_{\mathrm{epi}(w)}(\xi, t) \tag{7.9.7}$$

for $t \geq w(\xi)$, and this simply because $\mathcal{T}_{\mathrm{epi}(w)}(\xi, t) = \mathbb{R}^n \times \mathbb{R}$ whenever $w(\xi) < t$. On the other hand, since w is continuous, condition (C_1) is equivalent to the almost exact viability of epi (w) with respect to H. But the set epi (w) is closed in $\Omega \times \mathbb{R}$, and therefore the conclusion is an immediate consequence of Theorem 6.2.3.

(ii) Taking into account the second equality in Exercise 7.9.1, (7.9.4) states that $H(\xi, w(\xi)) \subseteq \mathcal{T}_{\text{hyp}(w)}(\xi, w(\xi))$. Since w is continuous, we have $H(\xi, t) \subseteq \mathcal{T}_{\text{hyp}(w)}(\xi, t)$ in case $w(\xi) > t$. On the other hand, condition (C_2) states that the set hyp (w) is invariant with respect to H. Since the set hyp (w) is closed in $\Omega \times \mathbb{R}$, by Corollary 2.2.1 we deduce that hyp (w) is proximal. Therefore, the "if" part follows from Corollary 6.7.1 and Problem 6.7.1. Finally, the "only if" part follows from Theorem 6.7.3. □

Problem 7.9.1. *Suppose that $w : \Omega \to \mathbb{R}$ is continuous. Prove that the condition (C_1) is equivalent to the following one below*

for every $\xi \in \Omega$, there exists an almost exact solution $(u, v) : [0, T) \to \mathbb{R}^n \times \mathbb{R}$ of the inclusion (7.9.2), with $(u(0), v(0)) = (\xi, w(\xi))$, such that, the function $s \mapsto w(u(s)) - v(s)$ is decreasing.

A natural question is whether we can weaken the continuity property of the function w in Theorem 7.9.3. This question arises from the fact that epi (w) is closed even in the case when w is l.s.c. and hyp (w) is closed in the case when w is u.s.c. The answer is in the negative as the following examples show. Consider first the differential inequation

$$\underline{D}w(\xi)(w(\xi)) - 1 \leq 0. \tag{7.9.8}$$

Here $H(x, y) = \{(y, 1)\}$, for all $(x, y) \in \mathbb{R} \times \mathbb{R}$

Exercise 7.9.2. *Prove that the l.s.c. function $w : \mathbb{R} \to \mathbb{R}$ given by $w(\xi) = 0$ for $\xi = 0$ and by $w(\xi) = 1$ for $\xi \neq 0$ is a solution to (7.9.8) but does not satisfy the corresponding condition (C_1).*

A second example is given by the differential inequation

$$0 \leq \overline{D}w(\xi)(w(\xi)) + 1. \tag{7.9.9}$$

Here $H(x, y) = \{(y, -1)\}$ for all $(x, y) \in \mathbb{R} \times \mathbb{R}$.

Exercise 7.9.3. *Prove that the u.s.c. function $w : \mathbb{R} \to \mathbb{R}$ given by $w(\xi) = 0$ for $\xi = 0$ and by $w(\xi) = -1$ for $\xi \neq 0$ is a solution to (7.9.9) but does not satisfy the corresponding condition (C_2).*

A condition which assures that (C_1) in part (i) of Theorem 7.9.3 holds for a l.s.c. function w, while (C_2) in part (ii) holds for an u.s.c. function w is given below.

Theorem 7.9.4. *Let $H : \Omega \times \mathbb{R} \rightsquigarrow \mathbb{R}^n \times \mathbb{R}$ be a nonempty and convex valued multi-function satisfying*

$$H(x, y_1) \subseteq H(x, y_2) \tag{7.9.10}$$

for each $x \in \Omega$ *and* $y_1, y_2 \in \mathbb{R}$, *with* $y_1 \le y_2$.

(i) *Assume that* H *is u.s.c. with compact values. Then, a l.s.c. function* $w : \Omega \to \mathbb{R}$ *is a solution to the differential inequation* (7.9.3) *if and only if it satisfies* (C_1)

(ii) *Assume that* H *is l.s.c. with closed values. Let* $w : \Omega \to \mathbb{R}$ *be u.s.c. such that* H *has the comparison property with respect to* $(\Omega \times \mathbb{R}, \mathrm{hyp}\,(w))$. *Then,* $w : \Omega \to \mathbb{R}$ *is a solution to the differential inequality* (7.9.4) *if and only if it satisfies* (C_2)

Proof. The proof follows the same steps as the one of Theorem 7.9.3. Indeed, in order to prove (i), as $\mathcal{T}_{\mathrm{epi}(w)}(x, w(x)) \subseteq \mathcal{T}_{\mathrm{epi}(w)}(x, t)$ if $w(x) \le t$, (7.9.10) implies that (7.9.6) is equivalent to (7.9.7). In its turn, (7.9.6) is equivalent to (7.9.3). On the other hand, since (7.9.10) is satisfied, condition (C_1) is equivalent to the almost exact viability of epi(w) with respect to H. Indeed, if $w(x) < t$, and $(u, v) : [0, T) \to \mathbb{R}^n \times \mathbb{R}$ is a solution to (7.9.2) with $(u(0), v(0)) = (\xi, w(\xi))$, and $w(u(s)) \le v(s)$ for all $s \in [0, T)$, then the function $s \mapsto (u(s), v(s) + t - w(x)) = (\overline{u}(s), \overline{v}(s))$ is a solution to (7.9.2) with $(\overline{u}(0), \overline{v}(0)) = (\xi, t)$ and satisfies $w(\overline{u}(s)) \le \overline{v}(s)$ for all $s \in [0, T)$. This completes the proof of (i). The proof of (ii) goes in the very same spirit, and therefore we do not give further details. □

Part 2

Evolution equations and inclusions

CHAPTER 8

Viability for single-valued semilinear evolutions

In this chapter we reconsider some problems already studied in the case of ordinary differential equations, within the more general frame of semilinear evolution equations governed by single-valued continuous perturbations of infinitesimal generators of C_0-semigroups. As in the previous case we just mentioned, we start with the autonomous case. So, after introducing the concept of mild viability and that one of A-tangent vector to a set at a given point, we prove a necessary condition for mild viability expressed in terms of a tangency condition which, whenever $A \equiv 0$, reduces to the Nagumo Tangency Condition. We next state and subsequently prove several necessary and sufficient conditions for mild viability. Further, we show how the quasi-autonomous case reduces to the autonomous one and we rephrase all the results already obtained in the autonomous case within this more general frame. We prove some necessary and sufficient conditions for mild viability in the specific case of a class of semilinear reaction-diffusion systems. We conclude the chapter with some results concerning the existence of noncontinuable, as well as of global mild solutions.

8.1. Necessary conditions for mild viability

We begin with the autonomous case. So, let X be a real Banach space, let $A : D(A) \subseteq X \to X$ be the infinitesimal generator of a C_0-semigroup, $\{S(t) : X \to X;\ t \geq 0\}$, K a nonempty subset in X and $f : K \to X$ a given function.

Definition 8.1.1. By a *mild solution* of the autonomous semilinear Cauchy problem

$$\begin{cases} u'(t) = Au(t) + f(u(t)) \\ u(0) = \xi, \end{cases} \tag{8.1.1}$$

on $[0,T]$, we mean a continuous function $u : [0,T] \to K$ such that the mapping $s \mapsto f(u(s))$ is Bochner integrable on $[0,T]$ and

$$u(t) = S(t)\xi + \int_0^t S(t-s)f(u(s))\,ds, \tag{8.1.2}$$

for each $t \in [0,T]$.

Remark 8.1.1. If u is a mild solution of the Cauchy problem (8.1.1), then u is a mild solution, in the sense of Definition 1.5.3, of the linear nonhomogeneous Cauchy problem

$$\begin{cases} u' = Au + g \\ u(0) = \xi, \end{cases}$$

where $g(s) = f(u(s))$ for each $s \in [0,T]$.

Definition 8.1.2. The set $K \subseteq X$ is *mild viable* with respect to $A + f$ if for each $\xi \in K$, there exists $T > 0$ such that the Cauchy problem (8.1.1) has at least one mild solution $u : [0,T] \to K$.

Definition 8.1.3. We say that $\eta \in X$ is *A-tangent to K at $\xi \in K$* if

$$\liminf_{h \downarrow 0} \frac{1}{h} \operatorname{dist}(S(h)\xi + h\eta; K) = 0.$$

In other words, $\eta \in X$ is A-tangent to K at $\xi \in K$ if for each $\delta > 0$ and each neighborhood V of 0 there exist $h \in (0,\delta)$ and $p \in V$ such that

$$S(h)\xi + h(\eta + p) \in K. \tag{8.1.3}$$

The set of all A-tangent elements to K at $\xi \in K$ is denoted by $\mathcal{T}_K^A(\xi)$. We notice that if $A \equiv 0$, then $\mathcal{T}_K^A(\xi)$ is the *contingent cone* at $\xi \in K$, i.e. $\mathcal{T}_K^0(\xi) = \mathcal{T}_K(\xi)$. See Definition 2.4.1.

Remark 8.1.2. Except for some particular cases, as for instance when $A = 0$, or ξ is an interior point of K, or $\xi = 0$, etc., $\mathcal{T}_K^A(\xi)$ is not a cone.

Proposition 8.1.1. *If $\eta \in \mathcal{T}_K^A(\xi)$ then, for every function $h \mapsto \eta_h$ from $(0,1)$ to X satisfying $\lim_{h\downarrow 0} \eta_h = \eta$, we have*

$$\liminf_{h \downarrow 0} \frac{1}{h} \operatorname{dist}(S(h)\xi + h\eta_h; K) = 0. \tag{8.1.4}$$

If there exists a function $h \mapsto \eta_h$ from $(0,1)$ to X satisfying both $\lim_{h\downarrow 0} \eta_h = \eta$ and (8.1.4), then $\eta \in \mathcal{T}_K^A(\xi)$.

Since the proof is completely similar with that one of Proposition 2.4.3, we do not enter into details.

Proposition 8.1.2. *A necessary and sufficient condition in order that* $\eta \in \mathcal{T}_K^A(\xi)$ *is*

$$\liminf_{h \downarrow 0} \frac{1}{h} \text{dist} \left(S(h)\xi + \int_0^h S(h-s)\eta \, ds; K \right) = 0.$$

Proof. Since, by (i) in Theorem 1.4.3, we have

$$\lim_{h \downarrow 0} \frac{1}{h} \int_0^h S(h-s)\eta \, ds = \eta,$$

the conclusion follows from Proposition 8.1.1. □

Theorem 8.1.1. *Let* X *be a Banach space,* $A : D(A) \subseteq X \to X$ *the infinitesimal generator of a* C_0*-semigroup,* $\{S(t) : X \to X;\ t \geq 0\}$, K *a nonempty subset in* X *and* $f : K \to X$ *a continuous function. If* K *is mild viable with respect to* $A + f$ *then, for each* $\xi \in K$, *we have*

$$\lim_{h \downarrow 0} \frac{1}{h} \text{dist} \left(S(h)\xi + hf(\xi); K \right) = 0. \tag{8.1.5}$$

Remark 8.1.3. Under the general hypotheses of Theorem 8.1.1, if K is mild viable with respect to $A + f$ then, for each $\xi \in K$, we have

$$f(\xi) \in \mathcal{T}_K^A(\xi).$$

Theorem 8.1.1 is an immediate consequence of the slightly more general Theorem 8.1.2 below.

Theorem 8.1.2. *Let* X *be a Banach space,* $A : D(A) \subseteq X \to X$ *the infinitesimal generator of a* C_0*-semigroup,* $\{S(t) : X \to X;\ t \geq 0\}$, K *a nonempty subset in* X *and* $f : K \to X$ *a given function. If* K *is mild viable with respect to* $A+f$, *then the tangency condition* (8.1.5) *is satisfied at each continuity point,* $\xi \in K$, *of* f.

Proof. Let $\xi \in K$ be a continuity point of f. Since K is mild viable with respect to $A + f$, there exist $T > 0$ and one mild solution $u : [0, T] \to K$ of (8.1.1). By virtue of (8.1.2), we have

$$\frac{1}{h} \text{dist} \left(S(h)\xi + hf(\xi); K \right) \leq \frac{1}{h} \| S(h)\xi + hf(\xi) - u(h) \|$$

$$= \frac{1}{h} \left\| \int_0^h [f(\xi) - S(h-s)f(u(s))] ds \right\| \leq \frac{1}{h} \int_0^h \| f(\xi) - S(h-s)f(u(s)) \| \, ds$$

$$\leq \sup_{s \in [0,h]} \| f(\xi) - S(h-s)f(u(s)) \|.$$

Since u is continuous, $u(0) = \xi$, $(t, x) \mapsto S(t)x$ is continuous from $\mathbb{R}_+ \times X$ to X, and ξ is a continuity point of f, we conclude that (8.1.5) holds true, as claimed. □

8.2. Sufficient conditions for mild viability

In order to handle several apparently different cases into a unitary frame, we introduce:

Definition 8.2.1. Let X be a Banach space, $A : D(A) \subseteq X \to X$ the infinitesimal generator of a C_0-semigroup, $\{S(t) : X \to X;\ t \geq 0\}$, and $f : K \to X$ a function. We say that $A + f$ is *locally of compact type* if f is continuous and, for each $\xi \in K$, there exist $\rho > 0$, a continuous function $\ell : \mathbb{R}_+ \to \mathbb{R}_+$ and a uniqueness function $\omega : \mathbb{R}_+ \to \mathbb{R}_+$, such that f is bounded on $D(\xi, \rho) \cap K$, and

$$\beta(S(t)f(C)) \leq \ell(t)\omega(\beta(C))$$

for each $t > 0$ and $C \subseteq D(\xi, \rho) \cap K$.

Several important instances when $A + f$ is locally of compact type are indicated in the remark below.

Remark 8.2.1. One may easily verify that $A+f$ is locally of compact type whenever:

(i) f is locally β-compact (see Definition 3.2.3);
(ii) f is continuous and the semigroup, $\{S(t) : X \to X;\ t \geq 0\}$, is compact (see Definition 1.5.5).

Indeed, if (i) holds, then we can take $\ell(t) = Me^{at}$, for $t \in \mathbb{R}_+$, where $M \geq 1$ and $a \geq 0$ are given by Theorem 1.4.1[1], while ω is given by Definition 3.2.3. As for the case (ii), we may take $\ell \equiv 0$ and $\omega \equiv 0$.

From this observation, and Remark 3.2.3, it readily follows that in each one of the next three situations below:

(iii) f is locally compact (see Definition 3.2.1);
(iv) f is continuous and K is locally compact;
(v) f is locally Lipschitz (see Definition 3.2.2),

$A + f$ is locally of compact type too.

Now we are ready to state the main sufficient conditions concerning the viability of a set K with respect to $A + f$.

Theorem 8.2.1. *Let X be a Banach space, $A : D(A) \subseteq X \to X$ the infinitesimal generator of a C_0-semigroup, $\{S(t) : X \to X;\ t \geq 0\}$, K a nonempty and locally closed subset in X and $f : K \to X$ a function such*

[1] In fact, by Theorem 1.4.1, we get $a \in \mathbb{R}$ but, here and thereafter, we may assume that $a \geq 0$. If not, then replace it by $|a|$ which satisfies also $\|S(t)\|_{\mathcal{L}(X)} \leq Me^{|a|t}$ for each $t \in \mathbb{R}_+$.

that $A+f$ is locally of compact type. Then K is mild viable with respect to $A+f$ if and only if, for each $\xi \in K$, we have

$$f(\xi) \in \mathcal{T}_K^A(\xi). \tag{8.2.1}$$

From Theorem 8.2.1 and (i) in Remark 8.2.1, we deduce

Theorem 8.2.2. *Let X be a Banach space, $A : D(A) \subseteq X \to X$ the infinitesimal generator of a C_0-semigroup, $\{S(t) : X \to X;\ t \geq 0\}$, K a nonempty and locally closed subset in X and $f : K \to X$ a locally β-compact function. Then K is mild viable with respect to $A+f$ if and only if, for each $\xi \in K$, the tangency condition* (8.2.1) *is satisfied.*

From Theorem 8.2.1 and (ii) in Remark 8.2.1, we get

Theorem 8.2.3. *Let X be a Banach space, $A : D(A) \subseteq X \to X$ the infinitesimal generator of a compact C_0-semigroup, $\{S(t) : X \to X;\ t \geq 0\}$, K a nonempty and locally closed subset in X and $f : K \to X$ a continuous function. Then K is mild viable with respect to $A+f$ if and only if, for each $\xi \in K$, the tangency condition* (8.2.1) *is satisfied.*

From Theorem 8.2.2 and (iii)$\sim$(v) in Remark 8.2.1, we obtain

Theorem 8.2.4. *Let X be a Banach space, $A : D(A) \subseteq X \to X$ the infinitesimal generator of a C_0-semigroup, $\{S(t) : X \to X;\ t \geq 0\}$, K a nonempty and locally closed subset in X and $f : K \to X$ a locally compact function*[2]*. Then K is mild viable with respect to $A+f$ if and only if, for each $\xi \in K$, the tangency condition* (8.2.1) *is satisfied.*

Theorem 8.2.5. *Let X be a Banach space, $A : D(A) \subseteq X \to X$ the infinitesimal generator of a C_0-semigroup, $\{S(t) : X \to X;\ t \geq 0\}$, K a nonempty and locally compact subset in X and $f : K \to X$ a continuous function. Then K is mild viable with respect to $A+f$ if and only if, for each $\xi \in K$, the tangency condition* (8.2.1) *is satisfied.*

Theorem 8.2.6. *Let X be a Banach space, $A : D(A) \subseteq X \to X$ the infinitesimal generator of a C_0-semigroup, $\{S(t) : X \to X;\ t \geq 0\}$, K a nonempty and locally closed subset in X and $f : K \to X$ a locally Lipschitz function. Then K is mild viable with respect to $A+f$ if and only if, for each $\xi \in K$, the tangency condition* (8.2.1) *is satisfied.*

Since the necessity of Theorems 8.2.1$\sim$8.2.6 is a simple consequence of Theorem 8.1.1, in what follows, we focus our attention only on the proof of the sufficiency. Finally, as Theorems 8.2.2$\sim$8.2.6 follow from Theorem 8.2.1 and Remark 8.2.1, we will prove only the sufficiency of Theorem 8.2.1.

---

[2]See Definition 3.2.1.

8.3. Existence of ε-approximate mild solutions

The proof of the sufficiency of Theorem 8.2.1 is divided into two steps. First, we will prove the existence of a family of approximate solutions for the Cauchy problem (8.1.1), all defined on one and the same interval $[0,T]$.

Lemma 8.3.1. *Let X be a Banach space, $A : D(A) \subseteq X \to X$ the infinitesimal generator of a C_0-semigroup, $\{S(t) : X \to X;\ t \geq 0\}$, K a nonempty and locally closed subset in X and $f : K \to X$ a continuous function satisfying the tangency condition* (8.2.1). *Let $\xi \in K$ be arbitrary and let $r > 0$ be such that $D(\xi, r) \cap K$ is closed. Then, there exist $\rho \in (0, r]$, and $T > 0$ such that, for each $\varepsilon \in (0,1)$, there exist $\sigma : [0,T] \to [0,T]$ nondecreasing, $\theta : \{(t,s);\ 0 \leq s < t \leq T\} \to [0,T]$ measurable, $g : [0,T] \to X$ Riemann integrable and $u : [0,T] \to X$ continuous such that:*

(i) $s - \varepsilon \leq \sigma(s) \leq s$ *for each* $s \in [0,T]$;
(ii) $u(\sigma(s)) \in D(\xi,\rho) \cap K$ *for each* $s \in [0,T]$ *and* $u(T) \in D(\xi,\rho) \cap K$;
(iii) $\|g(s)\| \leq \varepsilon$ *for each* $s \in [0,T]$;
(iv) $\theta(t,s) \leq t$ *for each* $0 \leq s < t \leq T$ *and* $t \mapsto \theta(t,s)$ *is nonexpansive on* $(s,T]$;
(v) $u(t) = S(t)\xi + \int_0^t S(t-s) f(u(\sigma(s)))\,ds + \int_0^t S(\theta(t,s))g(s)\,ds$ *for each* $t \in [0,T]$;
(vi) $\|u(t) - u(\sigma(t))\| \leq \varepsilon$ *for each* $t \in [0,T]$.

Definition 8.3.1. A 4-uple (σ, θ, g, u) as in Lemma 8.3.1 is called an *ε-approximate mild solution* of the Cauchy problem (8.1.1) on $[0,T]$.

Proof. Let $\xi \in K$ be arbitrary and let $r > 0$ be such that $D(\xi,r) \cap K$ be closed. Let us choose $\rho \in (0,r]$, $N > 0$, $M \geq 1$ and $a \geq 0$ such that

$$\|f(x)\| \leq N \tag{8.3.1}$$

for every $x \in D(\xi,\rho) \cap K$, and

$$\|S(t)\|_{\mathcal{L}(X)} \leq Me^{at} \tag{8.3.2}$$

for every $t \geq 0$.

The existence of $\rho > 0$ and $N > 0$ satisfying (8.3.1) is ensured by the fact that f is continuous and therefore locally bounded. Furthermore, the existence of $M \geq 1$ and $a \geq 0$ satisfying (8.3.2) follows from Theorem 1.4.1. Since $t \mapsto S(t)\xi$ is continuous at $t = 0$ and $S(0)\xi = \xi$, we may find a sufficiently small $T > 0$ such that

$$\sup_{t \in [0,T]} \|S(t)\xi - \xi\| + TMe^{aT}(N+1) \leq \rho. \tag{8.3.3}$$

Let $\varepsilon \in (0,1)$ be arbitrary but fixed. We begin by showing the existence of an ε-approximate solution on an interval $[0,\delta]$ with $\delta \in (0,T]$. As, for every $\xi \in K$, f satisfies the tangency condition (8.2.1), from Proposition 8.1.2, it follows that there exist $\delta \in (0,T]$ and $p \in X$ with $\|p\| \le \varepsilon$, such that

$$S(\delta)\xi + \int_0^\delta S(\delta - s) f(\xi)\, ds + \delta p \in K. \tag{8.3.4}$$

We continue by showing how to define the functions σ, θ, g, u. Namely, let $\sigma : [0,\delta] \to [0,\delta]$, $g : [0,\delta] \to X$ and $\theta : \{(t,s);\ 0 \le s < t \le \delta\} \to [0,\delta]$ be given by

$$\begin{aligned} \sigma(s) &= 0, \\ g(s) &= p, \\ \theta(t,s) &= 0 \end{aligned}$$

and let $u : [0,\delta] \to X$ be defined by

$$u(t) = S(t)\xi + \int_0^t S(t-s) f(\xi)\, ds + tp,$$

for each $t \in [0,\delta]$.

We will show that (σ, θ, g, u) is an ε-approximate mild solution to the Cauchy problem (8.1.1) on the interval $[0,\delta]$. Clearly σ is nondecreasing, g is Riemann integrable, θ is measurable and $t \mapsto \theta(t,s)$ is nonexpansive on $[0,\delta]$ and u is continuous. The conditions (i), (iii), (iv) and (v) are obviously fulfilled. To prove (ii), let us first observe that, as $u(\sigma(s)) = u(0) = \xi$ for each $s \in [0,\delta]$, we have $u(\sigma(s)) \in D(\xi,\rho) \cap K$. Moreover, since $\delta < T$, from (v), (8.3.1) and (8.3.2), we get

$$\|u(t) - \xi\| \le \|S(t)\xi - \xi\| + \int_0^t \|S(t-s)\|_{\mathcal{L}(X)} \| f(u(\sigma(s)))\| ds$$

$$+ \int_0^t \|S(\theta(t,s))\|_{\mathcal{L}(X)} \|g(s)\| ds \le \sup_{t \in [0,\delta]} \|S(t)\xi - \xi\| + \delta M e^{a\delta}(N+1) \tag{8.3.5}$$

for each $t \in [0,\delta]$. Thus, by (8.3.3), we have

$$\|u(\delta) - \xi\| \le \sup_{t \in [0,\delta]} \|S(t)\xi - \xi\| + \delta M e^{a\delta}(N+1) \le \rho.$$

Combining (8.3.4) with the last inequality, we get $u(\delta) \in D(\xi,\rho) \cap K$ and thus (ii) is satisfied. Diminishing $\delta > 0$, if necessary, by (8.3.5), we may assume that

$$\|u(t) - u(\sigma(t))\| \le \varepsilon$$

for each $t \in [0,\delta]$ and thus (vi) is also satisfied. We emphasize that we can do this because (8.3.5) is independent of p which, of course, may change with δ. Therefore (σ, θ, g, u) is an ε-approximate mild solution of (8.1.1) on $[0, \delta]$.

In the second step we will prove the existence of an ε-approximate solution for (8.1.1) defined on the whole interval $[0, T]$. To this aim we shall make use of Brezis–Browder Theorem 2.1.1. We denote by $D(c)$ the set

$$D(c) = [0, c] \times \{(t, s);\ 0 \leq s < t \leq c\} \times [0, c] \times [0, c],$$

with $c \geq 0$ and by $\mathcal{S}$ the set of all ε-approximate mild solutions to the problem (8.1.1), defined on $D(c)$, with $c \leq T$.

On $\mathcal{S}$ we introduce a preorder $\preceq$as follows: we say that $(\sigma_1, \theta_1, g_1, u_1)$, defined on $D(c_1)$, and $(\sigma_2, \theta_2, g_2, u_2)$, defined on $D(c_2)$, satisfy

$$(\sigma_1, \theta_1, g_1, u_1) \preceq (\sigma_2, \theta_2, g_2, u_2)$$

if $c_1 \leq c_2$, $\sigma_1(t) = \sigma_2(t)$, $g_1(t) = g_2(t)$, for $t \in [0, c_1]$ and $\theta_1(t, s) = \theta_2(t, s)$ for each $0 \leq s < t \leq c_1$.

Let $\mathcal{L}$ be an increasing sequence in $\mathcal{S}$,

$$\mathcal{L} = ((\sigma_i, \theta_i, g_i, u_i) : D(c_i) \to [0, c_i] \times [0, c_i] \times X \times X)_i.$$

We define an upper bound of $\mathcal{L}$ as follows. First, set

$$c^* = \sup\{c_i; i = 1, 2, \dots\}.$$

If $c^* = c_i$ for some $i = 1, 2, \dots$ then $(\sigma_i, \theta_i, g_i, u_i)$ is an upper bound for $\mathcal{L}$. If $c_i < c^*$ for $i = 1, 2, \dots$, we show that there exists $(\sigma^*, \theta^*, g^*, u^*)$ in $\mathcal{S}$, defined on $[0, c^*]$, and satisfying $(\sigma_m, \theta_m, g_m, u_m) \preceq (\sigma^*, \theta^*, g^*, u^*)$, for $m = 1, 2, \dots$.

First, we know that all the functions in the set $\{\sigma_m; m = 1, 2, \dots\}$ are nondecreasing, with values in $[0, c^*]$, and satisfy $\sigma_m(c_m) \leq \sigma_p(c_p)$ for $m, p = 1, 2, \dots$ with $m \leq p$. So, there exists

$$\lim_m \sigma_m(c_m) \in [0, c^*].$$

Now, let $m, k \in \{1, 2, \dots\}$ be arbitrary with $m \geq k$. For each $s \in [0, c_k)$, we have

$$|\theta_m(c_m, s) - \theta_k(c_k, s)|$$
$$\leq |\theta_m(c_m, s) - \theta_m(c_k, s)| + |\theta_m(c_k, s) - \theta_k(c_k, s)| \leq |c_m - c_k|,$$

because $t \mapsto \theta_m(t, s)$ is nonexpansive on (s, c^*). Accordingly, there exists

$$\lim_m \theta_m(c_m, s).$$

We are now ready to define the functions: $\sigma^* : [0, c^*] \to [0, c^*]$ by

$$\sigma^*(t) = \begin{cases} \sigma_m(t) & \text{if, for some } m = 1, 2, \dots,\ t \in [0, c_m] \\ \lim_m \sigma_m(c_m) & \text{if } t = c^*, \end{cases}$$

$\theta^* : \{(t,s); 0 \leq s < t \leq c^*\} \to [\,0, c^*\,]$ by

$$\theta^*(t,s) = \begin{cases} \theta_m(t,s) & \text{if, for some } m = 1,2,\ldots,\ 0 \leq s < t \leq c_m \\ \lim_m \theta_m(c_m, s) & \text{if } 0 \leq s < t = c^* \end{cases}$$

and $g^* : [\,0, c^*\,] \to X$ by

$$g^*(t) = \begin{cases} g_m(t) & \text{if, for some } m = 1,2,\ldots, t \in [\,0, c_m\,] \\ 0 & \text{if } t = c^*. \end{cases}$$

Obviously σ^* is nondecreasing, g^* is Riemann integrable, θ^* is measurable and $t \mapsto \theta^*(t,s)$ is nonexpansive on $[\,0, \delta\,]$. In order to define u^*, let us first prove that there exists $\lim_m u_m(c_m)$. To this aim, let us denote by $\chi_{[\,0,c_m\,]} : [\,0, T\,] \to \mathbb{R}_+$ the indicator function of $[\,0, c_m\,]$, i.e.

$$\chi_{[\,0,c_m\,]}(s) = \begin{cases} 1 & \text{if } s \in [\,0, c_m\,] \\ 0 & \text{if } s \in (c_m, T\,]. \end{cases}$$

By (v), we have

$$u_m(c_m) = S(c_m)\xi + \int_0^{c^*} \chi_{[\,0,c_m\,]}(s) S(c_m - s) f(u_m(\sigma_m(s)))\, ds$$

$$+ \int_0^{c^*} \chi_{[\,0,c_m\,]}(s) S(\theta_m(c_m, s)) g_m(s)\, ds.$$

Recalling (i), (ii), (iv) and (vi) and the fact that, by (iii), each g_m is Riemann integrable and bounded by ε, thanks to the Lebesgue Dominated Convergence Theorem 1.2.3, we deduce that there exists $\lim_m u_m(c_m)$.

Let us define $u^* : [\,0, c^*\,] \to X$ by

$$u^*(t) = \begin{cases} u_m(t) & \text{if, for some } m = 1,2,\ldots,\ t \in [\,0, c_m\,] \\ \lim_m u_m(c_m) & \text{if } t = c^*. \end{cases}$$

Clearly u^* is continuous on $[\,0, c^*\,]$. Moreover, $(\sigma^*, \theta^*, g^*, u^*)$ satisfies (i), (iii), (iv) and (v). Since u_m is an ε-approximate mild solution on $[\,0, c_m\,]$, by (ii) applied to u_m, we have $u_m(c_m) \in D(\xi, \rho) \cap K$ and since the latter is closed, we have $u^*(c^*) \in D(\xi, \rho) \cap K$. Similarly, by (vi) applied to u_m, we have

$$\|u_m(t) - u_m(\sigma_m(t))\| \leq \varepsilon$$

for each $m \in \mathbb{N}$ and $t \in [\,0, c_m\,]$ and since u^* is continuous, we deduce

$$\|u^*(t) - u^*(\sigma^*(t))\| \leq \varepsilon$$

for each $t \in [0, c^*]$. Hence u^* and σ^* satisfy (vi) and thus $(\sigma^*, \theta^*, g^*, u^*)$ is an ε-approximate mild solution of (8.1.1) on $[0, c^*]$ and

$$(\sigma_m, \theta_m, g_m, u_m) \preceq (\sigma^*, \theta^*, g^*, u^*),$$

for $m = 1, 2, \ldots$. Let us define the function $\mathcal{N} : \mathcal{S} \to \mathbb{R}$ by $\mathcal{N}((\sigma, \theta, g, u)) = c$, where $D(c)$ is the domain of definition of (σ, θ, g, u). Clearly $\mathcal{N}$ satisfies the hypotheses of Brezis-Browder Theorem 2.1.1. Then, $\mathcal{S}$ contains at least one $\mathcal{N}$-maximal element $(\overline{\sigma}, \overline{\theta}, \overline{g}, \overline{u})$ whose domain is $D(\overline{c})$. We will next show that $\overline{c} = T$. To this aim, let us assume by contradiction that $\overline{c} < T$. We know that $\overline{u}(\overline{c}) \in D(\xi, \rho) \cap K$. Moreover, by using (8.3.1), (8.3.2), (8.3.3), we get

$$\begin{aligned}
&\|\overline{u}(\overline{c}) - \xi\| \\
&= \left\| S(\overline{c})\xi + \int_0^{\overline{c}} S(\overline{c} - s) f(\overline{u}(\overline{\sigma}(s)))\, ds + \int_0^{\overline{c}} S(\overline{\theta}(\overline{c}, s))\overline{g}(s) ds - \xi \right\| \\
&\leq \sup_{t \in [0, \overline{c}]} \|S(t)\xi - \xi\| + \int_0^{\overline{c}} \|S(\overline{\theta}(\overline{c}, s))\overline{g}(s)\|\, ds \\
&\quad + \int_0^{\overline{c}} \|S(\overline{c} - s) f(\overline{u}(\overline{\sigma}(s)))\|\, ds \\
&\leq \sup_{t \in [0, \overline{c}]} \|S(t)\xi - \xi\| + \overline{c}\, M e^{aT}(N + 1) < \rho.
\end{aligned}$$

Then, as $\overline{u}(\overline{c}) \in K$ and $f(\overline{u}(\overline{c})) \in \mathcal{T}_K^A(\overline{u}(\overline{c}))$, there exist $\overline{\delta} \in (0, T - \overline{c})$, $\overline{\delta} \leq \varepsilon$ and $\overline{p} \in X$, $\|p\| \leq \varepsilon$ such that

$$S(\overline{\delta})\overline{u}(\overline{c}) + \int_0^{\overline{\delta}} S(\overline{\delta} - s) f(\overline{u}(\overline{c}))\, ds + \overline{\delta}\overline{p} \in D(\xi, \rho) \cap K. \tag{8.3.6}$$

Let us define the functions: $\widetilde{\sigma} : [0, \overline{c} + \overline{\delta}] \to [0, \overline{c} + \overline{\delta}]$, $\widetilde{g} : [0, \overline{c} + \overline{\delta}] \to X$ and $\widetilde{\theta} : \{(t, s);\ 0 \leq s < t \leq \overline{c} + \overline{\delta}\} \to [0, \overline{c} + \overline{\delta}]$ by

$$\widetilde{\sigma}(t) = \begin{cases} \overline{\sigma}(t) & \text{for } t \in [0, \overline{c}] \\ \overline{c} & \text{for } t \in (\overline{c}, \overline{c} + \overline{\delta}], \end{cases}$$

$$\widetilde{g}(t) = \begin{cases} \overline{g}(t) & \text{for } t \in [0, \overline{c}] \\ \overline{p} & \text{for } t \in (\overline{c}, \overline{c} + \overline{\delta}], \end{cases}$$

and

$$\widetilde{\theta}(t, s) = \begin{cases} \overline{\theta}(t, s) & \text{for } 0 \leq s < t \leq \overline{c} \\ t - \overline{c} + \overline{\theta}(\overline{c}, s) & \text{for } 0 \leq s < \overline{c} < t \leq \overline{c} + \overline{\delta} \\ 0 & \text{for } \overline{c} \leq s < t \leq \overline{c} + \overline{\delta}. \end{cases}$$

Clearly, $\widetilde{\sigma}$ is nondecreasing, $\widetilde{g}$ is Riemann integrable on $[0, \overline{c} + \overline{\delta}]$, $\widetilde{\theta}$ is measurable and they satisfy (i), (iii), and (iv).

Accordingly, we can define $\widetilde{u} : [0, \overline{c} + \overline{\delta}] \to X$ by

$$\widetilde{u}(t) = \begin{cases} \overline{u}(t) & \text{if } t \in [0, \overline{c}] \\ S(t - \overline{c})\overline{u}(\overline{c}) + \displaystyle\int_{\overline{c}}^{t} S(t-s)f(\overline{u}(\overline{c}))ds + (t - \overline{c})\overline{p} & \text{if } t \in [\overline{c}, \overline{c} + \overline{\delta}]. \end{cases}$$

A standard calculation, involving the form of $\widetilde{\theta}$, shows that $\widetilde{u}$ satisfies (v). From (8.3.6), we get $\widetilde{u}(\overline{c} + \overline{\delta}) \in D(\xi, \rho) \cap K$ and thus (ii) is satisfied.

Following the same arguments as in the first part of the proof, we conclude that, diminishing $\overline{\delta} > 0$ if necessary, we also get (vi). Thus $(\widetilde{\sigma}, \widetilde{\theta}, \widetilde{g}, \widetilde{u})$ is an element of $\mathcal{S}$ which satisfies

$$\mathcal{N}((\overline{\sigma}, \overline{\theta}, \overline{g}, \overline{u})) < \mathcal{N}((\widetilde{\sigma}, \widetilde{\theta}, \widetilde{g}, \widetilde{u}))$$

although

$$(\overline{\sigma}, \overline{\theta}, \overline{g}, \overline{u}) \preceq (\widetilde{\sigma}, \widetilde{\theta}, \widetilde{g}, \widetilde{u}).$$

This contradiction can be eliminated only if $\overline{c} = T$ and this completes the proof. □

Remark 8.3.1. Under the general hypotheses of Lemma 8.3.1, for each $\gamma > 0$, we can diminish both $\rho > 0$ and $T > 0$, such that $T < \gamma$, $\rho < \gamma$ and all the conditions (i) $\sim$ (vi) in Lemma 8.3.1 be satisfied.

8.4. Proof of the sufficiency of Theorem 8.2.1

In this section we will prove that, in the hypotheses of Theorem 8.2.1, there exists at least one sequence $\varepsilon_n \downarrow 0$ such that the corresponding sequence of ε_n-approximate mild solutions, $((\sigma_n, \theta_n, g_n, u_n))_n$, enjoys the property that $(u_n)_n$ is uniformly convergent on $[0, T]$ to some function $u : [0, T] \to K$ which is a mild solution of (8.1.1).

Proof. So, let $r > 0$, $\rho \in (0, r]$ and $T > 0$ as in Lemma 8.3.1. Since, by hypotheses, $A + f$ is locally of compact type, diminishing $\rho \in (0, r]$ and $T > 0$ if necessary, we can find a continuous function $\ell : \mathbb{R}_+ \to \mathbb{R}_+$ and a uniqueness function $\omega : \mathbb{R}_+ \to \mathbb{R}_+$ such that $f(D(\xi, \rho) \cap K)$ is bounded and

$$\beta(S(t)f(C)) \leq \ell(t)\omega(\beta(C)), \tag{8.4.1}$$

for each $t \in [0, T]$ and $C \subseteq D(\xi, \rho) \cap K$ and, in addition, all the conclusions of Lemma 8.3.1 be satisfied. See Remark 8.3.1 to conclude that we can do that.

Let $\varepsilon_n \downarrow 0$ be a sequence in $(0, 1)$ and let $((\sigma_n, \theta_n, g_n, u_n))_n$ be a sequence of ε_n-approximate mild solutions defined on $[0, T]$ whose existence is ensured by Lemma 8.3.1. From (v)[3], we have

$$u_n(t) = S(t)\xi + \int_0^t S(t-s)f(u_n(\sigma_n(s)))\,ds + \int_0^t S(\theta_n(t,s))g_n(s)\,ds \quad (8.4.2)$$

for each $n \in \mathbb{N}$ and $t \in [\,0, T\,]$.

We consider first the case when X is separable. Let $M \geq 1$ and $a \geq 0$ be the constants in (8.3.2). Let $t \in [\,0, T\,]$ be fixed. From Lemma 2.7.2, Problem 2.7.1, (iii), (vi), (8.4.2), (8.3.2), (8.4.1) and Remark 2.7.1, we get

$$\begin{aligned}
&\beta(\{u_n(t);\ n \geq k\}) \\
&\leq \beta\left(\left\{\int_0^t S(t-s)f(u_n(\sigma_n(s)))\,ds;\ n \geq k\right\}\right) \\
&\quad + \beta\left(\left\{\int_0^t S(\theta_n(t,s))g_n(s)\,ds\,;\ n \geq k\right\}\right) \\
&\leq \int_0^t \beta\left(\{S(t-s)f(u_n(\sigma_n(s)));\ n \geq k\}\right)ds \\
&\quad + \int_0^t \beta\left(\{S(\theta_n(t,s))g_n(s)\,;\ n \geq k\}\right)ds \\
&\leq \int_0^t \ell(t-s)\omega(\beta\left(\{u_n(\sigma_n(s));\ n \geq k\}\right))\,ds + TMe^{aT}\varepsilon_k \\
&\leq \int_0^t \sup_{\theta\in[\,0,T\,]} \ell(\theta)\omega(\beta\left(\{u_n(s);\ n \geq k\} + \{u_n(\sigma_n(s)) - u_n(s);\ n \geq k\}\right))\,ds \\
&\quad + TMe^{aT}\varepsilon_k \\
&\leq \int_0^t \sup_{\theta\in[\,0,T\,]} \ell(\theta)\omega(\beta\left(\{u_n(s);\ n \geq k\}\right) + \varepsilon_k)\,ds + TMe^{aT}\varepsilon_k.
\end{aligned}$$

Set $x_k(t) = \beta(\{u_n(t);\ n \geq k\}) + \varepsilon_k$, for $k = 1, 2, \ldots$ and $t \in [\,0, T\,]$, $\omega_0(x) = \sup_{\theta\in[\,0,T\,]} \ell(\theta)\omega(x)$, for $x \in \mathbb{R}_+$, and $\gamma_k = (TMe^{aT} + 1)\varepsilon_k$. We conclude

$$x_k(t) \leq \gamma_k + \int_0^t \omega_0(x_k(s))\,ds,$$

for $k = 1, 2, \ldots$ and $t \in [\,0, T\,]$.

By Remark 1.8.1, ω_0 is a uniqueness function. So, Lemma 1.8.2 shows that, diminishing $T > 0$ if necessary, we may assume that $\lim_k x_k(t) = 0$, which means that $\lim_k \beta(\{u_n(t);\ n \geq k\}) = 0$ uniformly for $t \in [\,0, T\,]$. From Lemma 2.7.3, it follows that, for each $t \in [\,0, T\,]$, $\{u_n(t);\ n = 1, 2, \ldots\}$ is relatively compact in X. At this point Theorem 1.5.2 comes into play and

[3]Throughout this proof, the references to (i)∼(vi) are to the corresponding items in Lemma 8.3.1.

shows that there exists $u \in C([0,T];X)$ such that, on a subsequence at least, we have $\lim_n u_n(t) = u(t)$ uniformly for $t \in [0,T]$. In view of (ii) and (vi), we get $\lim_n u_n(\sigma_n(t)) = u(t)$ uniformly for $t \in [0,T]$, and, since $D(\xi,\rho) \cap K$ is closed, $u(t) \in D(\xi,\rho) \cap K$ for each $t \in [0,T]$.

Passing to the limit in (8.4.2), for $n \to \infty$, and taking into account of (iii), we obtain

$$u(t) = S(t)\xi + \int_0^t S(t-s)f(u(s))\,ds,$$

for each $t \in [0,T]$, and this completes the proof in the case when X is separable.

If X is nonseparable, from Remark 2.7.3, it follows that there exists a separable and closed subspace, Y, of X such that

$$u_n(t), S(r)f(u_n(\sigma_n(s))), S(\theta_n(r,s))g_n(s) \in Y$$

for $n = 1, 2, \ldots$ and a.e. for $t, r, s \in [0,T]$. From Problem 2.7.2 and the monotonicity of ω, we deduce that

$$\beta_Y(S(t)f(C)) \le 2\beta(S(t)f(C)) \le 2\ell(t)\omega(\beta(C)) \le 2\ell(t)\omega(\beta_Y(C)),$$

for each $t > 0$ and each set $C \subseteq D(\xi,\rho) \cap K \cap Y$. From now on, we have to repeat the same routine as above by using the fact that the restriction of β_Y — as defined in Problem 2.7.2 — to $\mathcal{B}(Y)$ is the Hausdorff measure of noncompactness on Y. This completes the proof. □

8.5. The quasi-autonomous case

Let X be a real Banach space, $A : D(A) \subseteq X \to X$ the infinitesimal generator of a C_0-semigroup, $\mathcal{C}$ a nonempty subset in $\mathbb{R} \times X$, $f : \mathcal{C} \to X$ a given function. The goal of this section is to extend the necessary and sufficient conditions for viability already proved in the autonomous case to the general frame of quasi-autonomous semilinear Cauchy problems of the form

$$\begin{cases} u'(t) = Au(t) + f(t, u(t)) \\ u(\tau) = \xi. \end{cases} \tag{8.5.1}$$

Throughout, we denote by $\mathcal{X} = \mathbb{R} \times X$, which endowed with the usual norm $\|(t,u)\| = |t| + \|u\|$, is a Banach space.

Definition 8.5.1. Let $A : D(A) \subseteq X \to X$ be the infinitesimal generator of a C_0-semigroup, $\{S(t) : X \to X;\ t \ge 0\}$, and $f : \mathcal{C} \to X$ a function. We say that $A + f$ is *locally of compact type with respect to the second argument* if f is continuous and, for each $(\tau,\xi) \in \mathcal{C}$, there exist $\rho > 0$, a continuous function $\ell : \mathbb{R}_+ \to \mathbb{R}_+$ and a uniqueness function $\omega : \mathbb{R}_+ \to \mathbb{R}_+$, such that $f(D_{\mathcal{X}}((\tau,\xi),\rho) \cap \mathcal{C})$ is bounded and

$$\beta(S(t)f(C)) \leq \ell(t)\omega(\beta(\Pi_X C))$$

for each $t > 0$ and each $C \subseteq D_{\mathcal{X}}((\tau,\xi),\rho) \cap \mathcal{C}$.

Definition 8.5.2. By a *mild solution* of the problem (8.5.1) on $[\tau, T]$, we mean a continuous function $u : [\tau, T] \to X$ which satisfies:

(i) $(t, u(t)) \in \mathcal{C}$ for each $t \in [\tau, T]$;
(ii) the function $s \mapsto f(s, u(s))$ lies in $L^1(\tau, T; X)$;
(iii) u is a mild solution, in the sense of Definition 1.5.3, of the Cauchy problem

$$\begin{cases} u' = Au + g \\ u(\tau) = \xi, \end{cases}$$

where $g(s) = f(s, u(s))$ for a.a. $s \in [\tau, T]$, i.e.

$$u(t) = S(t-\tau)\xi + \int_\tau^t S(t-s)f(s,u(s))\,ds,$$

for each $t \in [\tau, T]$.

Definition 8.5.3. The set $\mathcal{C} \subseteq \mathbb{R} \times X$ is *mild viable* with respect to $A+f$ if for each $(\tau, \xi) \in \mathcal{C}$, there exists $T \in \mathbb{R}$, $T > \tau$ such that the Cauchy problem (8.5.1) has at least one mild solution $u : [\tau, T] \to X$.

Remark 8.5.1. In order to introduce the tangency concept we are going to use in the sequel, let us first observe that the quasi-autonomous Cauchy problem (8.5.1) can be equivalently rewritten as an autonomous one in the space $\mathcal{X}$, by setting $\mathcal{A} = (0, A)$, $z(s) = (t(s+\tau), u(s+\tau))$, $\mathcal{F}(z) = (1, f(z))$ and $\zeta = (\tau, \xi)$. Indeed, with the notations above, we have

$$\begin{cases} z'(s) = \mathcal{A}z(s) + \mathcal{F}(z(s)) \\ z(0) = \zeta. \end{cases}$$

It readily follows that $\mathcal{A}$ generates a C_0-semigroup, $\{\mathcal{S}(t) : \mathcal{X} \to \mathcal{X};\ t \geq 0\}$, on $\mathcal{X}$, where $\mathcal{S}(t) = (1, S(t))$ for each $t \geq 0$, $\{S(t) : X \to X;\ t \geq 0\}$ being the C_0-semigroup generated by A on X. So, the mild solution z of the problem above is given by the variation of constants formula

$$z(t) = \mathcal{S}(t)\zeta + \int_0^t \mathcal{S}(t-s)\mathcal{F}(z(s))\,ds$$

whose form, on components, is

$$z(t) = \left(\tau + t, S(t)\xi + \int_0^t S(t-s)f(z(s))\,ds\right).$$

Remark 8.5.2. One may easily see that $\mathcal{C}$ is mild viable with respect to $A+f$ in the sense of Definition 8.5.3 if and only if $\mathcal{C}$ is mild viable with respect to $\mathcal{A}+\mathcal{F}$ in the sense of Definition 8.1.2.

Theorem 8.5.1. *Let X be a Banach space, $A : D(A) \subseteq X \to X$ the infinitesimal generator of a C_0-semigroup, $\{S(t) : X \to X;\ t \geq 0\}$, $\mathcal{C}$ a nonempty subset in $\mathbb{R} \times X$ and $f : \mathcal{C} \to X$ a continuous function. If $\mathcal{C}$ is mild viable with respect to $A+f$ then, for each $(\tau,\xi) \in \mathcal{C}$, we have*

$$\lim_{h\downarrow 0} \frac{1}{h}\operatorname{dist}\left((\tau, S(h)\xi) + h(1, f(\tau,\xi)); \mathcal{C}\right) = 0. \tag{8.5.2}$$

Proof. From Remark 8.5.2 we know that $\mathcal{C}$ is mild viable with respect to $\mathcal{A}+\mathcal{F}$ in the sense of Definition 8.1.2 and thus, by Theorem 8.1.1, we conclude that, under the hypotheses of Theorem 8.5.1, for each $(\tau,\xi) \in \mathcal{C}$, we have

$$\lim_{h\downarrow 0} \frac{1}{h}\operatorname{dist}\left(\mathcal{S}(h)(\tau,\xi) + h\mathcal{F}(\tau,\xi); \mathcal{C}\right) = 0,$$

relation which is equivalent to (8.5.2). □

Remark 8.5.3. Under the general hypotheses of Theorem 8.5.1, if $\mathcal{C}$ is mild viable with respect to $A+f$ then, for each $(\tau,\xi) \in \mathcal{C}$, we have

$$(1, f(\tau,\xi)) \in \mathcal{T}_{\mathcal{C}}^{\mathcal{A}}(\tau,\xi).$$

As in the autonomous case, Theorem 8.5.1 is a direct consequence of a slightly more general necessary condition, i.e., Theorem 8.5.2 below.

Theorem 8.5.2. *Let X be a Banach space, $A : D(A) \subseteq X \to X$ the infinitesimal generator of a C_0-semigroup, $\{S(t) : X \to X;\ t \geq 0\}$, $\mathcal{C}$ a nonempty subset in $\mathbb{R} \times X$ and $f : \mathcal{C} \to X$ a given function. If $\mathcal{C}$ is mild viable with respect to $A+f$ then the tangency condition* (8.5.2) *is satisfied at each continuity point, $(\tau,\xi) \in \mathcal{C}$, of f.*

Proof. Recall Remark 8.5.2 and use Theorem 8.1.2. □

We will next prove the main sufficient conditions concerning the viability of a set $\mathcal{C}$ with respect to $A+f$.

Theorem 8.5.3. *Let X be a Banach space, $A : D(A) \subseteq X \to X$ the infinitesimal generator of a C_0-semigroup, $\{S(t) : X \to X;\ t \geq 0\}$, $\mathcal{C}$ a nonempty and locally closed subset in $\mathbb{R}\times X$ and $f : \mathcal{C} \to X$ a function such that $A+f$ is locally of compact type with respect to the second argument. Then $\mathcal{C}$ is mild viable with respect to $A+f$ if and only if for each $(\tau,\xi) \in \mathcal{C}$, we have*

$$\liminf_{h\downarrow 0} \frac{1}{h}\operatorname{dist}\left((\tau, S(h)\xi) + h(1, f(\tau,\xi)); \mathcal{C}\right) = 0. \tag{8.5.3}$$

Proof. We have $\beta(C) \geq \max\{\beta(\Pi_{\mathbb{R}}C), \beta(\Pi_X C)\} = \beta(\Pi_X C)$ for each bounded set C in $\mathfrak{X}$. Therefore, if $A + f$ is locally of compact type with respect to the second argument then $\mathcal{A} + \mathfrak{F}$ is locally of compact type (in the space $\mathfrak{X}$) in the sense of Definition 8.2.1. Thus the conclusion follows from Remark 8.5.2 and Theorem 8.2.1. $\square$

Problem 8.5.1. *Prove that, in the specific case when $\mathcal{C} = I \times K$, where I is an open (to the right) interval and K is a nonempty subset of X, the tangency condition* (8.5.3) *is equivalent to*

$$f(\tau, \xi) \in \mathcal{T}_K^A(\xi).$$

Theorem 8.5.4. *Let X be a Banach space, $A : D(A) \subseteq X \to X$ the infinitesimal generator of a compact C_0-semigroup, $\{S(t) : X \to X;\ t \geq 0\}$, $\mathcal{C}$ a nonempty and locally closed subset in $\mathbb{R} \times X$ and $f : \mathcal{C} \to X$ a continuous function. Then $\mathcal{C}$ is mild viable with respect to $A + f$ if and only if, for each $(\tau, \xi) \in \mathcal{C}$, the tangency condition* (8.5.3) *is satisfied.*

Proof. Use Remark 8.5.2 and Theorem 8.2.3. $\square$

The next results are simple corollaries of Theorem 8.5.3.

Theorem 8.5.5. *Let X be a Banach space $A : D(A) \subseteq X \to X$ the infinitesimal generator of a C_0-semigroup, $\{S(t) : X \to X;\ t \geq 0\}$, $\mathcal{C}$ a nonempty and locally closed subset in $\mathbb{R} \times X$ and $f : \mathcal{C} \to X$ a locally β-compact function. Then $\mathcal{C}$ is mild viable with respect to $A + f$ if and only if, for each $(\tau, \xi) \in \mathcal{C}$, the tangency condition* (8.5.3) *is satisfied.*

Theorem 8.5.6. *Let X be a Banach space $A : D(A) \subseteq X \to X$ the infinitesimal generator of a C_0-semigroup, $\{S(t) : X \to X;\ t \geq 0\}$, $\mathcal{C}$ a nonempty and locally closed subset in $\mathbb{R} \times X$ and $f : \mathcal{C} \to X$ a locally Lipschitz function. Then $\mathcal{C}$ is mild viable with respect to $A + f$ if and only if, for each $(\tau, \xi) \in \mathcal{C}$, the tangency condition* (8.5.3) *is satisfied.*

8.6. A class of reaction-diffusion systems

As direct applications of the results in Section 8.2, we will prove here two viability theorems referring to a class of semilinear reaction-diffusion systems. Let X and Y be two real Banach spaces[4].

For simplicity, in what follows, we denote by $\mathcal{G}(X)$ the class of all linear operators $A : D(A) \subseteq X \to X$ which generate C_0-semigroups on X. Moreover, if $A \in \mathcal{G}(X)$, then $\{S_A(t) : X \to X;\ \ t \geq 0\}$ denotes the C_0-semigroup

[4]Since there is no danger of confusion, we denote the norms on both spaces by the same symbol, $\|\cdot\|$.

generated by A. The class $\mathcal{G}(Y)$ and $\{S_B(t) : Y \to Y;\ \ t \geq 0\}$ are similarly defined.

Let $\mathcal{K} \subseteq X \times Y$ be nonempty, let $A \in \mathcal{G}(X)$, $B \in \mathcal{G}(Y)$, $F : \mathcal{K} \to X$, $G : \mathcal{K} \to Y$ and let us consider the Cauchy problem for the abstract semilinear reaction-diffusion system:

$$\begin{cases} u'(t) = Au(t) + F(u(t), v(t)) \\ v'(t) = Bv(t) + G(u(t), v(t)) \\ u(0) = \xi,\ \ v(0) = \eta. \end{cases} \tag{8.6.1}$$

Definition 8.6.1. We say that the set $\mathcal{K}$ is *mild viable* with respect to $(A + F, B + G)$ if for each $(\xi, \eta) \in \mathcal{K}$ there exists $T > 0$ such that (8.6.1) has at least one mild solution $(u, v) : [0, T] \to \mathcal{K}$ i.e.

$$\begin{cases} u(t) = S_A(t)\xi + \displaystyle\int_0^t S_A(t-s)F(u(s), v(s))\, ds \\ v(t) = S_B(t)\xi + \displaystyle\int_0^t S_B(t-s)G(u(s), v(s))\, ds \end{cases}$$

for each $t \in [0, T]$.

Remark 8.6.1. The system (8.6.1) can be rewritten as a semilinear autonomous equation in a product space. Namely, let $\mathcal{X} = X \times Y$ which, endowed with the norm $\|\cdot\|$, defined by $\|(x, y)\| = \|x\| + \|y\|$ for each $(x, y) \in \mathcal{X}$, is a real Banach space. Let $\mathcal{A} = (A, B) : D(\mathcal{A}) \subseteq \mathcal{X} \to \mathcal{X}$ be defined by: $D(\mathcal{A}) = D(A) \times D(B)$ and $\mathcal{A}(x, y) = (Ax, By)$ for each $(x, y) \in D(\mathcal{A})$. Let $\mathcal{F} : \mathcal{K} \to \mathcal{X}$ be defined by $\mathcal{F}(z) = (F(z), G(z))$ for all $z = (u, v) \in \mathcal{K}$ and let $\zeta = (\xi, \eta)$. So, (8.6.1) can be equivalently written as

$$\begin{cases} z'(t) = \mathcal{A}z(t) + \mathcal{F}(z(t)) \\ z(0) = \zeta. \end{cases} \tag{8.6.2}$$

We notice that, whenever both A and B generate C_0-semigroups, in its turn $\mathcal{A}$ is the infinitesimal generator of a C_0-semigroup $\{\mathcal{S}(t) : \mathcal{X} \to \mathcal{X};\ t \geq 0\}$, given by

$$\mathcal{S}(t)(\xi, \eta) = (S_A(t)\xi, S_B(t)\eta)$$

for each $t \geq 0$ and $(\xi, \eta) \in \mathcal{X}$.

However, from technical reasons, when necessary, we will use either the initial specific form of the system, i.e., (8.6.1), or the semilinear equation (8.6.2), as required by the necessities of the presentation. We notice that in either of the two forms, all metric concepts involved such as distance, measure of noncompactness, etc., are expressed in the terms of the norm on $\mathcal{X}$ introduced above.

Remark 8.6.2. Clearly $\mathcal{K}$ is mild viable with respect to $(A+F, B+G)$ in the sense of Definition 8.6.1 if and only if $\mathcal{K}$ is mild viable with respect to $\mathcal{A}+\mathcal{F}$ in the sense of Definition 8.1.2.

As usual, we denote by

$$\Pi_X C = \{u \in X;\ \exists v \in Y, (u, v) \in C\},$$
$$\Pi_Y C = \{v \in Y;\ \exists u \in X, (u, v) \in C\}.$$

Definition 8.6.2. Let $\mathcal{K} \subseteq \mathcal{X}$ and let A and F as above, where $\mathcal{X}$ is as in Remark 8.6.1. We say that $A+F$ is Y*-uniformly locally of compact type* if F is continuous and, for each $(\xi, \eta) \in \mathcal{K}$, there exist $\rho > 0$, a continuous function $\ell : \mathbb{R}_+ \to \mathbb{R}_+$ and a uniqueness function $\omega : \mathbb{R}_+ \to \mathbb{R}_+$ such that $F(D_{\mathcal{X}}((\xi, \eta), \rho) \cap \mathcal{K})$ is bounded and, for each set $C \subseteq D_{\mathcal{X}}((\xi, \eta), \rho) \cap \mathcal{K}$, with $\Pi_Y C$ relatively compact, we have

$$\beta(S_A(t)F(C)) \leq \ell(t)\omega(\beta(C))$$

for each $t > 0$.

Remark 8.6.3. If $A+F$ is Y-uniformly locally of compact type and ρ, ℓ and ω are as in Definition 8.6.2, then, for each set $C \subseteq D_{\mathcal{X}}((\xi, \eta), \rho) \cap \mathcal{K}$, with $\Pi_Y C$ relatively compact, we have

$$\beta(S_A(t)F(C)) \leq \ell(t)\omega(\beta(\Pi_X C))$$

for each $t > 0$. This follows from the simple observation that, due to the definition of the norm on $\mathcal{X}$, for each bounded subset C of $\mathcal{X}$, we have

$$\beta(C) \leq \beta(\Pi_X C) + \beta(\Pi_Y C).$$

Remark 8.6.4. Let $A \in \mathcal{G}(X)$. A simple example of a function F for which $A+F$ is Y-uniformly locally of compact type is that one in which $\mathcal{K} = K_X \times K_Y$, where $K_X \subseteq X$, $K_Y \subseteq Y$ and $F : \mathcal{K} \to X$ is of the special form $F(u, v) = f(u) + g(v)$, with $f : K_X \to X$ and $A+f$ is locally of compact type. In particular, if F is of the form specified above, where f is locally β-compact and $g : K_Y \to X$ is continuous, then $A+F$ is Y-uniformly locally of compact type.

The hypotheses which will be in effect throughout are:

(H_1) $A \in \mathcal{G}(X)$ *and* $B \in \mathcal{G}(Y)$;
(H_2) $\mathcal{K} \subseteq X \times Y$ *is nonempty and locally closed*;
(H_3) $(A+F, B+G) : \mathcal{K} \to X \times Y$ *is locally of compact type*[5];
(H_4) $A+F$ *is* Y*-uniformly locally of compact type*;
(H_5) $\{S_B(t) : Y \to Y;\ t \geq 0\}$ *is compact*;

[5]This means that $\mathcal{A}+\mathcal{F}$ is locally of compact type in the sense of Definition 8.2.1. For the definition of both $\mathcal{A}$ and $\mathcal{F}$ see Remark 8.6.1.

(H_6) *$G : \mathcal{K} \to Y$ is continuous.*

The main viability results referring to (8.6.1), are Theorems 8.6.1, which is an immediate consequence of Theorem 8.2.1, and Theorem 8.6.2, whose proof will be delivered in Section 8.7.

Theorem 8.6.1. *Assume that (H_1) $\sim$ (H_3) are satisfied. The necessary and sufficient condition in order that $\mathcal{K}$ be mild viable with respect to $(A+F, B+G)$ is that, for every $(\xi,\eta) \in \mathcal{K}$,*

$$\liminf_{h\downarrow 0} \frac{1}{h} \operatorname{dist}\left((u(h,\xi,\eta), v(h,\xi,\eta)); \mathcal{K}\right) = 0, \tag{8.6.3}$$

where

$$\begin{cases} u(h,\xi,\eta) = S_A(h)\xi + hF(\xi,\eta) \\ v(h,\xi,\eta) = S_B(h)\eta + hG(\xi,\eta). \end{cases} \tag{8.6.4}$$

Theorem 8.6.2. *Assume that (H_1), (H_2), (H_4) $\sim$ (H_6) are satisfied. The necessary and sufficient condition in order that $\mathcal{K}$ be mild viable with respect to $(A+F, B+G)$, is that, for all $(\xi,\eta) \in \mathcal{K}$, the tangency condition* (8.6.3) *is satisfied.*

The necessity part of both Theorems 8.6.1 and 8.6.2 follows from the simple result below.

Theorem 8.6.3. *Assume that (H_1) is satisfied, the set $\mathcal{K} \subseteq X \times Y$ is nonempty and $F : \mathcal{K} \to X$, $G : \mathcal{K} \to Y$ are continuous. If $\mathcal{K}$ is mild viable with respect to $(A+F, B+G)$ then, for each $(\xi,\eta) \in \mathcal{K}$, we have*

$$\lim_{h\downarrow 0} \frac{1}{h} \operatorname{dist}\left((u(h,\xi,\eta), v(h,\xi,\eta)); \mathcal{K}\right) = 0.$$

Proof. From Remark 8.6.1 it follows that we are in the hypotheses of Theorem 8.1.2, wherefrom the conclusion. □

We conclude with a nonautonomous version of Theorem 8.6.1 which results from it by using similar arguments as those in Section 8.5. We leave to the reader to formulate and to prove similar extensions to the nonautonomous case for Theorems 8.6.2 and 8.6.3.

Theorem 8.6.4. *Assume that (H_1) is satisfied, let $\mathcal{C} \subseteq \mathbb{R} \times \mathcal{X}$ be a locally closed set and let $(F,G) : \mathcal{C} \to \mathcal{X}$ be continuous. Let us assume that $(A+F, B+G)$ is locally of compact type with respect the second argument, i.e., with respect to $(u,v) \in \mathcal{X}$. Then a necessary and sufficient condition in order that $\mathcal{C}$ be mild viable with respect to $(A+F, B+G)$ is that*

$$\liminf_{h\downarrow 0} \frac{1}{h} \operatorname{dist}\left((\tau+h, u(\tau+h,\tau,\xi,\eta), v(\tau+h,\tau,\xi,\eta)); \mathcal{C}\right) = 0 \tag{8.6.5}$$

for each $(\tau, \xi, \eta) \in \mathcal{C}$, *where*

$$\begin{cases} u(\tau + h, \tau, \xi, \eta) = S_A(h)\xi + hF(\tau, \xi, \eta) \\ v(\tau + h, \tau, \xi, \eta) = S_B(h)\eta + hG(\tau, \xi, \eta). \end{cases}$$

8.7. Convergence in the case of Theorem 8.6.2

First, let us observe that, in view of Remark 8.6.1, Theorem 8.6.2 can be reformulated in the terms of the semilinear equation (8.6.2) as follows:

Theorem 8.7.1. *Assume that* (H_1), (H_2), $(H_4) \sim (H_6)$ *are satisfied. The necessary and sufficient condition in order that* $\mathcal{K}$ *be mild viable with respect to* $\mathcal{A} + \mathcal{F}$ *is that, for every* $\zeta \in \mathcal{K}$,

$$\liminf_{h \downarrow 0} \frac{1}{h} \mathrm{dist}\,(\mathcal{S}(h)\zeta + h\mathcal{F}(\zeta); \mathcal{K}) = 0. \tag{8.7.1}$$

To prove Theorem 8.7.1, we will use Lemma 8.3.1 to produce a sequence of ε_n-approximate solutions to the Cauchy problem

$$\begin{cases} z'(t) = \mathcal{A}z(t) + \mathcal{F}(z(t)) \\ z(0) = \zeta. \end{cases} \tag{8.7.2}$$

Proof. Let $\zeta = (\xi, \eta) \in \mathcal{K}$, let $\rho > 0$, $T > 0$, $M \geq 1$ and $N > 0$ as in Lemma 8.3.1. Since $A + F$ is Y-uniformly locally of compact type, diminishing both $\rho > 0$ and $T > 0$ if necessary, we conclude that there exist a continuous function $\ell : \mathbb{R}_+ \to \mathbb{R}_+$ and a uniqueness function $\omega : \mathbb{R}_+ \to \mathbb{R}_+$ such that $F(D_{\mathcal{X}}(\zeta, \rho) \cap \mathcal{K})$ is bounded and, for each $C \subseteq D_{\mathcal{X}}(\zeta, \rho) \cap \mathcal{K}$ and each $t > 0$, we have

$$\beta(S_A(t)F(C)) \leq \ell(t)\omega(\beta(C)). \tag{8.7.3}$$

Let $\varepsilon_n \downarrow 0$ be a sequence in $(0,1)$ and let $((\sigma_n, \theta_n, \mathcal{G}_n, z_n))_n$ be a sequence of ε_n-approximate mild solutions defined on $[0, T]$ whose existence is ensured by Lemma 8.3.1.

We consider first the case when X is separable.

From (v)[6], we have

$$z_n(t) = \mathcal{S}(t)\zeta + \int_0^t \mathcal{S}(t-s)\mathcal{F}(z_n(\sigma_n(s)))\,ds + \int_0^t \mathcal{S}(\theta_n(t,s))\mathcal{G}_n(s)\,ds, \tag{8.7.4}$$

for each $n \in \mathbb{N}$ and $t \in [0, T]$. Since $\|\mathcal{F}(z_n(\sigma_n(s)))\| \leq N$ and $\|\mathcal{G}_n(t)\| \leq \varepsilon_n$ for each $n \in \mathbb{N}$ and $t \in [0, T]$, thanks to the fact that F is Y-uniformly locally of compact type and to Theorem 1.5.3, we conclude that $\{z_n;\ n \in \mathbb{N}\}$ is relatively compact in $C([0, T]; \mathcal{X})$.

[6]Throughout this proof, the references to (i)∼(vi) are to the corresponding items in Lemma 8.3.1.

Indeed, $z_n = (u_n, v_n)$ satisfies $z_n(\sigma_n(t)) \in D_{\mathcal{X}}(\zeta,\rho) \cap \mathcal{K}$, for $n = 1, 2, \dots$ and each $t \in [0, T]$, where (u_n, v_n) satisfies the system

$$\begin{cases} u_n(t) = S_A(t)\xi + \displaystyle\int_0^t S_A(t-s)F(z_n(\sigma_n(s)))ds + \int_0^t S_A(\theta_n(t,s))\mathcal{G}_n^X(s)ds \\ v_n(t) = S_B(t)\eta + \displaystyle\int_0^t S_B(t-s)G(z_n(\sigma_n(s)))ds + \int_0^t S_B(\theta_n(t,s))\mathcal{G}_n^Y(s)ds. \end{cases}$$

We notice that here $\mathcal{G}_n(s) = (\mathcal{G}_n^X(s), \mathcal{G}_n^Y(s))$ for $n = 1, 2, \dots$ and $s \in [0, T]$. Since $\{G(z_n(\sigma_n(\cdot))); \ n \in \mathbb{N}\}$ is uniformly bounded on $[0, T]$, $\|\mathcal{G}_n^Y(s)\| \leq \varepsilon_n$, for $n = 1, 2, \dots$, and the C_0-semigroup $\{S_B(t) : Y \to Y; \ t \geq 0\}$ is compact, from Theorem 1.5.3, it follows that $\{v_n; \ n = 1, 2, \dots\}$ is relatively compact in $C([0, T]; Y)$. In particular, for each $t \in [0, T]$, $\{v_n(\sigma_n(t)); \ n = 1, 2, \dots\}$ is relatively compact in Y. Since

$$\Pi_X\{(u_n(\sigma_n(t)), v_n(\sigma_n(t))); \ n \geq k\} = \{u_n(\sigma_n(t)); \ n \geq k\},$$
$$\Pi_Y\{(u_n(\sigma_n(t)), v_n(\sigma_n(t))); \ n \geq k\} = \{v_n(\sigma_n(t)); \ n \geq k\},$$

for $k = 1, 2, \dots$, and $A + F$ is Y-uniformly locally of compact type, from (8.7.3) and Remark 8.6.3, we get

$$\beta(\{S_A(t-s)F(u_n(\sigma_n(s)), v_n(\sigma_n(s))); \ n \geq k\}) \leq \ell(t-s)\omega(\beta(\{u_n(\sigma_n(s)); \ n \geq k\})) \tag{8.7.5}$$

for $k = 1, 2, \dots$, each $t \in [0, T]$ and $s \in [0, t]$. We have

$$\beta(\{u_n(t); \ n \geq k\}) \leq \beta\left(\left\{\int_0^t S_A(t-s)F(u_n(\sigma_n(s)), v_n(\sigma_n(s)))\, ds; \ n \geq k\right\}\right) + \beta\left(\left\{\int_0^t S_A(\theta_n(t,s))\mathcal{G}_n^X(s)\, ds; \ n \geq k\right\}\right)$$

for $k = 1, 2, \dots$ and $t \in [0, T]$. From Theorem 1.4.1, there exist $M \geq 1$ and $a \geq 0$ such that

$$\|S_A(t)\|_{\mathcal{L}(X)} \leq Me^{at},$$

for each $t \geq 0$. By Lemma 2.7.2 and (8.7.5), we get

$$\beta(\{u_n(t); \ n \geq k\}) \leq \int_0^t \beta(\{S_A(t-s)F(z_n(\sigma_n(s))); \ n \geq k\})\, ds + \int_0^t \beta(\{S_A(\theta_n(t,s))\mathcal{G}_n^X(s); \ n \geq k\})\, ds$$
$$\leq \int_0^t \ell(t-s)\omega(\beta(\{u_n(\sigma_n(s)); \ n \geq k\}))ds + TMe^{aT}\varepsilon_k$$

$$\leq \int_0^t \ell(t-s)\omega(\beta(\{u_n(s);\ n\geq k\}+\{u_n(\sigma_n(s))-u_n(s);\ n\geq k\}))ds$$
$$+TMe^{aT}\varepsilon_k$$
$$\leq \int_0^t m\omega(\beta(\{u_n(s);\ n\geq k\})+\beta(\{u_n(\sigma_n(s))-u_n(s);\ n\geq k\}))ds$$
$$+TMe^{aT}\varepsilon_k,$$

where $m=\sup_{t\in[0,T]}\ell(t)$. Thus

$$\beta(\{u_n(t);\ n\geq k\})\leq \int_0^t m\omega(\beta(\{u_n(s);\ n\geq k\})+\varepsilon_k)\,ds+TMe^{aT}\varepsilon_k$$

for $k=1,2,\ldots$ and $t\in[0,T]$.

Set $x_k(t)=\beta(\{u_n(t);\ n\geq k\})+\varepsilon_k$ and $\gamma_k=(TMe^{aT}+1)\varepsilon_k$, for $k=1,2,\ldots$, and $\omega_0=m\omega$. Hence

$$x_k(t)\leq \gamma_k+\int_0^t \omega_0(x_k(s))ds$$

for $k=1,2,\ldots$ and $t\in[0,T]$.

From Remark 1.8.1, Lemma 1.8.2 and Lemma 2.7.3, we may assume without loss of generality that, for each $t\in[0,T]$, $\{u_n(t);\ n=1,2,\ldots\}$ is relatively compact in X. As, for each $t\in[0,T]$, $\{v_n(t);\ n=1,2,\ldots\}$ is relatively compact in Y, it follows that, for each $t\in[0,T]$, the family $\{z_n(t);\ n=1,2,\ldots\}$ is relatively compact in $\mathcal{X}$. Inasmuch as $(\mathcal{F}(z_n))_n$ is bounded, it is also uniformly integrable and so, by Theorem 1.5.2, there exists $z\in C([0,T];\mathcal{X})$ such that, on a subsequence at least,

$$\lim_n\left(z_n(t)-\int_0^t \mathcal{S}(\theta_n(t,s))\mathcal{G}_n(s)\,ds\right)=z(t)$$

uniformly for $t\in[0,T]$. But, by (iii),

$$\lim_n\int_0^t \mathcal{S}(\theta_n(t,s))\mathcal{G}_n(s)\,ds=0,$$

uniformly for $t\in[0,T]$, and therefore $\lim_n z_n(t)=z(t)$ uniformly for $t\in[0,T]$. Using (ii) and (vi), we deduce that $z(t)\in\mathcal{K}$ for each $t\in[0,T]$ and thus, passing to the limit for $n\to\infty$ in (8.7.4), we deduce that

$$z(t)=\mathcal{S}(t)\zeta+\int_0^t \mathcal{S}(t-s)\mathcal{F}(z(s))\,ds$$

for each $t\in[0,T]$. This means that $z=(u,v)$ is a mild solution of (8.7.2) or, equivalently, of (8.6.1). The proof of Theorem 8.7.1, and thus that one of Theorem 8.6.2, is therefore complete if X is separable.

If X is not separable, in view of Remark 2.7.3, there exists a separable and closed subspace Z of X such that

$$S_A(t)\xi, S_A(r)F(u_n(\sigma_n(s)), v_n(\sigma_n(s))), S_A(\theta_n(r,s))\mathcal{G}_n^X(s) \in Z$$

for $n = 1, 2, \ldots$ and a.e. for $t, s, r \in [\,0, T\,]$. In view of Problem 2.7.2 and of the monotonicity of ω, we have

$$\beta_Z(S_A(t)F(C)) \leq 2\beta(S_A(t)F(C)) \leq 2\ell(t)\omega(\beta(C)) \leq 2\ell(t)\omega(\beta_Z(C)),$$

for each $t > 0$ and each set $C \subseteq D_{\mathcal{X}}(\zeta, \rho) \cap \mathcal{K} \cap (Z \times Y)$ for which $\Pi_Y(C)$ is relatively compact. From now on, just repeat the arguments in the separable case with β replaced by β_Z and ω replaced by 2ω, by using Remark 2.7.3 instead of Lemma 2.7.2, and the fact that the restriction of β_Z — as defined in Problem 2.7.2 — to $\mathcal{B}(Z)$ is the Hausdorff measure of noncompactness on Z. This completes the proof. □

8.8. Global mild solutions

Let $A : D(A) \subseteq X \to X$ be the infinitesimal generator of a C_0-semigroup, let $\mathcal{C} \subseteq \mathbb{R} \times X$ be nonempty and let $f : \mathcal{C} \to X$. In this section we will state some results concerning the existence of noncontinuable, or even global mild solutions to the Cauchy problem

$$\begin{cases} u'(t) = Au(t) + f(t, u(t)) \\ u(\tau) = \xi. \end{cases} \tag{8.8.1}$$

We recall that a mild solution $u : [\,\tau, T) \to X$ to (8.8.1) is called *noncontinuable*, if there is no other mild solution $v : [\,\tau, \widetilde{T}) \to X$ of the same equation, with $T < \widetilde{T}$ and satisfying $u(t) = v(t)$ for all $t \in [\,\tau, T)$. The mild solution u is called *global* if $T = T_{\mathcal{C}}$, where $T_{\mathcal{C}}$ is defined by (3.6.2).

Theorem 8.8.1. *Let X be a Banach space, $A : D(A) \subseteq X \to X$ the infinitesimal generator of a C_0-semigroup, $\{S(t) : X \to X;\ t \geq 0\}$, $\mathcal{C}$ a nonempty subset in $\mathbb{R} \times X$ and $f : \mathcal{C} \to X$ a given function. The following conditions are equivalent:*

(i) *$\mathcal{C}$ is mild viable with respect to $A + f$;*
(ii) *for each $(\tau, \xi) \in \mathcal{C}$ there exists at least one noncontinuable mild solution $u : [\tau, T) \to X$ of* (8.8.1).

The proof, based on Brezis–Browder Theorem 2.1.1, is very similar with the one of Theorem 3.6.1 and therefore we omit it.

The next two results concern the existence of global solutions.

Theorem 8.8.2. *Let X be a Banach space, $A : D(A) \subseteq X \to X$ the infinitesimal generator of a C_0-semigroup and let $\mathcal{C} \subseteq \mathbb{R} \times X$ be an X-closed subset in $\mathbb{R} \times X$*[7]*. Let $f : \mathcal{C} \to X$ be a continuous function such that $\mathcal{C}$ is mild viable with respect to $A + f$. Let us assume that there exist two continuous functions $a : \mathbb{R} \to \mathbb{R}_+$ and $b : \mathbb{R} \to \mathbb{R}_+$ such that*

$$\|f(t, \xi)\| \leq a(t)\|\xi\| + b(t) \tag{8.8.2}$$

for each $(t, \xi) \in \mathcal{C}$. Then each mild solution of (8.8.1) *can be continued up to a global one, i.e., defined on $[\tau, T_{\mathcal{C}})$.*

Proof. Just apply the Gronwall Lemma 1.8.4 to show that whenever a noncontinuable mild solution u is defined on an interval $[\tau, T)$, with $T < T_{\mathcal{C}}$, it is bounded on its domain, and therefore it satisfies the Cauchy condition for the existence of $u^* = \lim_{t \uparrow T} u(t)$. But in this case, since $\mathcal{C}$ is X-closed, we have $(T, u^*) \in \mathcal{C}$, and thus u can be continued to the right of T which is absurd. Hence $T < T_{\mathcal{C}}$ is impossible and this completes the proof. □

In the specific case in which A generates a C_0-semigroup of contractions, i.e., when A is linear and dissipative, as stated in Theorem 1.4.8, the conclusion of Theorem 8.8.2 holds true under more general growth conditions on f. Namely, we have

Theorem 8.8.3. *Let X be a Banach space, $A : D(A) \subseteq X \to X$ a linear m-dissipative operator, $\mathcal{C}$ a nonempty subset in $\mathbb{R} \times X$ and $f : \mathcal{C} \to X$ a continuous function which is positively sublinear*[8]*. If f maps bounded subsets in $\mathcal{C}$ into bounded subsets in X and $\mathcal{C}$ is X-closed and mild viable with respect to $A + f$, then each mild solution of* (8.8.1) *can be continued up to a global one, i.e., defined on $[\tau, T_{\mathcal{C}})$.*

Since we will later prove a more general result, i.e. Theorem 10.6.2, allowing A to be nonlinear, we do not give further details.

---

[7]See Definition 3.6.2.

[8]See Definition 3.6.1.

CHAPTER 9

Viability for multi-valued semilinear evolutions

In this chapter we focus our attention on the case of semilinear evolution equations governed by multi-valued perturbations of infinitesimal generators of C_0-semigroups. We first consider the autonomous case and start with the definition of the A-quasi-tangent set at a point to a given set. Using this new concept, we next prove the main necessary condition for mild viability. Then, we state and prove several necessary and sufficient conditions for mild viability. We extend the previous results to the quasi autonomous case and we conclude with some facts concerning the existence of noncontinuable or even global mild solutions.

9.1. Necessary conditions for mild viability

Let X be a real Banach space, let $A : D(A) \subseteq X \to X$ be the infinitesimal generator of a C_0-semigroup, $\{S(t) : X \to X;\ t \geq 0\}$, K a nonempty subset in X and $F : K \rightsquigarrow X$ a given multi-function.

Definition 9.1.1. By a *mild solution* of the autonomous multi-valued semilinear Cauchy problem

$$\begin{cases} u'(t) \in Au(t) + F(u(t)) \\ u(0) = \xi, \end{cases} \tag{9.1.1}$$

on $[\,0, T\,]$, we mean a continuous function $u : [\,0, T\,] \to K$ for which there exists $f \in L^1(0, T; X)$ such that $f(s) \in F(u(s))$ a.e. for $s \in [\,0, T\,]$ and

$$u(t) = S(t)\xi + \int_0^t S(t-s)f(s)\,ds, \tag{9.1.2}$$

for each $t \in [\,0, T\,]$.

Definition 9.1.2. The set $K \subseteq X$ is *mild viable* with respect to $A + F$ if for each $\xi \in K$, there exists $T > 0$ such that the Cauchy problem (9.1.1) has at least one mild solution $u : [\,0, T\,] \to K$.

Let $E \subseteq X$ be nonempty. We denote by

$$\mathcal{E} = \{f \in L^1(\mathbb{R}_+; X);\ f(s) \in E \text{ a.e. for } s \in \mathbb{R}_+\}.$$

Definition 9.1.3. Let $A : D(A) \subseteq X \to X$ be the infinitesimal generator of a C_0-semigroup, $\{S(t) : X \to X;\ t \geq 0\}$, K a subset in X and $\xi \in K$. The set $E \subseteq X$ is *A-quasi-tangent* to the set K at the point $\xi \in K$ if for each $\rho > 0$, we have

$$\liminf_{h \downarrow 0} \frac{1}{h} \operatorname{dist}\left(S_{\mathcal{E}}(h)\xi; K \cap D(\xi, \rho)\right) = 0, \tag{9.1.3}$$

where

$$S_{\mathcal{E}}(h)\xi = \left\{S(h)\xi + \int_0^h S(h-s)f(s)\,ds;\ f \in \mathcal{E}\right\}.$$

We denote by $\mathcal{QTS}_K^A(\xi)$ the class of all A-quasi-tangent sets to K at $\xi \in K$.

Another A-tangency concept is introduced below.

Definition 9.1.4. We say that a set E is *A-tangent* to K at ξ if, for each $\rho > 0$, we have

$$\liminf_{h \downarrow 0} \frac{1}{h} \operatorname{dist}\left(S_E(h)\xi; K \cap D(\xi, \rho)\right) = 0, \tag{9.1.4}$$

where $S_E(h)\xi = \{S(h)\xi + \int_0^h S(h-s)\eta\,ds;\ \eta \in E\}$. We denote the class of all A-tangent sets to K at $\xi \in K$ by $\mathcal{TS}_K^A(\xi)$.

Since E can be identified with the subset of a.e. constant elements in $\mathcal{E}$, it readily follows that

$$\mathcal{TS}_K^A(\xi) \subseteq \mathcal{QTS}_K^A(\xi). \tag{9.1.5}$$

Remark 9.1.1. Let $K \subseteq X$, $\xi \in K$ and $E \subseteq X$. Then $E \in \mathcal{QTS}_K^A(\xi)$ if and only if for each $\varepsilon > 0$, $\rho > 0$ and $\delta > 0$ there exist $h \in (0, \delta)$, $p \in D(0, \varepsilon)$ and $f \in \mathcal{E}$ such that

$$S(h)\xi + \int_0^h S(h-s)f(s)\,ds + hp \in K \cap D(\xi, \rho).$$

Equivalently, $E \in \mathcal{QTS}_K^A(\xi)$ if and only if there exist three sequences, $(h_n)_n$ in $\mathbb{R}_+$ with $h_n \downarrow 0$, $(p_n)_n$ in X with $\lim_n p_n = 0$, and $(f_n)_n \in \mathcal{E}$ with $\lim_n \int_0^{h_n} S(h_n - s)f_n(s)\,ds = 0$, and such that

$$S(h_n)\xi + \int_0^{h_n} S(h_n - s)f_n(s)\,ds + h_n p_n \in K.$$

Remark 9.1.2. Let $K \subseteq X$, $\xi \in K$ and $E \in \mathcal{B}(X)$. Then $E \in \mathcal{QTS}_K^A(\xi)$ if and only if for each $\varepsilon > 0$ there exist $\delta \in (0, \varepsilon]$, $p \in X$ with $\|p\| \leq \varepsilon$ and $f \in \mathcal{E}$ such that

$$S(\delta)\xi + \int_0^\delta S(\delta - s)f(s)\,ds + \delta p \in K.$$

Equivalently, $E \in \mathcal{QTS}_K^A(\xi)$ if and only if there exist three sequences, $(h_n)_n$ in $\mathbb{R}_+$ with $h_n \downarrow 0$, $(f_n)_n$ in $\mathcal{E}$ and $(p_n)_n$ in X with $\lim_n p_n = 0$, and such that

$$S(h_n)\xi + \int_0^{h_n} S(h_n - s)f_n(s)\,ds + h_n p_n \in K$$

for $n = 1, 2, \ldots$.

Proposition 9.1.1. *Let $K \subseteq X$, $\xi \in K$ and $E \in \mathcal{B}(X)$. We have:*

(i) *if $S(t)K \subseteq K$ for each $t > 0$, then $\{0\} \in \mathcal{QTS}_K^A(\xi)$;*
(ii) *if $E \subseteq D$ and $E \in \mathcal{QTS}_K^A(\xi)$, then $D \in \mathcal{QTS}_K^A(\xi)$;*
(iii) *if $S(t)K \subseteq K$ for each $t > 0$ and $0 \in E$, then $E \in \mathcal{QTS}_K^A(\xi)$;*
(iv) *$E \in \mathcal{QTS}_K^A(\xi)$ if and only if $\overline{E} \in \mathcal{QTS}_K^A(\xi)$;*
(v) *Let $\eta \in X$. Then $\eta \in \mathcal{T}_K^A(\xi)$ if and only if $\{\eta\} \in \mathcal{QTS}_K^A(\xi)$;*
(vi) *if E is compact and convex then $E \in \mathcal{QTS}_K^A(\xi)$ if and only if there exists $\eta \in E$ such that $\eta \in \mathcal{T}_K^A(\xi)$.*

Proof. Except for (iv) and (vi) which will be proved below, the remaining properties are direct consequences of Remark 9.1.2. Let us observe that (iv) follows from the remark that each measurable function $f : \mathbb{R}_+ \to \overline{E}$ can be approximated uniformly with countably-valued functions taking values in E. See Theorem 1.2.1.

To prove (vi), let E be compact and convex and let $(f_n)_n$ in $\mathcal{E}$, $h_n \downarrow 0$ and $(p_n)_n$ with $\lim_n p_n = 0$ as in Remark 9.1.2, i.e., with

$$S(h_n)\xi + \int_0^{h_n} S(h_n - s)f_n(s)\,ds + h_n p_n \in K$$

for $n = 1, 2, \ldots$. For $n = 1, 2, \ldots$, let us define

$$\eta_n = \frac{1}{h_n}\int_0^{h_n} f_n(s)\,ds.$$

Since E is convex and closed, we have that $\eta_n \in E$ for $n = 1, 2, \ldots$. But E is compact and thus, we may assume with no loss of generality that there exists $\eta \in E$ such that

$$\lim_n \eta_n = \eta.$$

In view of Problem 1.3.1, we have

$$\lim_n \left\| \frac{1}{h_n} \int_0^{h_n} S(h_n - s) f_n(s)\, ds - \frac{1}{h_n} \int_0^{h_n} S(h_n - s)\eta\, ds \right\|$$

$$\leq \lim_n \frac{1}{h_n} \left\| \int_0^{h_n} [\, S(h_n - s) f_n(s) - f_n(s) \,]\, ds \right\| + \lim_n \|\eta_n - \eta\| = 0.$$

The conclusion follows from Remark 9.1.1 and (v). This completes the proof. □

Remark 9.1.3. From (vi) in Proposition 9.1.1 it follows that if E is convex and compact, then $E \in \mathcal{QTS}_K^A(\xi)$ if and only if $E \cap \mathcal{T}_K^A(\xi) \neq \emptyset$.

Problem 9.1.1. *Prove the result obtained by replacing* $\mathcal{QTS}_K^A(\xi)$ *in Proposition* 9.1.1 *by* $\mathcal{TS}_K^A(\xi)$ *with the special mention that in this case, in* (*vi*) *there is no need to assume that* E *is convex.*

Theorem 9.1.1. *Let* X *be a Banach space,* $A : D(A) \subseteq X \to X$ *the infinitesimal generator of a* C_0*-semigroup,* $\{S(t) : X \to X;\ t \geq 0\}$, K *a nonempty subset in* X *and* $F : K \rightsquigarrow X$ *a nonempty valued multi-function. If* K *is mild viable with respect to* $A + F$ *then, for each* $\xi \in K$ *at which* F *is u.s.c. and* $F(\xi)$ *is convex and quasi-weakly compact, we have*

$$F(\xi) \in \mathcal{QTS}_K^A(\xi). \tag{9.1.6}$$

Proof. Let $\xi \in K$. Since K is mild viable with respect to $A+F$ there exists at least one mild solution $u : [\,0, T\,] \to K$ of (9.1.1). Let $f \in L^1(0, T; X)$ be the function given by Definition 9.1.1. As u is continuous at $t = 0$ and F is u.s.c. at $u(0) = \xi$, it follows that for each $\rho > 0$ there exists $\delta(\rho) > 0$ such that

$$f(s) \in F(\xi) + D(0, \rho)$$

a.e. for $s \in [\,0, \delta(\rho)\,]$. Then, if $(\rho_n)_n$ is a sequence in $(0, 1)$, $\rho_n \downarrow 0$, there exists $h_n \downarrow 0$ such that

$$f(s) \in F(\xi) + D(0, \rho_n)$$

for $n = 1, 2, \ldots$ and a.e. for $s \in [\,0, h_n\,]$. So, for $n = 1, 2, \ldots$, there exist f_n and g_n with $f_n(s) \in F(\xi)$, $g_n(s) \in D(0, \rho_n)$ and

$$f(s) = f_n(s) + g_n(s)$$

a.e. for $s \in [\,0, h_n\,]$. Since $F(\xi)$ is convex and quasi-weakly compact, thanks to Lemma 6.1.1, we may assume without loss of generality that both f_n and g_n are integrable. Let us observe that

$$\left\| u(h_n) - S(h_n)\xi - \int_0^{h_n} S(h_n - s) f_n(s)\, ds \right\|$$

$$= \left\| S(h_n)\xi + \int_0^{h_n} S(h_n - s)f(s)\,ds - S(h_n)\xi - \int_0^{h_n} S(h_n - s)f_n(s)\,ds \right\|$$
$$\leq M h_n e^{a h_n} \rho_n,$$

for $n = 1, 2, \ldots$, where $M \geq 1$ and $a \geq 0$ are given by Theorem 1.4.1[1].

Now, let us denote by

$$p_n = \frac{1}{h_n}\left(u(h_n) - S(h_n)\xi - \int_0^{h_n} S(h_n - s)f_n(s)\,ds\right).$$

Since $\lim_n p_n = 0$ and

$$\lim_n \int_0^{h_n} S(h_n - s)f_n(s)\,ds = \lim_n (u(h_n) - S(h_n)\xi - h_n p_n) = 0$$

and

$$S(h_n)\xi + \int_0^{h_n} S(h_n - s)f_n(s)\,ds + h_n p_n = u(h_n) \in K$$

for $n = 1, 2, \ldots$, from Remark 9.1.1, it follows that $F(\xi) \in \mathfrak{QTS}_K^A(\xi)$ and this achieves the proof. □

Problem 9.1.2. *Prove that in Theorem* 9.1.1, *we can replace the assumption "quasi-weakly compact" by "quasi-weakly relatively compact" to obtain the very same conclusion.*

If X is reflexive, then each weakly closed and bounded set is quasi-weakly compact, and hence we have

Theorem 9.1.2. *Let X be reflexive, let $A : D(A) \subseteq X \to X$ be the infinitesimal generator of a C_0-semigroup, $\{S(t) : X \to X;\ t \geq 0\}$, K a nonempty subset in X and $F : K \rightsquigarrow X$ a nonempty valued multi-function. If K is mild viable with respect to $A + F$ then, for each $\xi \in K$ at which F is u.s.c. and $F(\xi)$ is convex and closed, we have* (9.1.6).

Coming back to general Banach spaces, if F is compact valued, then, instead of tangent sets, we can use tangent vectors to get a necessary condition in order for K to being mild viable with respect to $A + F$. Namely, we have

Theorem 9.1.3. *Let X be a Banach space, $A : D(A) \subseteq X \to X$ the infinitesimal generator of a C_0-semigroup, $\{S(t) : X \to X;\ t \geq 0\}$, K a nonempty subset in X and $F : K \rightsquigarrow X$ a nonempty valued multi-function. If K is mild viable with respect to $A + F$, then at each point $\xi \in K$, at which F is u.s.c. and $F(\xi)$ is convex and compact, we have*

$$F(\xi) \cap \mathfrak{T}_K^A(\xi) \neq \emptyset. \tag{9.1.7}$$

[1] In fact $a \in \mathbb{R}$, but, for our purposes, we may always assume that $a \geq 0$.

Proof. The conclusion follows from Theorem 9.1.1 and Remark 9.1.3. □

9.2. Sufficient conditions for mild viability

In order to handle several apparently different cases in a unitary frame, we introduce:

Definition 9.2.1. Let $A : D(A) \subseteq X \to X$ the infinitesimal generator of a C_0-semigroup $\{S(t) : X \to X;\ t \geq 0\}$ and $F : K \rightsquigarrow X$ a multi-function. We say that $A + F$ is *locally of compact type* if F is u.s.c. and, for each $\xi \in K$, there exist $\rho > 0$, a continuous function $\ell : \mathbb{R}_+ \to \mathbb{R}_+$ and a uniqueness function $\omega : \mathbb{R}_+ \to \mathbb{R}_+$, such that $F(D(\xi, \rho) \cap K)$ is bounded and, for each $C \subseteq D(\xi, \rho) \cap K$ and each $t > 0$, we have

$$\beta(S(t)F(C)) \leq \ell(t)\omega(\beta(C)).$$

Remark 9.2.1. As in the single-valued case, one may easily verify that $A + F$ is locally of compact type whenever:

(i) F is locally β-compact (see Definition 6.2.2);

(ii) F is u.s.c., has nonempty and closed values, is locally bounded[2], and the C_0-semigroup generated by A is compact (see Definition 1.5.5).

The main sufficient conditions for mild viability are:

Theorem 9.2.1. *Let X be a Banach space, $A : D(A) \subseteq X \to X$ the infinitesimal generator of a C_0-semigroup, $\{S(t) : X \to X;\ t \geq 0\}$, K a nonempty and locally closed subset in X and $F : K \rightsquigarrow X$ a nonempty, bounded, closed and convex valued multi-function such that $A+F$ is locally of compact type. If, for each $\xi \in K$, we have*

$$F(\xi) \in \mathfrak{QTS}_K^A(\xi), \tag{9.2.1}$$

then K is mild viable with respect to $A + F$.

Theorem 9.2.2. *Let X be a Banach space, $A : D(A) \subseteq X \to X$ the infinitesimal generator of a C_0-semigroup, $\{S(t) : X \to X;\ t \geq 0\}$, K a nonempty and locally closed subset in X and $F : K \rightsquigarrow X$ a nonempty, closed and convex valued, locally β-compact multi-function. Then K is mild viable with respect to $A + F$ if and only if, for each $\xi \in K$, the tangency condition $F(\xi) \cap \mathfrak{T}_K^A(\xi) \neq \emptyset$ is satisfied.*

---

[2]This happens for instance whenever X is reflexive, F is u.s.c. and has nonempty, closed, convex and bounded values. Indeed, F is strongly-weakly u.s.c. being strongly-strongly u.s.c. Furthermore, since X is reflexive, each bounded, closed convex set is weakly compact, and thus, by Lemma 2.6.1, F is locally bounded.

Theorem 9.2.3. *Let X be a Banach space, $A : D(A) \subseteq X \to X$ the infinitesimal generator of a compact C_0-semigroup, $\{S(t) : X \to X;\ t \geq 0\}$, K a nonempty and locally closed subset in X and $F : K \rightsquigarrow X$ a strongly-weakly u.s.c. multi-function with nonempty, weakly compact and convex values. If, for each $\xi \in K$, the tangency condition* (9.2.1) *is satisfied, then K is mild viable with respect to $A+F$.*

Theorem 9.2.4. *Let X be a Banach space, $A : D(A) \subseteq X \to X$ the infinitesimal generator of a C_0-semigroup, $\{S(t) : X \to X;\ t \geq 0\}$, K a nonempty and locally compact subset in X and $F : K \rightsquigarrow X$ a strongly-weakly u.s.c. multi-function with nonempty, weakly compact and convex values. If, for each $\xi \in K$, the tangency condition* (9.2.1) *is satisfied, then K is mild viable with respect to $A+F$.*

9.3. Existence of ε-approximate mild solutions

The proof of the sufficiency of Theorems 9.2.1 and 9.2.4 is based on the following existence result concerning ε-approximate solutions for the Cauchy problem (9.1.1).

Lemma 9.3.1. *Let X be a Banach space, $A : D(A) \subseteq X \to X$ the infinitesimal generator of a C_0-semigroup, $\{S(t) : X \to X;\ t \geq 0\}$, K a nonempty and locally closed subset in X and $F : K \rightsquigarrow X$ a nonempty-valued and locally bounded multi-function satisfying the tangency condition* (9.2.1)*. Let $\xi \in K$ be arbitrary and let $r > 0$ be such that $D(\xi, r) \cap K$ is closed. Then, there exist $\rho \in (0, r]$ and $T > 0$ such that, for each $\varepsilon \in (0,1)$, there exist $\sigma : [0,T] \to [0,T]$ nondecreasing, $\theta : \{(t,s);\ 0 \leq s < t \leq T\} \to [0,T]$ measurable, $g : [0,T] \to X$ and $f : [0,T] \to X$ Bochner integrable and $u : [0,T] \to X$ continuous such that:*

(i) *$s - \varepsilon \leq \sigma(s) \leq s$ for each $s \in [0,T]$;*
(ii) *$u(\sigma(s)) \in D(\xi,\rho) \cap K$ for each $s \in [0,T]$ and $u(T) \in D(\xi,\rho) \cap K$;*
(iii) *$\|g(s)\| \leq \varepsilon$ for each $s \in [0,T]$ and $f(s) \in F(u(\sigma(s)))$ a.e. for $s \in [0,T]$;*
(iv) *$\theta(t,s) \leq t$ for each $0 \leq s < t \leq T$ and $t \mapsto \theta(t,s)$ is nonexpansive on $(s,T]$;*
(v) *$u(t) = S(t)\xi + \int_0^t S(t-s)f(s)\,ds + \int_0^t S(\theta(t,s))g(s)\,ds$ for each $t \in [0,T]$;*
(vi) *$\|u(t) - u(\sigma(t))\| \leq \varepsilon$ for each $t \in [0,T]$.*

Definition 9.3.1. An *ε-approximate mild solution* of (9.1.1) on $[0,T]$ is a 5-uple $(\sigma, \theta, g, f, u)$ satisfying (i)∼(vi) in Lemma 9.3.1.

Proof. Let $\xi \in K$ be arbitrary and let $r > 0$ be such that $D(\xi, r) \cap K$ be closed. Let us choose $\rho \in (0, r\,]$, $N > 0$, $M \geq 1$ and $a \geq 0$ such that

$$\|y\| \leq N \tag{9.3.1}$$

for every $x \in D(\xi, \rho) \cap K$ and $y \in F(x)$, and

$$\|S(t)\|_{\mathcal{L}(X)} \leq Me^{at} \tag{9.3.2}$$

for every $t \geq 0$.

The existence of $\rho > 0$ and $N > 0$ satisfying (9.3.1) is ensured by the fact that F is locally bounded. Furthermore, the existence of $M \geq 1$ and $a \geq 0$ satisfying (9.3.2) follows from Theorem 1.4.1. Since $t \mapsto S(t)\xi$ is continuous at $t = 0$ and $S(0)\xi = \xi$, we may find a sufficiently small $T > 0$ such that

$$\sup_{t \in [\,0, T\,]} \|S(t)\xi - \xi\| + TMe^{aT}(N + 1) \leq \rho. \tag{9.3.3}$$

Let $\varepsilon \in (0, 1)$ be arbitrary but fixed. We begin by showing the existence of an ε-approximate mild solution on an interval $[\,0, \delta\,]$ with $\delta \in (0, T\,]$. As, for every $\xi \in K$, F satisfies the tangency condition (9.2.1), it follows that there exist $\delta \in (0, T\,]$, $f \in \mathcal{F}(\xi)$[3] and $p \in X$ with $\|p\| \leq \varepsilon$, such that

$$S(\delta)\xi + \int_0^\delta S(\delta - s)f(s)\,ds + \delta p \in K. \tag{9.3.4}$$

We continue by showing how to define the functions σ, θ, g, f and u. First, f is defined as above. Next, let $\sigma : [\,0, \delta\,] \to [\,0, \delta\,]$, $g : [\,0, \delta\,] \to X$ and $\theta : \{\,(t, s);\ 0 \leq s < t \leq \delta\,\} \to [\,0, \delta\,]$ be given by

$$\begin{aligned} \sigma(s) &= 0,\\ g(s) &= p,\\ \theta(t, s) &= 0 \end{aligned}$$

and let $u : [\,0, \delta\,] \to X$ be defined by

$$u(t) = S(t)\xi + \int_0^t S(t - s)f(s)\,ds + tp,$$

for each $t \in [\,0, \delta\,]$.

We will show that $(\,\sigma, \theta, g, f, u\,)$ is an ε-approximate mild solution to the Cauchy problem (9.1.1) on the interval $[\,0, \delta\,]$. Clearly σ is nondecreasing, g and f are Bochner integrable, θ is measurable and $t \mapsto \theta(t, s)$ is nonexpansive on $[\,0, \delta\,]$ and u is continuous. The conditions (i), (iii), (iv) and (v) are obviously fulfilled. To prove (ii), let us first observe that, as

[3] We recall that $\mathcal{F}(\xi) = \{f \in L^1(\mathbb{R}_+; X);\ f(s) \in F(\xi) \text{ a.e. for } s \in \mathbb{R}_+\}$.

$u(\sigma(s)) = u(0) = \xi$ for each $s \in [0, \delta]$, we have $u(\sigma(s)) \in D(\xi, \rho) \cap K$. Moreover, since $\varepsilon < 1$, from (9.3.1) and (9.3.2), we get

$$\|u(t) - \xi\| \leq \|S(t)\xi - \xi\| + \int_0^t \|S(t-s)\|_{\mathcal{L}(X)}\| f(s)\|ds$$

$$+ \int_0^t \|S(\theta(t,s))\|_{\mathcal{L}(X)}\|p\|ds \leq \sup_{t\in[0,\delta]} \|S(t)\xi - \xi\| + \delta\, Me^{a\delta}(N+1) \quad (9.3.5)$$

for each $t \in [0, \delta]$. Next, since $\delta < T$, by (9.3.3), we have

$$\|u(\delta) - \xi\| \leq \sup_{t\in[0,\delta]} \|S(t)\xi - \xi\| + \delta\, Me^{a\delta}(N+1) \leq \rho.$$

Combining (9.3.4) with the last inequality, we get $u(\delta) \in D(\xi, \rho) \cap K$ and thus (ii) is satisfied. Diminishing $\delta > 0$, if necessary, by (9.3.5), we may assume that

$$\|u(t) - u(\sigma(t))\| \leq \varepsilon$$

for each $t \in [0, \delta]$ and thus (vi) is also satisfied. We emphasize that the reason we can do this is because (9.3.5) is "independent" of p which, of course, may change with δ. Therefore $(\sigma, \theta, g, f, u)$ is an ε-approximate mild solution of (9.1.1) on $[0, \delta]$.

In the second step we will prove the existence of an ε-approximate mild solution for (9.1.1) defined on the whole interval $[0, T]$. To this aim we shall make use of Brezis–Browder Theorem 2.1.1. We denote by $D(c)$ the set

$$D(c) = [0, c] \times \{(t, s);\ 0 \leq s < t \leq c\} \times [0, c] \times [0, c] \times [0, c],$$

with $c > 0$, and by $\mathcal{S}$ the set of all ε-approximate mild solutions to the problem (9.1.1), defined on $D(c)$, with $c \leq T$.

On the set $\mathcal{S}$ we introduce a preorder relation $\preceq$as follows: we say that $(\sigma_1, \theta_1, g_1, f_1, u_1)$, defined on $D(c_1)$, and $(\sigma_2, \theta_2, g_2, f_2, u_2)$, defined on $D(c_2)$, satisfy

$$(\sigma_1, \theta_1, g_1, f_1, u_1) \preceq (\sigma_2, \theta_2, g_2, f_2, u_2)$$

if $c_1 \leq c_2$, $\sigma_1(t) = \sigma_2(t)$, $g_1(t) = g_2(t)$ and $f_1(t) = f_2(t)$ for $t \in [0, c_1]$ and $\theta_1(t, s) = \theta_2(t, s)$ for each $0 \leq s < t \leq c_1$.

Let $\mathfrak{L}$ be an increasing sequence in $\mathcal{S}$,

$$\mathfrak{L} = ((\sigma_i, \theta_i, g_i, f_i, u_i) : D(c_i) \to [0, c_i] \times [0, c_i] \times X \times X \times X)_i.$$

We define an upper bound of $\mathfrak{L}$ as follows. First, set

$$c^* = \sup\{c_i; i = 1, 2, \dots\}.$$

If $c^* = c_i$ for some $i = 1, 2, \dots$, then $(\sigma_i, \theta_i, g_i, f_i, u_i)$ is an upper bound for $\mathcal{L}$. If $c_i < c^*$ for $i = 1, 2, \dots$, we show that there exists $(\sigma^*, \theta^*, g^*, f^*, u^*)$ in $\mathcal{S}$, defined on $[0, c^*]$, and satisfying

$$(\sigma_m, \theta_m, g_m, f_m, u_m) \preceq (\sigma^*, \theta^*, g^*, f^*, u^*),$$

for $m = 1, 2, \dots$.

First, we know that all the functions in the set $\{\sigma_m; m = 1, 2, \dots\}$ are nondecreasing, with values in $[0, c^*]$, and satisfy $\sigma_m(c_m) \leq \sigma_p(c_p)$ for $m, p = 1, 2, \dots$ with $m \leq p$. So, there exists

$$\lim_m \sigma_m(c_m) \in [0, c^*].$$

Let now $m, k \in \{1, 2, \dots\}$ be arbitrary with $m \geq k$. For each $s \in [0, c_k)$, we have

$$\begin{aligned} &|\theta_m(c_m, s) - \theta_k(c_k, s)| \\ &\leq |\theta_m(c_m, s) - \theta_m(c_k, s)| + |\theta_m(c_k, s) - \theta_k(c_k, s)| \leq |c_m - c_k|, \end{aligned}$$

because $t \mapsto \theta_m(t, s)$ is nonexpansive on (s, c^*). Accordingly, there exists

$$\lim_m \theta_m(c_m, s).$$

We are now ready to define the functions: $\sigma^* : [0, c^*] \to [0, c^*]$ by

$$\sigma^*(t) = \begin{cases} \sigma_m(t) & \text{if, for some } m = 1, 2, \dots, \ t \in [0, c_m] \\ \lim_m \sigma_m(c_m) & \text{if } t = c^*, \end{cases}$$

$\theta^* : \{(t, s); 0 \leq s < t \leq c^*\} \to [0, c^*]$ by

$$\theta^*(t, s) = \begin{cases} \theta_m(t, s) & \text{if, for some } m = 1, 2, \dots, \ 0 \leq s < t \leq c_m \\ \lim_m \theta_m(c_m, s) & \text{if } 0 \leq s < t = c^*, \end{cases}$$

and $g^* : [0, c^*] \to X$ by

$$g^*(t) = \begin{cases} g_m(t) & \text{if, for some } m = 1, 2, \dots, t \in [0, c_m] \\ 0 & \text{if } t = c^*. \end{cases}$$

Obviously σ^* is nondecreasing, g^* is Bochner integrable, θ^* is measurable and $t \mapsto \theta^*(t, s)$ is nonexpansive on $[s, \delta]$. In order to define u^* and f^*, we will prove first that there exists $\lim_m u_m(c_m)$. To this aim, let us denote by $\chi_{[0, c_m]} : [0, T] \to \mathbb{R}_+$ the indicator function of $[0, c_m]$, i.e.

$$\chi_{[0, c_m]}(s) = \begin{cases} 1 & \text{if } s \in [0, c_m] \\ 0 & \text{if } s \in (c_m, T]. \end{cases}$$

By (v), we have

$$u_m(c_m) = S(c_m)\xi + \int_0^{c^*} \chi_{[0, c_m]}(s) S(c_m - s) f_m(s)\, ds$$

$$+\int_0^{c^*} \chi_{[0,c_m]}(s)S(\theta_m(c_m,s))g_m(s)\,ds.$$

Recalling (i), (ii), (iv) and (vi) and the fact that both g_m and f_m are Bochner integrable and, by (9.3.1) and (iii), both are bounded (the former by N and the latter by ε), thanks to the Lebesgue Dominated Convergence Theorem 1.2.3, we deduce that there exists $\lim_m u_m(c_m) = \xi^*$.

Let us define $u^* : [0, c^*] \to X$ by

$$u^*(t) = \begin{cases} u_m(t) & \text{if, for some } m = 1, 2, \ldots,\ t \in [0, c_m] \\ \xi^* & \text{if } t = c^*. \end{cases}$$

Furthermore, let us define $f^* : [0, c^*] \to X$ by

$$f^*(t) = \begin{cases} f_m(t) & \text{if, for some } m = 1, 2, \ldots, t \in [0, c_m] \\ \eta & \text{if } t = c^*, \end{cases}$$

where $\eta \in X$ is arbitrary but fixed.

Clearly u^* is continuous on $[0, c^*]$. Moreover, $(\sigma^*, \theta^*, g^*, f^*, u^*)$ satisfies (i), (iii), (iv) and (v). Since u_m is an ε-approximate mild solution on $[0, c_m]$, by (ii) applied to u_m, we have $u_m(c_m) \in D(\xi, \rho) \cap K$ and since the latter is closed, we have $u^*(c^*) \in D(\xi, \rho) \cap K$. Similarly, by (vi) applied to u_m, we have

$$\|u_m(t) - u_m(\sigma_m(t))\| \leq \varepsilon$$

for each $m \in \mathbb{N}$ and $t \in [0, c_m]$ and since u^* is continuous, we deduce

$$\|u^*(t) - u^*(\sigma^*(t))\| \leq \varepsilon$$

for each $t \in [0, c^*]$. Hence u^* and σ^* satisfy (ii) and (vi). Thus the quintuple $(\sigma^*, \theta^*, g^*, f^*, u^*)$ is an ε-approximate mild solution of (9.1.1) on $[0, c^*]$ and

$$(\sigma_m, \theta_m, g_m, f_m, u_m) \preceq (\sigma^*, \theta^*, g^*, f^*, u^*),$$

for $m = 1, 2, \ldots$. Let us define $\mathcal{N} : \mathcal{S} \to \mathbb{R}$ by $\mathcal{N}((\sigma, \theta, g, f, u)) = c$, where $D(c)$ is the domain of definition of $(\sigma, \theta, g, f, u)$. Clearly $\mathcal{N}$ satisfies the hypotheses of Brezis-Browder Theorem 2.1.1. Then, $\mathcal{S}$ contains at least one $\mathcal{N}$-maximal element $(\overline{\sigma}, \overline{\theta}, \overline{g}, \overline{f}, \overline{u})$ whose domain is $D(\overline{c})$. We will next show that $\overline{c} = T$. To this aim, let us assume by contradiction that $\overline{c} < T$. We know that $\overline{u}(\overline{c}) \in D(\xi, \rho) \cap K$. Moreover, by using (9.3.1), (9.3.2), (9.3.3), we get

$$\begin{aligned} &\|\overline{u}(\overline{c}) - \xi\| \\ &= \left\| S(\overline{c})\xi + \int_0^{\overline{c}} S(\overline{c} - s)\overline{f}(s)\,ds + \int_0^{\overline{c}} S(\overline{\theta}(\overline{c}, s))\overline{g}(s)ds - \xi \right\| \\ &\leq \sup_{t \in [0, \overline{c}]} \|S(t)\xi - \xi\| + \int_0^{\overline{c}} \|S(\overline{\theta}(\overline{c}, s))\overline{g}(s)\|\,ds + \int_0^{\overline{c}} \|S(\overline{c} - s)\overline{f}(s)\|\,ds \end{aligned}$$

$$\leq \sup_{t\in[0,\overline{c}]} \|S(t)\xi - \xi\| + \overline{c}Me^{aT}(N+1) < \rho.$$

Then, as $\overline{u}(\overline{c}) \in K$ and $F(\overline{u}(\overline{c})) \in \mathfrak{QTS}_K^A(\overline{u}(\overline{c}))$, there exist $\overline{\delta} \in (0, T-\overline{c})$, $\overline{\delta} \leq \varepsilon$, $\overline{f}_0 \in L^1(\mathbb{R}_+; F(\overline{u}(\overline{c})))$ and $\overline{p} \in X$, $\|\overline{p}\| \leq \varepsilon$ such that

$$S(\overline{\delta})\overline{u}(\overline{c}) + \int_0^{\overline{\delta}} S(\overline{\delta} - s)\overline{f}(s)\,ds + \overline{\delta}\overline{p} \in D(\xi, \rho) \cap K. \tag{9.3.6}$$

Let us define the functions: $\widetilde{\sigma} : [0, \overline{c}+\overline{\delta}] \to [0, \overline{c}+\overline{\delta}]$, $\widetilde{g} : [0, \overline{c}+\overline{\delta}] \to X$, $\widetilde{f} : [0, \overline{c}+\overline{\delta}] \to X$ and $\widetilde{\theta} : \{(t,s);\ 0 \leq s < t \leq \overline{c}+\overline{\delta}\} \to [0, \overline{c}+\overline{\delta}]$ by

$$\widetilde{\sigma}(t) = \begin{cases} \overline{\sigma}(t) & \text{for } t \in [0, \overline{c}] \\ \overline{c} & \text{for } t \in (\overline{c}, \overline{c}+\overline{\delta}], \end{cases}$$

$$\widetilde{g}(t) = \begin{cases} \overline{g}(t) & \text{for } t \in [0, \overline{c}] \\ \overline{p} & \text{for } t \in (\overline{c}, \overline{c}+\overline{\delta}], \end{cases}$$

$$\widetilde{f}(t) = \begin{cases} \overline{f}(t) & \text{for } t \in [0, \overline{c}] \\ \overline{f}_0(t) & \text{for } t \in (\overline{c}, \overline{c}+\overline{\delta}], \end{cases}$$

and

$$\widetilde{\theta}(t,s) = \begin{cases} \overline{\theta}(t,s) & \text{for } 0 \leq s < t \leq \overline{c} \\ t - \overline{c} + \overline{\theta}(\overline{c}, s) & \text{for } 0 \leq s < \overline{c} < t \leq \overline{c}+\overline{\delta} \\ 0 & \text{for } \overline{c} \leq s < t \leq \overline{c}+\overline{\delta}. \end{cases}$$

Clearly, $\widetilde{\sigma}$ is nondecreasing, $\widetilde{g}$ and $\widetilde{f}$ are Bochner integrable on $[0, \overline{c}+\overline{\delta}]$, $\widetilde{\theta}$ is measurable and they satisfy (i), (iii) and (iv).

Accordingly, we can define $\widetilde{u} : [0, \overline{c}+\overline{\delta}] \to X$ by

$$\widetilde{u}(t) = \begin{cases} \overline{u}(t) & \text{if } t \in [0, \overline{c}] \\ S(t-\overline{c})\overline{u}(\overline{c}) + \displaystyle\int_{\overline{c}}^t S(t-s)\overline{f}_0\,ds + (t-\overline{c})\overline{p} & \text{if } t \in [\overline{c}, \overline{c}+\overline{\delta}]. \end{cases}$$

A standard calculation, involving the form of $\widetilde{\theta}$, shows that $\widetilde{u}$ satisfies (v). From (9.3.6), we get $\widetilde{u}(\overline{c}+\overline{\delta}) \in D(\xi, \rho) \cap K$ and thus (ii) is satisfied.

Following the same arguments as in the first part of the proof, we conclude that, diminishing $\overline{\delta} > 0$ if necessary, we get also (vi). Thus $(\widetilde{\sigma}, \widetilde{\theta}, \widetilde{g}, \widetilde{f}, \widetilde{u})$ is an element of $\mathcal{S}$ which satisfies

$$\mathcal{N}((\overline{\sigma}, \overline{\theta}, \overline{g}, \overline{f}, \overline{u})) < \mathcal{N}((\widetilde{\sigma}, \widetilde{\theta}, \widetilde{g}, \widetilde{f}, \widetilde{u}))$$

although

$$(\overline{\sigma}, \overline{\theta}, \overline{g}, \overline{f}, \overline{u}) \preceq (\widetilde{\sigma}, \widetilde{\theta}, \widetilde{g}, \widetilde{f}, \widetilde{u}).$$

This contradiction can be eliminated only if $\overline{c} = T$ and this completes the proof of Lemma 9.3.1. $\square$

Remark 9.3.1. Under the general hypotheses of Lemma 9.3.1, for each $\gamma > 0$, we can diminish both $\rho > 0$ and $T > 0$, such that $T < \gamma$, $\rho < \gamma$ and all the conditions (i) $\sim$ (vi) in Lemma 9.3.1 be satisfied.

9.4. Proof of Theorem 9.2.1

In this section we will prove that, in the hypotheses of Theorem 9.2.1, there exists at least one sequence $\varepsilon_n \downarrow 0$ such that the corresponding sequence of ε_n-approximate mild solutions, $((\sigma_n, \theta_n, g_n, f_n, u_n))_n$, enjoys the property that $(u_n)_n$ is uniformly convergent on $[0, T]$ to some $u : [0, T] \to K$ which is a mild solution of (9.1.1).

Proof. Let $r > 0$, $\rho \in (0, r]$ and $T > 0$ as in Lemma 9.3.1. Since, by hypotheses, $A + F$ is locally of compact type, diminishing $\rho \in (0, r]$ and $T > 0$ if necessary, we can find a continuous function $\ell : \mathbb{R}_+ \to \mathbb{R}_+$ and a uniqueness function $\omega : \mathbb{R}_+ \to \mathbb{R}_+$ such that $F(D(\xi, \rho) \cap K)$ is bounded and, for each $t \in [0, T]$ and $C \subseteq D(\xi, \rho) \cap K$, we have

$$\beta(S(t)F(C)) \leq \ell(t)\omega(\beta(C)), \tag{9.4.1}$$

and all the conclusions of Lemma 9.3.1 be satisfied. See Remark 9.3.1.

Let $\varepsilon_n \downarrow 0$ be a sequence in $(0, 1)$ and let $((\sigma_n, \theta_n, g_n, f_n, u_n))_n$ be a sequence of ε_n-approximate mild solutions defined on $[0, T]$ whose existence is ensured by Lemma 9.3.1. From (v)[4], we have

$$u_n(t) = S(t)\xi + \int_0^t S(t - s)f_n(s)\,ds + \int_0^t S(\theta_n(t, s))g_n(s)\,ds \tag{9.4.2}$$

for each $n \in \mathbb{N}$ and $t \in [0, T]$.

Throughout, $M \geq 1$ and $a \geq 0$ denote the constants satisfying (9.3.2).

We consider first the case when X is separable. Let $t \in [0, T]$ be fixed. From Problem 2.7.1, Lemma 2.7.2, (iii), (vi), (9.4.2), (9.3.2), (9.4.1) and Remark 2.7.1, we deduce

$$\begin{aligned}
&\beta(\{u_n(t);\ n \geq k\}) \\
&\leq \beta\left(\left\{\int_0^t S(t - s)f_n(s)\,ds;\ n \geq k\right\}\right) \\
&+\beta\left(\left\{\int_0^t S(\theta_n(t, s))g_n(s)\,ds\,;\ n \geq k\right\}\right) \\
&\leq \int_0^t \beta\left(\{S(t - s)f_n(s);\ n \geq k\}\right) ds
\end{aligned}$$

[4]Throughout this proof, the references to (i)$\sim$(vi) are to the corresponding items in Lemma 9.3.1.

$$+\int_0^t \beta\left(\{S(\theta_n(t,s))g_n(s)\,;\; n\geq k\}\right) ds$$

$$\leq \int_0^t \ell(t-s)\omega(\beta\left(\{u_n(\sigma_n(s));\; n\geq k\}\right)) \,ds + TMe^{aT}\varepsilon_k$$

$$\leq \int_0^t \sup_{\theta\in[\,0,T\,]} \ell(\theta)\omega(\beta\left(\{u_n(s);\; n\geq k\} + \{u_n(\sigma_n(s)) - u_n(s);\; n\geq k\}\right)) \,ds$$

$$+TMe^{aT}\varepsilon_k$$

$$\leq \int_0^t \sup_{\theta\in[\,0,T\,]} \ell(\theta)\omega(\beta\left(\{u_n(s);\; n\geq k\}\right) + \varepsilon_k)\; ds + TMe^{aT}\varepsilon_k$$

Set $x_k(t) = \beta(\{u_n(t);\; n\geq k\}) + \varepsilon_k$, for $k = 1, 2, \ldots$ and $t\in[\,0,T\,]$. Set also $\omega_0(x) = \sup_{\theta\in[\,0,T\,]} \ell(\theta)\omega(x)$, for $x\in\mathbb{R}_+$ and $\gamma_k = (TMe^{aT}+1)\varepsilon_k$. We conclude

$$x_k(t) \leq \gamma_k + \int_0^t \omega_0(x_k(s))\,ds,$$

for $k = 1, 2, \ldots$ and $t\in[\,0,T\,]$.

By Remark 1.8.1, ω_0 is a uniqueness function. So, Lemma 1.8.2 shows that, diminishing $T>0$ if necessary, we may assume that $\lim_k x_k(t) = 0$, which means that $\lim_k \beta(\{u_n(t);\; n\geq k\}) = 0$ uniformly for $t\in[\,0,T\,]$. From Lemma 2.7.3, it follows that, for each $t\in[\,0,T\,]$, $\{u_n(t);\; n = 1, 2, \ldots\}$ is relatively compact in X. At this point Theorem 1.5.2 comes into play and shows that there exists $u\in C([\,0,T\,];X)$ such that, on a subsequence at least, we have $\lim_n u_n(t) = u(t)$ uniformly for $t\in[\,0,T\,]$. In view of (ii) and (vi), we get $\lim_n u_n(\sigma_n(t)) = u(t)$ uniformly for $t\in[\,0,T\,]$, and, since $D(\xi,\rho)\cap K$ is closed, $u(t)\in D(\xi,\rho)\cap K$ for each $t\in[\,0,T\,]$.

Next, since $f_n(s)\in F(u_n(\sigma_n(s)))$ for $n = 1, 2, \ldots$ and $s\in[\,0,T\,]$, and, by (9.3.1), $F(D(\xi,\rho)\cap K)$ is bounded, it follows that $\{f_n;\; n = 1, 2, \ldots\}$ is uniformly integrable. Further, since F is u.s.c. it is strongly-weakly u.s.c. too, and since $\{u_n(\sigma_n(s));\; n = 1, 2, \ldots,\; s\in[\,0,T\,]\}$ is relatively compact, from Lemma 2.6.1 and Theorem 1.3.2, it follows that the set

$$C = \overline{\mathrm{conv}} \bigcup_{n=1}^{\infty} \bigcup_{s\in[\,0,T\,]} F(u_n(\sigma_n(s)))$$

is weakly compact. As $f_n(s)\in C$ for $n = 1, 2, \ldots$ and a.e. for $s\in[\,0,T\,]$, we are in the hypotheses of Corollary 1.3.1 which, along with Theorem 1.3.4, shows that we may assume without loss of generality that $(f_n)_n$ is weakly convergent in $L^1(0,T;X)$ to some function f. Now Lemma 2.6.2 comes into play and shows that $f(s)\in F(u(s))$ a.e. for $s\in[\,0,T\,]$. As the graph of the mild solution operator $Q : L^1(0,T;X) \to C([0,T];X)$, defined by $(Qg)(t) = S(t)\xi + \int_0^t S(t-s)g(s)\,ds$, for each $g \in L^1(0,T;X)$, is weakly×strongly closed being strongly×strongly closed and convex, we may pass to the limit in (9.4.2), for $n \to \infty$. Taking into account of (iii), we obtain

$$u(t) = S(t)\xi + \int_0^t S(t-s)f(s)\,ds,$$

for each $t \in [0,T]$, and this concludes the proof in the case when X is separable.

If X is not separable, in view of Remark 2.7.3, it follows that there exists a separable and closed subspace, Y, of X such that

$$u_n(t), S(r)f_n(s), S(\theta_n(r,s))g_n(s) \in Y$$

for $n = 1,2,\ldots$ and a.e. for $t,r,s \in [0,T]$. From Problem 2.7.2 and the monotonicity of ω, we deduce that

$$\beta_Y(S(t)F(C)) \leq 2\beta(S(t)F(C)) \leq 2\ell(t)\omega(\beta(C)) \leq 2\ell(t)\omega(\beta_Y(C)),$$

for each $t > 0$ and each set $C \subseteq D(\xi,\rho) \cap K \cap Y$. From now on, we have to repeat the routine in the separable case, by using the fact that the restriction of β_Y — as defined in Problem 2.7.2 — to $\mathcal{B}(Y)$ is the Hausdorff measure of noncompactness on Y. This completes the proof. □

9.5. Proof of Theorem 9.2.3

We indicate briefly how to show that, in the hypotheses of Theorem 9.2.3, there exists at least one sequence $\varepsilon_n \downarrow 0$ such that the corresponding sequence of ε_n-approximate mild solutions, $((\sigma_n, \theta_n, g_n, f_n, u_n))_n$, enjoys the property that $(u_n)_n$ is uniformly convergent on $[0,T]$ to some function $u : [0,T] \to K$ which is a mild solution of (9.1.1).

Proof. As F is strongly-weakly u.s.c. and has weakly compact values, a simple argument by contradiction involving Lemma 2.6.1 shows that F is locally bounded and thus Lemma 9.3.1 applies. Let $r > 0$, $\rho \in (0,r]$ and $T > 0$ as in Lemma 9.3.1. Diminishing $\rho \in (0,r]$ and $T > 0$ if necessary, we may assume that $F(K \cap D(\xi,\rho))$ is bounded and, in addition, all the conclusions of Lemma 9.3.1 are satisfied. See Remark 9.3.1. The conclusion follows from (9.4.2) with the help of Theorem 1.5.3. □

9.6. Proof of Theorem 9.2.4

We prove that there exists at least one sequence $(\varepsilon_n)_n$, with $\varepsilon_n \downarrow 0$, and such that the corresponding sequence $((\sigma_n, \theta_n, g_n, f_n, u_n))_n$, of ε_n-approximate mild solutions, enjoys the property that $(u_n)_n$ is uniformly convergent on $[0,T]$ to some function $u:[0,T]\to K$ which is a mild solution of (9.1.1).

Proof. Let $r>0$, $\rho\in(0,r]$ and $T>0$ as in Lemma 9.3.1. Since, by hypotheses, K is locally compact and F is strongly-weakly u.s.c. and has weakly compact values, diminishing $\rho\in(0,r]$ and $T>0$ if necessary, we may assume with no loss of generality that both $K\cap D(\xi,\rho)$ is compact and $F(K\cap D(\xi,\rho))$ is weakly relatively compact and, in addition, all the conclusions of Lemma 9.3.1 are satisfied. See Remark 9.3.1. Let $\varepsilon_n\downarrow 0$ be a sequence in $(0,1)$ and let $((\sigma_n,\theta_n,g_n,f_n,u_n))_n$ be a sequence of ε_n-approximate mild solutions defined on $[0,T]$ whose existence is ensured by Lemma 9.3.1. From (v) in Lemma 9.3.1, we have

$$u_n(t)=S(t)\xi+\int_0^t S(t-s)f_n(s)\,ds+\int_0^t S(\theta_n(t,s))g_n(s)\,ds \tag{9.6.1}$$

for each $n\in\mathbb{N}$ and $t\in[0,T]$. From Lemma 1.3.1 it follows that, for each $t\in[0,T]$, the set $\left\{u_n(t)-\int_0^t S(\theta_n(t,s))g_n(s)\,ds;\ n=1,2\ldots\right\}$ is relatively compact. Since $\lim_n g_n(t)=0$, uniformly for $t\in[0,T]$, an appeal to Theorem 1.5.2 shows that $\{u_n;\ n=1,2,\ldots\}$ is relatively compact in $C([0,T];X)$. From now on the proof follows, except minor modifications, the very same lines as those of the proof of Theorem 9.2.1. □

9.7. The quasi-autonomous case

Let X be a Banach space, $A:D(A)\subseteq X\to X$ the infinitesimal generator of a C_0-semigroup, $\mathcal{C}$ a nonempty subset in $\mathbb{R}\times X$, $F:\mathcal{C}\rightsquigarrow X$ a multi-function and let us consider the Cauchy problem for the quasi-autonomous semilinear evolution inclusion

$$\begin{cases} u'(t)\in Au(t)+F(t,u(t))\\ u(\tau)=\xi.\end{cases} \tag{9.7.1}$$

Let $\mathfrak{X}=\mathbb{R}\times X$ endowed with the norm $\|(t,u)\|_{\mathfrak{X}}=|t|+\|u\|$.

Definition 9.7.1. Let $A:D(A)\subseteq X\to X$ be the infinitesimal generator of a C_0-semigroup, $\{S(t):X\to X;\ t\geq 0\}$, and $F:\mathcal{C}\rightsquigarrow X$ a multi-function. We say that $A+F$ is *locally of compact type with respect to the second argument* if F is u.s.c. and, for each $(\tau,\xi)\in\mathcal{C}$, there exist $\rho>0$, a continuous function $\ell:\mathbb{R}_+\to\mathbb{R}_+$ and a uniqueness function $\omega:\mathbb{R}_+\to\mathbb{R}_+$, such that

$$\beta(S(t)F(C))\leq \ell(t)\omega(\beta(C))$$

for each $t>0$ and each $C\subseteq D_{\mathfrak{X}}((\tau,\xi),\rho)\cap\mathcal{C}$.

Definition 9.7.2. By a *mild solution* of the quasi-autonomous multi-valued semilinear Cauchy problem (9.7.1), we mean a continuous function $u : [\tau, T] \to X$, with $(t, u(t)) \in \mathcal{C}$ for each $t \in [\tau, T]$, and for which there exists $f \in L^1(\tau, T; X)$ such that $f(s) \in F(s, u(s))$ a.e. for $s \in [\tau, T]$ and

$$u(t) = S(t-\tau)\xi + \int_\tau^t S(t-s)f(s)\,ds, \tag{9.7.2}$$

for each $t \in [\tau, T]$.

Definition 9.7.3. The set $\mathcal{C} \subseteq \mathbb{R} \times X$ is *mild viable* with respect to $A+F$ if for each $(\tau, \xi) \in \mathcal{C}$, there exists $T \in \mathbb{R}$, $T > \tau$ such that the Cauchy problem (9.7.1) has at least one mild solution $u : [\tau, T] \to X$.

Remark 9.7.1. The quasi-autonomous Cauchy problem (9.7.1) can be equivalently rewritten as an autonomous one in the space $\mathcal{X}$, by setting $\mathcal{A} = (0, A)$, $z(s) = (t(s+\tau), u(s+\tau))$, $\mathcal{F}(z) = (1, F(z))$[5] and $\zeta = (\tau, \xi)$. Indeed, with the notations above, we have

$$\begin{cases} z'(s) \in \mathcal{A}z(s) + \mathcal{F}(z(s)) \\ z(0) = \zeta. \end{cases}$$

It readily follows that $\mathcal{A}$ generates a C_0-semigroup $\{\mathcal{S}(t) : \mathcal{X} \to \mathcal{X};\ t \geq 0\}$ on $\mathcal{X}$, where $\mathcal{S}(t) = (1, S(t))$ for each $t \geq 0$, $\{S(t) : X \to X;\ t \geq 0\}$ being the C_0-semigroup generated by A on X.

Remark 9.7.2. One may easily see that $\mathcal{C}$ is mild viable with respect to $A+F$ in the sense of Definition 9.7.3 if and only if $\mathcal{C}$ is mild viable with respect to $\mathcal{A} + \mathcal{F}$ in the sense of Definition 9.1.2.

Theorem 9.7.1. *Let X be a Banach space and $A : D(A) \subseteq X \to X$ the infinitesimal generator of a C_0-semigroup, $\{S(t) : X \to X;\ t \geq 0\}$. Let $\mathcal{C}$ be a nonempty subset in $\mathbb{R} \times X$ and $F : \mathcal{C} \rightsquigarrow X$ a nonempty, quasi-weakly compact and convex valued, u.s.c. multi-function. If $\mathcal{C}$ is mild viable with respect to $A+F$ then, for each $(\tau, \xi) \in \mathcal{C}$, we have*

$$(1, F(\tau, \xi)) \in \mathcal{QTS}_{\mathcal{C}}^{\mathcal{A}}(\tau, \xi). \tag{9.7.3}$$

Proof. From Remark 9.7.2 we know that $\mathcal{C}$ is mild viable with respect to $\mathcal{A} + \mathcal{F}$ in the sense of Definition 9.1.2 and thus, by Theorem 9.1.1, we conclude that, under the hypotheses of Theorem 9.7.1, for each $z \in \mathcal{C}$, $z = (\tau, \xi)$, we have

$$\mathcal{F}(z) \in \mathcal{QTS}_{\mathcal{C}}^{\mathcal{A}}(z)$$

relation which is equivalent to (9.7.3). $\square$

[5]Here $(1, F(z)) = \{(1, \eta);\ \eta \in F(z)\}$.

Likewise, in the autonomous case, we have

Theorem 9.7.2. *Let X be a Banach space and $A : D(A) \subseteq X \to X$ the infinitesimal generator of a C_0-semigroup, $\{S(t) : X \to X;\ t \geq 0\}$. Let $\mathcal{C}$ be a nonempty subset in $\mathbb{R} \times X$ and $F : \mathcal{C} \leadsto X$ a given multi-function. If $\mathcal{C}$ is viable with respect to $A + F$ then the tangency condition*

$$(1, F(\tau, \xi)) \cap \mathcal{T}^{\mathcal{A}}_{\mathcal{C}}(\tau, \xi) \neq \emptyset \tag{9.7.4}$$

is satisfied at each upper semicontinuity point, $(\tau, \xi) \in \mathcal{C}$, of F, at which F is convex and compact.

Proof. Use Remark 9.7.2 and Theorem 9.1.3. □

From Theorem 9.7.1 and Remark 6.1.3, we deduce

Theorem 9.7.3. *Let X be reflexive and let $A : D(A) \subseteq X \to X$ be the infinitesimal generator of a C_0-semigroup, $\{S(t) : X \to X;\ t \geq 0\}$. Let $\mathcal{C}$ be a nonempty subset in $\mathbb{R} \times X$ and $F : \mathcal{C} \leadsto X$ a nonempty, closed and convex valued, u.s.c. multi-function. If $\mathcal{C}$ is mild viable with respect to $A + F$ then, for each $(\tau, \xi) \in \mathcal{C}$, the tangency condition* (9.7.3) *is satisfied.*

We can now pass to the main sufficient conditions concerning the viability of a set $\mathcal{C}$ with respect to $A + F$.

Theorem 9.7.4. *Let X be a Banach space, $A : D(A) \subseteq X \to X$ the infinitesimal generator of a C_0-semigroup, $\{S(t) : X \to X;\ t \geq 0\}$, $\mathcal{C}$ a nonempty and locally closed subset in $\mathbb{R} \times X$ and $F : \mathcal{C} \leadsto X$ a nonempty, bounded, closed and convex valued multi-function such that $A + F$ is locally of compact type with respect to the second argument. If, for each $(\tau, \xi) \in \mathcal{C}$, the tangency condition* (9.7.3) *is satisfied, then $\mathcal{C}$ is mild viable with respect to $A + F$.*

Proof. We have

$$\beta(\mathcal{S}(t)\mathcal{F}(C)) \leq \beta(S(t)F(C)) \leq \ell(t)\beta(C),$$

for each bounded set C in $\mathcal{X}$ and each $t > 0$. Thus, if $A + F$ is locally of compact type with respect to the second argument then $\mathcal{A} + \mathcal{F}$ is locally of compact type (in the space $\mathcal{X}$) in the sense of Definition 9.2.1. So the conclusion follows from Remark 9.7.2 and Theorem 9.2.1. □

Theorem 9.7.5. *Let X be a Banach space, $A : D(A) \subseteq X \to X$ the infinitesimal generator of a compact C_0-semigroup, $\{S(t) : X \to X;\ t \geq 0\}$, $\mathcal{C}$ a nonempty and locally closed subset in $\mathbb{R} \times X$ and $F : \mathcal{C} \leadsto X$ a nonempty, weakly compact and convex valued multi-function which is strongly-weakly u.s.c. If, for each $(\tau, \xi) \in \mathcal{C}$, the tangency condition* (9.7.3) *is satisfied, then $\mathcal{C}$ is mild viable with respect to $A + F$.*

Proof. Just use Remark 9.7.2 and Theorem 9.2.3. □

The next result is a simple corollary of Theorem 9.7.2, Remark 6.2.3 and Theorem 9.7.4.

Theorem 9.7.6. *Let X be a Banach space, $A : D(A) \subseteq X \to X$ the infinitesimal generator of a C_0-semigroup, $\{S(t) : X \to X;\ t \geq 0\}$, $\mathcal{C}$ a nonempty and locally closed subset in $\mathbb{R} \times X$ and $F : \mathcal{C} \rightsquigarrow X$ a nonempty, closed and convex valued, locally β-compact multi-function. Then $\mathcal{C}$ is mild viable with respect to $A+F$ if and only if, for each $(\tau,\xi) \in \mathcal{C}$, the tangency condition* (9.7.4) *is satisfied.*

9.8. Global mild solutions

Let $A : D(A) \subseteq X \to X$ be the infinitesimal generator of a C_0-semigroup, let $\mathcal{C} \subseteq \mathbb{R} \times X$ be nonempty and let $F : \mathcal{C} \rightsquigarrow X$ be a given multi-function. In this section we will state some results concerning the existence of noncontinuable, or even global mild solutions to the Cauchy problem

$$\begin{cases} u'(t) \in Au(t) + F(t,u(t)) \\ u(\tau) = \xi. \end{cases} \tag{9.8.1}$$

A mild solution $u : [\tau, T) \to X$ of (9.8.1) is called *noncontinuable*, if there is no other mild solution $v : [\tau, \widetilde{T}) \to X$ of the same equation, with $T < \widetilde{T}$ and satisfying $u(t) = v(t)$ for all $t \in [\tau, T)$. The mild solution u is called *global* if $T = T_{\mathcal{C}}$, where $T_{\mathcal{C}}$ is given by (3.6.2).

The next theorem follows from the Brezis–Browder Theorem 2.1.1.

Theorem 9.8.1. *Let X be a Banach space, let $A : D(A) \subseteq X \to X$ be the infinitesimal generator of a C_0-semigroup, $\mathcal{C}$ a nonempty subset in $\mathbb{R} \times X$ and $F : \mathcal{C} \rightsquigarrow X$ a given multi-function. The following conditions are equivalent:*

(i) *$\mathcal{C}$ is mild viable with respect to $A + F$;*
(ii) *for each $(\tau,\xi) \in \mathcal{C}$ there exists at least one noncontinuable mild solution $u : [\tau, T) \to X$ of* (9.8.1).

The proof, based on Brezis–Browder Theorem 2.1.1, is very similar with the one of Theorem 3.6.1 and therefore we will omit it.

We conclude with two results concerning the existence of global solutions.

Theorem 9.8.2. *Let X be a Banach space, let $A : D(A) \subseteq X \to X$ be the infinitesimal generator of a C_0-semigroup and let $\mathcal{C}$ be an X-closed subset in $\mathbb{R} \times X$*[6]*. Let $F : \mathcal{C} \rightsquigarrow X$ be a multi-function such that $\mathcal{C}$ is*

[6] See Definition 3.6.2.

mild viable with respect to $A + F$. *If there exist two continuous functions* $a : \mathbb{R} \to \mathbb{R}_+$ *and* $b : \mathbb{R} \to \mathbb{R}_+$ *such that*

$$\|\eta\| \leq a(t)\|\xi\| + b(t) \tag{9.8.2}$$

for each $(t, \xi) \in \mathcal{C}$ *and* $\eta \in F(t, \xi)$, *then each mild solution of* (9.8.1) *can be continued up to a global one, i.e., defined on* $[\,\tau, T_{\mathcal{C}})$.

Since the proof is similar with the one of Theorem 8.8.2, we do not enter into details.

Whenever A generates a C_0-semigroup of contractions, the conclusion of Theorems 9.8.2 holds true under more general growth conditions on F. Namely, we have

Theorem 9.8.3. *Let* X *be a Banach space, let* $A : D(A) \subseteq X \to X$ *be the infinitesimal generator of a* C_0*-semigroup of contractions and let* $\mathcal{C}$ *be an* X*-closed subset in* $\mathbb{R} \times X$. *Let* $F : \mathcal{C} \rightsquigarrow X$ *be a multi-function such that* $\mathcal{C}$ *is mild viable with respect to* $A+F$. *If* F *is positively sublinear*[7] *and maps bounded subsets in* $\mathcal{C}$ *into bounded subsets in* X, *then each mild solution of* (9.8.1) *can be continued up to a global one, i.e., defined on* $[\,\tau, T_{\mathcal{C}})$.

The proof of Theorem 9.8.3 is similar with that one of a more general result, i.e., Theorem 11.7.2, allowing A to be nonlinear and multi-valued. Therefore we do not enter into details.

[7]See Definition 6.6.1.

CHAPTER 10

Viability for single-valued fully nonlinear evolutions

In this chapter we reconsider some problems, already touched upon in Chapter 8 in the semilinear case, within the (partly) more general frame of fully nonlinear evolution equations governed by continuous perturbations of infinitesimal generators of nonlinear semigroups of contractions. We begin with the definition of the C^0-viability and with the one of A-tangent vector at a point to a given set, in the case of an m-dissipative, possibly nonlinear operator A. We prove a necessary condition for C^0-viability expressed in terms of this tangency concept and we continue with the statements and proofs of several necessary and sufficient conditions for C^0-viability. We extend the results to the quasi-autonomous case and next, we focus our attention on the problem of the existence of C^0-noncontinuable or even global solutions. We then consider a class of fully nonlinear reaction-diffusion systems and we prove several necessary and sufficient conditions for C^0-viability. We conclude with some sufficient conditions for C^0-local invariance.

10.1. Continuous perturbations of m-dissipative operators

Let X be a Banach space, let $A : D(A) \subseteq X \rightsquigarrow X$ be an m-dissipative operator generating a nonlinear semigroup of nonexpansive mappings, denoted by $\{S(t) : \overline{D(A)} \to \overline{D(A)};\ t \geq 0\}$, let K be a nonempty subset in $\overline{D(A)}$, and $f : K \to X$ a function. The goal in this chapter is to prove some necessary and sufficient conditions in order that a given subset K in X be C^0-viable with respect to $A + f$. We notice that the results in this chapter extend those in Chapter 8 to the fully nonlinear operators, but only in the m-dissipative case.

To begin with, let us consider the Cauchy problem

$$\begin{cases} u'(t) \in Au(t) + f(u(t)) \\ u(0) = \xi. \end{cases} \tag{10.1.1}$$

First, we introduce

Definition 10.1.1. By a C^0*-solution* of (10.1.1) on $[\,0,T\,]$ we mean a function $u:[\,0,T\,]\to K$ which renders $t\mapsto g(t)=f(u(t))$ integrable, satisfies $u(0)=\xi$ and it is a C^0-solution on $[\,0,T\,]$ of the equation

$$u'(t)\in Au(t)+g(t)$$

in the sense of Definition 1.6.2. By a C^0*-solution* of (10.1.1) on $[\,0,\widetilde{T})$ we mean a function $u:[\,0,\widetilde{T})\to K$ which is a C^0-solution of (10.1.1) on $[\,0,T\,]$ for each $0<T<\widetilde{T}$.

Definition 10.1.2. We say that K is C^0*-viable with respect to* $A+f$ if for each $\xi\in K$ there exist $T>0$ and a C^0-solution $u:[\,0,T\,]\to K$ of (10.1.1).

Let $\eta\in X$ and let us denote by $\{S_\eta(t):\overline{D(A)}\to\overline{D(A)};\ t\geq 0\}$ the semigroup of nonlinear contractions generated by $A_\eta=A+\eta$ on $\overline{D(A)}$. So, for each $\xi\in\overline{D(A)}$ and $t\geq 0$, $S_\eta(t)\xi$ is the value at $t\in\mathbb{R}_+$ of the unique C^0-solution u of the Cauchy problem

$$\begin{cases} u'(t)\in Au(t)+\eta\\ u(0)=\xi,\end{cases}$$

i.e., $S_\eta(t)\xi=u(t,0,\xi,\eta)$. Inspired by the semilinear case, we introduce the tangency concept we are going to use in the sequel.

Definition 10.1.3. Let $A:D(A)\subseteq X\rightsquigarrow X$ be an m-dissipative operator and K a nonempty subset in $\overline{D(A)}$. We say that the vector $\eta\in X$ is A*-tangent* to K at $\xi\in K$ if

$$\liminf_{h\downarrow 0}\frac{1}{h}\text{dist}\,(S_\eta(h)\xi;K)=0. \tag{10.1.2}$$

Clearly $\eta\in X$ is A*-tangent* to K at $\xi\in K$ if and only if for each $\delta>0$ and each $\rho>0$ there exist $h\in(0,\delta)$ and $p\in D(0,\rho)$ such that

$$S_\eta(h)\xi+hp\in K.$$

The set of all A-tangent elements to K at $\xi\in K$ is denoted by $\mathcal{T}_K^A(\xi)$.

Problem 10.1.1. *Show that whenever $A:D(A)\subseteq X\rightsquigarrow X$ is single-valued and $\xi\in K\cap D(A)$, the "nonlinear tangency condition" $\eta\in\mathcal{T}_K^A(\xi)$ is equivalent to the "Nagumo tangency condition"*

$$\liminf_{h\downarrow 0}\frac{1}{h}\text{dist}\,(\xi+h(A\xi+\eta);K)=0.$$

Problem 10.1.2. *Prove that, if $A : D(A) \subseteq X \to X$ is a linear m-dissipative operator, then $\eta \in X$ is A-tangent to K at $\xi \in K$ in the sense of Definition* 10.1.3 *if and only if it is A-tangent to K at $\xi \in K$ in the sense of Definition* 8.1.3.

We are now ready to state the main results of this chapter.

Theorem 10.1.1. *Let X be a Banach space, $A : D(A) \subseteq X \rightsquigarrow X$ an m-dissipative operator which generates a compact semigroup, let K be a nonempty and locally closed subset in $\overline{D(A)}$ and let $f : K \to X$ be a continuous function. Then a necessary and sufficient condition in order that K be C^0-viable with respect to $A + f$ is that*

$$f(\xi) \in \mathcal{T}_K^A(\xi), \tag{10.1.3}$$

for each $\xi \in K$.

Theorem 10.1.2. *Let X be a Banach space, $A : D(A) \subseteq X \rightsquigarrow X$ an m-dissipative operator, K a nonempty and locally closed subset in $\overline{D(A)}$ and let $f : K \to X$ be a locally Lipschitz function*[1]. *Then a necessary and sufficient condition in order that K be C^0-viable with respect to $A + f$ is* (10.1.3).

Theorem 10.1.3. *Let X be a Banach space, $A : D(A) \subseteq X \rightsquigarrow X$ an m-dissipative operator, K a nonempty and locally compact subset in $\overline{D(A)}$ and let $f : K \to X$ be a continuous function. Then a necessary and sufficient condition in order that K be C^0-viable with respect to $A + f$ is* (10.1.3).

The necessity of Theorems 10.1.1~10.1.3 is an immediate consequence of the next result which is interesting in itself.

Theorem 10.1.4. *Let X be a Banach space, $A : D(A) \subseteq X \rightsquigarrow X$ an m-dissipative operator, K a locally closed subset in X and $f : K \to X$ a given function. Then, a necessary condition in order that K be C^0-viable with respect to $A + f$ is*

$$\lim_{h \downarrow 0} \frac{1}{h} \text{dist}\,(S_{f(\xi)}(h)\xi; K) = 0, \tag{10.1.4}$$

at each point $\xi \in K$ of continuity for f.

Proof. Let $\xi \in K$ be a continuity point of f, and let z be a C^0-solution of (10.1.1) on $[0, T]$. We have

$$u(h, 0, \xi, f(z(\cdot))) \in K$$

---

[1] See Definition 3.2.2.

for each $h \in [\,0,T\,]$. On the other hand, by (1.6.2), we get

$$\|u(h,0,\xi,f(z(\cdot))) - S_{f(\xi)}(h)\xi\| \leq \int_0^h \|f(z(s)) - f(\xi)\|\,ds$$

for each $h \in [\,0,T\,]$. Since f is continuous at ξ and $\lim_{t\downarrow 0} z(t) = \xi$, it follows that

$$\lim_{h\downarrow 0} \frac{1}{h} \int_0^h \|f(z(s)) - f(\xi)\|\,ds = 0.$$

Therefore

$$\lim_{h\downarrow 0} \frac{1}{h} \text{dist}\,(S_{f(\xi)}(h)\xi; K) \leq \lim_{h\downarrow 0} \frac{1}{h} \|S_{f(\xi)}(h)\xi - u(h,0,\xi,f(z(\cdot)))\| = 0,$$

and this completes the proof. □

Remark 10.1.1. Let us define

$$\mathcal{F}_K^A(\xi) = \{\eta \in X\,;\ \lim_{h\downarrow 0} \frac{1}{h} \text{dist}\,(S_\eta(h)\xi; K) = 0\}.$$

Clearly $\mathcal{F}_K^A(\xi) \subseteq \mathcal{T}_K^A(\xi)$ and, in general, the inclusion is strict. In fact, Theorem 10.1.4 shows that, whenever f is continuous, a necessary condition for the viability of K with respect to $A+f$ is a tangency condition which is stronger than $f(\xi) \in \mathcal{T}_K^A(\xi)$ for each $\xi \in K$. More precisely, such a necessary condition is that, for each $\xi \in K$, we have

$$f(\xi) \in \mathcal{F}_K^A(\xi).$$

We notice that, if $A \equiv 0$, $\mathcal{F}_K^0(\xi)$ coincides with the Federer tangent cone to K at $\xi \in K$, i.e. $\mathcal{F}_K^0(\xi) = \mathcal{F}_K(\xi)$.

10.2. Existence of ε-approximate C^0-solutions

The proof of the sufficiency consists in showing that the tangency condition $f(\xi) \in \mathcal{T}_K^A(\xi)$ for each $\xi \in K$ along with Brezis-Browder Theorem 2.1.1 imply that, for each ξ in K, there exists at least one sequence of "approximate solutions" of (10.1.1), defined on the same interval, $\widetilde{u}_n : [\,0,T\,] \to X$, such that $(\widetilde{u}_n)_n$ converges, in some sense, to a C^0-solution of (10.1.1).

The next lemma is an existence result concerning ε-approximate C^0-solutions of (10.1.1) and it is an "m-dissipative plus continuous" version of Lemma 3.3.1. We notice that, bearing in mind our later purposes, although the problem is autonomous, we formulate our result on an arbitrary interval $[\,\tau,T\,]$ instead of $[\,0,T\,]$.

Lemma 10.2.1. *Let X be a real Banach space, $A : D(A) \subseteq X \rightsquigarrow X$ an m-dissipative operator, K a nonempty, locally closed subset in $\overline{D(A)}$ and $f : K \to X$ a continuous function satisfying the tangency condition*

$f(\xi) \in \mathcal{T}_K^A(\xi)$ for each $\xi \in K$. Then, for each $\xi \in K$ there exist $\rho > 0$, $T > \tau$ and $M > 0$ such that $D(\xi,\rho) \cap K$ is closed and, for each $\varepsilon > 0$, there exist three functions $\sigma : [\tau, T] \to [\tau, T]$ nondecreasing, $g : [\tau, T] \to X$ measurable, and $\widetilde{u} : [\tau, T] \to X$ continuous, satisfying:

(i) $t - \varepsilon \leq \sigma(t) \leq t$ *for each* $t \in [\tau, T]$, $\sigma(T) = T$;
(ii) $\widetilde{u}(\sigma(t)) \in D(\xi,\rho) \cap K$ *for each* $t \in [\tau, T]$, $\widetilde{u}([\tau, T])$ *is precompact;*
(iii) $g(t) = f(\widetilde{u}(\sigma(t)))$ *for each* $t \in [\tau, T]$;
(iv) $\|g(t)\| \leq M$ *a.e. for* $t \in [\tau, T]$;
(v) $\widetilde{u}(\tau) = \xi$ *and* $\|\widetilde{u}(s) - u(s, \sigma(\theta), \widetilde{u}(\sigma(\theta)), g)\| \leq (s - \sigma(\theta))\varepsilon$ *for each* $\theta \in [\tau, T)$ *and* $s \in [\sigma(\theta), T]$;
(vi) $\|\widetilde{u}(t) - \widetilde{u}(\sigma(t))\| \leq \varepsilon$ *for* $t \in [\tau, T]$.

Proof. We consider $\tau = 0$, the general case following by a standard translation argument, i.e., by the change of variable $t = \tau + s$. Let $\xi \in K$ be arbitrary and choose $\rho > 0$ and $M > 0$ such that $D(\xi,\rho) \cap K$ is closed (compact in the hypotheses of Theorem 10.1.3) and $\|f(u)\| \leq M$ for every $u \in D(\xi,\rho) \cap K$. This is always possible because K is locally closed and f continuous and thus locally bounded. Next, take $T > 0$ such that

$$\sup_{0 \leq t \leq T} \|S(t)\xi - \xi\| + T(M+1) \leq \rho. \tag{10.2.1}$$

We first prove that the conclusion of Lemma 10.2.1 remains true if we replace T as above with a possible smaller number $\mu \in (0, T]$ which, at this stage, is allowed to depend on $\varepsilon \in (0,1)$. Then, by using the Brezis-Browder Theorem 2.1.1, we will prove that we can take $\mu = T$ independent of ε.

Let $\varepsilon \in (0,1)$ be arbitrary. In view of (10.1.3), there exist $\delta \in (0, \varepsilon)$ and $p \in X$ with $\|p\| \leq \varepsilon$ such that

$$u(\delta, 0, \xi, f(\xi)) + \delta p \in K.$$

Let us define the functions $\sigma(s) = 0$ for $s \in [0, \delta)$, $\sigma(\delta) = \delta$, $g(s) = f(\xi)$ and $\widetilde{u}(s) = u(s, 0, \xi, f(\xi)) + sp$ for $s \in [0, \delta]$. We can easily see that, σ, g and $\widetilde{u}$ satisfy (i)~(v) with T substituted by δ. Diminishing $\delta > 0$ if necessary, we obtain also (vi).

In the next step, we will show that there exists at least one triplet $(\sigma, g, \widetilde{u})$ satisfying (i)~(vi). To this aim we shall make use of Brezis-Browder Theorem 2.1.1 as follows. Let $\mathcal{S}$ be the set of all quadruples $(\sigma, g, \widetilde{u}, \mu)$ with $\mu \leq T$ and satisfying (i)~(vi) with μ instead of T. This set is clearly nonempty, as we have already proved. On $\mathcal{S}$ we introduce a partial order $\preceq$ as follows. We say that

$$(\sigma_1, g_1, \widetilde{u}_1, \mu_1) \preceq (\sigma_2, g_2, \widetilde{u}_2, \mu_2)$$

if $\mu_1 \leq \mu_2$ and

$$(\sigma_1(s), g_1(s), \widetilde{u}_1(s)) = (\sigma_2(s), g_2(s), \widetilde{u}_2(s))$$

for each $s \in [0, \mu_1]$. Let us define the function $\mathcal{N} : \mathcal{S} \to \mathbb{R}$ by

$$\mathcal{N}((\sigma, g, \widetilde{u}, \mu)) = \mu.$$

It is clear that $\mathcal{N}$ is increasing on $\mathcal{S}$. Let us now take an increasing sequence

$$((\sigma_j, g_j, \widetilde{u}_j, \mu_j))_{j \in \mathbb{N}}$$

in $\mathcal{S}$ and let us show that it is bounded from above in $\mathcal{S}$. To define an upper bound, set

$$\mu^* = \sup\{\mu_j;\ j \in \mathbb{N}\}.$$

If $\mu^* = \mu_j$ for some $j \in \mathbb{N}$, $(\sigma_j, g_j, \widetilde{u}_j, \mu_j)$ is clearly an upper bound. If $\mu_j < \mu^*$ for each $j \in \mathbb{N}$, we define

$$\sigma(t) = \sigma_j(t), \quad g(t) = g_j(t), \quad \widetilde{u}(t) = \widetilde{u}_j(t)$$

for $j \in \mathbb{N}$ and every $t \in [0, \mu_j]$. This way, we define $(\sigma, g, \widetilde{u})$ on $[0, \mu^*)$. In order to extend the triplet to μ^*, we begin by checking that $\widetilde{u}([0, \mu^*))$ is precompact in X. By (iv), we know that $g \in L^1(0, \mu^*; X)$ and so, for each $j \in \mathbb{N}$, the function $u(\cdot, \mu_j, \widetilde{u}(\mu_j), g) : [\mu_j, \mu^*] \to \overline{D(A)}$ is continuous. As a consequence, the set $C_j = u([\mu_j, \mu^*], \mu_j, \widetilde{u}(\mu_j), g)$ is precompact. On the other hand, by (ii), we know that, for each $j \in \mathbb{N}$, $K_j = \widetilde{u}([0, \mu_j])$ is precompact too. By (v) we deduce that, for each $j \in \mathbb{N}$,

$$\widetilde{u}([0, \mu^*)) \subseteq C_j \cup K_j + (\mu^* - \mu_j) D(0, \varepsilon).$$

Let $\eta > 0$ be arbitrary and fix $j \in \mathbb{N}$ such that

$$(\mu^* - \mu_j)\varepsilon \leq \frac{\eta}{2}.$$

Since $C_j \cup K_j$ is precompact, there exists a finite family $\{\xi_1, \xi_2, \ldots, \xi_{n(\eta)}\}$ such that, for each $\xi \in C_j \cup K_j$, there exists $k \in \{1, 2, \ldots, n(\eta)\}$ such that

$$\|\xi - \xi_k\| \leq \frac{\eta}{2}.$$

The last two inequalities and the inclusion above yield

$$\widetilde{u}([0, \mu^*)) \subseteq \cup_{k=1}^{n(\eta)} D(\xi_k, \eta)$$

and accordingly $\widetilde{u}([0, \mu^*))$ is precompact. Now, take any limit point $\widetilde{u}_{\mu^*}$ of $\widetilde{u}(\mu_j)$ as j tends to ∞ and set $\widetilde{u}(\mu^*) = \widetilde{u}_{\mu^*}$. We define $\sigma(\mu^*) = \mu^*$ and so, $g(\mu^*) = f(\widetilde{u}(\sigma(\mu^*))) = \widetilde{u}_{\mu^*}$. Clearly $\widetilde{u}(\mu^*) \in D(\xi, \rho) \cap K$. So, with $\widetilde{u} : [0, \mu^*] \to X$, defined as above, we obviously have that $(\sigma, g, \widetilde{u}, \mu^*)$ satisfies (i)$\sim$(iv) on $[0, \mu^*]$. It is also easy to see that (v) holds for each $\theta \in [0, \mu^*)$ and each $s \in [\sigma(\theta), \mu^*)$. To check (v) for $s = \mu^*$, we have to fix any $\theta \in [0, \mu^*)$, to take $s = \mu_j$ with $\mu_j > \sigma(\theta)$ in (v) and to pass to the limit for j tending to ∞ both sides in (v) on that subsequence on which $(\widetilde{u}_j(\mu_j))_{j\in\mathbb{N}}$ tends to $\widetilde{u}_{\mu^*} = \widetilde{u}(\mu^*)$. So, $(\sigma, g, \widetilde{u}, \mu^*)$ is an upper bound for $((\sigma_j, g_j, \widetilde{u}_j, \mu_j))_{j\in\mathbb{N}}$ and consequently the set $\mathcal{S}$ endowed with the partial order $\preceq$ and the function $\mathcal{N}$ satisfy the hypotheses of Theorem 2.1.1. Accordingly there exists at least one element $(\sigma_\nu, g_\nu, \widetilde{u}_\nu, \nu)$ in $\mathcal{S}$ such that, if $(\sigma_\nu, g_\nu, \widetilde{u}_\nu, \nu) \preceq (\sigma_\mu, g_\mu, \widetilde{u}_\mu, \mu)$ then $\nu = \mu$.

We next show that $\nu = T$, where T satisfies (10.2.1). To this aim, let us assume by contradiction that $\nu < T$ and let $\xi_\nu = \widetilde{u}_\nu(\nu) = \widetilde{u}_\nu(\sigma_\nu(\nu))$ which belongs to $D(\xi, \rho) \cap K$. In view of (1.6.5), (v) and (iv), we have

$$
\begin{aligned}
\|\xi_\nu - \xi\| &\leq \|S(\nu)\xi - \xi\| + \|u(\nu, 0, \xi, g_\nu) - S(\nu)\xi\| \\
&\quad + \|\widetilde{u}_\nu(\nu) - u(\nu, 0, \xi, g_\nu)\| \\
&\leq \|S(\nu)\xi - \xi\| + \int_0^\nu \|g_\nu(s)\|\, ds + \nu\varepsilon \\
&\leq \sup_{0\leq t\leq \nu} \|S(t)\xi - \xi\| + \nu(M + \varepsilon).
\end{aligned}
$$

Recalling that $\nu < T$ and $\varepsilon < 1$, from (10.2.1), we get

$$
\|\xi_\nu - \xi\| < \rho. \tag{10.2.2}
$$

Next, we proceed as in the first part of the proof, with ν instead of 0 and with ξ_ν instead of ξ. So, from (10.1.3), combined with (10.2.2), we infer that there exist $\delta \in (0, \varepsilon]$ with $\nu + \delta \leq T$ and $p \in X$ satisfying $\|p\| \leq \varepsilon$, such that

$$
u(\nu + \delta, \nu, \xi_\nu, f(\xi_\nu)) + \delta p \in D(\xi, \rho) \cap K.
$$

We define $\sigma_{\nu+\delta} : [0, \nu + \delta] \to [0, \nu + \delta]$, $g_{\nu+\delta} : [0, \nu + \delta] \to X$ and $\widetilde{u}_{\nu+\delta} : [0, \nu + \delta] \to X$ by

$$
\sigma_{\nu+\delta}(t) = \begin{cases} \sigma_\nu(t) & \text{if } t \in [0, \nu] \\ \nu & \text{if } t \in (\nu, \nu + \delta) \\ \nu + \delta & \text{if } t = \nu + \delta, \end{cases}
$$

$$
g_{\nu+\delta}(t) = \begin{cases} g_\nu(t) & \text{if } t \in [0, \nu] \\ f(\xi_\nu) & \text{if } t \in (\nu, \nu + \delta], \end{cases}
$$

and respectively by

$$
\widetilde{u}_{\nu+\delta}(t) = \begin{cases} \widetilde{u}_\nu(t) & \text{if } t \in [0, \nu] \\ u(t, \nu, \xi_\nu, g_{\nu+\delta}) + (t - \nu)p & \text{if } t \in (\nu, \nu + \delta], \end{cases}
$$

Since $\widetilde{u}_{\nu+\delta}(\nu + \delta) \in K \cap D(\xi, \rho)$, $(\sigma_{\nu+\delta}, g_{\nu+\delta}, \widetilde{u}_{\nu+\delta})$ satisfies (i)~(iv) with T replaced by $\nu + \delta$. Obviously (v) holds for each t and s satisfying $\sigma_{\nu+\delta}(t) \leq s \leq \nu$, or $\nu \leq \sigma_{\nu+\delta}(t) \leq s$. The only case we have to check is that one in which $\sigma_{\nu+\delta}(t) \leq \nu < s \leq \nu+\delta$. To this aim, let us observe that, by virtue of the evolution property (1.6.4), i.e.

$$u(t,a,\xi,f) = u(t,\nu,u(\nu,a,\xi,f), f\,\big|_{[\nu,\nu+\delta]}), \tag{10.2.3}$$

we have

$$\begin{aligned}\widetilde{u}_{\nu+\delta}(s) - u(s,\sigma_{\nu+\delta}(t),\widetilde{u}_{\nu+\delta}(\sigma_{\nu+\delta}(t)), g_{\nu+\delta})\\ = u(s,\nu,\widetilde{u}_{\nu+\delta}(\nu), g_{\nu+\delta}) + (s-\nu)p\\ -u(s,\nu,u(\nu,\sigma_{\nu+\delta}(t),\widetilde{u}_{\nu+\delta}(\sigma_{\nu+\delta}(t)), g_{\nu+\delta}), g_{\nu+\delta}).\end{aligned}$$

Hence, in view of (v), we get

$$\begin{aligned}\|\widetilde{u}_{\nu+\delta}(s) - u(s,\sigma_{\nu+\delta}(t),\widetilde{u}_{\nu+\delta}(\sigma_{\nu+\delta}(t)), g_{\nu+\delta})\|\\ \leq \|\widetilde{u}_{\nu+\delta}(\nu) - u(\nu,\sigma_{\nu+\delta}(t),\widetilde{u}_{\nu+\delta}(\sigma_{\nu+\delta}(t)), g_{\nu+\delta})\| + (s-\nu)\|p\|\\ \leq (\nu-\sigma_{\nu+\delta}(t))\varepsilon + (s-\nu)\varepsilon = (s-\sigma_{\nu+\delta}(t))\varepsilon.\end{aligned}$$

So (v) holds for each $t \in [\,0,\nu+\delta)$ and each $s \in [\,\sigma_{\nu+\delta}(t),\nu+\delta\,]$. Diminishing δ is necessary, we get (vi).

We conclude that $(\sigma_{\nu+\delta}, g_{\nu+\delta}, \widetilde{u}_{\nu+\delta}, \nu+\delta) \in \mathcal{S}$,

$$(\sigma_{\nu}, g_{\nu}, \widetilde{u}_{\nu}, \nu) \preceq (\sigma_{\nu+\delta}, g_{\nu+\delta}, \widetilde{u}_{\nu+\delta}, \nu+\delta)$$

and $\nu < \nu+\delta$. This contradiction can be eliminated only if $\nu = T$ and this completes the proof of Lemma 10.2.1. □

Remark 10.2.1. Under the general hypotheses of Lemma 10.2.1, for each $\gamma > 0$, we can diminish both $\rho > 0$ and $T > \tau$, such that $T - \tau < \gamma$, $\rho < \gamma$ and all the conditions (i)∼(vi) in Lemma 10.2.1 be satisfied.

Definition 10.2.1. Let $\xi \in K$ and $\varepsilon > 0$. Let $T > \tau$ be (independent of $\varepsilon > 0$) as in Lemma 10.2.1. A triple $(\sigma, g, \widetilde{u})$ satisfying (i)∼(vi) is called an *ε-approximate C^0-solution* of (10.1.1) on $[\,\tau, T\,]$.

10.3. Convergence in the case of Theorems 10.1.1 and 10.1.3

In order to complete the proof of the sufficiency of Theorems 10.1.1 and 10.1.3, we will show that there exists a subsequence of ε_n-approximate C^0-solutions which is uniformly convergent on $[\,0,T\,]$ to a function u which is a C^0-solution of (10.1.1).

Proof. Let $(\varepsilon_n)_n$ be a sequence in $(0,1)$, strictly decreasing to 0, and let $((\sigma_n, g_n, \widetilde{u}_n))_n$ be a sequence of ε_n-approximate C^0-solutions of (10.1.1). From (iii), we know that $\{g_n;\ n = 1,2,\dots\}$ is uniformly bounded on $[\,0,T\,]$ and thus uniformly integrable in $L^1(0,T\,;X)$. See Remark 1.3.3.

Under the hypotheses of Theorem 10.1.1, since the semigroup of nonlinear contractions $\{S(t) : \overline{D(A)} \to \overline{D(A)};\ t \geq 0\}$ is compact, by Theorem 1.6.5, it follows that there exists $u \in C([0,T];X)$ such that, on a subsequence at least, we have

$$\lim_n u(t,0,\xi,g_n) = \widetilde{u}(t) \tag{10.3.1}$$

uniformly for $t \in [0,T]$. By (v) and (10.3.1), we also have

$$\lim_n \widetilde{u}_n(t) = \widetilde{u}(t)$$

uniformly for $t \in [0,T]$. From (vi), we conclude that

$$\lim_n \widetilde{u}_n(\sigma_n(t)) = \widetilde{u}(t) \tag{10.3.2}$$

uniformly for $t \in [0,T]$. By (ii), we know that $\widetilde{u}_n(\sigma_n(t)) \in D(\xi,\rho) \cap K$ for $n = 1,2,\ldots$ and $t \in [0,T]$. As $D(\xi,\rho) \cap K$ is closed, we conclude that $\widetilde{u}(t) \in D(\xi,\rho) \cap K$ for all $t \in [0,T]$. Further, since f is continuous and by (iii), we have $g_n(s) = f(\widetilde{u}_n(\sigma_n(s)))$ for $n = 1,2,\ldots$ and a.e. for $s \in [0,T]$, it follows that

$$\lim_n g_n(t) = f(\widetilde{u}(t))$$

uniformly for $t \in [0,T]$. Thus $\widetilde{u} = u(\cdot,0,\xi,f(\widetilde{u}(\cdot)))$ and therefore it is a C^0-solution of (10.1.1) on $[0,T]$. This completes the proof of Theorem 10.1.1. □

We can now pass to the proof of Theorem 10.1.3.

Proof. We begin by showing that, in the hypotheses of Theorem 10.1.3, (10.3.1) still holds true. First, let us remark that, in view of Remark 10.2.1, we can diminish $T > 0$ and $\rho > 0$, if necessary, such that $D(\xi,\rho) \cap K$ is compact and (i) $\sim$ (vi) in Lemma 10.2.1 be satisfied. So, by (v) and (vi) we have

$$\begin{aligned}\|u(t,0,\xi,g_n) - \widetilde{u}_n(\sigma_n(t))\| &\leq \|u(t,0,\xi,g_n) - \widetilde{u}_n(t)\| + \|\widetilde{u}_n(t) - \widetilde{u}_n(\sigma_n(t))\| \\ &\leq t\varepsilon_n + \varepsilon_n \leq (T+1)\varepsilon_n\end{aligned}$$

for $n = 1,2,\ldots$. Let $k = 1,2,\ldots$, and let us denote by

$$\begin{cases} C_k = \displaystyle\bigcup_{n=1}^{k} u([0,T],0,\xi,g_n) \\ C = \displaystyle\bigcup_{n=1}^{\infty} \widetilde{u}_n(\sigma_n([0,T])) \end{cases}$$

and let us observe that, in view of the inequality above, we have

$$\bigcup_{n=1}^{\infty} u([0,T],0,\xi,g_n) \subseteq C_k \cup C + (T+1)D(0,\varepsilon_k) \tag{10.3.3}$$

for each $k = 1,2,\ldots$. But, for $k = 1,2,\ldots$, C_k and C are precompact, C_k as a finite union of compact sets, i.e. $u([0,T],0,\xi,g_n)$ for $n = 1,2,\ldots,k$ and C as a subset of $D(\xi,\rho) \cap K$, which in its turn is compact. This remark along with (10.3.3) shows that, for each $t \in [0,T]$, $\{u(t,0,\xi,g_n);\ n = 1,2,\ldots\}$ is precompact too.

Thus Theorem 1.6.4 applies and so there exists $\widetilde{u} \in C([0,T];X)$ such that, on a subsequence at least, we have (10.3.1).

From now on, the proof follows the very same arguments as those in the last part of the proof of Theorem 10.1.1 and therefore we do not enter into details. □

10.4. Convergence in the case of Theorem 10.1.2

Under the hypotheses of Theorem 10.1.2, we will show that there exists a sequence of ε_n-approximate C^0-solutions which is uniformly convergent on $[0,T]$ to a C^0-solution of (10.1.1).

Proof. Let $\xi \in K$ and let $\rho > 0$ and $T > 0$ be given by Lemma 10.2.1. As f is locally Lipschitz, by Remark 10.2.1, we can diminish $T > 0$ and $\rho > 0$ if necessary, in order to assume that there exists $L > 0$ such that all the conditions in Lemma 10.2.1 be satisfied and, in addition,

$$\|f(u) - f(v)\| \leq L\|u - v\|$$

for all $u, v \in D(\xi,\rho) \cap K$.

Let $(\varepsilon_n)_n$ be a sequence strictly decreasing to 0 and let $((\sigma_n, g_n, \widetilde{u}_n))_n$ be a sequence of ε_n-approximate C^0-solutions of (10.1.1).

In view of (iii) in Lemma 10.2.1, we have

$$\|u(t,0,\xi,g_n) - u(t,0,\xi,g_k)\| \leq \int_0^t \|g_n(s) - g_k(s)\|\,ds$$

$$\leq \int_0^t \|f(\widetilde{u}_n(\sigma_n(s))) - f(\widetilde{u}_k(\sigma_k(s)))\|\,ds \leq L\int_0^t \|\widetilde{u}_n(\sigma_n(s)) - \widetilde{u}_k(\sigma_k(s))\|\,ds$$

and consequently

$$\|u(t,0,\xi,g_n) - u(t,0,\xi,g_k)\| \leq L\int_0^t \|\widetilde{u}_n(\sigma_n(s)) - u(s,0,\xi,g_n)\|\,ds$$

$$+L\int_0^t \|u(s,0,\xi,g_n)-u(s,0,\xi,g_k)\|\,ds+L\int_0^t \|u(s,0,\xi,g_k)-\widetilde{u}_k(\sigma_k(s))\|\,ds. \tag{10.4.1}$$

Using (v) and (vi) in Lemma 10.2.1 to estimate the first and the third integral on the right hand side in (10.4.1), we get

$$\|u(t,0,\xi,g_n)-u(t,0,\xi,g_k)\|$$
$$\leq L(T+1)(\varepsilon_n+\varepsilon_k)+L\int_0^t \|u(s,0,\xi,g_n)-u(s,0,\xi,g_k)\|\,ds.$$

From this inequality and Gronwall Lemma 1.8.4, we deduce

$$\|u(t,0,\xi,g_n)-u(t,0,\xi,g_k)\|\leq L(T+1)(\varepsilon_n+\varepsilon_k)e^{TL},$$

for $n,k=1,2,\ldots$ and $t\in[\,0,T\,]$. Since $\varepsilon_n\downarrow 0$, it follows that $(u(\cdot,0,\xi,g_n))_n$ is a Cauchy sequence in the sup-norm. Let u be the uniform limit of $(u(\cdot,0,\xi,g_n))_n$ on $[\,0,T\,]$. From now on, the proof follows the exactly the same steps as those in the last part of the proof of Theorem 10.1.1 and therefore we omit it. □

10.5. The quasi-autonomous noncylindrical case

Let $\mathcal{C}$ be a nonempty subset in $\mathbb{R}\times\overline{D(A)}$ and let $f:\mathcal{C}\to X$ be a continuous function. Let us consider the nonautonomous Cauchy problem

$$\begin{cases} u'(t)\in Au(t)+f(t,u(t))\\ u(\tau)=\xi. \end{cases} \tag{10.5.1}$$

Definition 10.5.1. By a C^0-*solution* of the problem (10.5.1) on $[\,\tau,T\,]$, we mean a continuous function $u:[\,\tau,T\,]\to\overline{D(A)}$ which satisfies:

(i) $(t,u(t))\in\mathcal{C}$ for each $t\in[\,\tau,T\,]$;

(ii) u is a C^0-solution, in the sense of Definition 1.6.2, of the Cauchy problem

$$\begin{cases} u'(t)\in Au(t)+g(t)\\ u(\tau)=\xi, \end{cases}$$

where $g(s)=f(s,u(s))$ for $s\in[\,\tau,T\,]$.

A C^0-*solution* of (10.5.1) on $[\,\tau,\widetilde{T})$ is a function $u:[\,\tau,\widetilde{T})\to\overline{D(A)}$ which, for each $T\in(\tau,\widetilde{T})$, is a C^0-solution of (10.5.1) on $[\,\tau,T\,]$ in the sense mentioned above.

Definition 10.5.2. The set $\mathcal{C}\subseteq\mathbb{R}\times\overline{D(A)}$ is C^0-*viable* with respect to $A+f$ if for each $(\tau,\xi)\in\mathcal{C}$, there exists $T\in\mathbb{R}$, $T>\tau$ such that the Cauchy problem (10.5.1) has at least one C^0-solution $u:[\,\tau,T\,]\to\overline{D(A)}$.

Remark 10.5.1. Let $\mathcal{X} = \mathbb{R} \times X$ with the norm $\|(t,u)\|_{\mathcal{X}} = |t| + \|u\|$ and let us observe that the quasi-autonomous Cauchy problem (10.5.1) can be equivalently rewritten as an autonomous one in the space $\mathcal{X}$, by setting $\mathcal{A}z = \{(0,v);\ v \in Au\}$ for $z = (t,u) \in D(\mathcal{A})$, where

$$D(\mathcal{A}) = \{(t,u);\ t \in \mathbb{R},\ u \in D(A)\},$$

$z(s) = (t(\tau+s), u(\tau+s))$, $\mathcal{F}(z) = (1, f(z))$ and $\zeta = (\tau, \xi)$. Indeed, with the notations above, we have

$$\begin{cases} z'(s) \in \mathcal{A}z(s) + \mathcal{F}(z(s)) \\ z(0) = \zeta. \end{cases}$$

It readily follows that $\mathcal{A}$ is m-dissipative and, in addition, that $z = (t,u)$ is a C^0-solution of the problem above if and only if

$$\begin{cases} t(s) = \tau + s \\ u(s) = u(\tau + s, \tau, \xi, f(\tau + \cdot, u(\cdot))). \end{cases}$$

Remark 10.5.2. One may easily see that $\mathcal{C}$ is C^0-viable with respect to $A + f$ in the sense of Definition 10.5.2 if and only if $\mathcal{C}$ is C^0-viable with respect to $\mathcal{A} + \mathcal{F}$ in the sense of Definition 10.1.2.

The next viability results are consequences of Theorems 10.1.1∼10.1.4.

Theorem 10.5.1. *Let X be a Banach space, $A : D(A) \subseteq X \rightsquigarrow X$ an m-dissipative operator which generates a compact semigroup, $\mathcal{C}$ a nonempty, locally closed subset in $\mathbb{R} \times \overline{D(A)}$ and $f : \mathcal{C} \to X$ a continuous function. Then a necessary and sufficient condition in order that $\mathcal{C}$ be C^0-viable with respect to $A + f$ is that*

$$(1, f(\tau,\xi)) \in \mathcal{T}^{\mathcal{A}}_{\mathcal{C}}(\tau,\xi), \tag{10.5.2}$$

for each $(\tau,\xi) \in \mathcal{C}$.

Theorem 10.5.2. *Let X be a Banach space, $A : D(A) \subseteq X \rightsquigarrow X$ an m-dissipative operator, $\mathcal{C}$ a nonempty, locally closed subset in $\mathbb{R} \times \overline{D(A)}$ and $f : \mathcal{C} \to X$ a locally Lipschitz function. Then a necessary and sufficient condition in order that $\mathcal{C}$ be C^0-viable with respect to $A + f$ is* (10.5.2).

Theorem 10.5.3. *Let X be a Banach space, $A : D(A) \subseteq X \rightsquigarrow X$ an m-dissipative operator, $\mathcal{C}$ a nonempty, locally compact subset in $\mathbb{R} \times \overline{D(A)}$ and $f : \mathcal{C} \to X$ a continuous function. Then a necessary and sufficient condition in order that $\mathcal{C}$ be C^0-viable with respect to $A + f$ is* (10.5.2).

Problem 10.5.1. Prove that, for each $(\tau,\xi) \in \mathcal{C}$, the condition (10.5.2) is equivalent to

$$\liminf_{h \downarrow 0} \frac{1}{h} \operatorname{dist}\left((\tau + h, u(\tau + h, \tau, \xi, f(\tau,\xi))); \mathcal{C}\right) = 0.$$

10.6. Noncontinuable C^0-solutions

In this section, we present some results concerning the existence of noncontinuable, or even global C^0-solutions to

$$\begin{cases} u'(t) \in Au(t) + f(t, u(t)) \\ u(\tau) = \xi, \end{cases} \tag{10.6.1}$$

where $A : D(A) \subseteq X \rightsquigarrow X$ is an m-dissipative operator, $\mathcal{C} \subseteq \mathbb{R} \times \overline{D(A)}$ is nonempty and $f : \mathcal{C} \to X$ is a given function. A C^0-solution $u : [\,\tau, T) \to X$ to (10.6.1) is called *noncontinuable*, if there is no other C^0-solution $v : [\,\tau, \widetilde{T}) \to X$ of the same equation, with $T < \widetilde{T}$ and satisfying $u(t) = v(t)$ for all $t \in [\,\tau, T)$. A C^0-solution $u : [\,\tau, T) \to X$ to (10.6.1) is called *global* if $T = T_{\mathcal{C}}$, where $T_{\mathcal{C}}$ is defined by (3.6.2). The next theorem follows from Brezis–Browder Theorem 2.1.1. Since its proof is almost identical with that of Theorem 3.6.1, we do not enter into details.

Theorem 10.6.1. *Let X be a Banach space, $A : D(A) \subseteq X \rightsquigarrow X$ an m-dissipative operator, let $\mathcal{C} \subseteq \mathbb{R} \times \overline{D(A)}$ be nonempty, and let $f : \mathcal{C} \to X$. Then, the following conditions are equivalent:*

(i) $\mathcal{C}$ *is C^0-viable with respect to $A + f$;*
(ii) *for each $(\tau, \xi) \in \mathcal{C}$ there exists at least one noncontinuable C^0-solution $u : [\,\tau, T) \to X$ of* (10.6.1).

The next result concerns the existence of global solutions.

Theorem 10.6.2. *Let X be a Banach space, $A : D(A) \subseteq X \rightsquigarrow X$ an m-dissipative operator, $\mathcal{C}$ a nonempty and X-closed subset*[2] *in $\mathbb{R} \times \overline{D(A)}$ and $f : \mathcal{C} \to X$ a continuous function which is positively sublinear*[3]. *If f maps bounded subsets in $\mathcal{C}$ into bounded subsets in X and $\mathcal{C}$ is C^0-viable with respect to $A + f$, then each C^0-solution of* (10.6.1) *can be continued up to a global one.*

Proof. Since $\mathcal{C}$ is C^0-viable with respect to $A + f$, for each $(\tau, \xi) \in \mathcal{C}$, there exists at least one noncontinuable C^0-solution $u : [\,\tau, T) \to X$ to (10.6.1). We will show that $T = T_{\mathcal{C}}$. To this aim, let us assume the contrary, i.e., that $T < T_{\mathcal{C}}$. In particular this means that $T < +\infty$. By using a translation argument if necessary, we may assume with no loss of generality that $0 \in D(A)$ and $0 \in A0$. From (1.6.2) with $\eta = 0$, $g \equiv 0$ and $v \equiv 0$, we get

$$\|u(t)\| \le \|\xi\| + \int_{E_t} [\,u(s), f(s, u(s))\,]_+ \, ds + \int_{H_t \setminus G_t} [\,u(s), f(s, u(s))\,]_+ \, ds$$

[2]See Definition 3.6.2.
[3]See Definition 3.6.1.

for each $t \in [\tau, T)$, where

$$E_t = \{s \in [0,t];\ [u(s), f(s,u(s))]_+ > 0 \text{ and } \|u(s)\| > c(s)\},$$
$$G_t = \{s \in [0,t];\ [u(s), f(s,u(s))]_+ \leq 0\},$$
$$H_t = \{s \in [0,t];\ \|u(s)\| \leq c(s)\}.$$

As $H_t \subseteq H_T$ and $[u,v]_+ \leq \|v\|$ for each $u, v \in X$, we get

$$\|u(t)\| \leq \|\xi\| + \int_{E_t} [a(s)\|u(s)\| + b(s)]\,ds + \int_{H_T} \|f(s,u(s))\|\,ds$$

for each $t \in [\tau, T)$. But f maps bounded subsets in $\mathcal{C}$ into bounded subsets in X and therefore there exists $M > 0$ such that $\|f(s,u(s))\| \leq M$ for all $s \in H_T$. Hence

$$\|u(t)\| \leq \|\xi\| + TM + \int_0^T b(s)\,ds + \int_0^t a(s)\|u(s)\|\,ds,$$

for each $t \in [\tau, T)$. By Gronwall Lemma 1.8.4, u is bounded on $[\tau, T)$.

Using once again the fact that f maps bounded subsets in $\mathcal{C}$ into bounded subsets in X, we deduce that $f(\cdot, u(\cdot))$ is bounded on $[\tau, T)$. So, there exists $\lim_{t\uparrow T} u(t) = u^*$. Since $\mathcal{C}$ is X-closed and $T < T_{\mathcal{C}}$, it follows that $(T, u^*) \in \mathcal{C}$. As $\mathcal{C}$ is C^0-viable with respect to f, we conclude that u can be continued to the right of T. But this is absurd, because u is non-continuable. This contradiction can be eliminated only if $T = T_{\mathcal{C}}$, and this achieves the proof. □

A useful consequence of Theorem 10.6.1 is

Theorem 10.6.3. *Let X be a Banach space, let $A : D(A) \subseteq X \rightsquigarrow X$ be an m-dissipative operator and let $\mathcal{C}$ be an X-closed subset in $\mathbb{R} \times \overline{D(A)}$. Let $f : \mathcal{C} \to X$ be a given function for which there exist two continuous functions $a : \mathbb{R} \to \mathbb{R}_+$ and $b : \mathbb{R} \to \mathbb{R}_+$ such that*

$$\|f(t,\xi)\| \leq a(t)\|\xi\| + b(t)$$

for each $(t,\xi) \in \mathcal{C}$. If $\mathcal{C}$ is C^0-viable with respect to $A + f$, then each C^0-solution of (10.6.1) *can be continued up to a global one, i.e., defined on* $[\tau, T_{\mathcal{C}})$.

10.7. A class of fully nonlinear reaction-diffusion systems

Let X and Y be two real Banach spaces[4] and let $A : D(A) \subseteq X \rightsquigarrow X$ and $B : D(B) \subseteq Y \rightsquigarrow Y$ be the infinitesimal generators of two nonlinear semigroups of contractions denoted by $\{S_A(t) : \overline{D(A)} \to \overline{D(A)};\ t \geq 0\}$ and

[4]Since there is no danger of confusion, we denote the norms on both spaces by the same symbol, $\|\cdot\|$.

by $\{S_B(t) : \overline{D(B)} \to \overline{D(B)};\ t \geq 0\}$ respectively. Let $\mathcal{C} \subseteq \mathbb{R} \times \overline{D(A)} \times \overline{D(B)}$ be nonempty and let $F : \mathcal{C} \to X$ and $G : \mathcal{C} \to Y$. Let us consider the Cauchy problem for the abstract nonlinear reaction-diffusion system

$$\begin{cases} x'(t) \in Ax(t) + F(t, x(t), y(t)) \\ y'(t) \in By(t) + G(t, x(t), y(t)) \\ x(\tau) = \zeta \ \text{ and } \ y(\tau) = \eta. \end{cases} \tag{10.7.1}$$

Definition 10.7.1. We say that the set $\mathcal{C}$ is *C^0-viable with respect to* $(A + F, B + G)$ if for each $(\tau, \zeta, \eta) \in \mathcal{C}$ there exists $T > \tau$ such that (10.7.1) has at least one C^0-solution $(x, y) : [\tau, T] \to \overline{D(A)} \times \overline{D(B)}$ with $(t, x(t), y(t)) \in \mathcal{C}$ for each $t \in [\tau, T]$.

Throughout, we denote by $\mathcal{X}$ the Banach space $X \times Y$, endowed with the norm $\|(x, y)\| = \|x\| + \|y\|$, for $(x, y) \in \mathcal{X}$.

Definition 10.7.2. A *C^0-solution* of (10.7.1) on $[\tau, T]$ is a continuous function $(x, y) : [\tau, T] \to \mathcal{X}$ such that:

(i) $(t, x(t), y(t)) \in \mathcal{C}$ for each $t \in [\tau, T]$;

(ii) x is a C^0-solution on $[\tau, T]$ of the Cauchy problem

$$\begin{cases} x'(t) \in Ax(t) + f(t) \\ x(\tau) = \zeta \end{cases}$$

and y is a C^0-solution on $[\tau, T]$ of the Cauchy problem

$$\begin{cases} y'(t) \in By(t) + g(t) \\ y(\tau) = \eta, \end{cases}$$

in the sense of Definition 1.6.2, where $f(t) = F(t, x(t), y(t))$ and $g(t) = G(t, x(t), y(t))$.

A *C^0-solution* of (10.7.1) on $[\tau, \widetilde{T})$ is a function $(x, y) : [\tau, \widetilde{T}) \to \mathcal{X}$ such that, for each $T \in (\tau, \widetilde{T})$, (x, y) is a C^0-solution of (10.7.1) on $[\tau, T]$ in the sense mentioned above.

Definition 10.7.3. Let $\mathcal{D} \subseteq \mathbb{R} \times \mathcal{X}$. The function $F : \mathcal{D} \to X$ is *locally Lipschitz with respect to $x \in X$* if it is continuous and, in addition, for each $(\tau, \zeta, \eta) \in \mathcal{D}$, there exist $\rho > 0$ and $L > 0$, such that

$$\|F(t, x, y) - F(t, \tilde{x}, y)\| \leq L\|x - \tilde{x}\|,$$

for all $(t, x, y), (t, \tilde{x}, y) \in D((\tau, \zeta, \eta), \rho) \cap \mathcal{D}$.

We notice that here and thereafter, $D((\tau, \zeta, \eta), \rho)$ denotes the closed ball with center $(\tau, \zeta, \eta) \in \mathbb{R} \times \mathcal{X}$ and radius $\rho > 0$.

The hypotheses which will be in effect throughout are:

(H_1) $A : D(A) \subseteq X \rightsquigarrow X$ and $B : D(B) \subseteq Y \rightsquigarrow Y$ are m-dissipative;

(H_2) $\mathcal{C} \subseteq \mathbb{R} \times \overline{D(A)} \times \overline{D(B)}$ is nonempty and locally closed;
(H_3) $\Pi_X \mathcal{C}$ is locally compact;
(H_4) $(F, G) : \mathcal{C} \to \mathcal{X}$ is continuous;
(H_5) $(F, G) : \mathcal{C} \to \mathcal{X}$ is locally Lipschitz;
(H_6) $F : \mathbb{R} \times X \times Y \to X$ is locally Lipschitz with respect to $x \in X$[5];
(H_7) $\{S_A(t) : \overline{D(A)} \to \overline{D(A)};\ \ t \geq 0\}$ is compact;
(H_8) $\{S_B(t) : \overline{D(B)} \to \overline{D(B)};\ \ t \geq 0\}$ is compact.

Let $(\tau, \zeta, \eta) \in \mathcal{C}$ and let us denote by $(x(\cdot, \tau, \zeta, \eta), y(\cdot, \tau, \zeta, \eta))$ the unique C^0-solution on $[\tau, \infty)$ of the system

$$\begin{cases} x'(t) \in Ax(t) + F(\tau, \zeta, \eta) \\ y'(t) \in By(t) + G(\tau, \zeta, \eta) \\ x(\tau) = \zeta \ \ y(\tau) = \eta. \end{cases}$$

The main viability results, referring to (10.7.1), are:

Theorem 10.7.1. *Assume that (H_1), (H_2), (H_4), (H_7) and (H_8) are satisfied. The necessary and sufficient condition in order that $\mathcal{C}$ be C^0-viable with respect to $(A + F, B + G)$ is*

$$\liminf_{h \downarrow 0} \frac{1}{h} \operatorname{dist}\left((\tau + h, x(\tau + h, \tau, \zeta, \eta), y(\tau + h, \tau, \zeta, \eta)); \mathcal{C}\right) = 0, \qquad (10.7.2)$$

for each $(\tau, \zeta, \eta) \in \mathcal{C}$.

Theorem 10.7.2. *Assume that (H_1), (H_2), (H_3), (H_4) and (H_8) are satisfied. The necessary and sufficient condition in order that $\mathcal{C}$ be C^0-viable with respect to $(A + F, B + G)$ is* (10.7.2).

Theorem 10.7.3. *Assume that (H_1), (H_2) and (H_5) are satisfied. The necessary and sufficient condition in order that $\mathcal{C}$ be C^0-viable with respect to $(A + F, B + G)$ is* (10.7.2).

Theorem 10.7.4. *Assume that (H_1), (H_2), (H_4), (H_6) and (H_8) are satisfied. The necessary and sufficient condition in order that $\mathcal{C}$ be C^0-viable with respect to $(A + F, B + G)$ is* (10.7.2).

The necessity of either Theorems 10.7.1~10.7.4 follows from the simple result below.

Theorem 10.7.5. *Assume that (H_1) and (H_4) are satisfied and let $\mathcal{C} \subseteq \mathbb{R} \times \mathcal{X}$ be nonempty. If $\mathcal{C}$ is C^0-viable with respect to $(A + F, B + G)$ then, for every $(\tau, \zeta, \eta) \in \mathcal{C}$, we have*

$$\lim_{h \downarrow 0} \frac{1}{h} \operatorname{dist}\left((\tau + h, x(\tau + h, \tau, \zeta, \eta), y(\tau + h, \tau, \zeta, \eta)); \mathcal{C}\right) = 0. \qquad (10.7.3)$$

---

[5]See Definition 10.7.3.

In order to simplify as much as possible the proofs, the following simple observation is needed.

Remark 10.7.1. Under the general hypotheses (H_1), (H_2) and (H_4), the problem (10.7.1) can be rewritten as an autonomous nonlinear evolution equation in the space $\mathcal{Z} = \mathbb{R} \times \mathcal{X}$ as follows. Let us define $\mathcal{A} : D(\mathcal{A}) \subseteq \mathcal{Z} \rightsquigarrow \mathcal{Z}$ by $\mathcal{A}u = (0, Ax, By)$ for each $u = (v, x, y) \in D(\mathcal{A})$, where,

$$D(\mathcal{A}) = \mathbb{R} \times D(A) \times D(B),$$

let us define $f : \mathcal{C} \to \mathcal{Z}$ by

$$f(u) = (1, F(u), G(u)) = (1, F(v, x, y), G(v, x, y))$$

for each $u = (v, x, y) \in \mathcal{C}$. Then (10.7.1), which is equivalent to the autonomous Cauchy problem

$$\begin{cases} v'(t) = 1 \\ x'(t) \in Ax(t) + F(v(t), x(t), y(t)) \\ y'(t) \in By(t) + G(v(t), x(t), y(t)) \\ v(\tau) = \tau, \quad x(\tau) = \zeta \ \text{ and } \ y(\tau) = \eta, \end{cases} \tag{10.7.4}$$

rewrites as

$$\begin{cases} u'(t) \in \mathcal{A}u(t) + f(u(t)) \\ u(\tau) = \xi, \end{cases} \tag{10.7.5}$$

where $\mathcal{A}$ and f are as above, $u(t) = (v(t), x(t), y(t))$ and $\xi = (\tau, \zeta, \eta)$.

Obviously, under the hypotheses (H_1) and (H_4), the operator $\mathcal{A}$ is m-dissipative on $\mathcal{X}$ and f is continuous. Moreover, $\mathcal{C}$ is C^0-viable with respect to $(A + F, B + G)$, in the sense of Definition 10.7.1, if it is C^0-viable with respect to $\mathcal{A} + f$, in the sense of Definition 10.1.2.

In view of Remark 10.7.1, Theorem 10.7.1 is a consequence of Theorem 10.1.1, Theorem 10.7.3 follows from Theorem 10.5.2. The remaining results have no correspondence with the viability results in Section 10.1. So, it remains to prove Theorems 10.7.2 and 10.7.4.

Remark 10.7.2. The tangency condition (10.7.2) is equivalent to

$$f(\xi) \in \mathcal{T}_{\mathcal{C}}^{\mathcal{A}}(\xi)$$

for each $\xi \in \mathcal{C}$, with $\xi = (\tau, \zeta, \eta)$. Therefore, under the hypotheses (H_1), (H_2) and (H_4), if the tangency condition (10.7.2) holds true, then $\mathcal{A}$, $\mathcal{C}$ and f in (10.7.5) satisfy the hypotheses of Lemma 10.2.1. Since (H_5) implies (H_4), it follows that the remark above applies in all Theorems 10.7.2$\sim$10.7.4.

10.8. Convergence in the case of Theorem 10.7.2

In view of Remarks 10.7.1 and 10.7.2, it suffices to show that, for each $\xi \in \mathcal{C}$, (10.7.5) has at least one C^0-solution $u : [\,\tau, T\,] \to \mathcal{Z}$.

Proof. Let $\xi = (\tau, \zeta, \eta) \in \mathcal{C}$ and let $\varepsilon_n \downarrow 0$ be a sequence in $(0,1)$. In view of Remark 10.7.2, we are in the hypotheses of Lemma 10.2.1 and therefore there exist $\rho > 0$, $T > \tau$, $M > 0$ and a sequence of ε_n-approximate C^0-solutions $((\sigma_n, g_n, \widetilde{u}_n))_n$ of (10.7.5). On components, $g_n = (1, g_n^X, g_n^Y)$ and $\widetilde{u}_n = (\widetilde{v}_n, \widetilde{x}_n, \widetilde{y}_n)$. By Remark 10.2.1, diminishing $\rho > 0$ and $T > \tau$ if necessary, we may assume that, in addition to the conditions in Lemma 10.2.1, $\Pi_X(D((\tau, \zeta, \eta), \rho) \cap \mathcal{C})$ is compact.

Next, from (iv), we know that $\{g_n;\ n = 1, 2, \dots\}$ is uniformly bounded on $[\,\tau, T\,]$, and thus uniformly integrable in $L^1(\tau, T\,; \mathcal{Z})$. See Remark 1.3.3. This means that the families $\{g_n^X;\ n = 1, 2, \dots\}$ and $\{g_n^Y;\ n = 1, 2, \dots\}$ are uniformly integrable in $L^1(\tau, T\,; X)$ and in $L^1(\tau, T\,; Y)$ respectively.

We next show that $\{u(\cdot, \tau, \xi, g_n);\ n = 1, 2, \dots\}$ is relatively compact in $C([\,\tau, T\,]; \mathcal{Z})$. We do this with the help of Theorem 1.6.5.

So, by (v) and (vi) in Lemma 10.2.1, we have

$$\begin{aligned}\|u(t, \tau, \xi, g_n) - \widetilde{u}_n(\sigma_n(t))\| &\leq \|u(t, \tau, \xi, g_n) - \widetilde{u}_n(t)\| + \|\widetilde{u}_n(\sigma_n(t)) - \widetilde{u}_n(t)\| \\ &\leq (t - \tau)\varepsilon_n + \varepsilon_n \leq (T - \tau + 1)\varepsilon_n\end{aligned}$$

for each $t \in [\,\tau, T\,]$.

For $k = 1, 2, \dots$, let us denote both

$$\begin{cases} \mathcal{K}_k = \displaystyle\bigcup_{n=1}^{k} \{u(t, \tau, \xi, g_n)\,;\ t \in [\,\tau, T\,]\}, \\ \mathcal{K} = \displaystyle\bigcup_{n=1}^{\infty} \{\widetilde{u}_n(\sigma_n(t))\,;\ t \in [\,\tau, T\,]\}. \end{cases}$$

Let us observe that, in view of the inequality above, we have

$$\bigcup_{n=1}^{\infty} \{u(t, \tau, \xi, g_n);\ t \in [\,\tau, T\,]\} \subseteq \mathcal{K}_k \cup \mathcal{K} + (T - \tau + 1) D_{\mathcal{Z}}(0, \varepsilon_k) \quad (10.8.1)$$

for $k = 1, 2, \dots$. Clearly $\mathcal{K}_k$ is relatively compact because it is the union of a finite number of compact sets, i.e., the ranges of the continuous functions $u(\cdot, \tau, \xi, g_n)$, $n = 1, 2, \dots k$, and $\mathcal{K}$ because $\mathcal{K} \subseteq \Pi_{\mathbb{R}}\mathcal{K} \times \Pi_X\mathcal{K} \times \Pi_Y\mathcal{K}$ and all factors $\Pi_{\mathbb{R}}\mathcal{K}$, $\Pi_X\mathcal{K}$ and $\Pi_Y\mathcal{K}$ are relatively compact. Indeed, $\Pi_{\mathbb{R}}\mathcal{K} = [\,\tau, T\,]$ is obviously compact. Further, $\Pi_X\mathcal{K}$ is relatively compact as a subset of a set enjoying the same property, i.e., of $\Pi_X(D(\xi, \rho) \cap \mathcal{C})$. See (H_3). Concerning the relative compactness of $\Pi_Y\mathcal{K}$, this follows from the fact that $\{S_B(t) : \overline{D(B)} \to \overline{D(B)},\ t \geq 0\}$, is compact. Indeed, let $(y_n)_n$ be an arbitrary sequence in $\Pi_Y\mathcal{K}$. This means that there exists $(\widetilde{u}_n(\sigma_n(t_n)))_n$, in $\mathcal{C}$, $\widetilde{u}_n(\sigma_n(t_n)) = (\widetilde{v}_n(\sigma_n(t_n)), \widetilde{x}_n(\sigma_n(t_n)), \widetilde{y}_n(\sigma_n(t_n)))$ and such that

$$y_n = \widetilde{y}_n(\sigma_n(t_n))$$

for $n = 1, 2, \ldots$. So, to conclude that $\Pi_Y\mathcal{K}$ is relatively compact, it suffices to show that $(\widetilde{y}_n)_n$ has at least one uniformly convergent subsequence in $C([\tau, T]; Y)$. Recalling that $\{g_n^Y;\ n = 1, 2, \ldots\}$ is uniformly integrable, by Theorem 1.6.5, it follows that there exists $\widetilde{y} \in C([\tau, T]; Y)$ such that, on a subsequence at least, we have

$$\lim_n y(t, \tau, \eta, g_n^Y) = \widetilde{y}(t)$$

uniformly for $t \in [\tau, T]$. By (v) in Lemma 10.2.1, we have

$$\|\widetilde{y}_n(t) - y(t, \tau, \eta, g_n^Y)\| \leq T\varepsilon_n$$

which shows that

$$\lim_n \widetilde{y}_n(t) = \widetilde{y}(t)$$

uniformly for $t \in [\tau, T]$. As $(t_n)_n$ is bounded, we may assume without loss of generality that there exists $t \in [\tau, T]$ such that $\lim_n t_n = t$. As $\lim_n \sigma_n(t_n) = t$, we conclude that $\lim_n y_n = \lim_n \widetilde{y}_n(\sigma_n(t_n)) = \widetilde{y}(t)$. Hence $\Pi_Y\mathcal{K}$ is relatively compact in Y. In view of (10.8.1), it follows that, for each $t \in [\tau, T]$, the set $\bigcup_{n=1}^{\infty} u(t, \tau, \xi, g_n)$ is relatively compact in $\mathcal{Z}$. An appeal to Theorem 1.6.4 shows that there exists $\widetilde{u} \in C([\tau, T]; \mathcal{Z})$ such that, at least on a subsequence, we have

$$\lim_n u(t, \tau, \xi, g_n) = \widetilde{u}(t) \tag{10.8.2}$$

uniformly for $t \in [\tau, T]$.

Recalling that $u(\cdot, \tau, \xi, g_n)$ is the C^0-solution of the Cauchy problem

$$\begin{cases} u'(t) \in \mathcal{A}u(t) + g_n(t) \\ u(\tau) = \xi, \end{cases} \tag{10.8.3}$$

where

$$g_n(t) = f(\widetilde{u}_n(\sigma_n(t))),$$

that $\lim_n \sigma_n(t) = t$ uniformly on $[\tau, T]$, $\lim_n \varepsilon_n = 0$ and, by (v) and (vi),

$$\|\widetilde{u}_n(\sigma_n(t)) - u(t, \tau, \xi, g_n)\| \leq (T + 1)\varepsilon_n$$

for $n = 1, 2, \ldots$, using (10.8.2), we get $\lim_n \widetilde{u}_n(\sigma_n(t)) = \widetilde{u}(t)$ uniformly for $t \in [\tau, T]$. But, by (ii), $\widetilde{u}_n(\sigma_n(t)) \in D(\xi, \rho) \cap \mathcal{C}$ for each $t \in [\tau, T]$. As $D(\xi, \rho) \cap \mathcal{C}$ is closed, it follows that $\widetilde{u}(t) \in D(\xi, \rho) \cap \mathcal{C}$ for each $t \in [\tau, T]$. Since f is continuous, we deduce $\lim_n g_n(t) = f(\widetilde{u}(t))$ uniformly for each $t \in [\tau, T]$. So, $\widetilde{u} = u(\cdot, \tau, \xi, f(\widetilde{u}(\cdot)))$, which means that $\widetilde{u} = (v, x, y)$ is a

C^0-solution of the problem (10.7.5), i.e., that (x, y) is a C^0-solution of the Cauchy problem (10.7.1) and the proof is complete. □

10.9. Convergence in the case of Theorem 10.7.4

Up to a certain point the proof of Theorem 10.7.4 is identical with the one of Theorem 10.7.2.

Proof. Let $\xi = (\tau, \zeta, \eta) \in \mathcal{C}$ and let $\varepsilon_n \downarrow 0$ be a sequence in $(0,1)$. In view of Remark 10.7.2 we are in the hypotheses of Lemma 10.2.1 and therefore there exist $\rho > 0$, $T > \tau$, $M > 0$ and a sequence of ε_n-approximate C^0-solutions $((\sigma_n, g_n, \widetilde{u}_n))_n$ of (10.7.5).

Since $F : \mathbb{R} \times \mathcal{X} \to X$ is locally Lipschitz with respect to $x \in X$, diminishing $\rho > 0$ and $T > \tau$ if necessary, we may assume that the conclusion of Lemma 10.2.1 holds and, in addition, there exists $L > 0$ such that

$$\|F(t, x, y) - F(t, \widetilde{x}, y)\| \leq L\|x - \widetilde{x}\| \tag{10.9.1}$$

for all $(t, x, y), (t, \tilde{x}, y) \in D((\tau, \zeta, \eta), \rho)$. See (H_6) and Definition 10.7.3.

We shall now prove that, under the hypotheses of Theorem 10.7.4, the sequence $(u(\cdot, \tau, \xi, g_n))_n$ has at least one uniformly convergent subsequence whose limit is a C^0-solution of (10.7.5).

Since by (H_8) the semigroup generated by B is compact, while by (iv)[6], the set $\{g_n^Y;\ n = 1, 2, \dots\}$ is uniformly integrable, by Theorem 1.6.5, we may assume that there exists $y \in C([\,\tau, T\,]; Y)$ such that, on a subsequence at least, $\lim_n y(t, \tau, \eta, g_n^Y) = y(t)$ uniformly for $t \in [\,\tau, T\,]$. In view of (v) and (vi), this shows that

$$\lim_n \widetilde{y}_n(\sigma_n(t)) = \lim_n \widetilde{y}_n(t) = y(t) \tag{10.9.2}$$

uniformly for $t \in [\,\tau, T\,]$.

Next, let us observe that (10.7.4) consists of two Cauchy problems, the first one,

$$\begin{cases} v'(t) = 1 \\ v(\tau) = \tau, \end{cases}$$

being decoupled from the second one whose unknown functions are x and y. Thus $v(t) = t$ for each $t \in [\,\tau, T\,]$. By Lemma 10.2.1, we conclude that

$$\lim_n \widetilde{v}_n(\sigma_n(t)) = \lim_n \widetilde{v}_n(t) = t \tag{10.9.3}$$

uniformly for $t \in [\,\tau, T\,]$.

---

[6] Here and thereafter within this section, (i)∼(vi) denotes the corresponding items in Lemma 10.2.1.

Since A is m-dissipative and the function $f : [\tau, T] \times X \to X$, defined by $f(t, x) = F(t, x, y(t))$ for $(t, x) \in [\tau, T] \times X$, is locally Lipschitz with respect to $x \in X$, there exists $T_0 \in (\tau, T]$ such that

$$\begin{cases} x'(t) \in Ax(t) + F(t, x(t), y(t)) \\ x(\tau) = \zeta, \end{cases} \tag{10.9.4}$$

has a unique C^0-solution $x : [\tau, T_0] \to X$. Taking into account of (10.2.1) and (10.9.1), we deduce that $T_0 = T$.

To conclude, it suffices to show that $t \mapsto u(t) = (v(t), x(t), y(t))$ is a solution of (10.7.5). To this aim, we shall prove that, on a subsequence at least, $(x(\cdot, \tau, \zeta, g_n^X))_n$ is uniformly convergent on $[\tau, T]$ to x.

Inasmuch as $g_n^X(t) = F(\widetilde{v}_n(\sigma_n(t)), \widetilde{x}_n(\sigma_n(t)), \widetilde{y}_n(\sigma_n(t)))$, we have

$$\|x(t, \tau, \zeta, g_n^X) - x(t)\|$$

$$\leq \int_\tau^t \|F(\widetilde{v}_n(\sigma_n(\theta)), \widetilde{x}_n(\sigma_n(\theta)), \widetilde{y}_n(\sigma_n(\theta))) - F(v(\theta), x(\theta), y(\theta))\| \, d\theta \tag{10.9.5}$$

for $n = 1, 2, \ldots$ and each $t \in [\tau, T]$.

Since, by (H_6), F is defined on $\mathbb{R} \times X \times Y$, we deduce

$$\|F(\widetilde{v}_n(\sigma_n(\theta)), \widetilde{x}_n(\sigma_n(\theta)), \widetilde{y}_n(\sigma_n(\theta))) - F(v(\theta), x(\theta), y(\theta))\|$$

$$\leq \|F(\widetilde{v}_n(\sigma_n(\theta)), \widetilde{x}_n(\sigma_n(\theta)), \widetilde{y}_n(\sigma_n(\theta))) - F(\widetilde{v}_n(\sigma_n(\theta)), x(\theta), \widetilde{y}_n(\sigma_n(\theta)))\|$$
$$+\|F(\widetilde{v}_n(\sigma_n(\theta)), x(\theta), \widetilde{y}_n(\sigma_n(\theta))) - F(v(\theta), x(\theta), y(\theta))\|$$
$$\leq L\|\widetilde{x}_n(\sigma_n(\theta)) - x(\theta)\| + \|F(\widetilde{v}_n(\sigma_n(\theta)), x(\theta), \widetilde{y}_n(\sigma_n(\theta))) - F(v(\theta), x(\theta), y(\theta))\|$$

for each $n = 1, 2, \ldots$ and $t \in [\tau, T]$. By (10.9.2), (10.9.3) and the continuity of F, we deduce that there exists $\gamma_n \downarrow 0$ such that

$$\|F(\widetilde{v}_n(\sigma_n(\theta)), x(\theta), \widetilde{y}_n(\sigma_n(\theta))) - F(v(\theta), x(\theta), y(\theta))\| \leq \gamma_n,$$

uniformly for $n = 1, 2, \ldots$ and $\theta \in [\tau, T]$. On the other hand, by (vi), we have

$$\|\widetilde{x}_n(\sigma_n(\theta)) - x(\theta)\| \leq \varepsilon_n$$

for $n = 1, 2, \ldots$. From (10.9.5), we deduce that

$$\|x(t, \tau, \zeta, g_n^X) - x(t)\| \leq (L\varepsilon_n + \gamma_n)(T - \tau)$$

for $n = 1, 2, \ldots$ and $t \in [\tau, T]$.

Consequently $\lim_n x(t, \tau, \zeta, g_n^X) = x(t)$ uniformly for $t \in [\tau, T]$. Clearly $(v(t), x(t), y(t)) \in \mathcal{C}$ for each $t \in [\tau, T]$. Passing to the limit in (v) for $\theta = \tau$ and using (iv), we deduce that (t, x, y) is a C^0-solution of the problem (10.7.4) and this completes the proof. □

10.10. Sufficient conditions for C^0-local invariance

Let X be a real Banach space, $A : D(A) \subseteq X \rightsquigarrow X$ an m-dissipative operator, D an open subset in X, K a nonempty and locally closed subset of $D \cap \overline{D(A)}$, $f : I \times D \to X$ a given function and let us consider the Cauchy problem for the nonautonomous evolution equation

$$\begin{cases} u'(t) \in Au(t) + f(t, u(t)) \\ u(\tau) = \xi. \end{cases} \tag{10.10.1}$$

Definition 10.10.1. The subset K is *locally C^0-invariant* with respect to $A+f$ if for each $(\tau, \xi) \in I \times K$ and each C^0-solution $u : [\tau, c] \to D \cap \overline{D(A)}$ of (10.10.1), there exists $T \in (\tau, c]$ such that $u(t) \in K$ for each $t \in [\tau, T]$. It is *C^0-invariant* if it satisfies the above condition of local C^0-invariance with $T = c$.

Problem 10.10.1. *Prove that if K is closed and locally C^0-invariant with respect to $A + f$, then it is C^0-invariant with respect to $A + f$.*

Theorem 10.10.1. *Let X be a Banach space, $A : D(A) \subseteq X \rightsquigarrow X$ an m-dissipative operator, D an open subset in X, K a nonempty and closed subset of $D \cap \overline{D(A)}$, I an open interval and $f : I \times D \to X$. Let us assume that $I \times K$ is C^0-viable with respect to $A + f$ and there exist an open neighborhood $V \subseteq D$ of K and one Carathéodory uniqueness function*[7] *$\omega : I \times \mathbb{R}_+ \to \mathbb{R}_+$, such that*

$$[\xi_1 - \xi_2, f(t, \xi_1) - f(t, \xi_2)]_+ \leq \omega(t, \|\xi_1 - \xi_2\|) \tag{10.10.2}$$

for each $t \in I$, $\xi_1 \in V \setminus K$ and $\xi_2 \in K$. Then K is locally C^0-invariant with respect to $A + f$.

Proof. Let $V \subseteq D$ be given by hypotheses and let $(\tau, \xi) \in I \times K$. If (10.10.1) has no C^0-solution leaving K, we have nothing to prove. Otherwise, let $u : [\tau, c] \to D$ be any C^0-solution to (10.10.1) which leaves K. Diminishing c if necessary, we may assume that $u(t) \in V$ for each $t \in [\tau, c]$. We prove that $u(t) \in K$ for each $t \in [\tau, c]$. To this end, let us assume by contradiction that there exists $t_1 \in [\tau, c]$ such that $u(t_1) \in V \setminus K$. Since K is closed and u is continuous, we can fix $\tau \leq t_0 < t_1$ such that $u(t) \in V \setminus K$ for every $t \in (t_0, t_1]$ and $u(t_0) \in K$. Let $v : [t_0, d] \to K$ be any C^0-solution of $v'(t) \in Av(t) + f(t, v(t))$ which satisfies $v(t_0) = u(t_0)$. Such C^0-solution exists because $I \times K$ is C^0-viable with respect to $A+f$. Let $t_2 = \min\{d, t_1\}$. Let $g : [t_0, t_2] \to \mathbb{R}_+$ be defined by $g(t) = \|u(t) - v(t)\|$ for each $t \in [t_0, t_2]$.

---

[7] See Definition 1.8.2.

From (1.6.2) in Theorem 1.6.2, we deduce

$$g(t) \leq \int_{t_0}^{t} [\,u(s) - v(s), f(s, u(s)) - f(s, v(s))\,]_+ \, ds.$$

Since V, K, and f satisfy (10.10.2), we get

$$g(t) \leq \int_{t_0}^{t} \omega(s, g(s)) \, ds$$

for $n = 1, 2, \ldots$ and $t \in [\,t_0, t_2\,]$. Applying Lemma 1.8.3 with an arbitrary sequence $\gamma_n \downarrow 0$, we deduce that there exists $\widetilde{t}_2 \in (t_0, t_2\,]$ such that $g(t) \equiv 0$, i.e., $u(t) = v(t)$ for all $t \in [\,t_0, \widetilde{t}_2\,]$. Since $u(t) \in V \setminus K$ and $v(t) \in K$ for each $t \in (t_0, \widetilde{t}_2\,]$, we arrived at a contradiction. This contradiction can be eliminated only if $u(t) \in K$ for all $t \in [\,\tau, c\,]$. The proof is complete. □

The viability assumption in Theorem 10.10.1 can be replaced by an appropriate tangency condition.

Theorem 10.10.2. *Let X be a Banach space, $A : D(A) \subseteq X \rightsquigarrow X$ an m-dissipative operator, D an open subset in X, K a nonempty and closed subset of $D \cap \overline{D(A)}$, I an open interval and $f : I \times D \to X$ a continuous function. Assume that, for each $(\tau, \xi) \in I \times K$, we have $f(\tau, \xi) \in \mathcal{T}_K^A(\xi)$, and there exist an open neighborhood $V \subseteq D$ of K and one Carathéodory uniqueness function $\omega : I \times \mathbb{R}_+ \to \mathbb{R}_+$, such that*

$$[\,\xi_1 - \xi_2, f(t, \xi_1) - f(t, \xi_2)\,]_+ \leq \omega(t, \|\xi_1 - \xi_2\|) \tag{10.10.3}$$

for each $t \in I$, $\xi_1 \in V \setminus K$ and $\xi_2 \in K$. Then K is locally C^0-invariant with respect to $A + f$.

Proof. Let $V \subseteq D$ be the open neighborhood of K such that f satisfies (10.10.3), let $(\tau, \xi) \in I \times K$ and let $u : [\,\tau, c\,] \to D$ be any solution to (10.10.1). Diminishing c if necessary, we may assume that $u(t) \in V$ for each $t \in [\,\tau, c\,]$. We prove that $u(t) \in K$ for each $t \in [\,\tau, c\,]$. To this end, let us assume by contradiction that there exists $t_1 \in [\,\tau, c\,]$ such that $u(t_1) \in V \setminus K$. Let $\tau \leq t_0 < t_1$ be such that $u(t) \in V \setminus K$ for every $t \in (t_0, t_1\,]$ and $u(t_0) \in K$. This is possible because K is closed and u is continuous. Let us consider the autonomous Cauchy problem

$$\begin{cases} v'(t) = 1 \\ z'(t) \in Az(t) + f(v(t), z(t)) \\ v(t_0) = t_0, \;\; z(t_0) = u(t_0). \end{cases} \tag{10.10.4}$$

Take $\varepsilon_n \downarrow 0$. By Lemma 10.2.1 we know that there exists at least one sequence $((\sigma_n, (1, g_n^X), (\widetilde{v}_n, \widetilde{z}_n)))_n$ of ε_n-approximate C^0-solutions of the Cauchy problem (10.10.4)[8], on $[\,t_0, T\,]$, with $T \in (t_0, t_1\,]$.

After some calculations involving (iv) and (v) in Lemma 10.2.1, we get

$$\|\widetilde{z}_n(\sigma_n(t)) - u(t)\| \leq \gamma_n + \int_{t_0}^{t} \omega(\theta, \|\widetilde{z}_n(\sigma_n(\theta)) - u(\theta)\|)\, d\theta,$$

where $\gamma_n \downarrow 0$. From Lemma 1.8.3, we deduce that there exists $T_0 \in (t_0, T\,]$ such that $\lim_n \|\widetilde{z}_n(\sigma_n(t)) - u(t)\| = 0$ uniformly for $t \in [\,t_0, T_0\,]$. Since $D(\xi, \rho) \cap K$ is closed and, by (ii) in Lemma 10.2.1, $\widetilde{z}_n(\sigma_n(t)) \in D(\xi, \rho) \cap K$ for each $t \in [\,t_0, T_0\,]$, we deduce that $u(t) \in K$ for each $t \in [\,t_0,\, T_0]$, thereby contradicting the definition of t_0. The contradiction can be eliminated only if $u(t) \in K$ for each $t \in [\,\tau, c\,]$ and this completes the proof. □

[8] We notice that, in the general hypotheses of Theorem 10.10.2, (10.10.4) may have no C^0-solution.

CHAPTER 11

Viability for multi-valued fully nonlinear evolutions

Here we extend our study in Chapter 10 to the more general case of nonlinear evolutions equations governed by multi-valued perturbations of m-dissipative operators. More precisely, we start with the autonomous case by defining the concept of A-quasi-tangent set at a point to a given set, in the case in which A is m-dissipative and possible nonlinear. Then, we prove the main necessary condition for C^0-viability expressed in terms of this new tangency concept. We next show that, under various natural extra-assumptions, the already established necessary condition is also sufficient. We extend the results to the quasi-autonomous case and we conclude with some results on the existence of noncontinuable or even global C^0-solutions.

11.1. Necessary conditions for C^0-viability

Let X be a Banach space, $A : D(A) \subseteq X \rightsquigarrow X$ the generator of a nonlinear semigroup of nonexpansive operators, $\{S(t) : \overline{D(A)} \to \overline{D(A)},\ t \geq 0\}$, K a nonempty subset in $\overline{D(A)}$ and $F : K \rightsquigarrow X$ a nonempty, closed and convex valued multi-function. We consider the Cauchy problem for the nonlinear perturbed differential inclusion

$$\begin{cases} u'(t) \in Au(t) + F(u(t)) \\ u(0) = \xi. \end{cases} \tag{11.1.1}$$

Definition 11.1.1. The function $u : [\,0, T\,] \to K$ is a C^0*-solution* of (11.1.1) if $u(0) = \xi$ and there exists $f \in L^1(0, T; X)$, with $f(t) \in F(u(t))$ a.e. for $t \in [\,0, T\,]$, and such that u is a C^0-solution on $[\,0, T\,]$ of the equation

$$u'(t) \in Au(t) + f(t)$$

in the sense of Definition 1.6.2.

Definition 11.1.2. We say that K is *C^0-viable with respect to $A+F$* if for each $\xi \in K$ there exist $T > 0$ and a C^0-solution $u : [0,T] \to K$ of (11.1.1).

The main goal of this section is to find necessary conditions in order that K be C^0-viable with respect to $A+F$. Let $E \subseteq X$ be a nonempty set. We recall that

$$\mathcal{E} = \{f \in L^1(\mathbb{R}_+; X);\ f(s) \in E \text{ a.e. for } s \in \mathbb{R}_+\}.$$

Definition 11.1.3. Let K be a subset in X and $\xi \in K$. We say that the set E is *A-quasi-tangent* to K at the point ξ if, for each $\rho > 0$, we have

$$\liminf_{h \downarrow 0} \frac{1}{h} \operatorname{dist}(S_{\mathcal{E}}(h)\xi; K \cap D(\xi,\rho)) = 0, \tag{11.1.2}$$

where $S_{\mathcal{E}}(h)\xi = \{u(h,0,\xi,f);\ f \in \mathcal{E}\}$. We denote by $\mathcal{QTS}_K^A(\xi)$ the class of all A-quasi-tangent sets to K at ξ.

Another A-tangency concept is introduced below.

Definition 11.1.4. We say that a set E is *A-tangent* to K at ξ if, for each $\rho > 0$, we have

$$\liminf_{h \downarrow 0} \frac{1}{h} \operatorname{dist}(S_E(h)\xi; K \cap D(\xi,\rho)) = 0, \tag{11.1.3}$$

where $S_E(h)\xi = \{S_\eta(h)\xi;\ \eta \in E\} = \{u(h,0,\xi,\eta);\ \eta \in E\}$. We denote by $\mathcal{TS}_K^A(\xi)$ the class of all A-tangent sets to K at $\xi \in K$.

As E can be identified with the subset of a.e. constant elements in $\mathcal{E}$, it readily follows that

$$\mathcal{TS}_K^A(\xi) \subseteq \mathcal{QTS}_K^A(\xi). \tag{11.1.4}$$

Remark 11.1.1. Let $K \subseteq X$ and $\xi \in K$. The following properties are simple consequences of Definition 11.1.3:

(i) if $E \in \mathcal{QTS}_K^A(\xi)$ and $E \subseteq D$ then $D \in \mathcal{QTS}_K^A(\xi)$;
(ii) if $E \in \mathcal{QTS}_K^A(\xi)$ and $D \subseteq E$ is dense in E, then $D \in \mathcal{QTS}_K^A(\xi)$.

In addition, by means of the correspondence $\eta \mapsto \{\eta\}$, $\mathcal{T}_K^A(\xi)$ can be identified with a subset in $\mathcal{TS}_K^A(\xi)$ and thus in $\mathcal{QTS}_K^A(\xi)$. See Definition 10.1.3.

Problem 11.1.1. *Prove that whenever E is compact, $E \in \mathcal{TS}_K^A(\xi)$ if and only if there exists $\eta \in E$ such that $\eta \in \mathcal{T}_K^A(\xi)$.*

Theorem 11.1.1. *Let X be a Banach space, $A : D(A) \subseteq X \rightsquigarrow X$ an m-dissipative operator, K a nonempty subset in $\overline{D(A)}$ and $F : K \rightsquigarrow X$ a* *nonempty, quasi-weakly compact*[1] *and convex valued, u.s.c. multi-function. Then a necessary condition in order that K be C^0-viable with respect to $A+F$ is that, for each $\xi \in K$, we have $F(\xi) \in \mathcal{QTS}_K^A(\xi)$.*

Proof. Let $\xi \in K$ and $\rho > 0$. Since K is C^0-viable with respect to $A+F$, there exists at least one C^0-solution $u : [\,0,T\,] \to K$ of (11.1.1). Let $f \in L^1(0,T;X)$ be the function given by Definition 11.1.1. As u is continuous at $t = 0$ and F is u.s.c. at $u(0) = \xi$, it follows that for each $\varepsilon > 0$ there exists $\delta(\varepsilon) > 0$ such that

$$f(s) \in F(\xi) + D(0,\varepsilon)$$

a.e. for $s \in [\,0,\delta(\varepsilon)\,]$. So, if $\varepsilon_n \downarrow 0$, there exists $h_n \downarrow 0$ such that

$$f(s) \in F(\xi) + D(0,\varepsilon_n) \quad \text{and} \quad u(h_n,0,\xi,f) \in D(\xi,\rho)$$

for $n = 1,2,\ldots$ and a.e. for $s \in [\,0,h_n\,]$. Then, for each $n = 1,2,\ldots$, there exist f_n and g_n with $f_n(s) \in F(\xi)$, $g_n(s) \in D(0,\varepsilon_n)$ and

$$f(s) = f_n(s) + g_n(s)$$

a.e. for $s \in [\,0,h_n\,]$. By Lemma 6.1.1, we may assume without loss of generality that both f_n and g_n are integrable. Let us observe that

$$\|u(h_n,0,\xi,f) - u(h_n,0,\xi,f_n)\| \le \int_0^{h_n} \|f(s) - f_n(s)\|\,ds \le h_n\varepsilon_n$$

for $n = 1,2,\ldots$.

Since $u(h_n,0,\xi,f) \in K \cap D(\xi,\rho)$ and $\varepsilon_n \downarrow 0$, we get $F(\xi) \in \mathcal{QTS}_K^A(\xi)$ and this completes the proof. □

Problem 11.1.2. *Prove that the conclusion of Theorem* 11.1.1 *remains unchanged if instead of quasi-weakly compact, we assume merely that, for each $\xi \in K$, $F(\xi)$ is quasi-weakly relatively compact.*

Arguing as in the proof of Theorem 11.1.1, we deduce

Theorem 11.1.2. *Let X be a Banach space, $A : D(A) \subseteq X \rightsquigarrow X$ an m-dissipative operator, K a nonempty subset in $\overline{D(A)}$ and $F : K \rightsquigarrow X$ a given multi-function. If K is C^0-viable with respect to $A+F$, then at each point $\xi \in K$ at which F is u.s.c. and $F(\xi)$ is quasi-weakly compact and convex, we have $F(\xi) \in \mathcal{QTS}_K^A(\xi)$.*

---

[1]See Definition 6.1.4.

11.2. Necessary and/or sufficient conditions for C^0-viability

The goal of this section is to find necessary and sufficient, or only sufficient conditions in order that K be C^0-viable with respect to $A+F$.

Definition 11.2.1. An m-dissipative operator $A : D(A) \subseteq X \rightsquigarrow X$ is called of *complete continuous type* if for each fixed $(\tau, \xi) \in \mathbb{R} \times \overline{D(A)}$, the graph of the C^0-solution operator, $f \mapsto u(\cdot, \tau, \xi, f)$, is weakly$\times$strongly sequentially closed in $L^1(\tau, T; X) \times C([\tau, T]; X)$.

Problem 11.2.1. *Prove that if A is linear and m-dissipative, then it is of complete continuous type. Show that if, in addition, A generates a compact C_0-semigroup, then $f \mapsto u(\cdot, \tau, \xi, f)$ is weakly-strongly sequentially continuous from $L^1(\tau, T; X)$ to $C([\tau, T]; X)$.*

Problem 11.2.2. *Prove that if X has uniformly convex dual and A (possibly nonlinear) generates a compact semigroup, then A is of complete continuous type. Prove that, in this case, the operator $f \mapsto u(\cdot, \tau, \xi, f)$ is weakly-strongly sequentially continuous from $L^1(\tau, T; X)$ to $C([\tau, T]; X)$.*

Remark 11.2.1. An example of an m-dissipative operator of complete continuous type, which is neither linear nor defined on a Banach space with uniformly convex dual, is offered by Theorem 1.7.9.

Theorem 11.2.1. *Let X be a Banach space, $A : D(A) \subseteq X \rightsquigarrow X$ an m-dissipative operator of complete continuous type which generates a compact semigroup of contractions, let K be a nonempty, locally closed subset in $\overline{D(A)}$ and $F : K \rightsquigarrow X$ a nonempty, weakly compact and convex valued strongly-weakly u.s.c. multi-function. Then a sufficient condition in order that K be C^0-viable with respect to $A+F$ is that, for each $\xi \in K$, $F(\xi) \in \mathfrak{QTS}_K^A(\xi)$.*

A consequence of Theorem 11.2.1 and of Problem 11.2.2 is

Theorem 11.2.2. *Let X be a Banach space with uniformly convex dual, $A : D(A) \subseteq X \rightsquigarrow X$ an m-dissipative operator which is the infinitesimal generator of a compact semigroup, K a nonempty, locally closed subset in $\overline{D(A)}$ and $F : K \rightsquigarrow X$ a nonempty, bounded, closed and convex valued strongly-weakly u.s.c. multi-function. Then a sufficient condition in order that K be C^0-viable with respect to $A+F$ is that, for each $\xi \in K$, $F(\xi) \in \mathfrak{QTS}_K^A(\xi)$.*

From Theorems 11.1.1 and 11.2.1 we deduce

Theorem 11.2.3. *Let X be a Banach space, $A : D(A) \subseteq X \rightsquigarrow X$ an m-dissipative operator of complete continuous type which generates a* *compact semigroup of contractions, K a nonempty, locally closed subset in $\overline{D(A)}$ and $F : K \rightsquigarrow X$ a nonempty, weakly compact and convex valued u.s.c. multi-function. Then a necessary and sufficient condition in order that K be C^0-viable with respect to $A+F$ is that, for each $\xi \in K$, $F(\xi) \in \mathfrak{QTS}_K^A(\xi)$.*

If A is only m-dissipative and of complete continuous type but the semigroup generated by A is not compact, we can prove

Theorem 11.2.4. *Let X be a Banach space, $A : D(A) \subseteq X \rightsquigarrow X$ an m-dissipative operator of complete continuous type, K a nonempty, locally compact subset in $\overline{D(A)}$ and $F : K \rightsquigarrow X$ a nonempty, weakly compact and convex valued strongly-weakly u.s.c. multi-function. Then a sufficient condition in order that K be C^0-viable with respect to $A + F$ is that, for each $\xi \in K$, $F(\xi) \in \mathfrak{QTS}_K^A(\xi)$.*

From Theorems 11.1.1 and 11.2.4, we get

Theorem 11.2.5. *Let X be a Banach space, $A : D(A) \subseteq X \rightsquigarrow X$ an m-dissipative operator of complete continuous type, K a nonempty, locally compact subset in $\overline{D(A)}$ and $F : K \rightsquigarrow X$ a nonempty, weakly compact and convex valued u.s.c. multi-function. Then a necessary and sufficient condition in order that K be C^0-viable with respect to $A + F$ is that, for each $\xi \in K$, $F(\xi) \in \mathfrak{QTS}_K^A(\xi)$.*

11.3. Existence of ε-approximate C^0-solutions

The proof of both Theorems 11.2.1 and 11.2.4 relies on showing that the tangency condition $F(\xi) \in \mathfrak{QTS}_K^A(\xi)$ for each $\xi \in K$ along with Brezis-Browder Theorem 2.1.1 imply that, for each ξ in K, there exists at least one sequence of "approximate solutions" of (11.1.1), defined on the same interval, $v_n : [0, T] \to X$, and such that $(v_n)_n$ converges uniformly to a C^0-solution of (11.1.1).

The next lemma is an existence result concerning ε-approximate C^0-solutions of (11.1.1).

Lemma 11.3.1. *Let X be a real Banach space, $A : D(A) \subseteq X \rightsquigarrow X$ an m-dissipative operator, K a nonempty and locally closed subset in $\overline{D(A)}$ and $F : K \rightsquigarrow X$ a nonempty-valued, locally bounded multi-function satisfying $F(\xi) \in \mathfrak{QTS}_K^A(\xi)$ for each $\xi \in K$. Then, for each $\xi \in K$, there exist $\rho > 0$, $T > 0$ and $M > 0$ such that $D(\xi, \rho) \cap K$ is closed and, for each $\varepsilon > 0$, there exist three functions: $\alpha : [0, T] \to [0, T]$ nondecreasing, $f : [0, T] \to X$ measurable and $v : [0, T] \to X$ continuous satisfying:*

(i) *$t - \varepsilon \le \alpha(t) \le t$ for all $t \in [0, T]$ and $\alpha(T) = T$;*
(ii) *$v(\alpha(t)) \in D(\xi, \rho) \cap K$ for all $t \in [0, T]$ and $v([0, T])$ is precompact;*

(iii) $f(t) \in F(v(\alpha(t)))$ *a.e. for* $t \in [0, T]$;
(iv) $\|f(t)\| \leq M$ *a.e. for* $t \in [0, T]$;
(v) $v(0) = \xi$ *and* $\|v(t) - u(t, \alpha(s), v(\alpha(s)), f)\| \leq (t - \alpha(s))\varepsilon$ *for all* $t, s \in [0, T]$, $0 \leq s \leq t \leq T$;
(vi) $\|v(t) - v(\alpha(t))\| \leq \varepsilon$ *for all* $t \in [0, T]$.

Definition 11.3.1. Let $\xi \in K$ and $\varepsilon > 0$. A triplet (α, f, v) satisfying (i)∼(vi) is called an *ε-approximate* C^0*-solution* of (11.1.1).

We can now proceed to the proof of Lemma 11.3.1.

Proof. Let $\xi \in K$ be arbitrary and choose $\rho > 0$ and $M > 0$ such that $D(\xi, \rho) \cap K$ is closed (compact in the hypotheses of Theorem 11.2.4) and $\|\eta\| \leq M$ for every $u \in D(\xi, \rho) \cap K$ and $\eta \in F(u)$. This is always possible because K is locally closed and F is locally bounded.

Next, take $T > 0$ such that

$$\sup_{0 \leq t \leq T} \|S(t)\xi - \xi\| + T(M + 1) \leq \rho. \tag{11.3.1}$$

We first prove that the conclusion of Lemma 11.3.1 remains true if we replace T as above with a possible smaller number $\delta \in (0, T]$ which, at this stage, is allowed to depend on $\varepsilon \in (0, 1)$. Then, by using the Brezis-Browder Theorem 2.1.1, we will prove that we can take $\delta = T$ independent of ε.

Let $\varepsilon \in (0, 1)$ be arbitrary. Since $F(\xi) \in \mathfrak{QTS}^A_K(\xi)$, it is easy to see that there exist $f \in \mathcal{F}(\xi)$[2], $\delta \in (0, \varepsilon)$ and $p \in X$ with $\|p\| \leq \varepsilon$ such that

$$u(\delta, 0, \xi, f) + \delta p \in K.$$

With f as above, let us define $\alpha : [0, \delta] \to [0, \delta]$ and $v : [0, \delta] \to X$ by $\alpha(t) = 0$ for $t \in [0, \delta)$, $\alpha(\delta) = \delta$, and respectively by

$$v(t) = u(t, 0, \xi, f) + tp \tag{11.3.2}$$

for each $t \in [0, \delta]$.

Let us observe that the functions α, f and v satisfy (i)∼(iv) with $T = \delta$. Clearly, $v(0) = \xi$. Moreover, since $\|p\| \leq \varepsilon$, we deduce

$$\|v(t) - u(t, \alpha(s), v(\alpha(s)), f)\| = \|v(t) - u(t, 0, v(0), f)\| = t\|p\| \leq (t - \alpha(s))\varepsilon$$

for all $0 \leq s \leq t \leq \delta$. Thus (v) is also satisfied. Next, diminishing $\delta > 0$ and redefining α if necessary, we get

$$\|v(t) - v(\alpha(t))\| = \|v(t) - v(0)\| \leq \|u(t, 0, \xi, f) - \xi\| + t\|p\|$$

$$\leq \|S(t)\xi - \xi\| + \int_0^t \|f(s)\|\, ds + t\varepsilon \leq \sup_{t \in [0, \delta]} \|S(t)\xi - \xi\| + \delta(M + \varepsilon) \leq \varepsilon$$

---

[2] We recall that $\mathcal{F}(\xi) = \{f \in L^1(\mathbb{R}_+; X);\ f(s) \in F(\xi) \text{ a.e. for } s \in \mathbb{R}_+\}$.

for each $t \in [0, \delta)$, and thus (vi) is also satisfied.

Next, we will show that there exists at least one triplet (α, f, v) satisfying (i)∼(vi). To this aim we shall use Brezis-Browder Theorem 2.1.1 as follows. Let $\mathcal{S}$ be the set of all triplets (α, f, v), defined on $[0, \mu]$, with $\mu \leq T$ and satisfying (i)∼(vi) with μ instead of T. This set is clearly nonempty, as we have already proved. On $\mathcal{S}$ we introduce a partial order $\preceq$ as follows. We say that

$$(\alpha_1, f_1, v_1) \preceq (\alpha_2, f_2, v_2)$$

if $\mu_1 \leq \mu_2$ and $\alpha_1(s) = \alpha_2(s)$, $f_1(s) = f_2(s)$ and $v_1(s) = v_2(s)$ for each $s \in [0, \mu_1]$.

Let us define the function $\mathcal{N} : \mathcal{S} \to \mathbb{R}$ by

$$\mathcal{N}(\alpha, f, v) = \mu.$$

It is clear that $\mathcal{N}$ is increasing on $\mathcal{S}$. Let us now take an increasing sequence

$$((\alpha_j, f_j, v_j))_j$$

in $\mathcal{S}$ and let us show that it is bounded from above in $\mathcal{S}$. We define an upper bound as follows. First, set

$$\mu^* = \sup\{\mu_j;\ j \in \mathbb{N}\}.$$

If $\mu^* = \mu_j$ for some $j \in \mathbb{N}$, (α_j, f_j, v_j) is clearly an upper bound. If $\mu_j < \mu^*$ for each $j \in \mathbb{N}$, let us define

$$\alpha(t) = \alpha_j(t), \quad f(t) = f_j(t), \quad v(t) = v_j(t)$$

for $j \in \mathbb{N}$ and every $t \in [0, \mu_j]$. To extend α, f and v to $t = \mu^*$, we proceed as follows.

First, we extend f at μ^* by setting $f(\mu^*) = \eta$, where $\eta \in X$ is arbitrary but fixed. Second, by (iii) and (iv) we know that $f \in L^\infty(0, \mu^*; X)$ and so, for each $j \in \mathbb{N}$, the function $u(\cdot, \mu_j, v(\mu_j), f) : [\mu_j, \mu^*] \to \overline{D(A)}$ is continuous. As a consequence, the set $C_j = u([\mu_j, \mu^*], \mu_j, v(\mu_j), f)$ is precompact. On the other hand, by (ii), for each $j \in \mathbb{N}$, we know that $K_j = v([0, \mu_j])$ is precompact too. By (v) we deduce that, for each $j \in \mathbb{N}$,

$$v([0, \mu^*)) \subseteq C_j \cup K_j + (\mu^* - \mu_j)D(0, \varepsilon).$$

Let $\nu > 0$ be arbitrary and fix $j \in \mathbb{N}$ such that

$$(\mu^* - \mu_j)\varepsilon \leq \frac{\nu}{2}.$$

Since $C_j \cup K_j$ is precompact, there exists a finite family $\{\xi_1, \xi_2, \ldots, \xi_{n(\nu)}\}$ such that, for each $\xi \in C_j \cup K_j$, there exists $k \in \{1, 2, \ldots, n(\nu)\}$ such that

$$\|\xi - \xi_k\| \leq \frac{\nu}{2}.$$

The last two inequalities and the inclusion above yield

$$v([0,\mu^*)) \subseteq \cup_{k=1}^{n(\nu)} D(\xi_k,\nu)$$

and accordingly $v([0,\mu^*))$ is precompact. Now, take any limit point v_{μ^*} of $v(\mu_j)$ as j tends to ∞ and set $v(\mu^*) = v_{\mu^*}$. Finally, we define $\alpha(\mu^*) = \mu^*$. Since $v(\mu_m) \in D(\xi,\rho) \cap K$, for each $m \in \mathbb{N}$, and the latter is closed, we have $v(\mu^*) \in D(\xi,\rho) \cap K$. Now let us observe that (α, f, v), where α, f and v are defined as above, satisfies (i)$\sim$(iv) with T replaced with μ^*. It is also easy to see that (v) holds for $0 \leq s \leq t < \mu^*$. To check (v) for $t = \mu^*$, we have to fix any $s \in [0,\mu^*)$, to take $t = \mu_j$ with $\mu_j > s$ in (v), and to pass to the limit for j tending to ∞ both sides in (v) on that subsequence on which $(v_j(\mu_j))_{j\in\mathbb{N}}$ tends to $v_{\mu^*} = v(\mu^*)$. To check (vi) we have to proceed similarly.

So, (α, f, v) is an upper bound for $((\alpha_j, f_j, v_j))_j$ and consequently the set $\mathcal{S}$ endowed with the partial order $\preceq$ and the function $\mathcal{N}$ satisfy the hypotheses of Theorem 2.1.1. Accordingly, there exists at least one element $(\alpha_\nu, f_\nu, v_\nu)$ in $\mathcal{S}$ such that, if $(\alpha_\nu, f_\nu, v_\nu) \preceq (\alpha_\sigma, f_\sigma, v_\sigma)$ then $\nu = \sigma$.

We next show that $\nu = T$, where T satisfies (11.3.1). To this aim, let us assume by contradiction that $\nu < T$ and let $\xi_\nu = v_\nu(\nu) = v_\nu(\alpha_\nu(\nu))$ which belongs to $D(\xi,\rho) \cap K$. In view of inequality (1.6.5), (iv) and (v), we have

$$\|\xi_\nu - \xi\| \leq \|S(\nu)\xi - \xi\| + \|u(\nu,0,\xi,f_\nu) - S(\nu)\xi\| + \|v_\nu(\nu) - u(\nu,0,\xi,f_\nu)\|$$

$$\leq \|S(\nu)\xi - \xi\| + \int_0^\nu \|f_\nu(s)\|\,ds + \nu\varepsilon \leq \sup_{0\leq t\leq\nu} \|S(t)\xi - \xi\| + \nu(M+\varepsilon).$$

Recalling that $\nu < T$ and $\varepsilon < 1$, from (11.3.1), we get

$$\|\xi_\nu - \xi\| < \rho. \tag{11.3.3}$$

At this point we act as at the beginning of the proof with ν instead of 0 and with ξ_ν instead of ξ. So, from (11.3.3), we infer that there exist $f \in \mathcal{F}(\xi_\nu)$, $\delta \in (0,\varepsilon]$ with $\nu + \delta \leq T$ and $p \in X$ satisfying $\|p\| \leq \varepsilon$, such that

$$u(\nu+\delta,\nu,\xi_\nu,f) + \delta p \in D(\xi,\rho) \cap K.$$

We define $\alpha_{\nu+\delta} : [0,\nu+\delta] \to [0,\nu+\delta]$, $f_{\nu+\delta} : [0,\nu+\delta] \to X$ and $v_{\nu+\delta} : [0,\nu+\delta] \to X$ by

$$\alpha_{\nu+\delta}(t) = \begin{cases} \alpha_\nu(t) & \text{if } t \in [0,\nu] \\ \nu & \text{if } t \in (\nu,\nu+\delta) \\ \nu+\delta & \text{if } t = \nu+\delta, \end{cases}$$

$$f_{\nu+\delta}(t) = \begin{cases} f_\nu(t) & \text{if } t \in [0,\nu] \\ f(t) & \text{if } t \in (\nu,\nu+\delta], \end{cases}$$

and

$$v_{\nu+\delta}(t) = \begin{cases} v_\nu(t) & \text{if } t \in [0, \nu] \\ u(t, \nu, \xi_\nu, f_{\nu+\delta}) + (t-\nu)p & \text{if } t \in (\nu, \nu+\delta]. \end{cases}$$

Since $v_{\nu+\delta}(\nu+\delta) \in K \cap D(\xi, \rho)$, $(\alpha_{\nu+\delta}, f_{\nu+\delta}, v_{\nu+\delta})$ satisfies (i)~(iv) with T replaced by $\nu + \delta$.

To check (v) we consider the following complementary cases: $s \le t \le \nu$, $\nu < s \le t$ and $s \le \nu \le t$.

Clearly (v) holds for each t, s satisfying $s \le t \le \nu$. If $\nu < s \le t$, we have

$$\|v_{\nu+\delta}(t) - u(t, \alpha_{\nu+\delta}(s), v_{\nu+\delta}(\alpha_{\nu+\delta}(s)), f_{\nu+\delta})\|$$

$$= \|u(t, \nu, \xi_\nu, f_{\nu+\delta}) + (t-\nu)p - u(t, \nu, \xi_\nu, f_{\nu+\delta})\| \le (t-\nu)\varepsilon = (t - \alpha_{\nu+\delta}(s))\varepsilon.$$

Let now $s < \nu \le t$ and let us observe that, by virtue of the evolution property (1.6.4), i.e.

$$u(t, a, \xi, f) = u(t, \nu, u(\nu, a, \xi, f), f\,|_{[\nu, \nu+\delta]}), \tag{11.3.4}$$

for $0 \le a \le \nu \le t \le \nu + \delta$, and of (v) (which is valid on both $[0, \nu]$ and $[\nu, \nu + \delta]$), we have

$$v_{\nu+\delta}(t) - u(t, \alpha_{\nu+\delta}(s), v_{\nu+\delta}(\alpha_{\nu+\delta}(s)), f_{\nu+\delta})$$

$$= u(t, \nu, v_{\nu+\delta}(\nu), f_{\nu+\delta}) + (t-\nu)p$$

$$-u(t, \nu, u(\nu, \alpha_{\nu+\delta}(s), v_{\nu+\delta}(\alpha_{\nu+\delta}(s)), f_{\nu+\delta}), f_{\nu+\delta})$$

and therefore

$$\|v_{\nu+\delta}(t) - u(t, \alpha_{\nu+\delta}(s), v_{\nu+\delta}(\alpha_{\nu+\delta}(s)), f_{\nu+\delta})\|$$

$$\le \|v_{\nu+\delta}(\nu) - u(\nu, \alpha_{\nu+\delta}(s), v_{\nu+\delta}(\alpha_{\nu+\delta}(s)), f_{\nu+\delta})\| + (t-\nu)\|p\| \le$$

$$\le (\nu - \alpha_{\nu+\delta}(s))\varepsilon + (t-\nu)\varepsilon = (t - \alpha_{\nu+\delta}(s))\varepsilon$$

which proves (v).

Similarly we deduce that (vi) is satisfied. So $(\alpha_{\nu+\delta}, f_{\nu+\delta}, v_{\nu+\delta}) \in \mathcal{S}$,

$$(\alpha_\nu, f_\nu, v_\nu) \preceq (\alpha_{\nu+\delta}, f_{\nu+\delta}, v_{\nu+\delta})$$

and $\nu < \nu + \delta$. This contradiction can be eliminated only if $\nu = T$ and this completes the proof of Lemma 11.3.1. □

Remark 11.3.1. Under the general hypotheses of Lemma 11.3.1, for each $\gamma > 0$, we can diminish both $\rho > 0$ and $T > 0$, such that $T < \gamma$, $\rho < \gamma$ and all the conditions (i)~(vi) in Lemma 11.3.1 be satisfied.

11.4. Convergence in the case of Theorem 11.2.1

In order to complete the proof of Theorem 11.2.1, we will show that a suitably chosen subsequence of ε_n-approximate C^0-solutions is uniformly convergent on $[0, T]$ to a function u which is a C^0-solution of (11.1.1).

Proof. A simple argument by contradiction involving Lemma 2.6.1 shows that F is locally bounded and thus Lemma 11.3.1 applies. Let $(\varepsilon_n)_n$ be a sequence in $(0,1)$ strictly decreasing to 0. Let $((\alpha_n, f_n, v_n))_n$ be a sequence of ε_n-approximate C^0-solutions of (11.1.1). By (iv)[3], the family $\{f_n;\ n = 1, 2, \ldots\}$ is uniformly bounded, and hence uniformly integrable. As the semigroup generated by A is compact, by Theorem 1.6.5, we deduce that there exists $\widetilde{u} \in C([0,T]; X)$ such that, on a subsequence at least, we have

$$\lim_n u(t, 0, \xi, f_n) = \widetilde{u}(t) \tag{11.4.1}$$

uniformly for $t \in [0, T]$. By (v) and (11.4.1), we also have $\lim_n v_n(t) = \widetilde{u}(t)$ uniformly for $t \in [0, T]$. From (vi), we conclude that

$$\lim_n v_n(\alpha_n(t)) = \widetilde{u}(t) \tag{11.4.2}$$

uniformly for $t \in [0, T]$. By (ii), we know that $v_n(\alpha_n(t)) \in D(\xi, \rho) \cap K$ for $n = 1, 2, \ldots$ and $t \in [0, T]$. As $D(\xi, \rho) \cap K$ is closed, we conclude that $\widetilde{u}(t) \in D(\xi, \rho) \cap K$ for all $t \in [0, T]$.

Next, by (iii), we have $f_n(s) \in F(v_n(\alpha_n(s)))$ for $n = 1, 2, \ldots$ and a.e. for $s \in [0, T]$. Since $\{v_n(\alpha_n(s));\ n = 1, 2, \ldots,\ s \in [0, T]\}$ is relatively compact and F is strongly-weakly u.s.c. with weakly compact values, from Lemma 2.6.1 combined with Theorem 1.3.2, we conclude that the set

$$\overline{\text{conv}} \bigcup_{n=1}^{\infty} F(v_n(\alpha_n([0, T])))$$

is weakly compact. Accordingly, from Corollary 1.3.1, we may assume with no loss of generality that there exists $f \in L^1(0, T; X)$ such that

$$\lim_n f_n = f, \tag{11.4.3}$$

weakly in $L^1(0, T; X)$. Since A is of complete continuous type, from (11.4.3) and (11.4.1), we conclude that $\widetilde{u}(t) = u(t, 0, \xi, f)$. From (11.4.3), (11.4.2) and Lemma 2.6.2, we obtain $f(s) \in F(u(s, 0, \xi, f))$ a.e. for $s \in [0, T]$. From (11.4.1), it follows that $\widetilde{u}$ is a C^0-solution of the problem (11.1.1) and this completes the proof. □

---

[3]Within this proof, references to (i)∼(vi) are to the corresponding items in Lemma 11.3.1.

11.5. Convergence in the case of Theorem 11.2.4

We will show that a suitably chosen subsequence of ε_n-approximate C^0-solutions is uniformly convergent on $[0,T]$ to a function u which is a C^0-solution of (11.1.1).

Proof. Let $(\varepsilon_n)_n$ be a sequence in $(0,1)$ strictly decreasing to 0. Take a sequence of ε_n-approximate C^0-solutions $((\alpha_n, f_n, v_n))_n$ of (11.1.1). First, let us observe that, in view of Remark 11.3.1, we can diminish $T>0$ and $\rho>0$, if necessary, such that $D(\xi,\rho)\cap K$ be compact and (i) $\sim$ (vi)[4] be satisfied. So, by (v) and (vi), we have

$$\begin{aligned}\|u(t,0,\xi,f_n)-v_n(\alpha_n(t))\| &\leq \|u(t,0,\xi,f_n)-v_n(t)\|+\|v_n(t)-v_n(\alpha_n(t))\|\\ &\leq t\varepsilon_n+\varepsilon_n\leq (T+1)\varepsilon_n\end{aligned}$$

for $n=1,2,\ldots$. Let $k=1,2,\ldots$, let us denote by

$$\begin{cases} C_k=\displaystyle\bigcup_{n=1}^{k} u([0,T],0,\xi,f_n)\\ C=\displaystyle\bigcup_{n=1}^{\infty} v_n(\alpha_n([0,T]))\end{cases}$$

and let us observe that, in view of the inequality above, we have

$$\bigcup_{n=1}^{\infty} u([0,T],0,\xi,f_n)\subseteq C_k\cup C+(T+1)D(0,\varepsilon_k) \tag{11.5.1}$$

for each $k=1,2,\ldots$. But, for $k=1,2,\ldots$, C_k and C are precompact, C_k as a finite union of compact sets, i.e. $u([0,T],0,\xi,f_n)$ for $n=1,2,\ldots,k$, and C as a subset of $D(\xi,\rho)\cap K$, which in its turn is compact. This remark along with (11.5.1) shows that, for each $t\in[0,T]$, $\{u(t,0,\xi,f_n);\ n=1,2,\ldots\}$ is precompact too.

Thus we can apply Theorem 1.6.4 and so there exists $\widetilde{u}\in C([0,T];X)$ such that, on a subsequence at least, we have

$$\lim_n u(t,0,\xi,f_n)=\widetilde{u}(t) \tag{11.5.2}$$

uniformly for $t\in[0,T]$. By (v) and (11.5.2), on the same subsequence, we also have

$$\lim_n v_n(t)=\widetilde{u}(t)$$

[4]Throughout this section, all the references to (i) $\sim$ (vi) are to the corresponding items in Lemma 11.3.1.

uniformly for $t \in [0, T]$. By (vi), we conclude that

$$\lim_n v_n(\alpha_n(t)) = \widetilde{u}(t) \tag{11.5.3}$$

uniformly for $t \in [0, T]$. Since, by (ii), $v_n(\alpha_n(s)) \in D(\xi, \rho) \cap K$ for each $s \in [0, T]$ and $D(\xi, \rho) \cap K$ is closed, it follows that $\widetilde{u}(s) \in D(\xi, \rho) \cap K$ for each $s \in [0, T]$.

From now on, the proof follows the very same arguments as those in the last part of the proof of Theorem 11.2.1 and therefore we do not enter into details. □

11.6. The fully nonlinear multi-valued quasi-autonomous case

Let X be a real Banach space, $A : D(A) \subseteq X \rightsquigarrow X$ an m-dissipative operator, $\mathcal{C}$ a nonempty subset in $\mathbb{R} \times \overline{D(A)}$, $F : \mathcal{C} \rightsquigarrow X$ a given multifunction. The goal of this section is to extend the necessary and sufficient conditions for C^0-viability already proved in the autonomous case to the general frame of quasi-autonomous nonlinear evolution inclusions of the form

$$\begin{cases} u'(t) \in Au(t) + F(t, u(t)) \\ u(\tau) = \xi. \end{cases} \tag{11.6.1}$$

We denote by $\mathfrak{X} = \mathbb{R} \times X$. Endowed with the norm

$$\|(t, u)\| = \sqrt{|t|^2 + \|u\|^2},$$

for each $(t, u) \in \mathfrak{X}$, the former is a Banach space. In addition, if X has uniformly convex dual, then $\mathfrak{X}$, endowed with the norm above, has uniformly convex dual too.

Definition 11.6.1. By a *C^0-solution* of the *quasi-autonomous multi-valued nonlinear Cauchy problem* (11.6.1), we mean a continuous function $u : [\tau, T] \to X$, with $(t, u(t)) \in \mathcal{C}$ for each $t \in [\tau, T]$, and for which there exists $f \in L^1(\tau, T; X)$ such that $f(s) \in F(s, u(s))$ a.e. for $s \in [\tau, T]$ and u is a C^0-solution of the evolution inclusion below

$$\begin{cases} u'(t) \in Au(t) + f(t) \\ u(\tau) = \xi, \end{cases} \tag{11.6.2}$$

in the sense of Definition 1.6.2.

Definition 11.6.2. The set $\mathcal{C} \subseteq \mathbb{R} \times \overline{D(A)}$ is *C^0-viable* with respect to $A + F$ if for each $(\tau, \xi) \in \mathcal{C}$, there exists $T \in \mathbb{R}$, $T > \tau$ such that the Cauchy problem (11.6.1) has at least one C^0-solution $u : [\tau, T] \to X$.

Remark 11.6.1. In order to introduce the tangency concept we are going to use in the sequel, let us first observe that the quasi-autonomous Cauchy problem (11.6.1) can be equivalently rewritten as an autonomous one in the space $\mathcal{X}$, by setting $\mathcal{A} = (0, A)$, $z(s) = (t(s+\tau), u(s+\tau))$, $\mathcal{F}(z) = (1, F(z))$[5] and $\zeta = (\tau, \xi)$. Indeed, with the notations above, we have

$$\begin{cases} z'(s) \in \mathcal{A}z(s) + \mathcal{F}(z(s)) \\ z(0) = \zeta. \end{cases}$$

It readily follows that $\mathcal{A}$ is m-dissipative. So, z is a C^0-solution for the problem above if it is given by

$$z(s) = (\tau + s, u(\tau + s, \tau, \xi, f(\tau + \cdot))),$$

where f is a function as in Definition 11.6.1.

Remark 11.6.2. One may easily see that $\mathcal{C}$ is C^0-viable with respect to $A+F$ in the sense of Definition 11.6.1 if and only if $\mathcal{C}$ is C^0-viable with respect to $\mathcal{A}+\mathcal{F}$ in the sense of Definition 11.1.2.

Theorem 11.6.1. *Let X be a Banach space, $A : D(A) \subseteq X \rightsquigarrow X$ an m-dissipative operator, $\mathcal{C}$ a nonempty subset in $\mathbb{R}\times\overline{D(A)}$ and let $F : \mathcal{C} \rightsquigarrow X$ a quasi-weakly compact and convex valued, u.s.c. multi-function. If $\mathcal{C}$ is C^0-viable with respect to $A+F$ then, for each $(\tau,\xi) \in \mathcal{C}$, we have*

$$(1, F(\tau, \xi)) \in \mathcal{QTS}^{\mathcal{A}}_{\mathcal{C}}(\tau, \xi). \tag{11.6.3}$$

Proof. By Remark 11.6.2 and Theorem 11.1.1, we conclude that, under the hypotheses of Theorem 11.6.1, since $\mathcal{C}$ is C^0-viable with respect to $\mathcal{A}+\mathcal{F}$ then, for each $z \in \mathcal{C}$, $z = (\tau, \xi)$, we have

$$\mathcal{F}(z) \in \mathcal{QTS}^{\mathcal{A}}_{\mathcal{C}}(z),$$

relation which is equivalent to (11.6.3). □

As in the autonomous case, Theorem 11.6.1 is a direct consequence of Theorem 11.6.2 below, which gives a necessary condition for C^0-viability taking place in a strictly more general frame than the one in Theorem 11.6.1.

Theorem 11.6.2. *Let X be a Banach space, $A : D(A) \subseteq X \rightsquigarrow X$ an m-dissipative operator, $\mathcal{C}$ a nonempty subset in $\mathbb{R}\times\overline{D(A)}$ and let $F : \mathcal{C} \rightsquigarrow X$ be a given multi-function. If $\mathcal{C}$ is viable with respect to $A+F$, then the tangency condition (11.6.3) is satisfied at each point $(\tau,\xi) \in \mathcal{C}$ at which F is u.s.c. and $F(\tau,\xi)$ is quasi-weakly compact and convex. If, in addition, $F(\tau,\xi)$ is compact, then $(1, F(\tau,\xi)) \cap \mathcal{T}^{\mathcal{A}}_{\mathcal{C}}(\tau,\xi) \neq \emptyset$.*

---

[5]Here $(1, F(z)) = \{(1, \eta);\ \eta \in F(z)\}$.

Proof. Use Theorem 11.1.2 and Problem 11.1.1. □

Now we can pass to the main sufficient conditions concerning the C^0-viability of a set $\mathcal{C}$ with respect to $A+F$. We have the following counterparts of Theorems 11.2.1, 11.2.3, 11.2.4 and 11.2.5.

Theorem 11.6.3. *Let X be a Banach space, $A : D(A) \subseteq X \rightsquigarrow X$ be an m-dissipative operator of complete continuous type which generates a compact semigroup of contractions, $\mathcal{C}$ a nonempty and locally closed subset in $\mathbb{R} \times \overline{D(A)}$ and let $F : \mathcal{C} \rightsquigarrow X$ be a nonempty, weakly compact and convex valued strongly-weakly u.s.c. multi-function. Then a sufficient condition in order that $\mathcal{C}$ be C^0-viable with respect to $A+F$ is that, for each $(\tau,\xi) \in \mathcal{C}$, we have $(1, F(\tau,\xi)) \in \mathfrak{QTS}_{\mathcal{C}}^{A}(\tau,\xi)$.*

Theorem 11.6.4. *Let X be a Banach space, $A : D(A) \subseteq X \rightsquigarrow X$ an m-dissipative operator of complete continuous type which generates a compact semigroup of contractions, $\mathcal{C}$ a nonempty and locally closed subset in $\mathbb{R} \times \overline{D(A)}$ and let $F : \mathcal{C} \rightsquigarrow X$ be a nonempty, weakly compact and convex valued u.s.c. multi-function. Then a necessary and sufficient condition in order that $\mathcal{C}$ be C^0-viable with respect to $A+F$ is that, for each $(\tau,\xi) \in \mathcal{C}$, we have $(1, F(\tau,\xi)) \in \mathfrak{QTS}_{\mathcal{C}}^{A}(\tau,\xi)$.*

Theorem 11.6.5. *Let X be a Banach space, $A : D(A) \subseteq X \rightsquigarrow X$ an m-dissipative operator of complete continuous type, $\mathcal{C}$ a nonempty and locally compact subset in $\mathbb{R}\times\overline{D(A)}$ and let $F : \mathcal{C} \rightsquigarrow X$ be a nonempty, weakly compact and convex valued strongly-weakly u.s.c. multi-function. Then a sufficient condition in order that $\mathcal{C}$ be C^0-viable with respect to $A+F$ is that, for each $(\tau,\xi) \in \mathcal{C}$, we have $(1, F(\tau,\xi)) \in \mathfrak{QTS}_{\mathcal{C}}^{A}(\tau,\xi)$.*

Theorem 11.6.6. *Let X be a Banach space, $A : D(A) \subseteq X \rightsquigarrow X$ an m-dissipative operator of complete continuous type, $\mathcal{C}$ a nonempty and locally compact subset in $\mathbb{R} \times \overline{D(A)}$ and let $F : \mathcal{C} \rightsquigarrow X$ be a nonempty, weakly compact and convex valued u.s.c. multi-function. Then a necessary and sufficient condition in order that $\mathcal{C}$ be C^0-viable with respect to $A+F$ is that, for each $(\tau,\xi) \in \mathcal{C}$, we have $(1, F(\tau,\xi)) \in \mathfrak{QTS}_{\mathcal{C}}^{A}(\tau,\xi)$.*

11.7. Noncontinuable C^0-solutions

Let $A : D(A) \subseteq X \rightsquigarrow X$ be an m-dissipative operator, let $\mathcal{C} \subseteq \mathbb{R} \times \overline{D(A)}$ be nonempty and let $F : \mathcal{C} \rightsquigarrow X$. Here, we present some results concerning the existence of noncontinuable, or even global C^0-solutions for the nonlinear evolution inclusion

$$\begin{cases} u'(t) \in Au(t) + F(t,u(t)) \\ u(\tau) = \xi. \end{cases} \tag{11.7.1}$$

A C^0-solution $u : [\tau, T) \to X$ to (11.7.1) is called *noncontinuable*, if there is no other C^0-solution $v : [\tau, \widetilde{T}) \to X$ of the same equation, with $T < \widetilde{T}$ and satisfying $u(t) = v(t)$ for all $t \in [\tau, T)$. The C^0-solution u is called *global* if $T = T_{\mathcal{C}}$, which is given by (3.6.2). The next theorem follows from the Brezis–Browder Theorem 2.1.1.

Theorem 11.7.1. *Let X be a Banach space, $A : D(A) \subseteq X \rightsquigarrow X$ an m-dissipative operator, let $\mathcal{C} \subseteq \mathbb{R} \times \overline{D(A)}$ be nonempty and let $F : \mathcal{C} \rightsquigarrow X$. Then the following conditions are equivalent*:

(i) *$\mathcal{C}$ is C^0-viable with respect to $A + F$;*
(ii) *for each $(\tau, \xi) \in \mathcal{C}$, (11.7.1) has at least one noncontinuable C^0-solution $u : [\tau, T) \to X$.*

The next result concerns the existence of global solutions.

Theorem 11.7.2. *Let X be a Banach space, $A : D(A) \subseteq X \rightsquigarrow X$ an m-dissipative operator, let $\mathcal{C} \subseteq \mathbb{R} \times \overline{D(A)}$ be a nonempty subset and let $F : \mathcal{C} \rightsquigarrow X$ be a given multi-function. If $\mathcal{C}$ is X-closed*[6]*, F is positively sublinear*[7] *and maps bounded subsets in $\mathcal{C}$ into bounded subsets in X and $\mathcal{C}$ is C^0-viable with respect to $A + F$, then each C^0-solution of* (11.7.1) *can be continued up to a global one, i.e., defined on $[\tau, T_{\mathcal{C}})$, where $T_{\mathcal{C}}$ is given by* (3.6.2).

Proof. Since $\mathcal{C}$ is C^0-viable with respect to $A + F$, for each $(\tau, \xi) \in \mathcal{C}$, there exists at least one noncontinuable C^0-solution $u : [\tau, T) \to X$ to (11.7.1). We will show that $T = T_{\mathcal{C}}$. To this aim, let us assume the contrary, i.e., that $T < T_{\mathcal{C}}$. By using a translation argument if necessary, we may assume with no loss of generality that $0 \in D(A)$ and $0 \in A0$. From (1.6.2) with $\eta = 0$, $g \equiv 0$ and $v \equiv 0$, we deduce

$$\|u(t)\| \leq \|\xi\| + \int_{E_t} [u(s), f(s)]_+ \, ds + \int_{H_t \setminus G_t} [u(s), f(s)]_+ \, ds$$

for each $t \in [\tau, T)$, where f is the function given by Definition 11.1.1, and

$$\begin{aligned} E_t &= \{s \in [\tau, t];\ [u(s), f(s)]_+ > 0 \text{ and } \|u(s)\| > c(s)\}, \\ G_t &= \{s \in [\tau, t];\ [u(s), f(s)]_+ \leq 0\}, \\ H_t &= \{s \in [\tau, t];\ \|u(s)\| \leq c(s)\}. \end{aligned}$$

Here $a, b, c : \mathbb{R} \to \mathbb{R}_+$ are the continuous functions in Definition 6.6.1. As $H_t \subseteq H_T$ and $[u, v]_+ \leq \|v\|$ for each $u, v \in X$, we get

$$\|u(t)\| \leq \|\xi\| + \int_{E_t} [a(s)\|u(s)\| + b(s)] \, ds + \int_{H_T} \|f(s)\| \, ds$$

---

[6] See Definition 3.6.2.
[7] See Definition 6.6.1.

for each $t \in [\tau, T)$. But F maps bounded subsets in $\mathcal{C}$ into bounded subsets in X and therefore there exists $m > 0$ such that $\|f(s)\| \leq m$ a.e. for $s \in H_T$. Hence

$$\|u(t)\| \leq \|\xi\| + Tm + \int_\tau^T b(s)\,ds + \int_\tau^t a(s)\|u(s)\|\,ds,$$

for each $t \in [\tau, T)$. By Gronwall Lemma 1.8.4, u is bounded on $[\tau, T)$.

Using once again the fact that F maps bounded subsets in $\mathcal{C}$ into bounded subsets in X, we deduce that f is bounded on $[\tau, T)$ and therefore, there exists $\lim_{t\uparrow T} u(t) = u^*$. Since $\mathcal{C}$ is X-closed, it follows that $(T, u^*) \in \mathcal{C}$. From this observation, recalling that $\mathcal{C}$ is C^0-viable with respect to $A + F$ and $T < +\infty$, we conclude that u can be continued to the right of T. But this is absurd, because u is noncontinuable. This contradiction can be eliminated only if $T = T_{\mathcal{C}}$, as claimed. □

CHAPTER 12

Carathéodory perturbations of m-dissipative operators

In this chapter we reconsider some problems, already touched upon in Chapter 5 and Chapter 10. More precisely, here, we deal with the (partly) more general frame of fully nonlinear evolutions equations governed by single-valued Carathéodory perturbations of m-dissipative operators in separable Banach spaces. We begin by proving a necessary condition for C^0-viability and continue with the statements and proofs of several necessary and sufficient conditions for C^0-viability. Finally, we focus our attention on the problem of the existence of C^0-noncontinuable or even global solutions.

12.1. Necessary and sufficient conditions for C^0-viability

The goal of this chapter is to prove a necessary and sufficient condition for a given subset in $\mathbb{R} \times X$, with X a Banach space, to be C^0-viable with respect to $A + f$, where A is an m-dissipative operator and f is a Carathéodory function. More precisely, let X be a Banach space, $A : D(A) \subseteq X \rightsquigarrow X$ an m-dissipative operator generating a nonlinear semigroup of nonexpansive mappings, $\{S(t) : \overline{D(A)} \to \overline{D(A)};\ t \geq 0\}$, K a nonempty subset in $\overline{D(A)}$, I a nonempty and open interval and $f : I \times K \to X$ a given function. We consider the Cauchy problem

$$\begin{cases} u'(t) \in Au(t) + f(t, u(t)) \\ u(\tau) = \xi. \end{cases} \tag{12.1.1}$$

Definition 12.1.1. By a C^0-*solution* of (12.1.1), on $[\,\tau, T\,]$, we mean a continuous function $u : [\,\tau, T\,] \to K$ such that $t \mapsto g(t) = f(t, u(t))$ for a.e. $t \in [\,\tau, T\,]$ belongs to $L^1(\tau, T; X)$ and u is a C^0-solution on $[\,\tau, T\,]$ of the problem

$$\begin{cases} u'(t) \in Au(t) + g(t) \\ u(\tau) = \xi, \end{cases}$$

in the sense of Definition 1.6.2. A C^0*-solution* of (12.1.1), on $[\tau,\widetilde{T})$, is a continuous function $u:[\tau,\widetilde{T})\to K$ such that, for each $\tau<T<\widetilde{T}$, u is a C^0-solution of (12.1.1), on $[\tau,T]$, in the sense just mentioned.

As we have already noticed, we are interested here in finding necessary and sufficient conditions in order that $I\times K$ be C^0*-viable* with respect to $A+f$ in the sense of the definition below.

Definition 12.1.2. We say that $I\times K$ is C^0*-viable with respect to* $A+f$ if for each $(\tau,\xi)\in I\times K$ there exist $T\in I$, $T>\tau$ and a C^0-solution $u:[\tau,T]\to K$ of (12.1.1).

We emphasize that we confined ourselves here only to the cylindrical case because, as we have already seen in Example 5.1.1, even for $A\equiv 0$, there is no hope to obtain viability of general noncylindrical domains, with respect to Carathéodory perturbations of m-dissipative operators, without any extra-assumptions.

We are now ready to state the main results of this chapter.

Theorem 12.1.1. *Let X be a Banach space, $A:D(A)\subseteq X\rightsquigarrow X$ an m-dissipative operator which is the infinitesimal generator of a compact semigroup, K a nonempty, locally closed subset in $\overline{D(A)}$, I a nonempty and open interval and $f:I\times K\to X$ a locally Carathéodory function*[1]*. Then a necessary and sufficient condition in order that $I\times K$ be C^0-viable with respect to $A+f$ is to exist a negligible subset $\mathcal{Z}$ in I such that, for each $(\tau,\xi)\in(I\setminus\mathcal{Z})\times K$, $f(\tau,\xi)\in\mathcal{T}_K^A(\xi)$.*

Theorem 12.1.2. *Let X be a Banach space, $A:D(A)\subseteq X\rightsquigarrow X$ an m-dissipative operator, K a nonempty, locally closed and separable subset in $\overline{D(A)}$, I a nonempty and open interval and $f:I\times K\to X$ a Lipschitz-Carathéodory function*[2]*. Then a necessary and sufficient condition in order that $I\times K$ be C^0-viable with respect to $A+f$ is to exist a negligible subset $\mathcal{Z}$ in I such that, for each $(\tau,\xi)\in(I\setminus\mathcal{Z})\times K$, $f(\tau,\xi)\in\mathcal{T}_K^A(\xi)$.*

Theorem 12.1.3. *Let X be a Banach space, $A:D(A)\subseteq X\rightsquigarrow X$ an m-dissipative operator, K a nonempty, locally compact and separable subset in $\overline{D(A)}$, I a nonempty and open interval and $f:I\times K\to X$ a locally Carathéodory function. Then a necessary and sufficient condition in order that $I\times K$ be C^0-viable with respect to $A+f$ is to exist a negligible subset $\mathcal{Z}$ in I such that, for each $(\tau,\xi)\in(I\setminus\mathcal{Z})\times K$, $f(\tau,\xi)\in\mathcal{T}_K^A(\xi)$.*

[1] See Definition 2.8.1.

[2] See Definition 5.2.1.

Remark 12.1.1. From Problem 1.6.1, it follows that, under the hypotheses of Theorem 12.1.1, $\overline{D(A)}$ is separable and hence K enjoys the same property.

One of the main tools in the proof of Theorems 12.1.1~12.1.3 is the following characterization of the tangency condition $f(\tau,\xi) \in \mathfrak{T}_K^A(\xi)$ for each $(\tau,\xi) \in (I \setminus \mathcal{Z}) \times K$. First, we recall that, for each $\tau \in \mathbb{R}$, $t \geq \tau$, $\xi \in \overline{D(A)}$ and $g \in L^1_{\mathrm{loc}}(\tau,\infty;X)$, $u(t,\tau,\xi,g(\cdot))$ denotes the value at t of the unique C^0-solution u of the Cauchy problem

$$\begin{cases} u'(t) \in Au(t) + g(t) \\ u(\tau) = \xi. \end{cases}$$

Of course, whenever g is constant, say $g = \eta$, $u(t,\tau,\xi,\eta) = S_\eta(t-\tau)\xi$.

Proposition 12.1.1. *Let X be a Banach space, $A : D(A) \subseteq X \rightsquigarrow X$ an m-dissipative operator, K a nonempty and separable subset in $\overline{D(A)}$, and let $f : I \times K \to X$ be a locally Carathéodory function. Then there exists a negligible subset $\mathcal{Z}$ in I such that, for each $(\tau,\xi) \in (I \setminus \mathcal{Z}) \times K$, we have $f(\tau,\xi) \in \mathfrak{T}_K^A(\xi)$ if and only if there exists a negligible subset $\mathcal{Z}_1$ in I such that, for each $(\tau,\xi) \in (I \setminus \mathcal{Z}_1) \times K$, we have*

$$\liminf_{h \downarrow 0} \frac{1}{h} \operatorname{dist}\left(u(\tau+h,\tau,\xi,f(\cdot,\xi));K\right) = 0. \tag{12.1.2}$$

Problem 12.1.1. *Prove Proposition* 12.1.1.

The necessity of Theorems 12.1.1~12.1.3 is an immediate consequence of the next result which is interesting by itself.

Theorem 12.1.4. *Let X be a Banach space, $A : D(A) \subseteq X \rightsquigarrow X$ an m-dissipative operator, K a locally closed and separable subset in X and let $f : I \times K \to X$ be a locally Carathéodory function. Then, a necessary condition in order that $I \times K$ be C^0-viable with respect to $A + f$ is to exist a negligible subset $\mathcal{Z}$ in I such that, for each $(\tau,\xi) \in (I \setminus \mathcal{Z}) \times K$, $f(\tau,\xi) \in \mathfrak{T}_K^A(\xi)$.*

Proof. Let $\mathcal{Z}$ be given by Theorem 2.8.5, let $\tau \in I \setminus \mathcal{Z}$, let $\xi \in K$, choose a solution u of (12.1.1) which is defined on a subinterval $[\,\tau,T\,]$ of I and take a continuous function $z : I \to K$ which coincides with u on $[\,\tau,T\,]$. We have

$$u(\tau+h,\tau,\xi,f(\cdot,z(\cdot))) \in K$$

for each $h \in [\,0,T-\tau\,]$. On the other hand, by (1.6.5), we get

$$\|u(\tau+h,\tau,\xi,f(\cdot,z(\cdot))) - S_{f(\tau,\xi)}(h)\xi\| \leq \int_\tau^{\tau+h} \|f(s,z(s)) - f(\tau,\xi)\|\,ds$$

for each $h \in [0, T - \tau]$. By Theorem 2.8.5 we know that

$$\lim_{h \downarrow 0} \frac{1}{h} \int_{\tau}^{\tau+h} \|f(s, z(s)) - f(\tau, \xi)\| \, ds = 0$$

and therefore

$$\lim_{h \downarrow 0} \frac{1}{h} \text{dist } (S_{f(\tau,\xi)}(h)\xi; K) \leq$$

$$\leq \lim_{h \downarrow 0} \frac{1}{h} \|S_{f(\tau,\xi)}(h)\xi - u(\tau + h, \tau, \xi, f(\cdot, z(\cdot)))\| = 0.$$

Hence $f(\tau, \xi) \in \mathcal{T}_K^A(\xi)$ for each $(\tau, \xi) \in (I \setminus \mathcal{Z}) \times K$ and this completes the proof. □

Let us pass now to the proof of the necessity of Theorems 12.1.1∼12.1.3. To complete this, we have merely to observe that, in either cases, K is separable and therefore the hypotheses of Theorem 12.1.4 are satisfied. This separability condition, required by Theorem 12.1.4, follows from Problem 1.6.1 – in the case of Theorem 12.1.1 – and by hypothesis – in the case of Theorems 12.1.2 and 12.1.3.

Remark 12.1.2. The proof of Theorem 12.1.4 shows that, even in a more general frame than that assumed in either Theorems 12.1.1∼12.1.3, a necessary condition for the C^0-viability of $I \times K$ with respect to $A + f$ is a tangency condition which is stronger than $f(\tau, \xi) \in \mathcal{T}_K^A(\xi)$ for each $(\tau, \xi) \in (I \setminus \mathcal{Z}) \times K$. More precisely, we have proved that such a necessary condition is to exist a negligible subset $\mathcal{Z}$ in I such that, for each $(\tau, \xi) \in (I \setminus \mathcal{Z}) \times K$, we have $f(\tau, \xi) \in \mathcal{F}_K^A(\xi)$, where $\mathcal{F}_K^A(\xi)$ is defined as in Remark 10.1.1.

12.2. Existence of ε-approximate C^0-solutions

The proof of the sufficiency consists in showing that the tangency condition $f(\tau, \xi) \in \mathcal{T}_K^A(\xi)$ for each $(\tau, \xi) \in (I \setminus \mathcal{Z}) \times K$ along with Brezis-Browder Theorem 2.1.1 imply that, for each (τ, ξ) in $I \times K$, there exists at least one sequence of "C^0-approximate solutions", $(v_n)_n$, of (12.1.1), defined on the same interval, $[\tau, T]$, such that $(v_n)_n$ converges uniformly to a C^0-solution of (12.1.1).

The next lemma is an existence result referring to ε-approximate C^0-solutions of (12.1.1) and it is an "m-dissipative plus locally Carathéodory" version of Lemma 5.3.1.

Lemma 12.2.1. *Let X be a real Banach space, $A : D(A) \subseteq X \rightsquigarrow X$ an m-dissipative operator, K a nonempty, locally closed and separable subset in $\overline{D(A)}$ and $f : I \times K \to X$ a locally Carathéodory function satisfying the tangency condition in Theorem* 12.1.1 *with the negligible set $\mathcal{Z}_1$ and* *thus the equivalent one in Proposition* 12.1.1 *with the negligible set* $\mathcal{Z}_2$. *Let* $\mathcal{Z} = \mathcal{Z}_1 \cup \mathcal{Z}_2 \cup \mathcal{Z}_3$, *where* $\mathcal{Z}_3$ *is the negligible set in* I *such that for each* $t \in I \setminus \mathcal{Z}_3$, $f(t, \cdot)$ *is continuous on* K. *Then, for each* $(\tau, \xi) \in I \times K$, *there exist* $\rho > 0$, $T \in I$, $T > \tau$, $\theta_0 \in I \setminus \mathcal{Z}$ *and* $\mathcal{M} \in L^1(\tau, T; \mathbb{R}_+)$ *such that* $D(\xi, \rho) \cap K$ *is closed and, for each* $\varepsilon \in (0,1)$ *and each open set* $\mathcal{L}$ *of* $\mathbb{R}$ *with* $\mathcal{Z} \subseteq \mathcal{L}$ *and* $\lambda(\mathcal{L}) < \varepsilon$, *there exist a family of nonempty and pairwise disjoint intervals,* $\mathcal{P}_T = \{[t_m, s_m);\ m \in \Gamma\}$, *with* Γ *finite or countable, and two functions,* $g : [\tau, T] \to X$ *measurable and* $v : [\tau, T] \to X$ *continuous, satisfying*:

(i) $\displaystyle\bigcup_{m\in\Gamma} [t_m, s_m) = [\tau, T)$ *and* $s_m - t_m \leq \varepsilon$ *for each* $m \in \Gamma$;

(ii) *if* $t_m \in \mathcal{L}$ *then* $[t_m, s_m) \subseteq \mathcal{L}$;

(iii) $v(t_m) \in D(\xi, \rho) \cap K$ *for each* $m \in \Gamma$, $v(T) \in D(\xi, \rho) \cap K$ *and* $v([\tau, T])$ *is precompact*;

(iv) $g(s) = \begin{cases} f(s, v(t_m)) \text{ a.e. on } [t_m, s_m) \text{ if } t_m \notin \mathcal{L} \\ f(\theta_0, v(t_m)) \text{ a.e. on } [t_m, s_m) \text{ if } t_m \in \mathcal{L}; \end{cases}$

(v) $\|g(t)\| \leq \mathcal{M}(t)$ *a.e. for* $t \in [\tau, T]$;

(vi) $v(\tau) = \xi$ *and* $\|v(t) - u(t, t_m, v(t_m), g)\| \leq (t - t_m)\varepsilon$ *for each* $m \in \Gamma$ *and* $t \in [t_m, T]$;

(vii) $\|v(t) - v(t_m)\| \leq \varepsilon$ *for each* $m \in \Gamma$ *and* $t \in [t_m, s_m)$.

Proof. Let $(\tau, \xi) \in I \times K$ be arbitrary and choose $\rho > 0$ and a locally integrable function $\ell(\cdot)$ such that $D(\xi, \rho) \cap K$ is closed (compact in the hypotheses of Theorem 12.1.3) and $\|f(t, u)\| \leq \ell(t)$ for almost every $t \in I$ and for every $u \in D(\xi, \rho) \cap K$. This is always possible because K is locally closed and f satisfies (C_4) in Definition 2.8.1.

Fix $\theta_0 \notin \mathcal{Z}$. Since $v \mapsto f(\theta_0, v)$ is continuous, diminishing $\rho > 0$ if necessary, we can find $M > 0$ such that

$$\|f(\theta_0, v)\| \leq M$$

for each $v \in D(\xi, \rho) \cap K$. Next, take $T \in I$, $T > \tau$ and let us define

$$\mathcal{M}(t) = \max\{M, \ell(t)\}$$

a.e. for $t \in [\tau, T]$. Obviously $\mathcal{M} \in L^1(\tau, T; \mathbb{R}_+)$ and consequently, taking a smaller $T > \tau$ if necessary, we may assume that

$$\sup_{\tau \leq t \leq T} \|S(t - \tau)\xi - \xi\| + \int_\tau^T \mathcal{M}(s)\, ds + T - \tau \leq \rho. \tag{12.2.1}$$

We first prove that the conclusion of Lemma 12.2.1 remains true if we replace T as above with a possible smaller number $\mu \in (\tau, T]$ which, at this stage, is allowed to depend on $\varepsilon \in (0, 1)$. Then, by using the Brezis-Browder Theorem 2.1.1, we will prove that we can take $\mu = T$, independent of ε.

Let $\varepsilon \in (0,1)$ be arbitrary but fixed and take an open set $\mathcal{L}$ of $\mathbb{R}$ with $\mathcal{Z} \subseteq \mathcal{L}$ and whose Lebesgue measure $\lambda(\mathcal{L}) < \varepsilon$.

Case 1. If $\tau \in \mathcal{L}$, since $f(\theta_0, \xi)$ is A-tangent to K at ξ, it is easy to see that there exist $\delta \in (0,\varepsilon)$ and $p \in X$ with $\|p\| \leq \varepsilon$ such that $[\tau, \tau+\delta) \subseteq \mathcal{L}$ and such that

$$u(\tau+\delta, \tau, \xi, f(\theta_0, \xi)) + \delta p \in K.$$

Now, let us define $g : [\tau, \tau+\delta] \to X$ and $v : [\tau, \tau+\delta] \to X$ by $g(t) = f(\theta_0, \xi)$ and respectively by

$$v(t) = u(t, \tau, \xi, g) + (t-\tau)p \tag{12.2.2}$$

for each $t \in [\tau, \tau+\delta]$.

Let us observe that the family $\mathcal{P}_{\tau+\delta} = \{[\tau, \tau+\delta)\}$ and the functions g and v satisfy (i)$\sim$(v) with T substituted by $\tau+\delta$ and $t_0 = \tau$, $s_0 = \tau+\delta$. Clearly, $v(\tau) = \xi$. Moreover, since $\|p\| \leq \varepsilon$, we deduce

$$\|v(t) - u(t, \tau, v(\tau), g)\| = (t-\tau)\|p\| \leq (t-\tau)\varepsilon.$$

Thus (vi) is also satisfied. Next, diminishing $\delta > 0$ if necessary, we get

$$\|v(t) - v(\tau)\| \leq \|u(t, \tau, \xi, g) - \xi\| + (t-\tau)\|p\|$$

$$\leq \|S(t-\tau)\xi - \xi\| + \int_\tau^t \|g(s)\|\, ds + (t-\tau)\varepsilon$$

$$\leq \sup_{t \in [0,\delta]} \|S(t)\xi - \xi\| + \int_\tau^t \mathcal{M}(s)\, ds + (t-\tau)\varepsilon \leq \varepsilon$$

for each $t \in [\tau, \tau+\delta)$, and thus (vii) is also satisfied.

Case 2. If $\tau \notin \mathcal{L}$, we have $\tau \notin \mathcal{Z}$ and in view of Proposition 12.1.1 there exist $\delta \in (0,\varepsilon)$ and $p \in X$ with $\|p\| \leq \varepsilon$ such that

$$u(\tau+\delta, \tau, \xi, f(\cdot, \xi)) + \delta p \in K.$$

Setting $g(s) = f(s, \xi)$ and defining v by (12.2.2), we can easily see that, again, the family $\mathcal{P}_{\tau+\delta} = \{[\tau, \tau+\delta)\}$ and the functions g and v satisfy (i)$\sim$(v) with T substituted by $\tau+\delta$. Moreover, by the very same arguments, diminishing $\delta > 0$ if necessary, we obtain (vi) and (vii).

Next, we will show that there exists at least one triplet $(\mathcal{P}_T, g, v)$ satisfying (i)$\sim$(vii). To this aim we shall use Brezis-Browder Theorem 2.1.1 as follows. Let $\mathcal{S}$ be the set of all triplets $(\mathcal{P}_\mu, g, v)$ with $\mu \leq T$ and satisfying (i)$\sim$(vii) with μ instead of T. This set is clearly nonempty, as we have already proved. On $\mathcal{S}$ we introduce a partial order $\preceq$ as follows. We say that

$$(\mathcal{P}_{\mu_1}, g_1, v_1) \preceq (\mathcal{P}_{\mu_2}, g_2, v_2),$$

where $\mathcal{P}_{\mu_k} = \{[t_m^k, s_m^k); m \in \Gamma_k\}$, $k = 1, 2$, if

(O_1) $\mu_1 \leq \mu_2$ and if $\mu_1 < \mu_2$ there exists $i \in \Gamma_2$ such that $\mu_1 = t_i^2$;
(O_2) for each $m_1 \in \Gamma_1$ there exists $m_2 \in \Gamma_2$ such that $t_{m_1}^1 = t_{m_2}^2$ and $s_{m_1}^1 = s_{m_2}^2$;
(O_3) $g_1(s) = g_2(s)$ and $v_1(s) = v_2(s)$ for each $s \in [\tau, \mu_1]$.

Let us define the function $\mathcal{N} : \mathcal{S} \to \mathbb{R}$ by

$$\mathcal{N}((\mathcal{P}_\mu, g, v)) = \mu.$$

It is clear that $\mathcal{N}$ is increasing on $\mathcal{S}$. Let us take now an increasing sequence

$$((\mathcal{P}_{\mu_j}, g_j, v_j))_j$$

in $\mathcal{S}$ and let us show that it is bounded from above in $\mathcal{S}$. We define an upper bound as follows. First, set

$$\mu^* = \sup\{\mu_j;\ j \in \mathbb{N}\}.$$

If $\mu^* = \mu_j$ for some $j \in \mathbb{N}$, $(\mathcal{P}_{\mu_j}, g_j, v_j)$ is clearly an upper bound. If $\mu_j < \mu^*$ for each $j \in \mathbb{N}$, let us observe that the family $\mathcal{P}_{\mu^*} = \cup_{j\in\mathbb{N}}\mathcal{P}_{\mu_j}$ is countable. Hence, relabelling if necessary, we may assume that

$$\mathcal{P}_{\mu^*} = \{[t_m, s_m); m \in \mathbb{N}\}.$$

We define

$$g(t) = g_j(t), \quad v(t) = v_j(t)$$

for $j \in \mathbb{N}$ and every $t \in [\tau, \mu_j]$. Now let us observe that $(\mathcal{P}_{\mu^*}, g, v)$, where $\mathcal{P}_{\mu^*}$, g and v are defined as above, satisfies (i), (ii), (iv) and (v) with T replaced with μ^*. Obviously we have $v(t_m) \in D(\xi, \rho) \cap K$ for each $m \in \mathbb{N}$. To see that $(\mathcal{P}_{\mu^*}, g, v)$ also satisfies (iii) we first have to check that $v([\tau, \mu^*))$ is precompact in X and next to show how to define $v(\mu^*)$. By (iv) and (v) we know that $g \in L^1(\tau, \mu^*; X)$ and so, for each $j \in \mathbb{N}$, the function $u(\cdot, \mu_j, v(\mu_j), g) : [\mu_j, \mu^*] \to \overline{D(A)}$ is continuous. As a consequence, the set $C_j = u([\mu_j, \mu^*], \mu_j, v(\mu_j), g)$ is precompact. On the other hand, by (iii), for each $j \in \mathbb{N}$, we know that $K_j = v([\tau, \mu_j])$ is precompact too. By (vi) and (O_1) we deduce that, for each $j \in \mathbb{N}$,

$$v([\tau, \mu^*)) \subseteq C_j \cup K_j + (\mu^* - \mu_j)D(0, \varepsilon).$$

Let $\eta > 0$ be arbitrary and fix $j \in \mathbb{N}$ such that

$$(\mu^* - \mu_j)\varepsilon \leq \frac{\eta}{2}.$$

Since $C_j \cup K_j$ is precompact, there exists a finite family $\{\xi_1, \xi_2, \ldots, \xi_{n(\eta)}\}$ such that, for each $\xi \in C_j \cup K_j$, there exists $k \in \{1, 2, \ldots, n(\eta)\}$, with

$$\|\xi - \xi_k\| \leq \frac{\eta}{2}.$$

The last two inequalities and the inclusion above yield

$$v([\,\tau,\mu^*)) \subseteq \cup_{k=1}^{n(\eta)} D(\xi_k,\eta),$$

and accordingly $v([\,\tau,\mu^*))$ is precompact. Now, take any limit point v_{μ^*} of $v(\mu_j)$ as j tends to ∞ and set $v(\mu^*) = v_{\mu^*}$. Clearly $v(\mu^*) \in D(\xi,\rho) \cap K$. So, with $v : [\,\tau,\mu^*\,] \to X$, defined as above, we obviously have that $(\mathcal{P}_{\mu^*}, g, v)$ satisfies (i)$\sim$(v). It is also easy to see that (vi) and (vii) hold for each $m \in \mathbb{N}$ and each $t \in [\,t_m,\mu^*)$. To check (vi) for $t = \mu^*$, we have to fix any $m \in \mathbb{N}$, to take $t = \mu_j$ with $\mu_j > t_m$ in (vi), and to pass to the limit for j tending to ∞ both sides in (vi) on that subsequence on which $(v_j(\mu_j))_j$ tends to $v_{\mu^*} = v(\mu^*)$. To check (vii), we have to proceed similarly. So $(\mathcal{P}_{\mu^*}, g, v)$ is an upper bound for $((\mathcal{P}_{\mu_j}, g_j, v_j))_j$, and consequently the set $\mathcal{S}$, endowed with the partial order $\preceq$, and the function $\mathcal{N}$ satisfy the hypotheses of Theorem 2.1.1. Accordingly, there exists at least one element $(\mathcal{P}_\nu, g_\nu, v_\nu)$ in $\mathcal{S}$ such that, if $(\mathcal{P}_\nu, g_\nu, v_\nu) \preceq (\mathcal{P}_\sigma, g_\sigma, v_\sigma)$ then $\nu = \sigma$.

We next show that $\nu = T$, where T satisfies (12.2.1). To this aim let us assume by contradiction that $\nu < T$, and let $\xi_\nu = v_\nu(\nu)$ which belongs to $D(\xi,\rho) \cap K$. In view of (1.6.5), (v) and (vi), we have

$$\begin{aligned}
\|\xi_\nu - \xi\| &\leq \|S(\nu-\tau)\xi - \xi\| + \|u(\nu,\tau,\xi,g_\nu) - S(\nu-\tau)\xi\| \\
&\quad + \|v_\nu(\nu) - u(\nu,\tau,\xi,g_\nu)\| \\
&\leq \|S(\nu-\tau)\xi - \xi\| + \int_\tau^\nu \|g_\nu(s)\|\,ds + (\nu-\tau)\varepsilon \\
&\leq \sup_{\tau \leq t \leq \nu} \|S(t-\tau)\xi - \xi\| + \int_\tau^\nu \mathcal{M}(s)\,ds + (\nu-\tau)\varepsilon.
\end{aligned}$$

Recalling that $\nu < T$ and $\varepsilon < 1$, from (12.2.1), we get

$$\|\xi_\nu - \xi\| < \rho. \tag{12.2.3}$$

There are two possibilities: either $\nu \in \mathcal{L}$, or $\nu \notin \mathcal{L}$.

If $\nu \in \mathcal{L}$, we proceed as in *Case* 1 above, with ν instead of τ, and with ξ_ν instead of ξ. So, since $f(\theta_0,\xi_\nu) \in \mathcal{T}_K^A(\xi_\nu)$, from (12.2.3), we infer that there exist $\delta \in (0,\varepsilon]$, with $\nu + \delta \leq T$, $[\,\nu,\nu+\delta) \subseteq \mathcal{L}$, and $p \in X$ satisfying $\|p\| \leq \varepsilon$ and such that

$$u(\nu+\delta,\nu,\xi_\nu,f(\theta_0,\xi_\nu)) + \delta p \in D(\xi,\rho) \cap K.$$

If $\nu \notin \mathcal{L}$ we proceed as in *Case* 2 above, with ν instead of τ, and with ξ_ν instead of ξ. So, from Proposition 12.1.1 combined with (12.2.3), we infer that there exist $\delta \in (0,\varepsilon]$, with $\nu + \delta \leq T$, and $p \in X$, satisfying $\|p\| \leq \varepsilon$ and such that

$$u(\nu+\delta,\nu,\xi_\nu,f(\cdot,\xi_\nu)) + \delta p \in D(\xi,\rho) \cap K.$$

We define successively $\mathcal{P}_{\nu+\delta} = \mathcal{P}_\nu \cup \{[\nu, \nu+\delta)\}$, $g_{\nu+\delta} : [\tau, \nu+\delta] \to X$ and $v_{\nu+\delta} : [\tau, \nu+\delta] \to X$ by

$$g_{\nu+\delta}(t) = \begin{cases} g_\nu(t) & \text{if } t \in [\tau, \nu] \\ f(\theta_0, \xi_\nu) & \text{if } t \in (\nu, \nu+\delta], \end{cases}$$

if $\nu \in \mathcal{L}$, and

$$g_{\nu+\delta}(t) = \begin{cases} g_\nu(t) & \text{if } t \in [\tau, \nu] \\ f(t, \xi_\nu) & \text{if } t \in (\nu, \nu+\delta], \end{cases}$$

if $\nu \notin \mathcal{L}$, and respectively by

$$v_{\nu+\delta}(t) = \begin{cases} v_\nu(t) & \text{if } t \in [\tau, \nu] \\ u(t, \nu, \xi_\nu, g_{\nu+\delta}) + (t-\nu)p & \text{if } t \in (\nu, \nu+\delta]. \end{cases}$$

Since $v_{\nu+\delta}(\nu+\delta) \in K \cap D(\xi, \rho)$, $(\mathcal{P}_{\nu+\delta}, g_{\nu+\delta}, v_{\nu+\delta})$ satisfies (i)~(v) with T replaced by $\nu+\delta$. Clearly (vi) holds for each t_m and t satisfying either $t_m \le t \le \nu$, or $\nu \le t_m \le t$. It remains to verify the case in which $t_m \le \nu < t \le \nu+\delta$. To this aim, let us observe that, by virtue of the evolution property (1.6.4) and (vi), we have

$$\|v_{\nu+\delta}(t) - u(t, t_m, v_{\nu+\delta}(t_m), g_{\nu+\delta})\|$$

$$= \|u(t, \nu, v_{\nu+\delta}(\nu), g_{\nu+\delta}) + (t-\nu)p - u(t, \nu, u(\nu, t_m, v_{\nu+\delta}(t_m), g_{\nu+\delta}), g_{\nu+\delta})\|$$

$$\le \|v_{\nu+\delta}(\nu) - u(\nu, t_m, v_{\nu+\delta}(t_m), g_{\nu+\delta})\| + (t-\nu)\|p\|$$

$$\le (\nu - t_m)\varepsilon + (t-\nu)\varepsilon = (t - t_m)\varepsilon.$$

So (vi) holds for each $m \in \mathbb{N}$ and each $t \in [t_m, \nu+\delta]$. Similarly, we conclude that (vii) is satisfied. We conclude that $(\mathcal{P}_{\nu+\delta}, g_{\nu+\delta}, v_{\nu+\delta}) \in \mathcal{S}$,

$$(\mathcal{P}_\nu, g_\nu, v_\nu) \preceq (\mathcal{P}_{\nu+\delta}, g_{\nu+\delta}, v_{\nu+\delta})$$

and $\nu < \nu+\delta$. This contradiction can be eliminated only if $\nu = T$. This completes the proof of Lemma 12.2.1. □

Remark 12.2.1. Under the general hypotheses of Lemma 12.2.1, for each $\gamma > 0$, we can diminish both $\rho > 0$ and $T > 0$, such that $T - \tau < \gamma$, $\rho < \gamma$ and all the conditions (i)~(vii) in Lemma 12.2.1 be satisfied.

Definition 12.2.1. Let $(\tau, \xi) \in I \times K$, $\varepsilon \in (0,1)$ and let $\mathcal{L}$ be the set as in Lemma 12.2.1. A triplet $(\mathcal{P}_T, g, v)$ satisfying (i)~(vii) is called an *ε-approximate C^0-solution* of (12.1.1).

12.3. Convergence in the case of Theorems 12.1.1 and 12.1.3

In order to complete the proof of the sufficiency of Theorems 12.1.1 and 12.1.3, we will show that a suitably chosen subsequence of ε_n-approximate C^0-solutions is uniformly convergent on $[\tau, T]$ to a function u which is a C^0-solution of (12.1.1).

Proof. Let $(\varepsilon_n)_n$ be a sequence strictly decreasing to 0 and let $(\mathcal{L}_n)$ be a decreasing sequence of open subsets in $\mathbb{R}$ such that $\mathcal{Z} \subseteq \mathcal{L}_n$ and $\lambda(\mathcal{L}_n) < \varepsilon_n$ for $n = 1, 2, \ldots$. Take $\mathcal{L} = \cap_{n \geq 1} \mathcal{L}_n$ and a sequence of ε_n-approximate C^0-solutions $((\mathcal{P}_T^n, g_n, v_n))_n$ of (12.1.1). From (v), we know that $\{g_n;\ n = 1, 2, \ldots\}$ is uniformly integrable in $L^1(\tau, T; X)$. See Remark 1.3.3.

Under the hypotheses of Theorem 12.1.1, since the semigroup of nonlinear contractions $\{S(t) : \overline{D(A)} \to \overline{D(A)};\ t \geq 0\}$, is compact, by Theorem 1.6.5, it follows that there exists $u \in C([\tau, T]; X)$ such that, on a subsequence at least, we have

$$\lim_n u(t, \tau, \xi, g_n) = u(t) \tag{12.3.1}$$

uniformly for $t \in [\tau, T]$.

We shall now prove that, under the hypotheses of Theorem 12.1.3, (12.3.1) still holds true. First, let us remark that, in view of Remark 12.2.1, we can diminish $T > \tau$ and $\rho > 0$, if necessary, such that $D(\xi, \rho) \cap K$ is compact and (i) $\sim$ (vii) in Lemma 12.2.1 be satisfied. So, by (vi) and (vii) we have

$$\begin{aligned}\|u(t, \tau, \xi, g_n) - v_n(t_m^n)\| &\leq \|u(t, \tau, \xi, g_n) - v_n(t)\| + \|v_n(t) - v_n(t_m^n)\| \\ &\leq (t - \tau)\varepsilon_n + \varepsilon_n \leq (T - \tau + 1)\varepsilon_n\end{aligned}$$

for $n = 1, 2, \ldots$, each $m \in \Gamma_n$ and $t \in [t_m^n, s_m^n)$. For each $k = 1, 2, \ldots$, let us denote by

$$\begin{cases} C_k = \bigcup_{n=1}^{k} \{u(t, \tau, \xi, g_n)\,;\ t \in [\tau, T]\} \\ C = \bigcup_{n=1}^{\infty} \{v_n(t_m^n)\,;\ m \in \Gamma_n\} \end{cases}$$

and let us observe that, in view of the inequality above, we have

$$\bigcup_{n=1}^{\infty} \{u(t, \tau, \xi, g_n)\,; t \in [\tau, T]\} \subseteq C_k \cup C + (T - \tau + 1)D(0, \varepsilon_k) \tag{12.3.2}$$

for each $k = 1, 2, \ldots$. But, for $k = 1, 2, \ldots$, C_k and C are precompact, C_k because for $n = 1, 2, \ldots$, $u(\cdot, \tau, \xi, g_n)$ is continuous on $[\tau, T]$, and C as a subset of $D(\xi, \rho) \cap K$ which, in its turn, is compact. This remark, along with (12.3.2), shows that, for each $t \in [\tau, T]$, $\{u(t, \tau, \xi, g_n);\ n = 1, 2, \dots\}$ is precompact too.

Thus Theorem 1.6.4 applies, and hence there exists $u \in C([\tau, T]; X)$ such that, on a subsequence at least, we have (12.3.1). So, under the hypotheses of either Theorems 12.1.1 or 12.1.3, (12.3.1) holds. By (vi) and (12.3.1), on the same subsequence, we also have

$$\lim_n v_n(t) = u(t) \tag{12.3.3}$$

uniformly for $t \in [\tau, T]$.

Now, let us observe that, by (i), for each $t \in [\tau, T)$ and each $n = 1, 2, \dots$, there exists $m \in \Gamma_n$ such that $t \in [t_m^n, s_m^n)$ and $s_m^n - t_m^n \leq \varepsilon_n$. As a consequence $\{t_m^n;\ n = 1, 2, \dots,\ m \in \Gamma_n\}$ is dense in $[\tau, T]$. In addition, $v_n(t_m^n)$ belongs to $D(\xi, \rho) \cap K$ for $n = 1, 2 \dots$ and $m \in \Gamma_n$. Therefore, from (12.3.1) and (12.3.3), we conclude that, for each $t \in [\tau, T]$, we have $u(t) \in D(\xi, \rho) \cap K$. Indeed, this is clearly the case if $t = \tau$. So, take $t \in (\tau, T]$, $n = 1, 2, \dots$, $m \in \Gamma_n$ and let us denote (for the sake of simplicity) $s = t_m^n$ if $t \in [t_m^n, s_m^n)$. Let us observe that, by virtue of (vi) and (1.6.5), we have

$$\|v_n(t) - v_n(s)\| \leq \|v_n(t) - u(t, s, v_n(s), g_n)\| + \|u(t, s, v_n(s), g_n) - S(t-s)v_n(s)\|$$

$$+ \|S(t-s)v_n(s) - v_n(s)\| \leq (t-s)\varepsilon_n + \int_s^t \|g_n(\theta)\| d\theta + \sup_{\eta \in C} \|S(t-s)\eta - \eta\|$$

$$\leq (t-s)\varepsilon_n + \int_s^t \mathcal{M}(\theta)\, d\theta + \sup_{\eta \in C} \|S(t-s)\eta - \eta\|,$$

where

$$C = \bigcup_{n=1}^{\infty} \{v_n(t)\,;\ t \in [\tau, T]\}.$$

Since, due to (12.3.3), C is precompact in X, by Problem 1.3.1, we have

$$\lim_{\delta \downarrow 0} \sup_{\eta \in C} \|S(\delta)\eta - \eta\| = 0.$$

Recalling that $\mathcal{M} \in L^1(\tau, T; \mathbb{R}_+)$, that s denotes a generic element t_m^n satisfying $0 \leq t - t_m^n < \varepsilon_n$, and using the relation above, we easily deduce that

$$u(t) \in \overline{\bigcup_{n=1}^{\infty} \{v_n(t_m^n)\,;\ m \in \Gamma_n\}},$$

where the latter is included in $D(\xi, \rho) \cap K$. As a consequence, we necessarily have $u(t) \in D(\xi, \rho) \cap K$ for each $t \in [\tau, T]$.

Next, let us observe that, if $s \notin \mathcal{L}$, there exists $n(s) \in \mathbb{N}^*$ such that, for each $n \geq n(s)$, $s \notin \mathcal{L}_n$. Hence, by (i) and (iv), we have $g_n(s) = f(s, v_n(t_m^n))$

for each $n \geq n(s)$ and for some $m \in \Gamma_n$ with $|s - t_m^n| \leq \varepsilon_n$. Therefore, we get

$$\lim_n g_n(s) = f(s, u(s))$$

a.e. for $s \in [\,\tau, T\,]$. From (v) and the Lebesgue Dominated Convergence Theorem 1.2.3, we deduce that

$$\lim_n g_n = f(\cdot, u(\cdot)) \tag{12.3.4}$$

in $L^1(\tau, T\,; X)$.

Let $\widetilde{u} : [\,\tau, T\,] \to \overline{D(A)}$ be the unique solution of the Cauchy problem

$$\begin{cases} \widetilde{u}'(t) \in A\widetilde{u}(t) + f(t, u(t)) \\ \widetilde{u}(\tau) = \xi. \end{cases}$$

In view of (12.3.1) and (1.6.5), we have

$$\|\widetilde{u}(t) - u(t, \tau, \xi, g_n)\| \leq \int_\tau^t \|f(s, u(s)) - g_n(s)\|\, ds$$

for $n = 1, 2, \ldots$ and each $t \in [\,\tau, T\,]$. Passing to the limit in the preceding inequality and using (12.3.4), we get $u(t) = \widetilde{u}(t) = u(t, \tau, \xi, f(\cdot, u(\cdot)))$ for each $t \in [\,\tau, T\,]$. But this means that u is a C^0-solution of (12.1.1), and this completes the proof. □

12.4. Convergence in the case of Theorem 12.1.2

Under the hypotheses of Theorem 12.1.2, we will show that there exists a sequence of ε_n-approximate solutions which is uniformly convergent on $[\,\tau, T\,]$ to a C^0-solution of (12.1.1).

Proof. Let $(\tau, \xi) \in I \times K$ and let $\rho > 0$ and $T \in I$, $T > \tau$ be given by Lemma 12.2.1. As f is a Lipschitz-Carathéodory function, by Remark 12.2.1, we can diminish $T > \tau$ and $\rho > 0$ if necessary, in order to assume that there exists $L \in L^1_{\text{loc}}(I)$ such that all the conditions in Lemma 12.2.1 be satisfied and, in addition,

$$\|f(t, u) - f(t, v)\| \leq L(t)\|u - v\|$$

for a.a. $t \in I$ and for all $u, v \in D(\xi, \rho) \cap K$. See (C_6) in Definition 5.2.1.

Let $(\varepsilon_n)_n$ be a sequence strictly decreasing to 0 and let $(\mathcal{L}_n)$ be a decreasing sequence of open subsets in $\mathbb{R}$ such that $\mathcal{Z} \subseteq \mathcal{L}_n$ and $\lambda(\mathcal{L}_n) < \varepsilon_n$ for $n = 1, 2, \ldots$. Take $\mathcal{L} = \cap_{n \geq 1} \mathcal{L}_n$ and a sequence of ε_n-approximate solutions $((\mathcal{P}_T^n, g_n, v_n))_n$ of (12.1.1).

For $n = 1, 2, \ldots$, let us denote by $\sigma_n : [\,\tau, T\,] \to [\,\tau, T\,]$ the a.e. defined function

$$\sigma_n(s) = t_m^n$$

a.e. for $s \in [t_m^n, s_m^n)$, where $\mathcal{P}_T^n = \{[t_m^n, s_m^n);\ m \in \Gamma_n\}$ is as in Lemma 12.2.1. Let $\Sigma_n = \{i \in \Gamma_n;\ t_i^n \in \mathcal{L}_n\}$ and let us denote by

$$\mathcal{E}_n = \bigcup_{i \in \Sigma_n} [t_i^n, s_i^n),$$

by $\mathcal{E}_n^t = \mathcal{E}_n \cap [\tau, t)$, by

$$\mathcal{H}_n = [\tau, T] \setminus \mathcal{E}_n$$

and by $\mathcal{H}_n^t = [\tau, t) \setminus \mathcal{E}_n^t$.

Since, by (ii) in Lemma 12.2.1, we have $[t_i^n, s_i^n) \subseteq \mathcal{L}_n$ whenever $t_i^n \in \mathcal{L}_n$, we deduce that $\mathcal{E}_n^t \subseteq \mathcal{L}_n$ and, since $\lambda(\mathcal{L}_n) < \varepsilon_n$, we conclude that

$$\lambda(\mathcal{E}_n^t) < \varepsilon_n$$

for each $t \in [\tau, T]$. In view of (iv) in Lemma 12.2.1, we have

$$\|u(t, \tau, \xi, g_n) - u(t, \tau, \xi, g_k)\| \leq \int_\tau^t \|g_n(s) - g_k(s)\|\, ds$$

$$\leq \int_{\mathcal{H}_n^t \cup \mathcal{H}_k^t} \|f(s, v_n(\sigma_n(s))) - f(s, v_k(\sigma_k(s)))\|\, ds + 2\int_{\mathcal{E}_n^t \cup \mathcal{E}_k^t} \mathcal{M}(s)\, ds$$

$$\leq \int_{\mathcal{H}_n^t \cup \mathcal{H}_k^t} L(s)\|v_n(\sigma_n(s)) - v_k(\sigma_k(s))\|\, ds + 2\int_{\mathcal{E}_n^t \cup \mathcal{E}_k^t} \mathcal{M}(s)\, ds.$$

Consequently

$$\|u(t, \tau, \xi, g_n) - u(t, \tau, \xi, g_k)\| \leq \int_{\mathcal{H}_n^t \cup \mathcal{H}_k^t} L(s)\|v_n(\sigma_n(s)) - u(s, \tau, \xi, g_n)\|\, ds$$

$$+ \int_{\mathcal{H}_n^t \cup \mathcal{H}_k^t} L(s)\|u(s, \tau, \xi, g_n) - u(s, \tau, \xi, g_k)\|\, ds$$

$$+ \int_{\mathcal{H}_n^t \cup \mathcal{H}_k^t} L(s)\|u(s, \tau, \xi, g_k) - v_k(\sigma_k(s))\|\, ds + 2\int_{\mathcal{E}_n^t \cup \mathcal{E}_k^t} \mathcal{M}(s)\, ds. \quad (12.4.1)$$

Using (vi) and (vii) in Lemma 12.2.1 to estimate the first and the third integral on the right hand side in (12.4.1), we get

$$\|u(t, \tau, \xi, g_n) - u(t, \tau, \xi, g_k)\|$$

$$\leq \int_{\mathcal{H}_n^t \cup \mathcal{H}_k^t} L(s)\left[(T - \tau + 1)\varepsilon_n + (T - \tau + 1)\varepsilon_k\right] ds$$

$$+2\int_{\mathcal{E}_n^t \cup \mathcal{E}_k^t} \mathcal{M}(s)\, ds + \int_{\mathcal{H}_n^t \cup \mathcal{H}_k^t} L(s)\|u(s, \tau, \xi, g_n) - u(s, \tau, \xi, g_k)\|\, ds.$$

Since $L, \mathcal{M} \in L^1_{\mathrm{loc}}(I)$, from this inequality we deduce

$$\|u(t, \tau, \xi, g_n) - u(t, \tau, \xi, g_k)\|$$

$$\leq \alpha(n,k) + \int_{\tau}^{t} L(s)\|u(s,\tau,\xi,g_n) - u(s,\tau,\xi,g_k)\|\,ds,$$

where $\alpha : \mathbb{N}\times\mathbb{N} \to \mathbb{R}_+$ satisfies

$$\lim_{n,k} \alpha(n,k) = 0.$$

From the Gronwall Lemma 1.8.4, we get

$$\|u(t,\tau,\xi,g_n) - u(t,\tau,\xi,g_k)\| \leq \alpha(n,k)e^{\int_{\tau}^{T} L(s)\,ds},$$

for each $n,k \in \mathbb{N}^*$ and $t \in [\,\tau, T\,]$. Thus $(u(\cdot,\tau,\xi,g_n))_n$ is a Cauchy sequence in the sup-norm. Let u be the uniform limit of $(u(\cdot,\tau,\xi,g_n))_n$ on $[\,\tau,T\,]$. By (vi) in Lemma 12.2.1, we have

$$\lim_{n} v_n(t) = u(t) \tag{12.4.2}$$

uniformly for $t \in [\,\tau,T\,]$. If $s \notin \mathfrak{L}$, there exists $n(s) \in \mathbb{N}$ such that, for each $n \geq n(s)$, we necessarily have $s \notin \mathfrak{L}_n$. By (i) and (iv) in Lemma 12.2.1, we have $g_n(s) = f(s, v_n(t_m^n))$ for each $n \geq n(s)$ and for some $m \in \Gamma_n$ satisfying $|s - t_m^n| \leq \varepsilon_n$. Clearly, by (iii), (vi) and (vii), we deduce that $u(t) \in D(\xi,\rho) \cap K$ for each $t \in [\,\tau,T\,]$. Hence, $\lim_n g_n(s) = f(s,u(s))$ for almost every $s \in [\,\tau,T\,]$. From (v) and the Lebesgue Dominated Convergence Theorem 1.2.3, we conclude that $\lim_n g_n = f(\cdot,u(\cdot))$ in $L^1(\tau,T\,;X)$. Now, repeating the same arguments as those following (12.3.4), we deduce that $u(t) = u(t,\tau,\xi,g)$ for each $t \in [\,\tau,T\,]$, where the function g is given by $g(s) = f(s,u(s))$ a.e. for $s \in [\,\tau,T\,]$. Consequently, u is a C^0-solution of (12.1.1) and this completes the proof. □

12.5. Noncontinuable C^0-solutions

In this section, we present some results concerning the existence of noncontinuable, or even global C^0-solutions to

$$\begin{cases} u'(t) \in Au(t) + f(t,u(t)) \\ u(\tau) = \xi. \end{cases} \tag{12.5.1}$$

A C^0-solution $u : [\,\tau,T) \to K$ to (12.5.1) is called *noncontinuable*, if there is no other C^0-solution $v : [\,\tau,\widetilde{T}) \to K$ of the same equation, with $T < \widetilde{T}$ and satisfying $u(t) = v(t)$ for all $t \in [\,\tau,T)$. The C^0-solution u is called *global* if $T = \sup I$. The next theorem follows from the Brezis–Browder Theorem 2.1.1. Since its proof is almost identical with that of Theorem 3.6.1, we do not enter into details.

Theorem 12.5.1. *Let X be a Banach space, $A : D(A) \subseteq X \rightsquigarrow X$ an m-dissipative operator, I a nonempty and open interval, let $K \subseteq \overline{D(A)}$*

be nonempty and let $f : I \times K \to X$. Then, the following conditions are equivalent:

(i) *$I \times K$ is C^0-viable with respect to $A + f$;*
(ii) *for each $(\tau, \xi) \in I \times K$ there exists at least one noncontinuable C^0-solution $u : [\tau, T) \to K$ of* (12.5.1).

The next result concerns the existence of global solutions.

Theorem 12.5.2. *Let X be a Banach space, $A : D(A) \subseteq X \rightsquigarrow X$ an m-dissipative operator, I a nonempty and open interval, let $K \subseteq \overline{D(A)}$ be nonempty and let $f : I \times K \to X$ be a Carathéodory function which is Carathéodory positively sublinear*[3]*. If K is closed and $I \times K$ is C^0-viable with respect to $A + f$, then each C^0-solution of* (12.5.1) *can be continued up to a global one, i.e., defined on* $[\tau, \sup I)$.

Proof. Since $I \times K$ is C^0-viable with respect to $A + f$, it follows that, for each $(\tau, \xi) \in I \times K$, there exists at least one noncontinuable C^0-solution $u : [\tau, T) \to K$ to (12.5.1). We will show that $T = \sup I$. To this aim, let us assume the contrary, i.e., that $T < \sup I$. In particular this means that $T < +\infty$. By using a translation argument if necessary, we may assume with no loss of generality that $0 \in D(A)$ and $0 \in A0$. From (1.6.2) with $\eta = 0$, $g \equiv 0$ and $v \equiv 0$, we get

$$\|u(t)\| \leq \|\xi\| + \int_{E_t} [\,u(s), f(s, u(s))\,]_+ \, ds + \int_{H_t \setminus G_t} [\,u(s), f(s, u(s))\,]_+ \, ds$$

for each $t \in [\tau, T)$, where

$$\begin{aligned} E_t &= \{s \in [\tau, t];\ [\,u(s), f(s, u(s))\,]_+ > 0 \text{ and } \|u(s)\| > c(s)\}, \\ G_t &= \{s \in [\tau, t];\ [\,u(s), f(s, u(s))\,]_+ \leq 0\}, \\ H_t &= \{s \in [\tau, t];\ \|u(s)\| \leq c(s)\}. \end{aligned}$$

As $H_t \subseteq H_T$ and $[\,u, v\,]_+ \leq \|v\|$ for each $u, v \in X$, we get

$$\|u(t)\| \leq \|\xi\| + \int_{E_t} [a(s)\|u(s)\| + b(s)] \, ds + \int_{H_T} \|f(s, u(s))\| \, ds$$

for each $t \in [\tau, T)$. But f is a Carathéodory function and therefore there exists $\ell \in L^1_{\rm loc}(I)$ such that $\|f(s, u(s))\| \leq \ell(s)$ for a.a. $s \in H_T$. See (C_3) in Definition 2.8.1. Hence

$$\|u(t)\| \leq \|\xi\| + \int_\tau^T \ell(s) \, ds + \int_\tau^T b(s) \, ds + \int_\tau^t a(s)\|u(s)\| \, ds,$$

for each $t \in [\tau, T)$. By the Gronwall Lemma 1.8.4, u is bounded on $[\tau, T)$.

---

[3]See Definition 5.5.1.

Using once again the fact that f is Carathéodory, we deduce that $f(\cdot, u(\cdot))$ is bounded on $[\tau, T)$ by a function in $L^1(\tau, T)$ and so, there exists $\lim_{t\uparrow T} u(t) = u^*$. Since K is closed and $T < \sup I$, it follows that $(T, u^*) \in I \times K$. From this observation, recalling that $I \times K$ is C^0-viable with respect to $A + f$, we conclude that u can be continued to the right of T. But this is absurd, because u is noncontinuable. This contradiction can be eliminated only if $T = \sup I$, and this completes the proof. □

CHAPTER 13

Applications

Here we collect several applications illustrating the effectiveness of the abstract developed theory. We begin with a sufficient condition for a set K, which is invariant with respect to the infinitesimal generator, A, of a C_0-semigroup, to be viable with respect to $A + f$ with $f : K \to X$ continuous. From this, we deduce the existence of orthogonal solutions of a first-order system of partial differential equations of hyperbolic type. Further, we deduce a necessary and sufficient condition in order that a first-order partial differential equation of hyperbolic type have a unique solution taking values in a certain closed subset in $\mathbb{R}$. Next, using viability techniques, we show how to get necessary and sufficient conditions for a pair of functions to be a Lyapunov pair for a semilinear evolution equation. We notice that the existence of such a pair implies the asymptotic stability of the null solution of the semilinear evolution equation in question. We continue with several comparison results: for a semilinear diffusion equation, for a semilinear pray-predator system, for a nonlinear diffusion inclusion and for a fully nonlinear reaction-diffusion system. We next prove a null controllability result for a class of semilinear evolution equations, and we conclude with an existence result for periodic solutions to a fully nonlinear evolution equation.

13.1. Viability in the first approximation

Let X be a Banach space and let $A : D(A) \subseteq X \to X$ be the infinitesimal generator of a C_0-semigroup, $\{S(t) : X \to X;\ t \geq 0\}$. Let K be a nonempty subset in X, invariant with respect to A, in the sense that $S(t)K \subseteq K$ for each $t \in \mathbb{R}_+$, and let $f : K \to X$ be a continuous function. Here, we intend to find appropriate sufficient conditions on f in order that K be mild viable with respect to $A + f$.

Lemma 13.1.1. *Let X be a Banach space, $A : D(A) \subseteq X \to X$ the infinitesimal generator of a C_0-semigroup, $\{S(t) : X \to X;\ t \geq 0\}$ and K a nonempty subset in X. Assume that K is invariant with respect to A, i.e., $S(t)K \subseteq K$ for each $t \in \mathbb{R}_+$. Then $\mathcal{T}_K(\xi) \subseteq \mathcal{T}_K^A(\xi)$ for each $\xi \in K$.*

Proof. Let $\eta \in \mathcal{T}_K(\xi)$. By Proposition 8.1.1, it suffices to check that

$$\liminf_{h\downarrow 0} \frac{1}{h}\text{dist}\,(S(h)\xi + hS(h)\eta; K) = 0.$$

Let $M \geq 1$ and $a \in \mathbb{R}$ be given by Theorem 1.4.1, i.e., $\|S(t)\| \leq Me^{at}$ for each $t \geq 0$. Since $S(t)K \subseteq K$ for each $t \geq 0$, we have

$$\text{dist}\,(S(h)\xi + hS(h)\eta; K)$$

$$\leq \text{dist}\,(S(h)\xi + hS(h)\eta; S(h)K) \leq Me^{ah}\text{dist}\,(\xi + h\eta; K).$$

Thus

$$\liminf_{h\downarrow 0} \frac{1}{h}\text{dist}\,(S(h)\xi + hS(h)\eta; K) \leq \liminf_{h\downarrow 0} \frac{1}{h}Me^{ah}\text{dist}\,(\xi + h\eta; K) = 0$$

and this completes the proof. □

Problem 13.1.1. *Show that, if in Lemma* 13.1.1, *instead of a C_0-semigroup, we consider a C_0-group of isometries, $\{G(t) : X \to X;\ t \in \mathbb{R}\}$, satisfying $G(t)K = K$ for each $t \in \mathbb{R}$, then*

$$\mathcal{T}_K(\xi) = \mathcal{T}_K^A(\xi).$$

Theorem 13.1.1. *Let X be a Banach space, $A : D(A) \subseteq X \to X$ the infinitesimal generator of a C_0-semigroup, $\{S(t) : X \to X;\ t \geq 0\}$, K a nonempty and locally closed subset in X, and $f : K \to X$ a continuous function such that $A + f$ is locally of compact type. If $S(t)K \subseteq K$ for each $t \geq 0$ and*

$$f(\xi) \in \mathcal{T}_K(\xi) \tag{13.1.1}$$

for each $\xi \in K$, then K is mild viable with respect to $A + f$.

Proof. The conclusion follows from Lemma 13.1.1 and Theorem 8.2.1. The proof is complete. □

Theorem 13.1.2. *Let X be a Banach space, $A : D(A) \subseteq X \to X$ the infinitesimal generator of a C_0-group of isometries $\{G(t) : X \to X;\ t \in \mathbb{R}\}$, K a nonempty and locally closed subset in X, and $f : K \to X$ a continuous function such that $A + f$ is locally of compact type*[1]. *If $G(t)K = K$ for each $t \in \mathbb{R}$, then a necessary and sufficient condition in order that K be mild viable with respect to $A + f$ is* (13.1.1).

Proof. The conclusion follows from Problem 13.1.1 and Theorem 8.2.1. □

[1]See Definition 8.2.1.

Problem 13.1.2. *Prove that a locally closed set K is invariant with respect to the C_0-semigroup $\{S(t) : X \to X\,;\ t \geq 0\}$ if and only if $0 \in \mathcal{T}_K^A(\xi)$ for each $\xi \in K$.*

Example 13.1.1. As a first application of Theorem 13.1.2, we prove a sufficient condition for a first-order nonlinear hyperbolic system to have solutions with orthogonal components.

Let $f_1 : \mathbb{R} \times \mathbb{R} \to \mathbb{R}$, $f_2 : \mathbb{R} \times \mathbb{R} \to \mathbb{R}$ and let $\xi, \eta \in L^2(\mathbb{R}^n)$. Let $a \in \mathbb{R}^n$ and let us consider the hyperbolic system[2]

$$\begin{cases} u_t = a\nabla v + f_1(u,v) \\ v_t = a\nabla u + f_2(u,v) \\ u(0,x) = \xi(x) \\ v(0,x) = \eta(x). \end{cases} \tag{13.1.2}$$

We are looking for mild solutions, $(u,v) : [\,0,T\,] \to L^2(\mathbb{R}^n) \times L^2(\mathbb{R}^n)$, of (13.1.2), satisfying

$$\langle u(t,\cdot), v(t,\cdot)\rangle = 0 \tag{13.1.3}$$

for each $t \in [\,0,T\,]$, whenever the initial datum, $(\xi,\eta) \in X$, satisfies

$$\langle \xi, \eta\rangle = 0. \tag{13.1.4}$$

Here and thereafter, $\langle\cdot,\cdot\rangle$ denotes the inner product on $L^2(\mathbb{R}^n)$, i.e.,

$$\langle u, v\rangle = \int_{\mathbb{R}^n} u(x)v(x)\,dx$$

for each $u, v \in L^2(\mathbb{R}^n)$. Roughly speaking, by a *mild solution* of (13.1.2) we mean a mild solution of the Cauchy problem for the abstract semilinear evolution equation

$$\begin{cases} u'(t) = Au(t) + f(u(t)) \\ u(0) = \xi, \end{cases} \tag{13.1.5}$$

where $X = L^2(\mathbb{R}^n) \times L^2(\mathbb{R}^n)$, $A : D(A) \subseteq X \to X$ is defined by

$$\begin{cases} D(A) = \{(u,v) \in X;\ (a\nabla v, a\nabla u) \in X\} \\ A(u,v) = (a\nabla v, a\nabla u) \ \text{ for all } \ (u,v) \in D(A), \end{cases} \tag{13.1.6}$$

$f : X \to X$ is given by

$$f(u,v)(x) = (f_1(u(x),v(x)), f_2(u(x),v(x))), \tag{13.1.7}$$

for each $(u,v) \in X$ and a.e. for $x \in \mathbb{R}^n$.

---

[2]Throughout, if $w : \mathbb{R}^n \to \mathbb{R}$ is continuous, we denote by, $a\nabla w = \sum_{i=1}^{n} a_i \frac{\partial w}{\partial x_i}$ in the sense of distributions over $\mathbb{R}^n$.

On X we consider the usual Hilbert space norm

$$\|(u,v)\| = \sqrt{\langle u,u\rangle + \langle v,v\rangle},$$

for each $(u,v) \in X$.

Theorem 13.1.3. *Let $f_i : \mathbb{R} \times \mathbb{R} \to \mathbb{R}$, $i = 1,2$, be globally Lipschitz. Then, a necessary and sufficient condition in order that for each initial datum $(\xi,\eta) \in L^2(\mathbb{R}^n) \times L^2(\mathbb{R}^n)$, satisfying* (13.1.4), *to exist a unique mild solution $(u,v) : \mathbb{R}_+ \to L^2(\mathbb{R}^n) \times L^2(\mathbb{R}^n)$ of* (13.1.2), *satisfying* (13.1.3) *for each $t \in \mathbb{R}_+$, is*

$$\langle \xi, f_2(\xi,\eta)\rangle + \langle \eta, f_1(\xi,\eta)\rangle = 0, \tag{13.1.8}$$

for each $(\xi,\eta) \in L^2(\mathbb{R}^n) \times L^2(\mathbb{R}^n)$ satisfying (13.1.4).

Proof. We will apply Theorem 13.1.3 as follows. First, let us observe that the linear operator A, defined by (13.1.6), generates a C_0-group of isometries, $\{G(t) : X \to X;\ t \in \mathbb{R}\}$, given by

$$[G(t)(u,v)](x) = \frac{1}{2}\begin{pmatrix} u(x+ta) + u(x-ta) + v(x+ta) - v(x-ta) \\ u(x+ta) - u(x-ta) + v(x+ta) + v(x-ta) \end{pmatrix}^{\mathcal{T}},$$

where $B^{\mathcal{T}}$ denotes the transpose of the matrix B. Second, since f_i, $i = 1,2$, are globally Lipschitz, the function $f : X \to X$, given by (13.1.7), is well-defined and globally Lipschitz on X.

Next, let us define

$$K = \{(\xi,\eta) \in X;\ \xi \text{ and } \eta \text{ satisfy } (13.1.4)\}$$

and let us remark that K is nonempty and closed in X. Moreover, a simple computational argument based on the fact that the Lebesgue measure on $\mathbb{R}^n$ is translation invariant, shows that $G(t)K = K$ for each $t \in \mathbb{R}$. Thanks to Theorem 13.1.2, K is mild viable with respect to $A + f$ if and only if $f(\xi,\eta) \in \mathcal{T}_K(\xi,\eta)$ for each $(\xi,\eta) \in K$. By virtue of Corollary 2.4.1, the last condition is equivalent to the existence of two sequences, $(h_n)_n$ in $\mathbb{R}_+$ and $((p_n,q_n))_n$ in X, with $h_n \downarrow 0$, $\lim_n (p_n,q_n) = (0,0)$ and such that

$$(\xi,\eta) + h_n(f_1(\xi,\eta), f_2(\xi,\eta)) + h_n(p_n,q_n) \in K$$

for $n = 1,2,\ldots$. Equivalently,

$$\langle \xi + h_n f_1(\xi,\eta) + h_n p_n, \eta + h_n f_2(\xi,\eta) + h_n q_n\rangle = 0$$

for $n = 1,2,\ldots$. A simple calculation using the fact that $\langle \xi,\eta\rangle = 0$, $h_n \downarrow 0$ and $\lim_n p_n = \lim_n q_n = 0$, shows that the last relation is equivalent to (13.1.8), and this completes the proof. □

Example 13.1.2. Let $\Sigma \subseteq \mathbb{R}$ be nonempty and closed, $a \in \mathbb{R}^n$, $n \geq 1$, $g : \Sigma \to \mathbb{R}$ a continuous function and $\eta : \mathbb{R}^n \to \Sigma$ a bounded and uniformly continuous function. We consider the initial-value problem for the *transport equation in* $\mathbb{R}^n$,

$$\begin{cases} u_t = a\nabla u + g(u) & (t,x) \in [\,0,T\,] \times \mathbb{R}^n \\ u(0,x) = \eta(x) & x \in \mathbb{R}^n, \end{cases} \tag{13.1.9}$$

and we are interested in finding sufficient conditions for (13.1.9) to have at least one mild solution $u : [\,0,T\,] \to C_{ub}(\mathbb{R}^n)$ satisfying $u(t,x) \in \Sigma$ for all $(t,x) \in [\,0,T\,] \times \mathbb{R}^n$. Here $C_{ub}(\mathbb{R}^n)$ denotes the space of all uniformly continuous and bounded functions from $\mathbb{R}^n$ to $\mathbb{R}$, endowed with the sup-norm. Finally, by a *mild solution* of (13.1.9), we understand a mild solution of (13.1.5), with $X = C_{ub}(\mathbb{R}^n)$, $A : D(A) \subseteq X \to X$ given by

$$\begin{cases} D(A) = \{u \in X;\ a\nabla u \in X\} \\ Au = a\nabla u \ \text{ for each } u \in D(A), \end{cases}$$

$$K = \{\xi \in X;\ \xi(x) \in \Sigma \ \text{ for all } \ x \in \mathbb{R}^n\}$$

and $f : K \to X$ defined by $f(u)(x) = g(u(x))$ for each $u \in K$ and each $x \in \mathbb{R}$.

The main result concerning (13.1.9) is

Theorem 13.1.4. *Let $g : \Sigma \to \mathbb{R}$ be a continuous function which is globally Lipschitz on Σ. Assume that, for each $\xi \in \Sigma$, we have*

$$\liminf_{h \downarrow 0} \frac{1}{h} \operatorname{dist}(\xi + hg(\xi); \Sigma) = 0. \tag{13.1.10}$$

Then, for each $\eta \in K$ and each $T > 0$, the problem (13.1.9) *has a unique mild solution $u : [\,0,T\,] \to K$.*

Proof. We will show that we are in the hypotheses of Theorem 13.1.2. First, let us observe that A is the infinitesimal generator of a C_0-group of isometries $\{G(t) : X \to X;\ t \in \mathbb{R}\}$ defined by

$$[G(t)\xi(\cdot)](x) = \xi(x + ta)$$

for each $\xi \in X$. Since Σ is nonempty and closed, it follows that K is nonempty and closed too. It is a simple exercise to verify that f is globally Lipschitz on K and that $G(t)K = K$ for each $t \in \mathbb{R}$. In order to be able to use Theorem 13.1.2, we only have to verify the tangency condition (13.1.1). We will proceed indirectly. Namely, we will prove that K is viable with respect to f and then we will apply Theorem 3.2.3. To prove that K is viable with respect to f, let $\eta \in K$ and let $x \in \mathbb{R}^n$. Since g satisfies the tangency condition (13.1.10), for each $x \in \mathbb{R}^n$, the problem

$$\begin{cases} v'(t,x) = g(v(t,x)) \\ v(0)(x) = \eta(x) \end{cases}$$

has a unique noncontinuable solution $v(\cdot, x) : [0, T(x)) \to \Sigma$. Since g is globally Lipschitz on Σ, it maps bounded subsets in Σ into bounded subsets in $\mathbb{R}$ and is positively sublinear. See Definition 3.6.1 and Remark 3.6.2. By Theorem 3.6.3 it follows that $T(x) = \infty$ for each $x \in \mathbb{R}^n$.

Now, using once again the fact that g is globally Lipschitz on Σ, we deduce that

$$|v(t,x) - v(t,y)| \leq |\eta(x) - \eta(y)| + \int_0^t L|v(s,x) - v(s,y)|\,ds$$

for each $t \in [0,T]$ and $x, y \in \mathbb{R}^n$. Let $T > 0$ be arbitrary but fixed. By virtue of the Gronwall Lemma 1.8.4, we have

$$|v(t,x) - v(t,y)| \leq e^{TL}|\eta(x) - \eta(y)|$$

for each $x, y \in \mathbb{R}^n$. Therefore $v(t,\cdot)$ is uniformly continuous and bounded, i.e., $v(t,\cdot) \in X$. But $v(t,x) \in \Sigma$ for each $(t,x) \in [0,T] \times \mathbb{R}^n$ and consequently $v(t,\cdot) \in K$ for each $t \in [\tau, T]$. Hence, the function $u : [0,T] \to K$, $u(t)(x) = v(t,x)$, is a C^1-solution of the Cauchy problem

$$\begin{cases} u'(t) = f(u(t)) \\ u(0) = \eta. \end{cases}$$

Therefore, K is viable with respect to f and, in view of Theorem 3.2.3, it satisfies the tangency condition (13.1.1). Consequently, all the hypotheses of Theorem 13.1.2 are satisfied, and this achieves the conclusion. □

13.2. Lyapunov pairs

Let X be a Banach space, let A be the infinitesimal generator of a C_0-semigroup, $\{S(t) : X \to X; t \geq 0\}$, and let $f : X \to X$ be a globally Lipschitz function. Given $\xi \in X$, the Cauchy problem

$$\begin{cases} u'(t) = Au(t) + f(u(t)) \\ u(0) = \xi \end{cases} \tag{13.2.1}$$

has a unique mild solution $u : [0, \infty) \to X$.

Definition 13.2.1. Let $V : X \to (-\infty, +\infty]$ be a *proper* function, i.e., a function V whose *effective domain*, $D(V) = \{\xi \in X;\ V(\xi) < +\infty\}$, is nonempty. The *A-contingent derivative* $\underline{D}V(\xi)(u)$ of V, at $\xi \in D(V)$ in the direction $u \in X$, is defined by

$$\underline{D}V(\xi)(u) = \liminf_{\substack{h\downarrow 0\\ w\to 0}} \frac{1}{h}\Big[V(S(h)\xi + h(u+w)) - V(\xi)\Big].$$

Definition 13.2.2. Let $V, g : X \to (-\infty, +\infty]$ be two lower semicontinuous functions. We say that (V, g) is a *Lyapunov pair* for the problem (13.2.1) if

$$V(u(t)) + \int_0^t g(u(s))ds \le V(\xi), \tag{13.2.2}$$

for each $\xi \in X$ and each $t \ge 0$, where, as already mentioned, $u : [0, \infty) \to X$ denotes the unique mild solution of (13.2.1).

Let us define $\mathcal{A} : D(\mathcal{A}) \subseteq X \times \mathbb{R} \to X \times \mathbb{R}$ by $D(\mathcal{A}) = D(A) \times \mathbb{R}$ and $\mathcal{A}(\xi, t) = (A\xi, 0)$ for each $(\xi, t) \in D(\mathcal{A})$. Recalling that

$$\mathrm{epi}(V) = \{(x, t) \in X \times \mathbb{R}; V(x) \le t, x \in D(V)\},$$

it is easy to see that $\mathcal{T}^{\mathcal{A}}_{\mathrm{epi}(V)}(\xi, \mu)$ coincides with the set of all $(u, \lambda) \in X \times \mathbb{R}$ with the property that there exist three sequences $(h_n)_n$ and $(\theta_n)_n$ in $\mathbb{R}$, $(w_n)_n$ in X with $h_n \downarrow 0$, $\lim_n w_n = 0$, $\lim_n \theta_n = 0$ such that

$$(S(h_n)\xi + h_n(u + w_n), \mu + h_n(\lambda + \theta_n)) \in \mathrm{epi}(V),$$

for $n = 1, 2, \ldots$. Notice that $\mathcal{T}^{\mathcal{A}}_{\mathrm{epi}(V)}(\xi, \mu)$ is the set of $\mathcal{A}$-tangent vectors to $\mathrm{epi}(V)$, as introduced in Definition 8.1.3.

The following result describes the set of all $\mathcal{A}$-tangent vectors to the epigraph at one of its points.

Lemma 13.2.1. *Let $V : X \to (-\infty, +\infty]$ be a function and $\xi \in D(V)$. Then*

$$\mathcal{T}^{\mathcal{A}}_{\mathrm{epi}(V)}(\xi, V(\xi)) = \bigcap_{V(\xi)\le\mu} \mathcal{T}^{\mathcal{A}}_{\mathrm{epi}(V)}(\xi, \mu).$$

Problem 13.2.1. *Prove Lemma* 13.2.1.

Furthermore, we have the following geometric property.

Lemma 13.2.2. *For all $\xi \in D(V)$, we have*

$$\mathrm{epi}(\underline{D}V(\xi)) = \mathcal{T}^{\mathcal{A}}_{\mathrm{epi}(V)}(\xi, V(\xi)).$$

Proof. Let $(u, \lambda) \in \mathrm{epi}(\underline{D}V(\xi))$. In view of Definition 13.2.1, this is equivalent to

$$\liminf_{\substack{h\downarrow 0\\ w\to 0}} \frac{1}{h}\Big[V(S(h)\xi + h(u+w)) - V(\xi)\Big] \le \lambda.$$

The above inequality ensures that there exist $(h_n)_n$ in $\mathbb{R}$ and $(w_n)_n$ in X such that $h_n \downarrow 0$, $\lim_n w_n = 0$ and

$$V(S(h_n)\xi + h_n(u + w_n)) \leq V(\xi) + h_n\left(\lambda + \frac{1}{n}\right),$$

for $n = 1, 2, \ldots$. This means that $(u, \lambda) \in \mathcal{T}^{\mathcal{A}}_{\mathrm{epi}(V)}(\xi, V(\xi))$.

Conversely, let $(u, \lambda) \in \mathcal{T}^{\mathcal{A}}_{\mathrm{epi}(V)}(\xi, V(\xi))$. Then there exist $(h_n)_n$ and $(\theta_n)_n$ in $\mathbb{R}$, $(w_n)_n$ in X with $h_n \downarrow 0$, $\lim_n \theta_n = 0$, and $\lim_n w_n = 0$ satisfying

$$\frac{1}{h_n}\left(V(S(h_n)\xi + h_n(u + w_n)) - V(\xi)\right) \leq \lambda + \theta_n,$$

for $n = 1, 2, \ldots$.

Passing to the limit, we obtain

$$\limsup_n \frac{1}{h_n}\left(V(S(h_n)\xi + h_n(u + w_n)) - V(\xi)\right) \leq \lambda\,,$$

which shows that $(u, \lambda) \in \mathrm{epi}(\underline{D}V(\xi))$. See Definition 13.2.1. □

Before stating the main result of this section, let us present an auxiliary result providing suitable approximations of a lower semicontinuous function by locally Lipschitz functions.

Lemma 13.2.3. *Let X be a Banach space and let $g : X \to (-\infty, +\infty]$ be a proper, lower semicontinuous function which satisfies the unilateral growth condition*

$$g(x) \geq -C(1 + \|x\|^p), \tag{13.2.3}$$

for some constants $C > 0$ and $p \geq 1$, and for every $x \in X$. Then there exists a sequence $(g_n)_n$ of functions, $g_n : X \to \mathbb{R}$, such that every g_n is Lipschitz continuous on bounded subsets of X and $g_n \uparrow g$ pointwise on X as $n \to \infty$.

Problem 13.2.2. *Give a proof of Lemma* 13.2.3 *by using the approximation*

$$g_n(x) = \inf_{y \in X}\{g(y) + n\|x - y\|^p\},$$

for $n = 1, 2, \ldots$ and each $x \in X$.

Let us now formulate a characterization of a Lyapunov pair for (13.2.1) by using the A-contingent derivative introduced in Definition 13.2.1.

Theorem 13.2.1. *Let X be a Banach space, let $V : X \to (-\infty, +\infty]$ be a proper, lower semicontinuous function and let $g : X \to (-\infty, +\infty]$ be a* *proper, lower semicontinuous function which satisfies* (13.2.3). *Then* (V, g) *is a Lyapunov pair for problem* (13.2.1) *if and only if*

$$\underline{D}V(\xi)(f(\xi)) + g(\xi) \le 0, \tag{13.2.4}$$

for all $\xi \in D(V)$.

Proof. Let us first notice that $K = \mathrm{epi}(V)$ is closed in $X \times \mathbb{R}$ simply because $V : X \to (-\infty, +\infty]$ is lower semicontinuous. Corresponding to the lower semicontinuous function $g : X \to (-\infty, +\infty]$, Lemma 13.2.3 provides a sequence of functions $g_n : X \to \mathbb{R}$ with the properties given therein. In particular, due to the pointwise convergence $g_n \uparrow g$, one may easily see that (13.2.4) is equivalent to

$$\underline{D}V(\xi)(f(\xi)) + g_n(\xi) \le 0,$$

for all $\xi \in D(V)$ and $n = 1, 2, \ldots$. By Lemmas 13.2.1 and 13.2.2, we deduce that the given inequality is satisfied if and only if

$$(f(\xi), -g_n(\xi)) \in \mathcal{T}^{\mathcal{A}}_{\mathrm{epi}(V)}(\xi, \mu), \tag{13.2.5}$$

for all $(\xi, \mu) \in \mathrm{epi}(V)$ and $n = 1, 2, \ldots$. We now apply Theorem 8.2.6 on the space $X \times \mathbb{R}$, with the subset $K = \mathrm{epi}(V)$ and the evolution equation

$$(u'(t), z'(t)) = \mathcal{A}(u(t), z(t)) + (f(u(t)), -g_n(u(t))) \tag{13.2.6}$$

on $X \times \mathbb{R}$. We conclude that (13.2.5) holds if and only if for every $\xi \in D(V)$ there exists $T_\xi > 0$ such that

$$V(u(t)) \le V(\xi) - \int_0^t g_n(u(s))ds, \tag{13.2.7}$$

for all $n = 1, 2, \ldots$ and each $t \in [0, T_\xi]$, where $u(\cdot)$ is the solution of (13.2.1). Due to the special form of the differential equation in (13.2.6), we can choose $T_\xi = T$, independent of $n = 1, 2, \ldots$. Passing to the limit as $n \to \infty$ in (13.2.7) and taking into account the properties of the sequence $(g_n)_n$ given by Lemma 13.2.3, we conclude that

$$V(u(t)) + \int_0^t g(u(s))ds \le V(\xi),$$

for all $t \in [0, T]$. Then, a standard continuation argument shows that (13.2.2) holds true for all $t \ge 0$. This completes the proof. □

Problem 13.2.3. *Let* $\{S(t) : X \to X; t \ge 0\}$ *be a semigroup of contractions. Let* $r > 0$ *and consider the Lipschitz function* $f : X \to X$ *given by* $f(x) = -x/r$ *if* $\|x\| \le r$ *and* $f(x) = -x/\|x\|$ *if* $\|x\| \ge r$. *Show that, for every* $\xi \in X$ *with* $\|\xi\| > r$, *the solution of* (13.2.1) *reaches the closed ball* $D(0, r)$ *in a finite time which is less or equal than* $\|\xi\|$.

13.3. A comparison result for a semilinear diffusion equation

Throughout, by a *domain* we mean a nonempty, open and connected subset in $\mathbb{R}^n$. So, let $\Omega \subseteq \mathbb{R}^n$, $n = 1, 2, \ldots$ be a bounded domain with C^2 boundary Γ, and let us consider the Cauchy problem for the semilinear diffusion equation

$$\begin{cases} u_t = \Delta u + f(t, x, u) & \text{in } Q_{\tau,T} \\ u = 0 & \text{on } \Sigma_{\tau,T} \\ u(\tau, x) = \xi(x) & \text{in } \Omega, \end{cases} \tag{13.3.1}$$

where, if $0 \le \tau < T \le \infty$, $Q_{\tau,T} = (\tau, T) \times \Omega$, $\Sigma_{\tau,T} = (\tau, T) \times \Gamma$, Δ is the usual Laplace operator, i.e., $\Delta u = \sum_{i=1}^{n} \frac{\partial^2 u}{\partial x_i^2}$, $f : \mathbb{R}_+ \times \overline{\Omega} \times \mathbb{R} \to \mathbb{R}$ is continuous and bounded[3], and $\xi \in L^2(\Omega)$, $\xi(x) \ge 0$ a.e. for $x \in \Omega$.

Definition 13.3.1. A *mild solution* of the problem (13.3.1), on $[\tau, T]$, is a function $u : [\tau, T] \to L^2(\Omega)$ with $t \mapsto f(t, \cdot, u(t, \cdot)) \in L^1(\tau, T; L^2(\Omega))$ and such that u is a mild solution in the sense of Definition 1.5.3 of the problem

$$\begin{cases} u'(t) = Au(t) + f_0(t) \\ u(\tau) = \xi, \end{cases}$$

where A is the Laplace operator Δ with homogeneous Dirichlet boundary conditions on $L^2(\Omega)$ as in Theorem 1.7.2 and $f_0(t)(x) = f(t, x, u(t, x))$ a.e. for $t \in [\tau, T]$ and $x \in \Omega$. By a *mild solution* of the problem (13.3.1), on $[\tau, \widetilde{T})$, $\tau < \widetilde{T} \le \infty$, we mean a function $u \in C([\tau, \widetilde{T}); L^2(\Omega))$ such that for each $\tau < T < \widetilde{T}$, u is a mild solution of (13.3.1), on $[\tau, T]$ in the sense stated above.

Let $g : \mathbb{R}_+ \times \overline{\Omega} \times \mathbb{R} \to \mathbb{R}$ be continuous, bounded and satisfying

$$f(t, x, u) \le g(t, x, u)$$

for each $(t, x, u) \in \mathbb{R}_+ \times \overline{\Omega} \times \mathbb{R}$. In this section we will prove that, under some appropriate conditions on g, for each $u_0 \in L^2(\Omega)$, with $u_0(x) \ge 0$ a.e. for $x \in \Omega$, and for every global mild solution $\widetilde{u} : \mathbb{R}_+ \times \overline{\Omega} \to \mathbb{R}_+$ of the semilinear diffusion equation

$$\begin{cases} u_t = \Delta u + g(t, x, u) & \text{in } Q_{0,\infty} \\ u = 0 & \text{on } \Sigma_{0,\infty} \\ u(0, x) = u_0(x) & \text{in } \Omega, \end{cases} \tag{13.3.2}$$

[3] In fact, we may merely assume that f is only continuous, and has sublinear growth with respect to the last argument.

the semilinear diffusion equation (13.3.1) has at least one global mild solution which, for each $t \in [\tau, \infty)$, lies almost everywhere "between" 0 and $\widetilde{u}(t, \cdot)$, provided $0 \leq \xi(x) \leq \widetilde{u}(\tau, x)$ a.e. for $x \in \Omega$.

More precisely, since g is continuous and, in view of Theorem 1.7.2, the Laplace operator with homogeneous boundary conditions in $L^2(\Omega)$, i.e., Δ, generates a compact C_0-semigroup, the problem (13.3.2) has at least one noncontinuable mild solution $\widetilde{u} : [0, T_m) \to L^2(\Omega)$. Since g is bounded, it readily follows that $T_m = \infty$. Let $\mathcal{C} \subseteq \mathbb{R} \times L^2(\Omega)$ be the infinite tube defined by

$$\mathcal{C} = \left\{(t, u) \in \mathbb{R}_+ \times L^2(\Omega);\ 0 \leq u(x) \leq \widetilde{u}(t, x) \text{ a.e. for } x \in \Omega\right\}. \tag{13.3.3}$$

So, our goal is to show that, for each $(\tau, \xi) \in \mathcal{C}$, the Cauchy problem (13.3.1) has at least one mild solution $u : [\tau, \infty) \to L^2(\Omega)$ with $(t, u(t)) \in \mathcal{C}$ for each $t \in [\tau, \infty)$. Namely, we will prove

Theorem 13.3.1. *Let $\Omega \subseteq \mathbb{R}^n$, $n = 1, 2, \ldots$, be a bounded domain with C^2 boundary Γ, let $f, g : \mathbb{R}_+ \times \overline{\Omega} \times \mathbb{R} \to \mathbb{R}_+$ be continuous, with g bounded, nondecreasing with respect to its last argument, and $f(t, x, u) \leq g(t, x, u)$ for each $(t, x, u) \in \mathbb{R} \times \overline{\Omega} \times \mathbb{R}$. Let $u_0 \in L^2(\Omega)$, $u_0(x) \geq 0$ a.e. for $x \in \Omega$ and let $\mathcal{C}$ be defined by* (13.3.3)*, where $\widetilde{u} : \mathbb{R}_+ \times \Omega \to \mathbb{R}_+$ is a global solution of* (13.3.2)*. Then, for each $(\tau, \xi) \in \mathcal{C}$ there exists at least one global solution $u : [\tau, \infty) \times \Omega \to \mathbb{R}_+$ of* (13.3.1) *satisfying for each $\tau < \delta < T$:*

(i) $u \in C([\tau, T]; L^2(\Omega)) \cap L^2(\delta, T; H^2(\Omega)) \cap W^{1,1}(\delta, T; H_0^1(\Omega))$;
(ii) *for each $t \in [\tau, \infty)$, we have $(t, u(t)) \in \mathcal{C}$.*

Proof. Since (i) follows from a classical regularity result in the theory of parabolic equations — see Vrabie [**175**], Theorem 11.6.1, p. 265 —, it remains to prove (ii). In other words, we have to show first that $\mathcal{C}$ is mild viable, in $X = L^2(\Omega)$, with respect to $\Delta + f$ and second that every mild solution $u : [\tau, T) \to L^2(\Omega)$, with $(t, u(t)) \in \mathcal{C}$ for each $t \in [\tau, T)$, can be extended to a global one satisfying the very same constraint. We will first make use of Theorem 8.5.4 and second of Theorem 8.8.2. To this aim, we denote by $X = L^2(\Omega)$, and we rewrite both (13.3.1) and (13.3.2) as evolution equations in X.

First, let us define $A : D(A) \subseteq X \to X$ by

$$\begin{cases} D(A) = H_0^1(\Omega) \cap H^2(\Omega) \\ Au = \Delta u \quad \text{for } u \in D(A), \end{cases}$$

and $F, G : \mathbb{R} \times X \to X$ by

$$F(t, u)(x) = f(t, x, u(x)) \quad G(t, u)(x) = g(t, x, u(x))$$

for each $t \in \mathbb{R}$, each $u \in X$ and a.e. for $x \in \Omega$. Since both f and g are continuous and bounded, both F and G are well-defined, continuous and bounded. With the notations above, the problem (13.3.1) can be rewritten as the abstract Cauchy problem for an evolution equation, i.e.,

$$\begin{cases} u'(t) = Au(t) + F(t, u(t)) \\ u(\tau) = \xi. \end{cases} \tag{13.3.4}$$

Similarly, (13.3.2) can be rewritten as

$$\begin{cases} u'(t) = Au(t) + G(t, u(t)) \\ u(0) = u_0. \end{cases} \tag{13.3.5}$$

At this point, let us recall that A generates a compact C_0-semigroup. See Theorem 1.7.2. Furthermore, $\mathcal{C}$ is nonempty and locally closed, and F is continuous. Since both ξ and f are nonnegative, in view of Theorem 1.7.5, all solutions of (13.3.1) are a fortiori nonnegative. Therefore, in order to check the viability of $\mathcal{C}$ with respect to $A + F$ it suffices to show that the set

$$\widetilde{\mathcal{C}} = \left\{(t, u) \in \mathbb{R}_+ \times L^2(\Omega);\ u(x) \leq \widetilde{u}(t, x) \text{ a.e. for } x \in \Omega\right\},$$

which obviously is closed, is mild viable with respect to $A+F$. So, in order to apply Theorem 8.5.4 it suffices to verify the tangency condition

$$\liminf_{h \downarrow 0} \frac{1}{h} \operatorname{dist}\left((\tau, S(h)\xi) + h(1, F(\tau, \xi)); \widetilde{\mathcal{C}}\right) = 0, \tag{13.3.6}$$

for each $(\tau, \xi) \in \widetilde{\mathcal{C}}$, where $\{S(t) : X \to X,\ t \geq 0\}$ is the C_0-semigroup of contractions generated by A. So, let $(\tau, \xi) \in \widetilde{\mathcal{C}}$ be arbitrary and let us observe that, in order to prove (13.3.6), it suffices to show that, for each $h > 0$, there exists $u_h \in L^2(\Omega)$ with $(\tau + h, u_h) \in \widetilde{\mathcal{C}}$ and

$$\liminf_{h \downarrow 0} \frac{1}{h} \|S(h)\xi + hF(\tau, \xi) - u_h\| = 0. \tag{13.3.7}$$

Let us define

$$u_h = S(h)\xi + \int_\tau^{\tau+h} S(\tau + h - s)F(s, \xi)\, ds$$

$$+ \int_\tau^{\tau+h} S(\tau + h - s)\big[\,G(s, \widetilde{u}(s)) - G(s, \widetilde{u}(\tau))\,\big]\, ds.$$

Since $\xi \leq \widetilde{u}(\tau)$, $F(s, \xi) \leq G(s, \xi)$ for each $s \in \mathbb{R}_+$, and $u \mapsto G(s, u)$ is nondecreasing, in view of Theorem 1.7.5, we get

$$u_h \leq S(h)\widetilde{u}(\tau) + \int_\tau^{\tau+h} S(\tau + h - s)G(s, \widetilde{u}(\tau))\, ds$$

$$+ \int_\tau^{\tau+h} S(\tau + h - s)\big[\,G(s, \widetilde{u}(s)) - G(s, \widetilde{u}(\tau))\,\big]\, ds = \widetilde{u}(\tau + h).$$

Therefore, $(\tau + h, u_h) \in \widetilde{\mathcal{C}}$. To complete the proof, it remains to check (13.3.7). Since G and $\widetilde{u}$ are continuous, we first observe that

$$\lim_{h\downarrow 0} \frac{1}{h} \int_{\tau}^{\tau+h} S(\tau + h - s)[\,G(s, \widetilde{u}(s)) - G(s, \widetilde{u}(\tau))\,]\, ds = 0.$$

Similarly, we have

$$\lim_{h\downarrow 0} \frac{1}{h} \left\| S(h)\xi + hF(\tau, \xi) - S(h)\xi - \int_{\tau}^{\tau+h} S(\tau + h - s)F(s, \xi)\, ds \right\| = 0,$$

which proves (13.3.7). Thus $\mathcal{C}$ is mild viable with respect to $A+F$. Since F is bounded, we are in the hypotheses of Theorem 8.8.2. So each mild solution $u : [\,\tau, T) \to L^2(\Omega)$ of (13.3.1) whose graph lies in $\mathcal{C}$ can be continued to a mild solution defined on $[\,\tau, \infty)$ and whose graph is in $\mathcal{C}$ too, and this completes the proof. □

13.4. A comparison result for a predator–pray system

Let $\Omega \subseteq \mathbb{R}^n$, $n = 1, 2, \ldots$, be a bounded domain with C^2 boundary Γ, let $\delta_i \geq 0$, $i = 1, 2$, $a > 0$, $r > 0$, let $f : \mathbb{R}_+ \times \mathbb{R} \to \mathbb{R}_+$ and $g : \mathbb{R}_+ \times \mathbb{R} \to \mathbb{R}_-$ be two continuous functions and let us consider the following general *predator-pray system*[4]

$$\begin{cases} u_t = \delta_1 \Delta u - au + f(u, v) & \text{in } Q_{\tau,\infty} \\ v_t = \delta_2 \Delta v + rv + g(u, v) & \text{in } Q_{\tau,\infty} \\ u = v = 0 & \text{on } \Sigma_{\tau,\infty} \\ u(\tau, x) = \xi(x) \quad v(\tau, x) = \eta(x) & \text{in } \Omega. \end{cases} \tag{13.4.1}$$

Here, for $0 \leq \tau < T \leq \infty$, we denote by $Q_{\tau,T} = (\tau, T) \times \Omega$, $\Sigma_{\tau,T} = (\tau, T) \times \Gamma$. Moreover, $\xi, \eta \in L^2(\Omega)$, $\xi(x) \geq 0$ and $\eta(x) \geq 0$ a.e. for $x \in \Omega$.

Let $\widetilde{f} : \mathbb{R}_+ \times \mathbb{R} \to \mathbb{R}_+$ and $\widetilde{g} : \mathbb{R}_+ \times \mathbb{R} \to \mathbb{R}_-$ be two continuous functions such that

$$\begin{cases} f(u, v) \leq \widetilde{f}(u, v) \\ g(u, v) \geq \widetilde{g}(u, v) \end{cases} \tag{13.4.2}$$

for each $(u, v) \in \mathbb{R} \times \mathbb{R}$. Let us consider also the comparison predator-pray system

$$\begin{cases} u_t = \delta_1 \Delta u - au + \widetilde{f}(u, v) & \text{in } Q_{0,\infty} \\ v_t = \delta_2 \Delta v + rv + \widetilde{g}(u, v) & \text{in } Q_{0,\infty} \\ u = v = 0 & \text{on } \Sigma_{0,\infty}, \\ u(0, x) = u_0(x), \quad v(0, x) = v_0(x) & \text{in } \Omega \end{cases} \tag{13.4.3}$$

---

[4]In order to simplify the notation, throughout this section, whenever no confusion may occur, we will write u, $\widetilde{u}$, v, $\widetilde{v}$, instead of $u(t, x)$, $\widetilde{u}(t, x)$, $v(t, x)$ and $\widetilde{v}(t, x)$ respectively.

and let $(\widetilde{u}, \widetilde{v}) : \mathbb{R}_+ \times \Omega \to \mathbb{R}_+ \times \mathbb{R}_+$ be a mild solution of (13.4.3).

In this section, by using the viability results presented in Section 8.6, we will prove a sufficient condition such that, for each $(\xi, \eta) \in L^2(\Omega) \times L^2(\Omega)$, with

$$\begin{cases} 0 \le \xi(x) \le \widetilde{u}(\tau, x) \\ \widetilde{v}(\tau, x) \le \eta(x) \end{cases} \tag{13.4.4}$$

a.e. for $x \in \Omega$, the predator-pray system (13.4.1) has at least one solution $(u, v) : \mathbb{R}_+ \times \Omega \to \mathbb{R}_+ \times \mathbb{R}_+$, such that, for each $t \in [\tau, \infty)$, we have

$$\begin{cases} 0 \le u(t, x) \le \widetilde{u}(t, x) \\ \widetilde{v}(t, x) \le v(t, x) \end{cases} \tag{13.4.5}$$

a.e. for $x \in \Omega$.

Let $\mathcal{C} \subseteq \mathbb{R} \times L^2(\Omega) \times L^2(\Omega)$ be defined by

$$\mathcal{C} = \left\{(t, u, v) \in \mathbb{R}_+ \times L^2(\Omega) \times L^2(\Omega);\ (u, v) \text{ satisfy (13.4.7) below}\right\} \tag{13.4.6}$$

$$\begin{cases} 0 \le u(x) \le \widetilde{u}(t, x) \\ \widetilde{v}(t, x) \le v(x) \end{cases} \tag{13.4.7}$$

a.e. for $x \in \Omega$. We intend to show that, for each $(\tau, \xi, \eta) \in \mathcal{C}$, the problem (13.4.1) has at least one mild solution $(u, v) : [\tau, \infty) \to L^2(\Omega) \times L^2(\Omega)$ with $(t, u(t), v(t)) \in \mathcal{C}$ for each $t \in [\tau, \infty)$. To this aim, let us assume that there exist the constants $c_i \ge 0$, $i = 1, \ldots, 5$, such that

$$\begin{cases} |\widetilde{f}(u, v)| \le c_1|u| + c_2 \\ |\widetilde{g}(u, v)| \le c_3|u| + c_4|v| + c_5 \end{cases} \tag{13.4.8}$$

for each $(u, v) \in \mathbb{R}_+ \times \mathbb{R}_+$. Now we are ready to prove

Theorem 13.4.1. *Let $\Omega \subseteq \mathbb{R}^n$, $n = 1, 2, \ldots$, be a bounded domain with C^2 boundary Γ, let $f : \mathbb{R} \times \mathbb{R} \to \mathbb{R}_+$ and $g : \mathbb{R} \times \mathbb{R} \to \mathbb{R}_-$ be continuous and let $\widetilde{f} : \mathbb{R} \times \mathbb{R} \to \mathbb{R}_+$ and $\widetilde{g} : \mathbb{R} \times \mathbb{R} \to \mathbb{R}_-$ be continuous and such that, for each $(u_0, v_0) \in \mathbb{R}_+ \times \mathbb{R}_+$, $u \mapsto \widetilde{f}(u, v_0)$ and $v \mapsto \widetilde{g}(u_0, v)$ are nondecreasing, $u \mapsto \widetilde{g}(u, v_0)$ and $v \mapsto \widetilde{f}(u_0, v)$ are nonincreasing and satisfy* (13.4.2) *and* (13.4.8). *Let $(u_0, v_0) \in L^2(\Omega) \times L^2(\Omega)$ with $u_0(x) \ge 0$ and $v_0(x) \ge 0$ a.e. for $x \in \Omega$ and let $(\widetilde{u}, \widetilde{v}) : \mathbb{R}_+ \to L^2(\Omega) \times L^2(\Omega)$ be a global mild solution*[5] *of* (13.4.3) *with $\widetilde{u} \ge 0$ for each $t \ge 0$ and a.e. for $x \in \Omega$. Let $\mathcal{C}$ be defined by* (13.4.6). *Then, for each $(\tau, \xi, \eta) \in \mathcal{C}$, the problem* (13.4.1) *has at least one global mild solution $(u, v) : [\tau, \infty) \to L^2(\Omega) \times L^2(\Omega)$ satisfying for each $\tau < \delta < T$:*

[5]In view of Definition 13.3.1 it is clear to see what a mild solution in this context ought to be. In our specific case, such a solution exists since both $\delta_1 \Delta - aI$ and $\delta_2 \Delta + rI$ generate compact C_0-semigroups and $\widetilde{f}$, $\widetilde{g}$ are continuous and have sublinear growth.

(i) $u, v \in C([\tau, T]; L^2(\Omega)) \cap L^2(\delta, T; H^2(\Omega)) \cap W^{1,1}(\delta, T; H_0^1(\Omega))$;
(ii) *for each* $t \in [\tau, \infty)$, *we have* $(t, u(t), v(t)) \in \mathcal{C}$.

Proof. We observe that (i) is a classical regularity result in the theory of parabolic equations in a $L^2(\Omega)$-setting. See Vrabie [**175**], Theorem 11.6.1, p. 265. It remains to prove (ii). So, we have to show first that $\mathcal{C}$ is mild viable with respect to $(\delta_1\Delta - aI + f, \delta_2\Delta + rI + g)$ and second that every mild solution $(u, v) : [\tau, T) \to L^2(\Omega) \times L^2(\Omega)$, satisfying $(t, u(t), v(t)) \in \mathcal{C}$ for each $t \in [\tau, T)$, can be extended to a global one obeying the very same constraints. To this aim, we will first make use of Theorem 8.6.4 and second of Theorem 8.8.2.

Let $\widetilde{\mathcal{C}} \subseteq \mathbb{R}_+ \times L^2(\Omega) \times L^2(\Omega)$ be defined by

$$\widetilde{\mathcal{C}} = \{(t, u, v) \in \mathbb{R}_+ \times L^2(\Omega) \times L^2(\Omega);\ (u, v) \text{ satisfy (13.4.10) below}\} \tag{13.4.9}$$

$$\begin{cases} u(x) \leq \widetilde{u}(t, x) \\ \widetilde{v}(t, x) \leq v(x) \end{cases} \tag{13.4.10}$$

a.e. for $x \in \Omega$. Since f and ξ are nonnegative, by Theorem 1.7.5, u is nonnegative and therefore, to prove that $\mathcal{C}$ is mild viable with respect to $(\delta_1\Delta - aI + f, \delta_2\Delta + rI + g)$, it suffices to show that $\widetilde{\mathcal{C}}$ is mild viable with respect to $(\delta_1\Delta - aI + f, \delta_2\Delta + rI + g)$.

Let us denote by $X = L^2(\Omega)$, and $\mathcal{X} = X \times X$ which, endowed with the usual norm $\|(u, v)\| = \|u\|_X + \|v\|_X$, is a Banach space. We rewrite (13.4.1) as an evolution system in X or, equivalently, as an evolution equation in $\mathcal{X}$. To this aim, let us define $A : D(A) \subseteq X \to X$ and $B : D(B) \subseteq X \to X$ by

$$\begin{cases} D(A) = H_0^1(\Omega) \cap H^2(\Omega) \\ Au = \delta_1\Delta u - au \quad \text{for } u \in D(A) \end{cases}$$

and respectively by

$$\begin{cases} D(B) = H_0^1(\Omega) \cap H^2(\Omega) \\ Bv = \delta_2\Delta v + rv \quad \text{for } v \in D(B). \end{cases}$$

Further, let us define $F : \mathcal{X} \to X$ and $G : \mathcal{X} \to X$ by

$$F(u, v)(x) = f(u(x), v(x))$$

and respectively by

$$G(u, v)(x) = g(u(x), v(x))$$

for each $(u, v) \in X \times X$ and a.e. for $x \in \Omega$. With the notations above, the problem (13.4.1) can be rewritten as the abstract evolution system

$$\begin{cases} u'(t) = Au(t) + F(u(t), v(t)) \\ v'(t) = Bv(t) + G(u(t), v(t)), \end{cases} \tag{13.4.11}$$

while (13.4.2) takes the abstract form

$$\begin{cases} u'(t) = Au(t) + \widetilde{F}(u(t), v(t)) \\ v'(t) = Bv(t) + \widetilde{G}(u(t), v(t)), \end{cases} \tag{13.4.12}$$

where $\widetilde{F}$ and $\widetilde{G}$ are defined likewise F and G but with $\widetilde{f}$ instead of f and $\widetilde{g}$ instead of g. Since $\widetilde{f}$ and $\widetilde{g}$ are continuous and have sublinear growth — see (13.4.8) —, $\widetilde{F}$ and $\widetilde{G}$ are well-defined, continuous and have sublinear growth. From the very same reason F and G are well-defined, continuous and have sublinear growth too. Since both A and B generate compact semigroups, in view of Theorem 8.6.4, to show that $\widetilde{\mathcal{C}}$ is mild viable with respect to $(A+F, B+G)$, we have merely to check the tangency condition

$$\liminf_{h\downarrow 0} \frac{1}{h} \operatorname{dist}\left((\tau, S_A(h)\xi, S_B(h)\eta) + h(1, F(\xi,\eta), G(\xi,\eta)); \widetilde{\mathcal{C}}\right) = 0, \tag{13.4.13}$$

for each $(\tau, \xi, \eta) \in \widetilde{\mathcal{C}}$, where $\{S_A(t) : X \to X,\ t \geq 0\}$ is the C_0-semigroup generated by A and $\{S_B(t) : X \to X,\ t \geq 0\}$ is the C_0-semigroup generated by B. To do this, it suffices to prove that for each $(\tau, \xi, \eta) \in \widetilde{\mathcal{C}}$ and each $h > 0$ there exists $(u_h, v_h) \in \mathcal{X}$ with $(\tau + h, u_h, v_h) \in \widetilde{\mathcal{C}}$ and

$$\begin{cases} \liminf\limits_{h\downarrow 0} \dfrac{1}{h} \|S_A(h)\xi + hF(\xi,\eta) - u_h\| = 0 \\ \liminf\limits_{h\downarrow 0} \dfrac{1}{h} \|S_B(h)\eta + hG(\xi,\eta) - v_h\| = 0. \end{cases} \tag{13.4.14}$$

So, let $(\tau, \xi, \eta) \in \widetilde{\mathcal{C}}$, and let us define u_h and v_h by

$$u_h = S_A(h)\xi + \int_\tau^{\tau+h} S_A(\tau + h - s)F(\xi,\eta)\, ds$$

$$+ \int_\tau^{\tau+h} S_A(\tau + h - s)[\, \widetilde{F}(\widetilde{u}(s), \widetilde{v}(s)) - \widetilde{F}(\widetilde{u}(\tau), \widetilde{v}(\tau))\,]\, ds$$

and respectively by

$$v_h = S_B(h)\eta + \int_\tau^{\tau+h} S_B(\tau + h - s)G(\xi,\eta)\, ds$$

$$+ \int_\tau^{\tau+h} S_B(\tau + h - s)[\, \widetilde{G}(\widetilde{u}(s), \widetilde{v}(s)) - \widetilde{G}(\widetilde{u}(\tau), \widetilde{v}(\tau))\,]\, ds.$$

Now let us observe that, inasmuch as $\xi \leq \widetilde{u}(\tau)$ and $\eta \geq \widetilde{v}(\tau)$ a.e. on Ω, in view of Theorem 1.7.5, we have both

$$S_A(h)\xi \leq S_A(h)\widetilde{u}(\tau) \ \text{ and } \ S_B(h)\eta \geq S_B(h)\widetilde{v}(\tau).$$

Since $f \leq \widetilde{f}$, taking into account of the monotonicity properties of $\widetilde{f}$, we get

$$F(\xi,\eta) \leq \widetilde{F}(\xi,\eta) \leq \widetilde{F}(\widetilde{u}(\tau),\widetilde{v}(\tau)).$$

Similarly, using the fact that $g \geq \widetilde{g}$ and the monotonicity properties of $\widetilde{g}$, we deduce

$$G(\xi,\eta) \geq \widetilde{G}(\xi,\eta) \geq \widetilde{G}(\widetilde{u}(\tau),\widetilde{v}(\tau)).$$

These inequalities and Theorem 1.7.5, show that both $u_h \leq \widetilde{u}(\tau+h)$ and $v_h \geq \widetilde{v}(\tau+h)$, and thus $(\tau+h, u_h, v_h) \in \widetilde{\mathcal{C}}$. On the other hand

$$\|S_A(h)\xi + hF(\xi,\eta) - u_h\| \leq \int_\tau^{\tau+h} \|S_A(\tau+h-s)F(\xi,\eta) - F(\xi,\eta)\|\,ds$$

$$+Me^{ah}\int_\tau^{\tau+h} \|\widetilde{F}(\widetilde{u}(s),\widetilde{v}(s)) - \widetilde{F}(\widetilde{u}(\tau),\widetilde{v}(\tau))\|\,ds,$$

where $M \geq 1$ and $a \in \mathbb{R}$ are the growth constants of the C_0-semigroup $\{S_A(t) : X \to X,\ t \geq 0\}$ given by Theorem 1.4.1. Consequently, the first equality in (13.4.14) holds. Similarly, we get the second equality, and this completes the proof of the viability part. As $\widetilde{f}$ and $\widetilde{g}$ have sublinear growth, $0 \leq f \leq \widetilde{f}$ and $\widetilde{g} \leq g \leq 0$, it follows that f and g have sublinear growth too. Further, since $\widetilde{\mathcal{C}}$ is X-closed[6], by Theorem 8.8.2 it follows that each mild solution $(u,v) : [\,\tau,T\,] \to \mathfrak{X}$ of (13.4.1) satisfying $(t,u(t),v(t)) \in \widetilde{\mathcal{C}}$ for each $t \in [\,\tau,T\,]$ can be continued up to a global one $(u^*,v^*) : [\,\tau,T_{\mathcal{C}}) \to \mathfrak{X}$ satisfying the very same condition on $[\,\tau,T_{\mathcal{C}})$. Finally, as $(\widetilde{u},\widetilde{v})$ is defined on $\mathbb{R}_+$, it readily follows that $T_{\mathcal{C}} = \infty$[7] and this completes the proof. □

13.5. A comparison result for a nonlinear diffusion inclusion

Let $\Omega \subseteq \mathbb{R}^n$, $n = 1,2,\ldots$, be a bounded domain with C^2 boundary Γ, let $\varphi : \mathbb{R} \to \mathbb{R}$ be a continuous and strictly increasing function with $\varphi(0) = 0$ and let us consider the Cauchy problem for the nonlinear diffusion equation

$$\begin{cases} u_t \in \Delta\varphi(u) + [\,f_1(t,x,u), f_2(t,x,u)\,] & \text{in } Q_{\tau,T} \\ u = 0 & \text{on } \Sigma_{\tau,T} \\ u(\tau,x) = \xi(x) & \text{in } \Omega, \end{cases} \tag{13.5.1}$$

where $Q_{\tau,T} = (\tau,T) \times \Omega$, $\Sigma_{\tau,T} = (\tau,T) \times \Gamma$, $\Delta\varphi$ is the nonlinear diffusion operator, $f_i : \mathbb{R} \times \overline{\Omega} \times \mathbb{R} \to \mathbb{R}_+$ for $i = 1,2$. We assume that f_1 is an l.s.c.

---

[6]See Definition 3.6.2.

[7]For the definition of $T_{\mathcal{C}}$ see (3.6.2).

function[8] and f_2 is an u.s.c. function on $\mathbb{R}\times\overline{\Omega}\times\mathbb{R}$. Further we assume that

$$0 \leq f_1(t,x,u) \leq f_2(t,x,u)$$

for each $(t,x,u)\in\mathbb{R}\times\overline{\Omega}\times\mathbb{R}$. We notice that here, unlike in Section 3, we have to use an $L^1(\Omega)$-setting simply because only in this space the equation above can be rewritten as an u.s.c. perturbed m-dissipative-type evolution inclusion.

Definition 13.5.1. By a C^0*-solution* of the problem (13.5.1), on $[\tau,T]$, we mean a continuous function $u:[\tau,T)\to L^1(\Omega)$, for which there exists $f\in L^1(0,T;L^1(\Omega))$ with $f(t,x)\in[f_1(t,x,u(t,x)),f_1(t,x,u(t,x))]$ a.e. for $t\in[\tau,T]$ and $x\in\Omega$, such that u is a C^0-solution in the sense of Definition 1.6.2 of the problem

$$\begin{cases} u'(t) = Au(t) + f(t) \\ u(\tau) = \xi, \end{cases}$$

where $A=\Delta\varphi$ is the m-dissipative operator in Theorem 1.7.7. By a C^0-*solution* of the problem (13.5.1), on $[\tau,\widetilde{T})$, $\tau<\widetilde{T}\leq\infty$, we mean a function $u\in C([\tau,\widetilde{T});L^1(\Omega))$ such that for each $\tau<T<\widetilde{T}$, u is a C^0-solution of (13.5.1) on $[\tau,T]$ in the sense stated before.

Next let us consider another function $g:\mathbb{R}\times\overline{\Omega}\times\mathbb{R}\to\mathbb{R}_+$ which is continuous, bounded, nondecreasing with respect to its last argument and such that $f_1(t,x,u)\leq g(t,x,u)$ for each $(t,x,u)\in\mathbb{R}\times\overline{\Omega}\times\mathbb{R}$, and we intend to show that, for each $u_0\in L^1(\Omega)$, with $u_0(x)\geq 0$ a.e. for $x\in\Omega$, and for every global C_0-solution $\widetilde{u}:\mathbb{R}_+\times\overline{\Omega}\to\mathbb{R}_+$ of the nonlinear diffusion equation

$$\begin{cases} u_t = \Delta\varphi(u) + g(t,x,u) & \text{in } Q_{0,\infty} \\ u = 0 & \text{on } \Sigma_{0,\infty} \\ u(0,x) = u_0(x) & \text{in } \Omega, \end{cases} \tag{13.5.2}$$

the nonlinear diffusion equation (13.5.1) has at least one global C_0-solution which, for each $t\in[\tau,\infty)$, lies almost everywhere "between" 0 and $\widetilde{u}(t,\cdot)$, provided $0\leq\xi(x)\leq\widetilde{u}(\tau,x)$ a.e. for $x\in\Omega$.

Namely, as g is continuous and, in view of Theorem 1.7.7, the operator $\Delta\varphi$ with homogeneous boundary conditions in $L^1(\Omega)$ generates a compact semigroup of contractions, the problem (13.5.2) has at least one noncontinuable C_0-solution $\widetilde{u}:[0,T_m)\to L^1(\Omega)$. Recalling that g is bounded, thanks

[8] Here the concept of l.s.c. function, (u.s.c. function) has a different meaning from the one introduced for multi-valued functions. Namely, a real function f defined on a topological space Y is called l.s.c. (u.s.c.) at a point $y\in Y$ if $\liminf_{x\to y} f(x) = f(y)$ ($\limsup_{x\to y} f(x) = f(y)$).

to Theorem 11.7.2, we readily conclude that $T_m = \infty$. Let $\mathcal{C} \subseteq \mathbb{R} \times L^1(\Omega)$ be the infinite tube defined by

$$\mathcal{C} = \left\{(t,u) \in \mathbb{R}_+ \times L^1(\Omega);\ 0 \le u(x) \le \widetilde{u}(t,x) \text{ a.e. for } x \in \Omega\right\}. \qquad (13.5.3)$$

The main result in this section is the following nonlinear $L^1(\Omega)$-version of Theorem 13.3.1.

Theorem 13.5.1. *Let $\Omega \subseteq \mathbb{R}^n$, $n = 1, 2, \ldots$, be a bounded domain with C^2 boundary Γ and let $\varphi : \mathbb{R} \to \mathbb{R}$ be continuous on $\mathbb{R}$ and C^1 on $\mathbb{R} \setminus \{0\}$, with $\varphi(0) = 0$, and for which there exist $C > 0$ and $a > 0$ if $n \le 2$ and $a > (n-2)/n$ if $n \ge 3$ such that*

$$\varphi'(r) \ge C|r|^{a-1}$$

for each $r \in \mathbb{R} \setminus \{0\}$. Let $f_i, g : \mathbb{R} \times \overline{\Omega} \times \mathbb{R} \to \mathbb{R}_+$, $i = 1, 2$, with f_1 l.s.c., f_2 u.s.c. and g continuous, bounded and nondecreasing with respect of its last argument. Let us assume also that

$$0 \le f_1(t,x,u) \le \min\{f_2(t,x,u), g(t,x,u)\},$$

for each $(t,x,u) \in \mathbb{R} \times \overline{\Omega} \times \mathbb{R}$. Let $u_0 \in L^1(\Omega)$, $u_0(x) \ge 0$ a.e. for $x \in \Omega$, and let $\mathcal{C}$ be defined by (13.5.3), where $\widetilde{u} : \mathbb{R}_+ \to L^1(\Omega)$ is a global C_0-solution of (13.5.2). Then, for each $(\tau, \xi) \in \mathcal{C}$ there exists at least one global C^0-solution $u : [\,\tau, \infty) \to L^1(\Omega)$ of (13.5.1) satisfying

$$0 \le u(t,x) \le \widetilde{u}(t,x),$$

for each $t \ge \tau$ and a.e. for $x \in \Omega$.

The proof of Theorem 13.5.1 rests heavily upon a nonlinear version of Theorem 1.7.5, which is mainly based on Theorem 1.7.4 and is interesting in itself. So, we postpone for a moment the proof of Theorem 13.5.1 in favor of

Lemma 13.5.1. *Let Ω be a bounded domain in $\mathbb{R}^n$, $n = 1, 2, \ldots$, with C^2 boundary Γ, let $\varphi : \mathbb{R} \to \mathbb{R}$ be strictly increasing*[9] *with $\varphi(0) = 0$, let $u_0, v_0 \in L^1(\Omega)$, $f_0, g_0 \in L^1(\tau, T; L^1(\Omega))$ and let $u : [\,\tau, T\,] \to L^1(\Omega)$ be the unique C^0-solution of the Cauchy problem*

$$\begin{cases} u_t = \Delta\varphi(u) + f_0(t,x) & \text{in } Q_{\tau,T} \\ u = 0 & \text{on } \Sigma_{\tau,T} \\ u(\tau,x) = u_0(x) & \text{in } \Omega \end{cases}$$

and $v : [\,\tau, T\,] \to L^1(\Omega)$ the unique C^0-solution of the very same Cauchy problem but with f_0 replaced by g_0 and u_0 replaced by v_0. If $u_0(x) \le v_0(x)$

[9] In fact, the conclusion of Lemma 13.5.1 remains unchanged if we assume merely that φ is nondecreasing.

a.e. for $x \in \Omega$ *and* $f_0(s,x) \leq g_0(s,x)$ *for each* $s \in [\tau, T]$ *and a.e. for* $x \in \Omega$, *then*

$$u(s,x) \leq v(s,x)$$

for each $s \in [0,T]$ *and a.e. for* $x \in \Omega$.

Proof. Firstly, from Theorem 1.7.10, it follows that, for each $\lambda > 0$ and each $\xi_1, \xi_2 \in L^1(\Omega)$ with $\xi_1(x) \leq \xi_2(x)$ a.e. for $x \in \Omega$, we have

$$[(I - \lambda\Delta\varphi)^{-1}\xi_1](x) \leq [(I - \lambda\Delta\varphi)^{-1}\xi_2](x) \tag{13.5.4}$$

a.e. for $x \in \Omega$. Secondly, let $\varepsilon > 0$ and let $u_\varepsilon, v_\varepsilon : [\tau, t_n] \to L^1(\Omega)$ be two ε-*DS*-solutions corresponding to u and v satisfying

$$\begin{cases} \|u(t) - u_\varepsilon(t)\| \leq \varepsilon \\ \|v(t) - v_\varepsilon(t)\| \leq \varepsilon \end{cases}$$

for each $t \in [\tau, t_n]$ with $T - t_n \leq \varepsilon$. See Definition 1.6.2. Let us recall that $u_\varepsilon(t) = u_k$ and $v_\varepsilon(t) = v_k$ for $t \in [t_{k-1}, t_k)$, $k = 1, 2, \ldots, n$, where

$$\frac{u_k - u_{k-1}}{t_k - t_{k-1}} = \Delta\varphi(u_k) + f_k \quad \text{and} \quad \frac{v_k - v_{k-1}}{t_k - t_{k-1}} = \Delta\varphi(v_k) + g_k$$

for $k = 1, 2, \ldots, n$, with

$$\sum_{i=1}^{n} \int_{t_{k-1}}^{t_k} \|f_0(t) - f_k\| \, dt \leq \varepsilon \quad \text{and} \quad \sum_{i=1}^{n} \int_{t_{k-1}}^{t_k} \|g_0(t) - g_k\| \, dt \leq \varepsilon.$$

Notice that, since both f_0 and g_0 are continuous, we may assume with no loss of generality that both $f_k = f_0(t_k)$ and $g_k = g_0(t_k)$ for $k = 1, 2, \ldots, n$.

We use an inductive argument to evaluate v_i by means on u_i, taking into account of $g_i \leq f_i$ and (13.5.4). So, if we assume that $v_{k-1} \leq u_{k-1}$, we get

$$\begin{aligned} v_k &= (I - (t_k - t_{k-1})\Delta\varphi)^{-1}(v_{k-1} + (t_k - t_{k-1})g_k) \\ &\leq (I - (t_k - t_{k-1})\Delta\varphi)^{-1}(u_{k-1} + (t_k - t_{k-1})f_k) = u_k. \end{aligned}$$

Passing to the limit for $\varepsilon \to 0$, we get the conclusion for $t \in [\tau, T)$. The case $t = T$ follows by simply passing to the limit. □

We are now prepared to continue with the proof of Theorem 13.5.1.

Proof. To get the conclusion, we will first make use of Theorem 11.6.3, to show that $\mathcal{C}$ is C^0-viable with respect to $A + F$, and second of Theorem 11.7.2 to prove that each C^0-solution $u : [\tau, T) \to L^1(\Omega)$ of (13.5.1) whose graph lies in $\mathcal{C}$ can be continued to a C^0-solution defined on $\mathbb{R}_+$ and whose graph is in $\mathcal{C}$ too. To this aim, we denote by $X = L^1(\Omega)$ and we rewrite (13.5.1) as an evolution equation in X. First, since in our case φ is single-valued, let us define $A : D(A) \subseteq X \to X$ by

$$\begin{cases} D(A) = \{u \in L^1(\Omega);\ \varphi(u) \in W^{1,1}(\Omega),\ \Delta\varphi(u) \in L^1(\Omega)\} \\ Au = \Delta\varphi(u) \quad \text{for } u \in D(A), \end{cases}$$

$F : \mathbb{R} \times X \rightsquigarrow X$ by defining $F(t,u)$ as the set of all functions $f(t,\cdot) : \Omega \to \mathbb{R}$ such that $f(t,\cdot)$ is measurable and $f(t,x) \in [\,f_1(t,x,u(x)), f_2(t,x,u(x))\,]$ a.e. for $x \in \Omega$, and $G : \mathbb{R} \times X \to X$ by

$$G(t,u)(x) = g(t,x,u(x))$$

for each $t \in \mathbb{R}$, each $u \in X$ and a.e. for $x \in \Omega$. Since f_1 is l.s.c. and f_2 is u.s.c. and both are bounded, from Problem 2.6.1, we conclude that F is strongly-weakly u.s.c. with nonempty, convex and weakly compact values. Moreover, since g is continuous and bounded, G is well-defined, continuous and bounded. With the notations above, the problem (13.5.1) can be rewritten as a Cauchy problem for an evolution inclusion, i.e.

$$\begin{cases} u'(t) \in Au(t) + F(t,u(t)) \\ u(\tau) = \xi, \end{cases} \tag{13.5.5}$$

while (13.5.2) takes the abstract form

$$\begin{cases} u'(t) = Au(t) + G(t,u(t)) \\ u(0) = u_0. \end{cases} \tag{13.5.6}$$

At this point, let us recall that, in view of Theorem 1.7.7, A generates a compact semigroup of nonexpansive mappings. Since ξ and all elements in $F(\tau,\xi)$ are nonnegative, by Lemma 13.5.1 it follows that each solution of (13.5.2) is necessarily nonnegative. Therefore, to prove that $\mathcal{C}$ is viable with respect to $A+F$ it suffices to show that the larger set

$$\widetilde{\mathcal{C}} = \big\{(t,u) \in \mathbb{R}_+ \times L^1(\Omega);\ u(x) \le \widetilde{u}(t,x) \text{ a.e. for } x \in \Omega\big\}$$

is C^0-viable with respect to $A+F$. Clearly, $\widetilde{\mathcal{C}}$ is nonempty and closed, and F is strongly-weakly u.s.c. with nonempty, convex and weakly compact values. Moreover, let us observe that, by Theorem 1.7.9, A is of complete continuous type. So, in order to apply Theorem 11.6.3, it remains to check the tangency condition $(1, F(\tau,\xi)) \in \mathcal{QTS}^A_{\widetilde{\mathcal{C}}}(\tau,\xi)$ for each $(\tau,\xi) \in \widetilde{\mathcal{C}}$. Since $\mathcal{TS}^A_{\widetilde{\mathcal{C}}}(\tau,\xi) \subseteq \mathcal{QTS}^A_{\widetilde{\mathcal{C}}}(\tau,\xi)$, it suffices to show that

$$\liminf_{h\downarrow 0} \frac{1}{h} \operatorname{dist}\left((\tau+h, u(\tau+h,\tau,\xi,F(\tau,\xi)));\widetilde{\mathcal{C}}\right) = 0, \tag{13.5.7}$$

for each $(\tau,\xi) \in \widetilde{\mathcal{C}}$. We recall that

$$u(\tau+\cdot,\tau,\xi,F(\tau,\xi)) = \{u(\tau+\cdot,\tau,\xi,f);\ f \in F(\tau,\xi)\},$$

where $s \mapsto u(s, \tau, \xi, f)$ is the unique C^0-solution of the Cauchy problem

$$\begin{cases} u'(t) = Au(t) + f \\ u(\tau) = \xi. \end{cases} \tag{13.5.8}$$

So, let $(\tau, \xi) \in \widetilde{\mathcal{C}}$ be arbitrary and let us observe that, to prove (13.5.7), it suffices to show that, for each $h > 0$, there exist both $u_h \in L^1(\Omega)$ and $f_h \in F(\tau, \xi)$ such that $(\tau + h, u_h) \in \widetilde{\mathcal{C}}$ and

$$\liminf_{h \downarrow 0} \frac{1}{h} \|u(\tau + h, \tau, \xi, f_h) - u_h\| = 0. \tag{13.5.9}$$

Let us define $f_h(x) = f_1(\tau, x, \xi(x))$ a.e. for $x \in \Omega$ and

$$u_h = u(\tau + h, \tau, \xi, f_h)$$

$$+u(\tau + h, \tau, \widetilde{u}(\tau), G(\cdot, \widetilde{u}(\cdot))) - u(\tau + h, \tau, \widetilde{u}(\tau), G(\tau, \widetilde{u}(\tau))).$$

Since $\xi \le \widetilde{u}(\tau)$, $f_h \le G(\tau, \xi)$ and $u \mapsto G(\tau, u)$ is nondecreasing, in view of Lemma 13.5.1, we deduce

$$u_h(x) \le u(\tau + h, \tau, \widetilde{u}(\tau), G(\tau, \widetilde{u}(\tau)))$$

$$+u(\tau + h, \tau, \widetilde{u}(\tau), G(\cdot, \widetilde{u}(\cdot))) - u(\tau + h, \tau, \widetilde{u}(\tau), G(\tau, \widetilde{u}(\tau))) = \widetilde{u}(\tau + h).$$

Thus $(\tau + h, u_h) \in \widetilde{\mathcal{C}}$. On the other hand, from (1.6.5) and the continuity of both G and $\widetilde{u}$, we get

$$\lim_{h \downarrow 0} \frac{1}{h} \|u(\tau + h, \tau, \widetilde{u}(\tau), G(\cdot, \widetilde{u}(\cdot))) - u(\tau + h, \tau, \widetilde{u}(\tau), G(\tau, \widetilde{u}(\tau)))\|$$

$$\le \lim_{h \downarrow 0} \frac{1}{h} \int_\tau^{\tau + h} \|G(s, \widetilde{u}(s)) - G(\tau, \widetilde{u}(\tau))\| \, ds = 0$$

and so (13.5.9) is satisfied. Consequently, we have

$$\lim_{h \downarrow 0} \frac{1}{h} \operatorname{dist}\left((\tau + h, u(\tau + h, \tau, \xi, f_h)); \widetilde{\mathcal{C}}\right) = 0.$$

In view of Theorem 11.6.3, $\widetilde{\mathcal{C}}$, and hence $\mathcal{C}$, is C^0-viable with respect to $A + F$. As F is bounded, an appeal to Theorem 11.7.2, shows that each C^0-solution $u : [\tau, T) \to L^1(\Omega)$, whose graph lies in $\mathcal{C}$, can be continued to a C^0-solution defined on $[\tau, \infty)$, and whose graph is in $\mathcal{C}$ too. The proof is complete. □

13.6. Comparison for a fully nonlinear reaction-diffusion system

Let $\Omega \subseteq \mathbb{R}^n$, $n = 1, 2, \ldots$, be a bounded domain with C^2 boundary Γ, let $\varphi : \mathbb{R} \to \mathbb{R}$ and $\psi : \mathbb{R} \to \mathbb{R}$ be two continuous and nondecreasing functions with $\varphi(0) = \psi(0) = 0$ and let us consider the nonlinear reaction diffusion system

$$\begin{cases} u_t = \Delta\varphi(u) + f(u,v) & \text{in } Q_{\tau,T} \\ v_t = \Delta\psi(v) + g(u,v) & \text{in } Q_{\tau,T} \\ u = v = 0 & \text{in } \Sigma_{\tau,T} \\ u(\tau,x) = \xi(x), \;\; v(\tau,x) = \eta(x) & \text{on } \Omega, \end{cases} \tag{13.6.1}$$

where, for $0 \leq \tau < T \leq \infty$, $Q_{\tau,T} = (\tau,T) \times \Omega$, $\Sigma_{\tau,T} = (\tau,T) \times \Gamma$. Further, $\Delta\varphi$ and $\Delta\psi$ are nonlinear diffusion operators, both $f : \mathbb{R} \times \mathbb{R} \to \mathbb{R}_+$ and $g : \mathbb{R} \times \mathbb{R} \to \mathbb{R}_-$ are continuous and $\xi, \eta \in L^1(\Omega)$ are nonnegative. We notice that, if either φ or ψ is not strictly increasing, (13.6.1) is degenerate.

Let $\widetilde{f} : \mathbb{R} \times \mathbb{R} \to \mathbb{R}_+$ and $\widetilde{g} : \mathbb{R} \times \mathbb{R} \to \mathbb{R}_-$ be two continuous functions such that

$$\begin{cases} f(u,v) \leq \widetilde{f}(u,v) \\ g(u,v) \geq \widetilde{g}(u,v) \end{cases} \tag{13.6.2}$$

for each $(u,v) \in \mathbb{R} \times \mathbb{R}$. Further, let us consider the comparison reaction-diffusion system

$$\begin{cases} u_t = \Delta\varphi(u) + \widetilde{f}(u,v) & \text{in } Q_{0,\infty} \\ v_t = \Delta\psi(v) + \widetilde{g}(u,v) & \text{in } Q_{0,\infty} \\ u = v = 0 & \text{on } \Sigma_{0,\infty} \\ u(0,x) = u_0(x), \;\; v(0,x) = v_0(x) & \text{in } \Omega, \end{cases} \tag{13.6.3}$$

and let $(\widetilde{u}, \widetilde{v}) : \mathbb{R}_+ \times \Omega \to \mathbb{R}_+ \times \mathbb{R}_+$ be a C^0-solution of (13.6.3).

Here, by using the viability results in Section 10.7, we will prove a sufficient condition in order that, for each $(\xi, \eta) \in L^1(\Omega) \times L^1(\Omega)$, with

$$\begin{cases} 0 \leq \xi(x) \leq \widetilde{u}(\tau,x) \\ \widetilde{v}(\tau,x) \leq \eta(x) \end{cases} \tag{13.6.4}$$

a.e. for $x \in \Omega$, the reaction-diffusion system (13.6.1) has at least one solution $(u,v) : \mathbb{R}_+ \times \Omega \to \mathbb{R}_+ \times \mathbb{R}_+$, such that, for each $t \in [\tau, \infty)$, we have

$$\begin{cases} 0 \leq u(t,x) \leq \widetilde{u}(t,x) \\ \widetilde{v}(t,x) \leq v(t,x) \end{cases} \tag{13.6.5}$$

a.e. for $x \in \Omega$.

Let $\mathcal{C} \subseteq \mathbb{R}_+ \times L^1(\Omega) \times L^1(\Omega)$ be defined by

$$\mathcal{C} = \big\{(t,u,v) \in \mathbb{R}_+ \times L^1(\Omega) \times L^1(\Omega);\; (u,v) \text{ satisfy (13.6.7) below}\big\} \tag{13.6.6}$$

$$\begin{cases} 0 \le u(x) \le \widetilde{u}(t,x) \\ \widetilde{v}(t,x) \le v(x) \end{cases} \tag{13.6.7}$$

a.e. for $x \in \Omega$.

We want to prove that, for each $(\tau, \xi, \eta) \in \mathcal{C}$, the Cauchy problem (13.6.1) has at least one mild solution $(u, v) : [\tau, \infty) \to L^1(\Omega) \times L^1(\Omega)$ with $(t, u(t), v(t)) \in \mathcal{C}$ for each $t \in [\tau, \infty)$. To this end, let us assume that there exist $c_i \ge 0$, $i = 1, \ldots, 5$, such that

$$\begin{cases} |\widetilde{f}(u,v)| \le c_1|u| + c_2 \\ |\widetilde{g}(u,v)| \le c_3|u| + c_4|v| + c_5 \end{cases} \tag{13.6.8}$$

for each $(u, v) \in \mathbb{R}_+ \times \mathbb{R}_+$. Namely, we will prove

Theorem 13.6.1. *Let $\Omega \subseteq \mathbb{R}^n$, $n = 1, 2, \ldots$, be a bounded domain with C^2 boundary Γ and let $\varphi : \mathbb{R} \to \mathbb{R}$ and $\psi : \mathbb{R} \to \mathbb{R}$ be two continuous and nondecreasing functions with $\varphi(0) = \psi(0) = 0$ and such that ψ is C^1 on $\mathbb{R} \setminus \{0\}$ and there exist $C > 0$ and $a > 0$ if $n \le 2$ and $a > (n-2)/n$ if $n \ge 3$ such that*

$$\psi'(r) \ge C|r|^{a-1}$$

for each $r \in \mathbb{R} \setminus \{0\}$. Let $f : \mathbb{R} \times \mathbb{R} \to \mathbb{R}_+$ and $g : \mathbb{R} \times \mathbb{R} \to \mathbb{R}_-$ be continuous and let $\widetilde{f} : \mathbb{R} \times \mathbb{R} \to \mathbb{R}_+$ and $\widetilde{g} : \mathbb{R} \times \mathbb{R} \to \mathbb{R}_-$ be continuous and such that, for each $(u_0, v_0) \in \mathbb{R}_+ \times \mathbb{R}_+$, $u \mapsto \widetilde{f}(u, v_0)$ and $v \mapsto \widetilde{g}(u_0, v)$ are nondecreasing, $u \mapsto \widetilde{g}(u, v_0)$ and $v \mapsto \widetilde{f}(u_0, v)$ are nonincreasing and satisfy (13.6.2) *and* (13.6.8). *Let us assume that there exist $L > 0$ and $\widetilde{L} > 0$ such that*

$$\begin{cases} |f(u_1,v) - f(u_2,v)| \le L|u_1 - u_2| \\ |\widetilde{f}(u_1,v) - \widetilde{f}(u_2,v)| \le \widetilde{L}|u_1 - u_2| \end{cases}$$

for each $u_1, u_2, v \in \mathbb{R}$. Let $(u_0, v_0) \in L^1(\Omega) \times L^1(\Omega)$ with $u_0(x) \ge 0$ and $v_0(x) \ge 0$ a.e. for $x \in \Omega$, let $(\widetilde{u}, \widetilde{v}) : \mathbb{R}_+ \times \Omega \to \mathbb{R}_+ \times \mathbb{R}_+$ be a global solution[10] *of* (13.6.3) *and let $\mathcal{C}$ be defined by* (13.6.6). *Then, for each $(\tau, \xi, \eta) \in \mathcal{C}$,* (13.6.1) *has at least one global C^0-solution $(u, v) : [\tau, \infty) \to L^1(\Omega) \times L^1(\Omega)$ satisfying $(t, u(t), v(t)) \in \mathcal{C}$ for each $t \in [\tau, \infty)$.*

Proof. We begin by showing that $\mathcal{C}$, given by (13.6.6), is C^0-viable with respect $(\Delta\varphi + f, \Delta\psi + g)$. Secondly, we will show that every C^0-solution $(u, v) : [\tau, T) \to L^1(\Omega) \times L^1(\Omega)$, satisfying $(t, u(t), v(t)) \in \mathcal{C}$ for each $t \in [\tau, T)$, can be extended to a global one obeying the very same

[10] In view of Definition 13.5.1, it is clear what a C^0-solution ought to be in this context. Such a solution exists since $\Delta\psi$ generates a compact semigroup of contractions, $\widetilde{f}$, $\widetilde{g}$ are continuous and have sublinear growth and $\widetilde{f}$ is globally Lipschitz with respect to u, uniformly with respect to $v \in \mathbb{R}$. For the existence see Theorems 10.7.4, and for the global continuation see Theorem 10.6.2.

constraints. To this aim, we will first make use of Theorem 10.7.4 and then of Theorem 10.6.3.

Now, for $t \in \mathbb{R}_+$, let us consider the constraints

$$\begin{cases} u(x) \leq \widetilde{u}(t,x) \\ \widetilde{v}(t,x) \leq v(x), \end{cases} \tag{13.6.9}$$

a.e. for $x \in \Omega$, which are less restrictive than (13.6.7), and let us define $\widetilde{\mathcal{C}} \subseteq \mathbb{R} \times L^1(\Omega) \times L^1(\Omega)$ by

$$\widetilde{\mathcal{C}} = \left\{(t,u,v) \in \mathbb{R}_+ \times L^1(\Omega) \times L^1(\Omega);\ (u,v) \text{ satisfy } (13.6.9)\right\}. \tag{13.6.10}$$

Since f and ξ are nonnegative, by Lemma 13.5.1, it follows that u is nonnegative too, and so, to prove that $\mathcal{C}$ is C^0-viable with respect to $(\Delta\varphi + f, \Delta\psi + g)$, it suffices to show that $\widetilde{\mathcal{C}}$ is C^0-viable with respect to $(\Delta\varphi + f, \Delta\psi + g)$.

Let us denote by $X = L^1(\Omega)$, and by $\mathcal{X} = X \times X$. We endow $\mathcal{X}$ with the usual norm $\|(u,v)\| = \|u\| + \|v\|$, and we rewrite (13.6.1) as an evolution system in X or, equivalently, as an evolution equation in $\mathcal{X}$. To this aim, let us define the operators $A : D(A) \subseteq X \to X$ and $B : D(B) \subseteq X \to X$ by

$$\begin{cases} D(A) = \{u \in L^1(\Omega);\ \varphi(u) \in W_0^{1,1}(\Omega),\ \Delta\varphi(u) \in L^1(\Omega)\} \\ Au = \Delta\varphi(u) \quad \text{for } u \in D(A) \end{cases}$$

and by

$$\begin{cases} D(B) = \{u \in L^1(\Omega);\ \psi(u) \in W_0^{1,1}(\Omega),\ \Delta\psi(u) \in L^1(\Omega)\} \\ Bv = \Delta\psi(v) \quad \text{for } v \in D(B), \end{cases}$$

respectively. Further, let us define $F : \mathcal{X} \to X$ and $G : \mathcal{X} \to X$ by

$$F(u,v)(x) = f(u(x), v(x))$$

and respectively by

$$G(u,v)(x) = g(u(x), v(x))$$

for each $(u,v) \in X \times X$ and a.e. for $x \in \Omega$. With the notations above, the problem (13.6.1) can be rewritten as the abstract evolution system

$$\begin{cases} u'(t) = Au(t) + F(u(t), v(t)) \\ v'(t) = Bv(t) + G(u(t), v(t)), \end{cases} \tag{13.6.11}$$

while (13.6.2) takes the abstract form

$$\begin{cases} u'(t) = Au(t) + \widetilde{F}(u(t), v(t)) \\ v'(t) = Bv(t) + \widetilde{G}(u(t), v(t)), \end{cases} \tag{13.6.12}$$

where $\widetilde{F}$ and $\widetilde{G}$ are defined just like F and G but with $\widetilde{f}$ instead of f and $\widetilde{g}$ instead of g. Since $\widetilde{f}$ and $\widetilde{g}$ are continuous and have sublinear growth —

see (13.6.8) —, $\widetilde{F}$ and $\widetilde{G}$ are well-defined, continuous and have sublinear growth. From the very same reason F and G are well-defined, continuous and have sublinear growth too.

Throughout, if $(\tau,\xi,\eta)\in\widetilde{\mathcal{C}}$, we denote by

$$\begin{cases} u(\tau+h,\tau,\xi,\eta)=u(\tau+h,\tau,\xi,F(\xi,\eta))\\ v(\tau+h,\tau,\xi,\eta)=v(\tau+h,\tau,\eta,G(\xi,\eta)).\end{cases}$$

Since B generates a compact semigroup and F is locally Lipschitz with respect to $u\in X$ — see Definition 10.7.3 —, in view of Theorem 10.7.4, to show that $\widetilde{\mathcal{C}}$ is C^0-viable with respect to $(A+F,B+G)$, we have merely to prove that the tangency condition

$$\liminf_{h\downarrow 0}\frac{1}{h}\operatorname{dist}\left((\tau+h,u(\tau+h,\tau,\xi,\eta),v(\tau+h,\tau,\xi,\eta));\widetilde{\mathcal{C}}\right)=0 \tag{13.6.13}$$

holds for each $(\tau,\xi,\eta)\in\widetilde{\mathcal{C}}$. To this aim, it suffices to show that, for each $(\tau,\xi,\eta)\in\widetilde{\mathcal{C}}$ and each $h>0$, there exists $(u_h,v_h)\in\mathcal{X}$ with $(\tau+h,u_h,v_h)\in\widetilde{\mathcal{C}}$ and

$$\begin{cases} \liminf\limits_{h\downarrow 0}\dfrac{1}{h}\|u(\tau+h,\tau,\xi,\eta)-u_h\|=0\\ \liminf\limits_{h\downarrow 0}\dfrac{1}{h}\|v(\tau+h,\tau,\xi,\eta)-v_h\|=0.\end{cases} \tag{13.6.14}$$

So, let $(\tau,\xi,\eta)\in\widetilde{\mathcal{C}}$, and let us define u_h and v_h by

$$\begin{aligned} u_h&=u(\tau+h,\tau,\xi,\eta)\\ &+u(\tau+h,\tau,\widetilde{u}(\tau),\widetilde{F}(\widetilde{u}(\cdot),\widetilde{v}(\cdot)))-u(\tau+h,\tau,\widetilde{u}(\tau),\widetilde{F}(\widetilde{u}(\tau),\widetilde{v}(\tau)))\end{aligned}$$

and respectively by

$$\begin{aligned} v_h&=v(\tau+h,\tau,\xi,\eta)\\ &+v(\tau+h,\tau,\widetilde{v}(\tau),\widetilde{G}(\widetilde{u}(\cdot),\widetilde{v}(\cdot)))-v(\tau+h,\tau,\widetilde{v}(\tau),\widetilde{G}(\widetilde{u}(\tau),\widetilde{v}(\tau))).\end{aligned}$$

From $f\le\widetilde{f}$ and from the monotonicity properties of the latter, we get

$$F(\xi,\eta)\le\widetilde{F}(\xi,\eta)\le\widetilde{F}(\widetilde{u}(\tau),\widetilde{v}(\tau)).$$

Similarly, we deduce

$$G(\xi,\eta)\ge\widetilde{G}(\xi,\eta)\ge\widetilde{G}(\widetilde{u}(\tau),\widetilde{v}(\tau)).$$

Now let us observe that, inasmuch as $\xi\le\widetilde{u}(\tau)$ and $\eta\ge\widetilde{v}(\tau)$ a.e. on Ω, in view of Lemma 13.5.1, we have both

$$\begin{cases} u(\tau+h,\tau,\xi,\eta)\le u(\tau+h,\tau,\widetilde{u}(\tau),\widetilde{F}(\widetilde{u}(\tau),\widetilde{v}(\tau)))\\ v(\tau+h,\tau,\xi,\eta)\ge v(\tau+h,\tau,\widetilde{v}(\tau),\widetilde{G}(\widetilde{u}(\tau),\widetilde{v}(\tau))).\end{cases}$$

From these inequalities we get both $u_h\le\widetilde{u}(\tau+h)$ and $v_h\ge\widetilde{v}(\tau+h)$ and thus $(\tau+h,u_h,v_h)\in\widetilde{\mathcal{C}}$. On the other hand

$$\|u(\tau+h,\tau,\xi,\eta)-u_h\| \leq \int_\tau^{\tau+h} \left\|\widetilde{F}(\widetilde{u}(s),\widetilde{v}(s))-\widetilde{F}(\widetilde{u}(\tau),\widetilde{v}(\tau))\right\| ds.$$

Consequently the first equality in (13.6.14) holds. Similarly, we get the second equality, and this completes the proof of the viability part. Since f and g have sublinear growth and $\widetilde{\mathcal{C}}$ is X-closed[11] being closed, by Theorem 10.6.3 it follows that each C^0-solution $(u,v):[\tau,T]\to\mathcal{X}$ of (13.6.1) satisfying $(t,u(t),v(t))\in\widetilde{\mathcal{C}}$ for each $t\in[\tau,T]$ can be continued up to a global one $(u^*,v^*):[\tau,T_{\mathcal{C}})\to\mathcal{X}$ satisfying the same condition on $[\tau,T_{\mathcal{C}})$[12]. Finally, as $(\widetilde{u},\widetilde{v})$ is defined on $\mathbb{R}_+$, it readily follows that $T_{\mathcal{C}}=\infty$ and this completes the proof. □

13.7. A controllability problem

Let X be a Banach space, $A:D(A)\subseteq X\to X$ the infinitesimal generator of a C_0-semigroup, $\{S(t):X\to X;\ t\geq 0\}$, $g:X\to X$ a given function, $\xi\in X$, and $c(\cdot)$ a measurable control taking values in $D(0,1)$. Here, the problem we consider is how to find a control $c(\cdot)$ in order to reach the origin starting from the initial point ξ in some time T, by mild solutions of the state equation

$$\begin{cases} u'(t)=Au(t)+g(u(t))+c(t) \\ u(0)=\xi. \end{cases} \tag{13.7.1}$$

Let us consider $G:X\rightsquigarrow X$, defined by $G(x)=g(x)+D(0,1)$. We can rewrite the above problem as follows. For a given $\xi\in X$, find $T>0$ and a mild solution of multi-valued semilinear Cauchy problem

$$\begin{cases} u'(t)\in Au(t)+G(u(t)) \\ u(0)=\xi, \end{cases} \tag{13.7.2}$$

that satisfies $u(T)=0$.

The main result of this section is given by the following theorem.

Theorem 13.7.1. *Let $g:X\to X$ be a continuous function such that for some $L>0$ we have*

$$\|g(x)\|\leq L\|x\|, \tag{13.7.3}$$

for every $x\in X$. Assume that the semigroup $\{S(t):X\to X;t\geq 0\}$ is compact and satisfies the condition

$$\|S(t)x\|\leq e^{at}\|x\|, \tag{13.7.4}$$

[11] See Definition 3.6.2.

[12] For the definition of $T_{\mathcal{C}}$ see (3.6.2).

for every $x \in X$. *Then, for every* $\xi \in X$ *with* $\xi \neq 0$ *there exists a mild solution* $u : [0, \infty) \to X$ *of* (13.7.2) *which satisfies the inequation*

$$\|u(t)\| \leq \|\xi\| - t + (L + a) \int_0^t \|u(s)\| ds \tag{13.7.5}$$

for every $t \geq 0$ *for which* $u(t) \neq 0$.

Corollary 13.7.1. *Under the hypotheses of Theorem* 13.7.1, *the following properties hold.*

(i) *In case* $L + a \leq 0$, *for any* $\xi \in X$, $\xi \neq 0$, *there exist a control* $c(\cdot)$ *and a mild solution of* (13.7.1) *that reaches the origin of* X *in some time* $T \leq \|\xi\|$ *and satisfies*

$$\|u(t)\| \leq \|x\| - t \tag{13.7.6}$$

for any $0 \leq t \leq T$.

(ii) *In case* $L + a > 0$, *for every* $\xi \in X$ *satisfying* $0 < \|\xi\| < 1/(L + a)$, *there exist a control* $c(\cdot)$ *and a mild solution of* (13.7.1) *that reaches the origin of* X *in some time*

$$T \leq \frac{1}{L + a} \log \frac{1}{1 - (L + a)\|\xi\|},$$

and satisfies

$$\|u(t)\| \leq e^{(L+a)t} \left(\|\xi\| - \frac{1}{L + a} \right) + \frac{1}{L + a} \tag{13.7.7}$$

for any $0 \leq t \leq T$.

We notice that, in view of Theorem 1.4.2, the condition (13.7.4) can always be satisfied if we replace the initial norm with an equivalent one. We begin with the proof of Corollary 13.7.1.

Proof. In the case (i), since $L + a \leq 0$, by (13.7.5) we deduce that there exists a mild solution $u : [0, \infty) \to X$ of (13.7.2) which satisfies the inequality $\|u(t)\| \leq \|\xi\| - t$ for every $t \geq 0$ for which $u(t) \neq 0$. This implies that there exists $T > 0$ with $T \leq \|\xi\|$ such that $u(T) = 0$. By Definition 9.1.1, there exists $c \in L^1(0, T; X)$ such that $c(s) \in D(0, 1)$ a.e. for $s \in [\,0, T\,]$ and u is a mild solution of (13.7.1). This completes the proof of (i). To prove (ii) we proceed similarly, by observing that (13.7.7) comes from (13.7.5), via the Gronwall Lemma 1.8.4, with $x(\cdot) = \|u(\cdot)\| - \frac{1}{L+a}$. □

We now proceed with the proof of Theorem 13.7.1

Proof. We consider the space $\mathbb{R} \times X$, the operator $\mathcal{A} = (0, A)$ which generates the C_0-semigroup $(1, S(t))$ on $\mathbb{R} \times X$, the locally closed set

$$K = \{(\lambda, x) \in \mathbb{R}_+ \times X \setminus \{0\};\ \|x\| \leq \lambda\},$$

and the multi-function $F : \mathbb{R} \times X \rightsquigarrow \mathbb{R} \times X$ defined by

$$F(t, x) = ((L + a)\|x\| - 1, g(x) + D(0, 1)),$$

for every $(t, x) \in \mathbb{R} \times X$. We show that

$$((L + a)\|\xi\| - 1, g(\xi) + D(0, 1)) \in \mathfrak{QTS}_K^{\mathcal{A}}(\lambda, \xi), \tag{13.7.8}$$

for every $(\lambda, \xi) \in K$. In view of Remark 9.1.1, to prove this it suffices to check that, for each $\xi \in X$, $\xi \neq 0$, there exist $(h_n)_n$, $(\theta_n)_n$ both in $\mathbb{R}$, and $(g_n)_n \in \mathcal{G}(\xi)$[13], with $h_n \downarrow 0$ and $\lim_n \theta_n = 0$ and such that

$$\begin{aligned} &\left\| S(h_n)\xi + \int_0^{h_n} S(h_n - s) g_n(\xi) ds \right\| \\ &\leq \|\xi\| + h_n((L + a)\|\xi\| - 1) + h_n \theta_n. \end{aligned}$$

We will consider $a \neq 0$, the case $a = 0$ following by simpler arguments. Namely, let us first observe that

$$\begin{aligned} &\left\| S(h)\xi + \int_0^h S(h - s) g(\xi) ds - \int_0^h S(h - s) S(s) e^{-as} \frac{\xi}{\|\xi\|} ds \right\| \\ &= \left\| S(h)\xi + \int_0^h S(h - s) g(\xi) ds + \frac{1}{a\|\xi\|}(e^{-ah} - 1) S(h)\xi \right\| \\ &\leq \|S(h)\xi\| (1 + \frac{1}{a\|\xi\|}(e^{-ah} - 1)) + \left\| \int_0^h S(h - s) g(\xi) ds \right\| \\ &\leq e^{ah}\|\xi\| - \frac{1}{a}(e^{ah} - 1) + \left\| \int_0^h S(h - s) g(\xi) ds \right\|, \end{aligned}$$

for h sufficiently small. Further,

$$\lim_{h \downarrow 0} \left(\frac{e^{ah} - 1}{h}\|\xi\| - \frac{e^{ah} - 1}{ah} + \left\| \frac{1}{h} \int_0^h S(h - s) g(\xi) ds \right\| \right) \leq (L + a)\|\xi\| - 1.$$

From these inequalities, it is easy to see that for any arbitrary sequence $(h_n)_n$, with $h_n \downarrow 0$, there exist $(\theta_n)_n$ in $\mathbb{R}$, $\theta_n \downarrow 0$, and $(g_n)_n$, defined by

$$g_n(s) = g(\xi) - S(s) e^{-as} \frac{\xi}{\|\xi\|} \in g(\xi) + D(0, 1),$$

for $n = 1, 2, \ldots$ and a.e. for $s \geq 0$, and such that $(h_n)_n$, $(\theta_n)_n$ and $(g_n)_n$ satisfy the conditions in Remark 9.1.1. Thus, we get (13.7.8). Moreover,

---

[13] We recall that $\mathcal{G}(\xi) = \{g \in L^1(\mathbb{R}_+; X);\ g(s) \in G(\xi) \text{ a.e. for } s \in \mathbb{R}_+\}$. See also Definition 9.1.3.

from Theorems 9.2.1 and 9.8.2, we deduce that for each $\xi \in X$, $\xi \neq 0$, there exist $T > 0$ and a noncontinuable mild solution $(z, u) : [0, T) \to \mathbb{R} \times X$ of the Cauchy problem

$$\begin{cases} z'(t) = (L + a)\|u(t)\| - 1 \\ u'(t) \in Au(t) + G(u(t)) \\ z(0) = \|\xi\| \quad \text{and} \quad u(0) = \xi, \end{cases} \tag{13.7.9}$$

which satisfies $(z(t), u(t)) \in K$ for every $t \in [0, T)$. This means that (13.7.5) is satisfied for every $t \in [0, T)$.

Now, let us observe that u, as a solution of (13.7.2), can be continued to $\mathbb{R}_+$ simply because G has sublinear growth. So, $u(T)$ exists, even though the solution (z, u) of (13.7.9) is defined merely on $[0, T)$. Clearly $u(T)$ must be 0 since otherwise, (z, u) can be continued to the right of T, thereby contradicting the fact that (z, u) is noncontinuable. This completes the proof. □

13.8. Periodic solutions

Let X be a Banach space, $A : D(A) \subseteq X \rightsquigarrow X$ an m-dissipative operator, $f : \mathbb{R} \times \overline{D(A)} \to X$ a given continuous function which is T-periodic with respect to its first argument and let us consider the periodic problem

$$\begin{cases} u'(t) \in Au(t) + f(t, u(t)) \\ u(0) = u(T). \end{cases} \tag{13.8.1}$$

Theorem 13.8.1. *Let X be a Banach space, $A : D(A) \subseteq X \rightsquigarrow X$ an m-dissipative operator and let $f : \mathbb{R} \times \overline{D(A)} \to X$ be continuous. If the semigroup of contractions generated by A is compact, $\overline{D(A)}$ is convex*[14], *$0 \in D(A)$, $0 \in A0$, $f(\cdot, x)$ is T-periodic for each $x \in \overline{D(A)}$, i.e.,*

$$f(t + T, x) = f(t, x),$$

for each $(t, x) \in \mathbb{R} \times \overline{D(A)}$, and there exists $r > 0$ such that

$$[x, f(t, x)]_+ \leq 0$$

for each $(t, x) \in \mathbb{R} \times \overline{D(A)}$ with $\|x\| = r$, and f is bounded on $\mathbb{R} \times (\overline{D(A)} \cap D(0, r))$, then (13.8.1) has at least one T-periodic solution.

Proof. Let us denote by $K = \overline{D(A)} \cap D(0, r)$ which is nonempty, bounded, closed and convex. We first suppose that, in addition to the hypotheses assumed, f is locally Lipschitz on $\mathbb{R} \times \overline{D(A)}$ and there exists $\delta > 0$

[14]This happens, for instance, whenever X^* is uniformly convex. See Barbu [**10**], (c) in Proposition 3.6, p. 77.

such that

$$[x, f(t,x)]_+ \leq -\delta$$

for each $(t,x) \in \mathbb{R} \times K$, with $\|x\| = r$. Then, for each $\xi \in \overline{D(A)} \cap D(\xi, r)$, the Cauchy problem

$$\begin{cases} u'(t) \in Au(t) + f(t, u(t)) \\ u(0) = \xi \end{cases} \tag{13.8.2}$$

has a unique noncontinuable solution $u : [0, T_m) \to \overline{D(A)}$. We will next show that $T_m = \infty$. To this aim, we begin by observing that K is invariant with respect to $A + f$. Indeed, since $0 \in D(A)$ and $0 \in A0$, we have that $u(\tau + h, 0, 0, 0) = 0$. Therefore, an appeal to (1.6.2) with $\eta = 0$, $g \equiv 0$ and $\widetilde{v} \equiv 0$, yields

$$\|u(\tau + h, \tau, \xi, f(\tau, \xi))\| \leq \|\xi\| + \int_\tau^{\tau+h} [u(s, \tau, \xi, f(\tau, \xi)), f(\tau, \xi)]_+ \, ds,$$

for each $h > 0$. Inasmuch as $\limsup_{(u,v)\to(x,y)} [u, v]_+ = [x, y]_+$, we conclude that, there exists $h_0 > 0$ such that for each $h \in (0, h_0)$,

$$\|u(\tau + h, \tau, \xi, f(\tau, \xi))\| \leq \|\xi\| - h\frac{\delta}{2} < r.$$

But this shows that, for each $(\tau, \xi) \in \mathbb{R} \times K$, we have

$$\liminf_{h \downarrow 0} \frac{1}{h} \operatorname{dist}(u(\tau + h, \tau, \xi, f(\tau, \xi)); K) = 0.$$

In view of Theorem 10.1.2, it follows that K is C^0-viable with respect to $A + f$. Since f is locally Lipschitz, the Cauchy problem (13.8.2) has the uniqueness property and accordingly K is invariant with respect to $A + f$. Since f is bounded on $\mathbb{R} \times K$, in view of Theorem 10.6.2 combined with (i) in Remark 3.6.2, it follows that $T_m = \infty$. So, we can define the Poincaré map $P : K \to K$ by

$$P(\xi) = u(T),$$

where $u : [0, T] \to K$ is the unique solution of (13.8.2). The idea is to show that P has at least one fixed point $\xi \in K$ which, by means of (13.8.2) will produce a T-periodic solution for (13.8.1). We prove this with the help of Schauder Fixed Point Theorem 1.3.3. So, as K is invariant with respect to $A + f$, it follows that P maps K into itself. To prove that P is continuous, let $\xi \in K$ and let $(\xi_n)_n$ in K with $\lim_n \xi_n = \xi$. From Theorem 1.6.6, we deduce that $\{P(\xi_n);\ n = 1, 2, \dots\}$ is relatively compact. On the other hand, each convergent subsequence of $(P(\xi_n))_n$ converges to $P(\xi)$. Indeed, let us assume for simplicity that $(P(\xi_n))_n$ is itself convergent, and let us denote by u_n the unique C^0-solution of (13.8.2) corresponding to the initial datum ξ_n. By (1.6.5), we have

$$\|u_n(t) - u(t)\| \leq \|\xi_n - \xi\| + \int_0^t \|f(s, u_n(s)) - f(s, u(s))\| \, ds \tag{13.8.3}$$

for $n = 1, 2, \ldots$ and $t \in [\,0, T\,]$.

At this point let us recall that f is locally Lipschitz. Then, for each $\tau \in [\,0, T\,]$, there exist $\rho_\tau > 0$ and $L_\tau > 0$ such that

$$\|f(t, x) - f(t, y)\| \leq L_\tau \|x - y\|$$

for each $(t, x), (t, y) \in [\,\tau - \rho_\tau, \tau + \rho_\tau\,] \times D(\xi, \rho_\tau)$. Since $[\,0, T\,]$ is compact, there exists a finite family, $\{\tau_i;\ i = 1, 2, \ldots, n\}$ in $[\,0, T\,]$, such that

$$[\,0, T\,] \subseteq \cup_{i=1}^n [\,\tau_i - \rho_{\tau_i}, \tau_i + \rho_{\tau_i}\,].$$

Let $\rho = \min\{\rho_{\tau_i};\ i = 1, 2, \ldots, n\}$ and $L = \max\{L_{\tau_i};\ i = 1, 2, \ldots, n\}$. We then have

$$\|f(t, x) - f(t, y)\| \leq L\|x - y\|$$

for each $(t, x), (t, y) \in [\,0, T\,] \times D(\xi, \rho)$. As $\lim_n \xi_n = \xi$, we may assume without loss of generality that $\|\xi_n - \xi\| \leq \rho e^{-TL}$ for $n = 1, 2, \ldots$. Fix $n = 1, 2, \ldots$ and let us remark that, on a right neighborhood $[\,0, a\,]$ of 0 we have $\|u_n(s) - u(s)\| \leq \rho$. Let a be the greatest number in $(0, T\,]$ with the property above. We will show that $a = T$. Indeed, if we assume by contradiction that $a < T$, from (13.8.3) and the Lipschitz condition, we get

$$\|u_n(t) - u(t)\| \leq \|\xi_n - \xi\| + \int_0^t L\|u_n(s) - u(s)\| \, ds$$

for $n = 1, 2, \ldots$ and $t \in [\,0, a\,]$. By Gronwall Lemma 1.8.4, we have

$$\|u_n(t) - u(t)\| \leq \|\xi_n - \xi\| e^{aL} \leq \rho e^{-(T-a)L}$$

for $n = 1, 2, \ldots$ and $t \in [\,0, a\,]$. Since both u_n and u are continuous, this inequality contradicts the maximality of a. This contradiction can be eliminated only if $a = T$. Thus we have

$$\|u_n(t) - u(t)\| \leq \|\xi_n - \xi\| e^{TL}$$

for $n = 1, 2, \ldots$ and $t \in [\,0, T\,]$. Hence each convergent subsequence of $(P(\xi_n))_n$ converges to $P(\xi)$. As $\{P(\xi_n);\ n = 1, 2, \ldots\}$ is relatively compact, this shows that even $\lim_n P(\xi_n) = P(\xi)$, and thus P is continuous. Finally, again by Theorem 1.6.6, we conclude that $P(K)$ is relatively compact. From the Schauder Fixed Point Theorem 1.3.3 it follows that P has at least one fixed point $\xi \in K$. Clearly the solution u of (13.8.2), with initial datum $\xi = P(\xi)$, is T-periodic.

We may now pass to the general case, i.e., f merely continuous. Let $\varepsilon > 0$ be arbitrary. By a well-known approximation result, there exists a locally Lipschitz function $f_\varepsilon : \mathbb{R} \times K \to X$ such that

$$\|f(t,x) - \varepsilon x - f_\varepsilon(t,x)\| \leq \frac{\varepsilon}{2}$$

for each $(t,x) \in \mathbb{R} \times K$. By the inequality above and (viii) in Exercise 1.6.1, we get

$$[x, f_\varepsilon(t,x)]_+ \leq -\frac{\varepsilon}{2}$$

for each $(t,x) \in \mathbb{R} \times K$, with $\|x\| = r$. In view of the first part of the proof, we know that the periodic problem

$$\begin{cases} u_\varepsilon'(t) \in Au_\varepsilon(t) + f_\varepsilon(t, u_\varepsilon(t)) \\ u_\varepsilon(0) = u_\varepsilon(T) \end{cases}$$

has at last one T-periodic solution $u_\varepsilon : \mathbb{R} \to K$. Let $(\varepsilon_n)_n$ be a sequence in $(0,1)$ with $\varepsilon_n \downarrow 0$ and let us denote by $u_n = u_{\varepsilon_n}$. Again by Theorem 1.6.6, we deduce that $\{u_n;\ n = 1,2,\dots\}$ is relatively compact in $C([a,T];X)$ for each $a \in (0,T)$. Then, $\{u_n(T);\ n = 1,2,\dots\}$ is relatively compact in X. Taking into account that $u_n(0) = u_n(T)$, again by Theorem 1.6.6, we conclude that $\{u_n;\ n = 1,2,\dots\}$ is relatively compact even in $C([0,T];X)$. So, we may assume with no loss of generality that there exists a continuous function $u : [0,T] \to K$ such that $\lim_n u_n(t) = u(t)$ uniformly for $t \in [0,T]$. Let us denote by $f_n = f_{\varepsilon_n}$. Since $\lim_n f_n(t, u_n(t)) = f(t, u(t))$ uniformly for $t \in [0,T]$, we deduce that u is a T-periodic solution of (13.8.1). The proof is complete. □

Problem 13.8.1. *Let $X = \ell_2$, $r > 0$ and let $f : X \to X$ be defined by*

$$f_n((x_k)_k) = a_n x_n + b_n(\|(x_k)_k\| - r)^2$$

for $n = 1,2,\dots$, where $a_n < 0$, $\lim_n a_n = 0$, $(b_k)_k \in \ell_2$ and $(b_k a_k^{-1})_k \notin \ell_2$. Prove that f is compact, $\langle f((x_k)_k), (x_k)_k\rangle < 0$ for each $(x_k)_k \in \ell_2$ with $\|(x_k)_k\| = r$, but, for each $T > 0$, $u'(t) = f(u(t))$ has no T-periodic solution.

Solutions to the proposed problems

Solutions to Chapter 1

Problem 1.3.1. The conclusion is straightforward if C is finite. The general case follows from the preceding one by taking ε-nets and using (i).

Problem 1.3.2. Let $(x_n)_n$ be a sequence in $B+C$ with $\lim_n x_n = x$. We have $x_n = y_n + z_n$ with $y_n \in B$ and $z_n \in C$, for $n = 1, 2, \ldots$. In view of Theorem 1.3.5 both B and C are weakly relatively sequentially compact. So, we may assume with no loss of generality that $\lim_n y_n = y$ and $\lim_n z_n = z$ weakly in X. Thanks to Theorem 1.1.1, we get $y \in B$ and $z \in C$. Thus $x \in B + C$ and this completes the proof.

Problem 1.3.3. Since $\varepsilon_n \downarrow 0$, for each $\varepsilon > 0$ there exists $n(\varepsilon) \in \mathbb{N}$ such that $\varepsilon_n \leq \frac{\varepsilon}{2}$ for each $n \in \mathbb{N}$, $n \geq n(\varepsilon)$. Since $\lim_{h\downarrow 0} m(h) = m(0) = 0$, for the same $\varepsilon > 0$, there exists $\delta_1(\varepsilon) > 0$ such that $m(h) \leq \frac{\varepsilon}{2}$ for each $h \in [\,0, \delta_1(\varepsilon)]$. Finally, since $\{u_1, u_2, \ldots, u_{n(\varepsilon)}\}$ is a finite set of continuous functions, it is equi-uniformly continuous on $[\,\tau, T\,]$. Therefore, for the very same $\varepsilon > 0$, there exists $\delta_2(\varepsilon) > 0$ such that $\|u_k(t) - u_k(\widetilde{t})\| \leq \varepsilon$, for $k = 1, 2, \ldots, n(\varepsilon)$ and each $t, \widetilde{t} \in [\,\tau, T\,]$, with $|t - \widetilde{t}| \leq \delta_2(\varepsilon)$. Set $\delta(\varepsilon) = \min\{\delta_1(\varepsilon), \delta_2(\varepsilon)\}$. Then, for each $n \in \mathbb{N}$ and each $t, \widetilde{t} \in [\,\tau, T\,]$, with $|t - \widetilde{t}| \leq \delta(\varepsilon)$, we have

$$\|u_n(t) - u_n(\widetilde{t})\| \leq \frac{\varepsilon}{2} + \frac{\varepsilon}{2}$$

which shows that $\{u_n;\ n \in \mathbb{N}\}$ is equi-uniformly continuous on $[\,\tau, T\,]$, as claimed.

Problem 1.3.4. Take $\varepsilon = 1$. Then, there exists $\delta > 0$ such that, for each interval E in $[\,\tau, T\,]$ whose length, $\lambda(E)$, is less than δ, we have

$$\int_E \|f(s)\|\, ds \leq 1$$

for each $f \in \mathcal{F}$. Since $[\,\tau, T\,]$ is compact, it has a finite covering of nonoverlaping intervals $E_1, E_2, \ldots, E_k$ with $\lambda(E_i) \leq \delta$ for $i = 1, 2, \ldots, k$. Then

$$\int_\tau^T \|f(s)\|\, ds = \sum_{i=1}^{k} \int_{E_i} \|f(s)\|\, ds \leq k$$

for each $f \in \mathcal{F}$. Thus $\mathcal{F}$ is norm bounded by k.

Exercise 1.6.1. To check (i)$\sim$(viii) just recall the definition of both directional derivatives $[\,x, y\,]_\pm$ and $(x, y)_\pm$ and apply the triangle inequality. Finally, (ix) is obtained in the very same manner as the classical chain rule, while the first implication in (x) follows from (iv) recalling that, whenever $\|\cdot\|$ is Gâteaux differentiable on $X \setminus \{0\}$, we have $[\,x, y\,]_+ = [\,x, y\,]_-$ for each $x \in X \setminus \{0\}$ and $y \in X$. Conversely, if $[\,x, y\,]_+ = -[\,-x, y\,]_+$ for each $x \in X \setminus \{0\}$ and $y \in X$, by (iv), (v) and (vi), we deduce that $[\,x, \cdot\,]$ is linear and thus $\|\cdot\|$ is Gâteaux differentiable on $X \setminus \{0\}$.

Exercise 1.6.2. Let $((t_n, x_n))_n$ and (t, x) in $\mathbb{R}_+ \times C$ be such that

$$\lim_n (t_n, x_n) = (t, x).$$

Using the semigroup properties, we get

$$\|S(t_n)x_n - S(t)x\| \leq \|S(t_n)x_n - S(t_n)x\| + \|S(t_n)x - S(t)x\|$$

$$\leq \begin{cases} \|x_n - x\| + \|S(t_n - t)x - x\| & \text{if } t_n \geq t \\ \|x_n - x\| + \|S(t - t_n)x - x\| & \text{if } t_n < t. \end{cases}$$

The conclusion follows from the continuity of $\tau \mapsto S(\tau)x$ at $\tau = 0$.

Exercise 1.6.3. We will check (i)~(iv) in Definition 1.6.3. First, let us observe that $S(0)\xi = u(0, 0, \xi, 0) = \xi$ for each $\xi \in \overline{D(A)}$ which proves (i). From the evolution property (1.6.4), we get

$$S(t + s)\xi = u(t + s, 0, \xi, 0) = u(t, 0, u(s, 0, \xi, 0), 0) = S(t)S(s)\xi$$

for each $t, s \geq 0$, wherefrom (ii). Moreover, as $u(\cdot, 0, \xi, 0)$ is continuous at $\tau = 0$ and $u(0, 0, \xi, 0) = \xi$, we deduce $\lim_{t \downarrow 0} S(t)\xi = \lim_{t \downarrow 0} u(t, 0, \xi, 0) = \xi$. Thus, we get (iii). Finally, from (1.6.2), we have

$$\|S(t)\xi - S(t)\eta\| = \|u(t, 0, \xi, 0) - u(t, 0, \eta, 0)\| \leq \|\xi - \eta\|$$

for each $t \geq 0$ and each $\xi, \eta \in \overline{D(A)}$, wherefrom (iv).

Exercise 1.6.4. Apply (1.6.2) and then use (ii) in Exercise 1.6.1 to evaluate $[\,\widetilde{u}(s) - \widetilde{v}(s), f(s) - g(s)\,]_+$ under the integral sign.

Problem 1.6.1. Let $(t_n)_n$ in $\mathbb{R}_+$ with $t_n \downarrow 0$, and let $n_0 \in \mathbb{N}$ be such that, for each $n \geq n_0$, $D(0, n) \cap C \neq \emptyset$. Since, for each $n \geq n_0$, $S(t_n)(D(0, n) \cap C)$ is precompact there exists a finite family of points C_n in $D(0, n) \cap C$ such that for every $\xi \in D(0, n) \cap C$ there exists $\xi_n \in C_n$ satisfying

$$\|S(t_n)\xi - S(t_n)\xi_n\| \leq t_n.$$

Let $\xi \in C$ and $\varepsilon > 0$ and choose $n \in \mathbb{N}$ such that $t_n \leq \varepsilon$, $\|\xi - S(t_n)\xi\| \leq \varepsilon$ and $\|\xi\| \leq n$. Taking $\xi_n \in C_n$ as above, we have

$$\|\xi - S(t_n)\xi_n\| \leq \|\xi - S(t_n)\xi\| + \|S(t_n)\xi - S(t_n)\xi_n\| \leq 2\varepsilon.$$

So $D = \cup_{n \geq n_0} S(t_n)C_n$ (which obviously is countable) is dense in C and this completes the proof.

Problem 1.8.1. Since

$$x(t) \leq \frac{1}{n} + \int_0^t \omega(x(s))\, ds$$

for $n = 1, 2, \ldots$ and each $t \in [\,0, T\,]$, an appeal to Lemma 1.8.2 shows that there exists $T_0 \in (0, T\,]$ such that $x \equiv 0$ on $[\,0, T_0\,]$. Let T_m be the greatest number T_0 in $(0, T\,]$ with the property above. If we assume that $T_m < T$, by applying Lemma 1.8.2 on $[\,T_m, T\,]$ we get a contradiction, i.e., that T_m is not maximal. This contradiction can be eliminated only if $T_m = T$ and this completes the proof.

Problem 1.8.2. Just repeat the proof of Problem 1.8.1 by using Lemma 1.8.3 instead of Lemma 1.8.2.

Problem 1.8.3. Let $y : [\tau, T) \to \mathbb{R}$ be defined by

$$y(t) = m + \int_\tau^t k(s)x(s)\,ds$$

for every $t \in [\tau, T)$. Clearly y is absolutely continuous on $[\tau, T)$ and

$$y'(t) = k(t)x(t) \le k(t)y(t)$$

a.e. for $t \in [\tau, T)$. Multiplying both sides by $e^{-\int_\tau^t k(\theta)\,d\theta}$, after some simple calculations, we get

$$y(t) \le me^{\int_\tau^t k(\theta)\,d\theta}$$

for every $t \in [\tau, T)$. But $x(t) \le y(t)$ for each $t \in [\tau, T)$ and this completes the proof.

Solutions to Chapter 2

Problem 2.1.1. Let $x \in K$ and $t > 0$. Let us fix $\varepsilon > 0$. On $\mathcal{S} = K \times [0, t]$ we introduce the binary relation $\preceq$ defined by $(v_1, s_1) \preceq (v_2, s_2)$ if $s_1 \le s_2$ and $\|S(s_2 - s_1)v_1 - v_2\| \le \varepsilon(s_2 - s_1)$. It is easy to see that $\preceq$ is a preorder on $\mathcal{S}$. The reflexivity is obvious. To prove the transitivity, let $(v_1, t_1) \preceq (v_2, t_2)$ and $(v_2, t_2) \preceq (v_3, t_3)$. We have $\|S(t_2 - t_1)v_1 - v_2\| \le \varepsilon(t_2 - t_1)$ and $\|S(t_3 - t_2)v_2 - v_3\| \le \varepsilon(t_3 - t_2)$. But

$$\begin{aligned}\|S(t_3 - t_1)v_1 - v_3\| &\le \|S(t_3 - t_2)S(t_2 - t_1)v_1 - S(t_3 - t_2)v_2\| + \|S(t_3 - t_2)v_2 - v_3\| \\ &\le \|S(t_2 - t_1)v_1 - v_2\| + \|S(t_3 - t_2)v_2 - v_3\| \le \varepsilon(t_3 - t_1),\end{aligned}$$

as claimed. Let us define the function $\mathcal{N} : \mathcal{S} \to \mathbb{R}$ by $\mathcal{N}((v, s)) = s$. Clearly, $\mathcal{N}$ is increasing. Let us now prove that it satisfies condition (i) in Theorem 2.1.1. To this end, let $((v_n, t_n))_n$ any increasing sequence in $\mathcal{S}$. It is obvious that the sequence $(t_n)_n$ is convergent, say, to t_0, and $t_n \le t_0$ for every $n \in \mathbb{N}$. We prove that the sequence $(v_n)_n$ is Cauchy and hence convergent. First, let us observe that, for each $n \in \mathbb{N}$ and each $k \in \mathbb{N}$, we have

$$\|S(t_{n+k} - t_n)v_n - v_{n+k}\| \le \varepsilon(t_{n+k} - t_n). \tag{14.2.1}$$

Since $(t_n)_n$ is convergent too, for every $\eta > 0$ there exists $n_\eta \in \mathbb{N}$ such that for $n \ge n_\eta$ we have $\varepsilon(t_{n+k} - t_n) < \eta$. For $n \ge n_\eta$ and $k, l \in \mathbb{N}$ we get

$$\|v_{n+k} - v_{n+l}\| \le \|S(t_{n+k} - t_n)v_n - S(t_{n+l} - t_n)v_n\| + 2\eta.$$

Fix $n \ge n_\eta$. Letting $k \to \infty$, $l \to \infty$, we deduce that the right hand side of the above inequality tends to 2η. Therefore, there exists n'_η such that for $k, l \ge n'_\eta$ we have

$$\|v_{n+k} - v_{n+l}\| \le 3\eta,$$

which shows that $(v_n)_n$ is a Cauchy sequence. Let $v_0 = \lim_n v_n$. Letting $k \to \infty$ in (14.2.1), we conclude

$$\|S(t_0 - t_n)v_n - v_0\| \le \varepsilon(t_0 - t_n),$$

which shows that $(v_n, t_n) \preceq (v_0, t_0)$ for every $n \in \mathbb{N}$. We apply Theorem 2.1.1 and deduce the existence of an $\mathcal{N}$-maximal element $(\widetilde{v}, \widetilde{t}) \in \mathcal{S}$ such that $(x, 0) \preceq (\widetilde{v}, \widetilde{t})$. We show that $\widetilde{t} = t$. Indeed, let us assume by contradiction that $\widetilde{t} < t$. By (2.1.3), there exists $0 < t' < t - \widetilde{t}$ such that

$$\operatorname{dist}(S(t')\widetilde{v}; K) < \varepsilon t'.$$

Therefore, there exists $v \in K$ such that $\|S(t')\widetilde{v} - v\| < \varepsilon t'$, which implies that $(\widetilde{v}, \widetilde{t}) \preceq (v, \widetilde{t} + t')$. But this contradicts the fact that $(\widetilde{v}, \widetilde{t})$ is $\mathcal{N}$-maximal. Hence we have proved that there exists $\widetilde{v} \in K$ such that $(x, 0) \preceq (\widetilde{v}, t)$, which means that

$$\|S(t)x - \widetilde{v}\| \leq \varepsilon t.$$

Since $\widetilde{v} \in K$, we deduce that $\operatorname{dist}(S(t)x; K) \leq \varepsilon t$. As ε is arbitrary and K is closed, we get the conclusion.

The hypothesis that $S(t)$ is nonexpansive is not essential. More precisely, if there exists $a \in \mathbb{R}_+$ such that $\|S(t)v_1 - S(t)v_2\| \leq e^{at}\|v_1 - v_2\|$ for each $t \geq 0$ and $v_1, v_2 \in M$, then we can get the same conclusion repeating the proof above by considering the preorder $(v_1, t_1) \preceq (v_2, t_2)$ if $t_1 \leq t_2$ and $\|S(t_2 - t_1)v_1 - v_2\| \leq (\varepsilon/a)(e^{a(t_2 - t_1)} - 1)$.

Problem 2.2.1. Let us first assume that K is closed relative to D. Then there exists a closed set K_0 such that $K = D \cap K_0$. Since D is open, for each $x \in D$ there exists $r_x > 0$ such that $D(x, r_x) \subseteq D$. As $D(x, r_x) \cap K_0 = D(x, r_x) \cap K$ and $D(x, r_x) \cap K_0$ is closed, it follows that $D(x, r_x) \cap K$ is closed, as claimed. Conversely, if K is locally closed, it follows that for each $x \in K$ there exists $r_x > 0$ such that $K \cap D(x, r_x)$ is closed. Let us denote by $B(x, r_x)$ the open ball with center x and radius r_x and let us observe that $D = \cup_{x \in K} B(x, r_x)$ is open and $K \subseteq D$. We will show that K is closed relative to D. To this aim, we will prove that $K = D \cap \overline{K}$. Since the inclusion $K \subseteq D \cap \overline{K}$ is obvious, it remains to show that $D \cap \overline{K} \subseteq K$. Let $\xi \in D \cap \overline{K}$ be arbitrary. As $\xi \in D$, there exists $x \in K$ such that $\xi \in B(x, r_x)$. On the other hand, since $\xi \in \overline{K}$, there exists $(\xi_n)_n$ with $\xi_n \in K$ for $n = 1, 2, \ldots$ and $\lim_n \xi_n = \xi$. Since $B(x, r_x)$ is open, we may assume without loss of generality that $\xi_n \in B(x, r_x)$ for $n = 1, 2, \ldots$. Accordingly, $\xi_n \in K \cap D(x, r_x)$ for $n = 1, 2, \ldots$. Taking into account that $K \cap D(x, r_x)$ is closed, we conclude that $\xi \in K \cap D(x, r_x)$ and hence $\xi \in K$, as claimed.

Problem 2.3.1. Let us observe that, by Definition 2.3.1, $E \in \mathcal{TS}_K(\xi)$ if and only if, for each $\rho > 0$, we have

$$\liminf_{h \downarrow 0} \frac{1}{h} \operatorname{dist}(\xi + hE; K \cap D(\xi, \rho)) = 0,$$

relation equivalent to

$$\sup_{\delta > 0} \inf_{h \in (0, \delta)} \frac{1}{h} \operatorname{dist}(\xi + hE; K \cap D(\xi, \rho)) = 0.$$

In its turn, this relation is equivalent to: for each $\delta > 0$ and each $\varepsilon > 0$, there exists $h \in (0, \delta)$ such that

$$\operatorname{dist}(\xi + hE; K \cap D(\xi, \rho)) < \varepsilon h.$$

Since dist $(C; D) < \alpha$ if and only if there exist $x \in C$ and $y \in D(0, \alpha)$ such that $x + y \in D$, we finally deduce that $E \in \mathcal{TS}_K(\xi)$ if and only if for each $\rho > 0$, each $\delta > 0$ and each $\varepsilon > 0$, there exist $h \in (0, \delta)$, $\eta \in E$ and $p \in D(0, \varepsilon)$ such that

$$\xi + h\eta + hp \in K \cap D(\xi, \rho),$$

and so (i) is equivalent to (iii).

Now, to prove that (iii) implies (iv), let us take in (iii) $\rho = \delta = \varepsilon = \frac{1}{n}$, for $n = 1, 2, \ldots$. It follows that there exist $(h_n)_n$, in $\mathbb{R}_+$, with $h_n \downarrow 0$, $(\eta_n)_n$ in E and $(p_n)_n$, in X, with $\lim_n p_n = 0$, and such that

$$\xi + h_n\eta_n + h_n p_n \in K \cap D\left(\xi, \frac{1}{n}\right)$$

for $n = 1, 2, \ldots$. From $\|h_n\eta_n + h_n p_n\| \leq \frac{1}{n}$ and $h_n p_n \to 0$ we get $h_n\eta_n \to 0$, and thus (iv) is fulfilled.

Conversely, let us assume that (iv) holds. Let $\rho > 0$, $\delta > 0$, and $\varepsilon > 0$. Then, there exists $n_0 \in \mathbb{N}$ such that $h_{n_0} \in (0, \delta)$, $p_{n_0} \in D(0, \varepsilon)$, $\xi + h_{n_0}\eta_{n_0} + h_{n_0}p_{n_0} \in K$ and

$$\|h_{n_0}\eta_{n_0} + h_{n_0}p_{n_0}\| < \rho.$$

Therefore $\xi + h_{n_0}\eta_{n_0} + h_{n_0}p_{n_0} \in K \cap D(\xi, \rho)$, hence (iii) is satisfied. Since the equivalence between (i) and (ii) is straightforward, the proof is complete.

Problem 2.3.2. Arguing as in the proof of Problem 2.3.1, we deduce that conditions (ii), (iii) and (iv) are equivalent. On the other hand, since $A \subseteq B$ implies dist $(x; A) \geq$ dist $(x; B)$, it follows that whenever $E \in \mathcal{TS}_K(\xi)$ we have (ii). Next we show that, for each bounded set E, the converse implication is true. Let $E \in \mathcal{B}(X)$ satisfying (ii). But (ii) and (v) are equivalent and thus there exist three sequences, $(h_n)_n$ in $\mathbb{R}_+$, with $h_n \downarrow 0$, $(\eta_n)_n$ in E and $(p_n)_n$ in X, with $\lim_n p_n = 0$, such that

$$\xi + h_n\eta_n + h_n p_n \in K.$$

Since E is bounded, it follows that $\lim_n h_n\eta_n = 0$ and, in view of the equivalence between (i) and (iv) in Problem 2.3.1, we get $E \in \mathcal{TS}_K(\xi)$. Since the equivalence between (iv) and (v) is obvious, the proof is complete.

Problem 2.3.3. Let $\varepsilon > \text{dist}_{HP}(B; C)$. Then, $\varepsilon > e(B; C)$ and $\varepsilon > e(C; B)$. Therefore, $B \subseteq C + D(0, \varepsilon)$ and $C \subseteq B + D(0, \varepsilon)$. Consequently

$$\inf\{\varepsilon > 0; B \subseteq C + D(0, \varepsilon),\ C \subseteq B + D(0, \varepsilon)\} \leq \text{dist}_{HP}(B; C).$$

To prove the converse inequality, we first show that

$$e(B; C) = \inf\{\varepsilon > 0;\ B \subseteq C + D(0, \varepsilon)\}. \tag{14.2.2}$$

Indeed, $\varepsilon > e(B; C)$ if and only if $\varepsilon > \sup_{x \in B} \text{dist}(x; C)$ which is equivalent to $B \subseteq C + D(0, \varepsilon)$. So, $\inf\{\varepsilon > 0;\ B \subseteq C + D(0, \varepsilon)\} \leq e(B; C)$. Next, if $\varepsilon > 0$ is such that $B \subseteq C + D(0, \varepsilon)$, then $e(B; C) \leq \varepsilon$. Thus

$$e(B; C) \leq \inf\{\varepsilon > 0;\ B \subseteq C + D(0, \varepsilon)\}$$

and this completes the proof of (14.2.2). Let

$$d = \inf\{\varepsilon > 0;\ B \subseteq C + D(0,\varepsilon),\ C \subseteq B + D(0,\varepsilon)\}.$$

By (14.2.2), we have both $e(C;B) \leq d$ and $e(B;C) \leq d$ and so $\mathrm{dist}_{HP}(B;C) \leq d$, and this completes the proof.

Problem 2.3.4. Let us denote by $d = \sup_{x\in X}(\mathrm{dist}\,(x;C) - \mathrm{dist}\,(x;B))$ and by $\mathrm{dist}\,(x;y) = \|x-y\|$. We begin by checking that $e(B;C) \leq d$. Let $x \in B$ and $\varepsilon > 0$. We have

$$\mathrm{dist}\,(x;C) = \mathrm{dist}\,(x;C) - \mathrm{dist}\,(x;B) \leq d < d+\varepsilon$$

which shows that $x \in C + D(0, d+\varepsilon)$. Hence $B \subseteq C + D(0, d+\varepsilon)$ and, since $\varepsilon > 0$ is arbitrary, in view of (14.2.2), we get $e(B;C) \leq d$.

To prove the converse inequality, i.e., $e(B;C) \geq d$, let $\varepsilon > 0$, let $x \in X$ and choose $b \in B$ such that $\mathrm{dist}\,(x;b) < \mathrm{dist}\,(x;B) + \varepsilon/2$. Next, choose $c \in C$ such that $\mathrm{dist}\,(b;c) < \mathrm{dist}\,(b;C) + \varepsilon/2 \leq e(B;C) + \varepsilon/2$. We have

$$\mathrm{dist}\,(x;C) \leq \mathrm{dist}\,(x;c) \leq \mathrm{dist}\,(x;b) + \mathrm{dist}\,(b;c) < \mathrm{dist}\,(x;B) + e(B;C) + \varepsilon.$$

Therefore

$$\mathrm{dist}\,(x;C) - \mathrm{dist}\,(x;B) \leq e(B;C) + \varepsilon.$$

But both $\varepsilon > 0$ and $x \in X$ were arbitrary, and thus $d \leq e(B;C)$. But

$$\max\left\{\sup_{x\in X}\{\mathrm{dist}\,(x;C) - \mathrm{dist}\,(x;B)\}, \sup_{x\in X}\{\mathrm{dist}\,(x;B) - \mathrm{dist}\,(x;C)\}\right\}$$
$$= \sup_{x\in X}\left|\mathrm{dist}\,(x;C) - \mathrm{dist}\,(x;B)\right|.$$

The proof is complete.

Problem 2.3.5. Since (i)~(v) are simple consequences of Definition 2.3.1, we confine ourselves only to the proof of (vi). To check (vi), let $E \in \mathcal{B}(X)$ and let $(E_n)_n$ be such that $E_n \in \mathcal{TS}_K(\xi)$ for $n = 1,2,\ldots$ and

$$\lim_n e(E_n;E) = 0.$$

Let $\varepsilon > 0$ and fix $n = 1,2,\ldots$ such that

$$e(E_n;E) \leq \varepsilon.$$

Since $E_n \in \mathcal{TS}_K(\xi)$, in view of the equivalence between (i) and (ii) in Problem 2.3.2, there exist $\widetilde{\eta}_n \in E_n$ and $h_n \in (0,\varepsilon)$ such that

$$\mathrm{dist}\ (\xi + h_n\widetilde{\eta}_n; K) \leq h_n\varepsilon.$$

Since $e(E_n;E) \leq \varepsilon$, there exists $\eta_n \in E$ such that $\|\eta_n - \widetilde{\eta}_n\| \leq 2\varepsilon$. We then have

$$\mathrm{dist}\ (\xi + h_n\eta_n; K) \leq \mathrm{dist}\ (\xi + h_n\widetilde{\eta}_n; K) + h_n\|\widetilde{\eta}_n - \eta_n\| \leq 3h_n\varepsilon.$$

But this inequality combined with (iii) in Problem 2.3.2 shows that $E \in \mathcal{TS}_K(\xi)$ and this completes the proof.

Problem 2.4.1. If $\eta \in K$ and K is a cone, we have $h\eta \in K$ for each $h > 0$. Therefore $\liminf_{h\downarrow 0}\frac{1}{h}\mathrm{dist}\,(0 + h\eta; K) = 0$, which proves that $\eta \in \mathcal{T}_K(0)$. Hence $K \subseteq \mathcal{T}_K(0)$. To prove the converse inclusion, we have merely to observe that, whenever $\eta \notin K$ we have also $\eta \notin \mathcal{T}_K(0)$. So let $\eta \notin K$. Since K is closed, $\operatorname{dist}(\eta; K) > 0$. On the other hand

$$\frac{1}{h}\operatorname{dist}(0 + h\eta; K) = \operatorname{dist}(\eta; \frac{1}{h}K) = \operatorname{dist}(\eta; K),$$

which shows that $\eta \notin \mathcal{T}_K(0)$. The proof is complete.

Problem 2.4.2. Let $E \in \mathcal{TS}_K(\xi)$ be compact. Then, we may assume with no loss of generality that $(\eta_n)_n$ in (v), Problem 2.3.2, is convergent to some $\eta \in E$. We then have $\xi + h_n\eta_n + h_np_n \in K$ if and only if $\xi + h_n\eta + h_nq_n \in K$, where $q_n = p_n + \eta_n - \eta \to 0$. In view of Problem 2.3.2, $\{\eta\} \in \mathcal{TS}_K(\xi)$, i.e. $\eta \in \mathcal{T}_K(\xi)$.

Problem 2.4.3. Clearly K is a closed cone and therefore $\mathcal{T}_K(0) = K$. See Problem 2.4.1. Let $\Omega = X \setminus K$. Clearly

$$\Omega = \{f \in C([\,0,1\,])\,; \text{for each } t \in [\,0,1\,],\ f(t) > 0\}.$$

For $f \in \Omega$, we define

$$\widetilde{f}(t) = f(t) - \min_{[\,0,1\,]} f(t) = f(t) - f(t_{\min}).$$

We have $\widetilde{f}(t_{\min}) = 0$ and therefore $\widetilde{f} \in K$ and $\|\widetilde{f} - f\| = \min_{[\,0,1\,]} f(t)$. So,

$$\operatorname{dist}(hE; K) \le \|hf - h\widetilde{f}\| = \min_{[\,0,1\,]} hf(t) = h \min_{[\,0,1\,]} f(t)$$

for each $f \in E$ and $h > 0$. Hence

$$\operatorname{dist}(hE; K) \le h \inf_{f \in E} \min_{[\,0,1\,]} f(t) = 0$$

and consequently $E \in \mathcal{TS}_K(0)$. Nevertheless $E \cap \mathcal{T}_K(0) = \emptyset$ simply because $E \subseteq \Omega$.

Problem 2.4.4. Let $(\eta_n)_n$ in $\mathcal{T}_K(\xi)$ with $\lim_n \eta_n = \eta$. Then $(\{\eta_n\})_n$ is in $\mathcal{TS}_K(\xi)$ and $\lim_n e(\{\eta_n\}; \{\eta\}) = 0$. In view of (vi) in Problem 2.3.5, $\{\eta\} \in \mathcal{TS}_K(\xi)$ which means that $\eta \in \mathcal{T}_K(\xi)$, as claimed.

Problem 2.4.5. Let $\eta \in \mathcal{T}_\Sigma(\xi)$. If $\eta = 0$ we have nothing to prove. So let $\eta \neq 0$. In view of Corollary 2.4.1, there exist $(h_n)_n$, $h_n \downarrow 0$ and $(p_n)_n$, with $\lim_n p_n = 0$, such that $\xi + h_n\eta + h_np_n \in \Sigma$, i.e., $\|\xi + h_n\eta + h_np_n\| = r$ for $n = 1, 2, \ldots$. Hence

$$0 = \frac{\|\xi + h_n\eta + h_np_n\|^2 - \|\xi\|^2}{2h_n}$$

and consequently,

$$0 = \lim_n \frac{\|\xi + h_n\eta + h_np_n\|^2 - \|\xi\|^2}{2h_n} = \lim_n \frac{\|\xi + h_n\eta\|^2 - \|\xi\|^2}{2h_n} = (\xi, \eta)_+.$$

Thus, if $\eta \in \mathcal{T}_\Sigma(\xi)$, we have $(\xi, \eta)_+ = 0$. Conversely, if $(\xi, \eta)_+ = 0$, we have either $\eta = 0$ which shows that $\eta \in \mathcal{T}_\Sigma(\xi)$, or $\eta \neq 0$. In the latter case, in view of (i) in Exercise 1.6.1, we have $[\,\xi, \eta\,]_+ = 0$ which means

$$\lim_{h \downarrow 0} \frac{\|\xi + h\eta\| - \|\xi\|}{h} = 0.$$

Since $\|\xi\| = r$, for each $h > 0$, we have

$$\frac{1}{h}\text{dist}\,(\xi + h\eta; \Sigma) \leq \frac{1}{h}\left\|\xi + h\eta - r\frac{\xi + h\eta}{\|\xi + h\eta\|}\right\| = \left|\frac{\|\xi + h\eta\| - \|\xi\|}{h}\right|$$

which proves that $\lim_{h\downarrow 0}\frac{1}{h}\text{dist}\,(\xi + h\eta; \Sigma) = 0$. Thus $\eta \in \mathcal{F}_{\Sigma}(\xi) \subseteq \mathcal{T}_{\Sigma}(\xi)$ and this completes the proof of (i).

To prove the second assertion just repeat, with minor modifications, the arguments above.

Problem 2.5.1. Let $\eta \in \mathcal{C}_K(\xi_0)$ and let $(\xi_m)_m$ in K with $\lim_m \xi_m = \xi_0$. The idea is to find $(\eta_m)_m$ in $\mathcal{T}_K(\xi_m)$ with $\lim_m \eta_m = \eta$. This would imply that $\eta \in \liminf_{\substack{\xi\to\xi_0\\ \xi\in K}} \mathcal{T}_K(\xi)$. To this end, let us observe that, since $\eta \in \mathcal{C}_K(\xi_0)$, for every $\varepsilon > 0$, there exist $m_\varepsilon \in \mathbb{N}$ and $h_\varepsilon > 0$ such that, for all $0 < h < h_\varepsilon$ and $m \geq m_\varepsilon$, we have

$$\text{dist}\,(\xi_m + h\eta; K) < h\varepsilon.$$

Fix m as above and take $\mu_m^h \in K$ with $\|\mu_m^h - \xi_m - h\eta\| < h\varepsilon$. Let us consider

$$\eta_m^h = \frac{1}{h}(\mu_m^h - \xi_m).$$

Since $\|\eta_m^h - \eta\| \leq \varepsilon$, $\{\eta_m^h;\ 0 < h < h_\varepsilon\}$ is bounded and therefore it has a limit point η_m as $h \downarrow 0$. In its turn, η_m satisfies $\|\eta_m - \eta\| \leq \varepsilon$. A simple computational argument shows that $\eta_m \in \mathcal{T}_K(\xi_m)$. Since $\mathcal{T}_K(\xi) \subseteq \mathcal{B}_K(\xi)$ for each $\xi \in K$, we have

$$\liminf_{\substack{\xi\to\xi_0\\ \xi\in K}} \mathcal{T}_K(\xi) = \liminf_{\substack{\xi\to\xi_0\\ \xi\in K}} \mathcal{B}_K(\xi)$$

for each $\xi_0 \in K$. If, in addition, K is locally closed, by Lemma 2.2.1, it is proximal. Then, by Lemma 2.5.1, we have

$$\liminf_{\substack{\xi\to\xi_0\\ \xi\in K}} \mathcal{B}_K(\xi) = \mathcal{C}_K(\xi_0),$$

and this completes the proof.

Problem 2.6.1. Clearly F is nonempty, closed and convex values. In addition, since f_i, $i = 1, 2$, are bounded it follows that, for each $\xi \in L^1(\Omega)$, $F(\xi)$ lies and is bounded in $L^\infty(\Omega)$ and thus in $L^2(\Omega)$. As a consequence, $F(\xi)$ is weakly compact in $L^2(\Omega)$. But $L^2(\Omega) \subseteq L^1(\Omega)$, and therefore $F(\xi)$ is weakly compact in $L^1(\Omega)$. To prove that F is strongly-weakly u.s.c. on $L^1(\Omega)$, we proceed by contradiction. So, let us assume that there exists $\xi \in L^1(\Omega)$ such that F is not strongly-weakly u.s.c. at ξ. Then, there would exist an open halfspace E in $L^1(\Omega)$ containing $F(\xi)$ and a sequence $(\xi_n)_n$ in $L^1(\Omega)$ with $\lim_n \xi_n = \xi$ and $F(\xi_n) \nsubseteq E$. This means that there exists a sequence $(\eta_n)_n$ with $\eta_n \in F(\xi_n)$ and $\eta_n \in L^1(\Omega) \setminus E$. Clearly $(\eta_n)_n$ is uniformly bounded and thus, we may assume with no loss of generality that it is weakly convergent in $L^1(\Omega)$ to some function η. By Corollary 1.1.1, there exists a sequence $(g_n)_n$ with $g_n \in \text{conv}\,\{\eta_k;\ k \geq n\}$, for $n = 1, 2, \ldots$ and such that $\lim_n g_n = \eta$ strongly in $L^1(\Omega)$. We may assume further that $\lim_n g_n = \eta$ a.e. on Ω. Since f_1 is l.s.c. and f_2 is u.s.c., we conclude that $\eta \in F(\xi) \subseteq E$. On the other hand, we have $\eta \in L^1(\Omega)\setminus E$ simply because the latter is closed. This contradiction can be eliminated only if F is strongly-weakly u.s.c. on $L^1(\Omega)$, as claimed.

Problem 2.7.1. Except for (ii), which is nothing but an equivalent formulation of Theorem 1.3.1, (i)~(vii) are simple consequences of Definitions 2.7.1 and 2.7.2. It remains to check (viii), i.e., that γ is Lipschitz continuous with respect to the Hausdorff-Pompeiu distance with Lipschitz constant 1 in the case of β and 2 in the case of α. We will confine ourselves only to the proof of

$$|\beta(B) - \beta(C)| \leq \mathrm{dist}_{HP}(B; C). \tag{14.2.3}$$

In view of Problem 2.3.3, we have

$$\mathrm{dist}_{HP}(B; C) = \inf\{\varepsilon > 0;\ B \subseteq C + D(0,\varepsilon),\ C \subseteq B + D(0,\varepsilon)\}$$

and therefore

$$\begin{cases} \beta(B) \leq \mathrm{dist}_{HP}(B; C) + \beta(C) \\ \beta(C) \leq \mathrm{dist}_{HP}(B; C) + \beta(B), \end{cases}$$

which completes the proof of (14.2.3).

Problem 2.7.2. If $\{x_1, x_2, \ldots, x_n\} \subseteq X$ is such that $B \subseteq \cup_{i=1}^n D(x_i,\varepsilon)$, then $B = \cup_{i=1}^n (B \cap D(x_i,\varepsilon))$ and $B \cap D(x_i,\varepsilon) \in \mathcal{B}_{2\varepsilon}(X)$, for $i = 1, 2, \ldots, n$. Therefore $\alpha(B) \leq 2\beta(B)$. Clearly $\beta(B) \leq \beta_Y(B)$. Finally let $B \in \mathcal{B}(Y)$, $B \subseteq \cup_{i=1}^n B_i$, with $B_i \in \mathcal{B}_\varepsilon(X)$ and $B_i \cap B \neq \emptyset$ for $i = 1, 2, \ldots, n$. Let $x_i \in B_i \cap B$, for $i = 1, 2, \ldots, n$. As $x_i \in Y$ for $i = 1, 2, \ldots, n$ and $B \subseteq \cup_{i=1}^n D(x_i,\varepsilon)$, we conclude that $\beta_Y(B) \leq \alpha(B)$.

Solutions to Chapter 3

Problem 3.4.1. Let us consider a sequence $(\varepsilon_n)_n$ in $(0,1)$, decreasing to 0, and let $((\sigma_n, g_n, u_n))_n$ be a sequence of ε_n-approximate solutions of (3.3.1) on $[\,0,T\,]$ as given by Lemma 3.3.1. We will show that $(u_n)_n$ has at least one convergent subsequence in the sup-norm.

First, let us observe that, in view of (iv) (i), (ii) in Lemma 3.3.1 and of (3.3.2), we have

$$\|u_n(t) - \xi\| \leq \int_0^t \|f(u_n(\sigma_n(s)))\|\,ds + \int_0^t \|g_n(s)\|\,ds \leq T(M+1)$$

for each $n \in \mathbb{N}$ and $t \in [\,0,T\,]$. Thus $\{u_n;\ n \in \mathbb{N}\}$ is uniformly bounded on $[\,0,T\,]$. Similarly, we get

$$\|u_n(t) - u_n(\widetilde{t})\| \leq \left|\int_t^{\widetilde{t}} \|f(u_n(\sigma_n(s)))\|\,ds\right| + \left|\int_t^{\widetilde{t}} \|g_n(s)\|\,ds\right| \leq (M+1)|t - \widetilde{t}|$$

for every $n \in \mathbb{N}$ and every $t, \widetilde{t} \in [\,0,T\,]$. Consequently $\{u_n;\ n \in \mathbb{N}\}$ is equicontinuous on $[\,0,T\,]$. By virtue of Arzelá-Ascoli Theorem 1.3.6 – we recall that X is finite dimensional and therefore the uniform boundedness of $\{u_n;\ n \in \mathbb{N}\}$ implies the condition (ii) in Theorem 1.3.6 –, we conclude that there exists a continuous function $u : [\,0, T\,] \to X$ such that, on a subsequence, denoted for simplicity also by $(u_n)_n$, we have

$$\lim_n u_n(t) = u(t)$$

uniformly for $t \in [\,0, T\,]$. From this point the proof runs exactly as that one of Theorem 3.2.2.

Problem 3.4.2. Let us consider a sequence $(\varepsilon_n)_n$ in $(0, 1)$, decreasing to 0, and let $((\sigma_n, g_n, u_n))_n$ be a sequence of ε_n-approximate solutions of (3.3.1) on $[\,0, T\,]$ as given by Lemma 3.3.1. We will show that $(u_n)_n$ is a Cauchy sequence in the sup-norm.

We have

$$\|u_n(\sigma_n(t)) - u_m(\sigma_m(t))\|$$

$$\leq \|u_n(\sigma_n(t)) - u_n(t)\| + \|u_n(t) - u_m(t)\| + \|u_m(t) - u_m(\sigma_m(t))\|. \qquad (14.3.1)$$

At this point let us observe that, for each $n \in \mathbb{N}$ and each $t \in [\,0, T\,]$, we have

$$\|u_n(t) - u_n(\sigma_n(t))\| \leq \int_{\sigma_n(t)}^{t} \|f(u_n(\sigma_n(s)))\| \, ds + \int_{\sigma_n(t)}^{t} \|g_n(s)\| \, ds.$$

In view of (i) and (ii), we deduce that

$$\|u_n(t) - u_n(\sigma_n(t))\| \leq (M + \varepsilon_n)(t - \sigma_n(t)) \leq (M + 1)\varepsilon_n \qquad (14.3.2)$$

for each $n \in \mathbb{N}$ and each $t \in [\,\tau, T\,]$. From the Lipschitz condition (3.2.1)

$$\|u_n(t) - u_m(t)\| \leq L \int_0^t \|u_n(\sigma_n(s)) - u_m(\sigma_m(s))\| \, ds + \int_0^t \|g_n(s) - g_m(s)\| \, ds$$

for every $n, m \in \mathbb{N}$ and every $t \in [\,0, T\,]$. From (ii) in Lemma 3.3.1, (14.3.1) and (14.3.2), we get

$$\|u_n(t) - u_m(t)\| \leq T[L(M+1) + 1](\varepsilon_n + \varepsilon_m) + L \int_0^t \|u_n(s) - u_m(s)\| \, ds \quad (14.3.3)$$

for every $n, m \in \mathbb{N}$ and every $t \in [\,0, T\,]$. By Gronwall's Lemma 1.8.4, we deduce

$$\|u_n(t) - u_m(t)\| \leq T[L(M+1) + 1](\varepsilon_n + \varepsilon_m)e^{LT}$$

for all $n, m \in \mathbb{N}$ and $t \in [\,0, T\,]$.

But this inequality shows that $(u_n)_n$ is a Cauchy sequence in the sup-norm. So, there exists a function $u : [\,0, T\,] \to X$ such that

$$\lim_n u_n(s) = u(s)$$

uniformly for $s \in [\,0, T\,]$. Taking into account of (iii) and of the fact that $D(\xi, \rho) \cap K$ is closed, we deduce that $u(t) \in D(\xi, \rho) \cap K$ for every $t \in [\,0, T\,]$. Finally, passing to the limit for $n \to \infty$ both sides in the approximate equations

$$u_n(t) = \xi + \int_0^t f(u_n(\sigma_n(s))) \, ds + \int_0^t g_n(s) \, ds,$$

and observing that $\lim_n \sigma_n(t) = t$ uniformly for $t \in [\,0, T\,]$, we conclude that

$$u(t) = \xi + \int_0^t f(u(s))\, ds$$

for every $t \in [\,0, T\,]$, which completes the proof of Theorem 3.2.3.

Problem 3.4.3. We proceed as in the case of Problem 3.4.1 with the mention that $\rho > 0$ should be chosen small enough in order that $f(D(\xi, \rho) \cap K)$ be relatively compact. Then use Lemma 1.3.1 and the approximate equations (iv) in Lemma 3.3.1, to deduce the compactness of the cross-sections $\{u_n(t);\ n \in \mathbb{N}\}$ for $t \in [\,0, T\,]$. From now on, the proof continues as in the case of Problem 3.4.1.

Problem 3.5.1. We have

$$\frac{1}{h}\text{dist}\,(\xi + hf(\tau, \xi); K) \le \frac{1}{h}\text{dist}\,\left(\xi + \int_\tau^{\tau+h} f(\theta, \xi)\, d\theta; K\right)$$
$$+\frac{1}{h}\left\| hf(\tau, \xi) - \int_\tau^{\tau+h} f(\theta, \xi)\, d\theta \right\|$$
$$\le \frac{1}{h}\text{dist}\,\left(\xi + \int_\tau^{\tau+h} f(\theta, \xi)\, d\theta; K\right) + \frac{1}{h}\int_\tau^{\tau+h} \|f(\theta, \xi) - f(\tau, \xi)\|\, d\theta.$$

Since $t \mapsto f(t, \xi)$ is continuous from the right at $\tau \in I$, it follows that the last term in the inequality above tends to 0 when $h \downarrow 0$. So, if f satisfies (3.5.3) at (τ, ξ) then $f(\tau, \xi) \in \mathcal{T}_K(\xi)$. Conversely, if $f(\tau, \xi) \in \mathcal{T}_K(\xi)$, a similar argument shows that

$$\frac{1}{h}\text{dist}\,\left(\xi + \int_\tau^{\tau+h} f(\theta, \xi)\, d\theta; K\right) \le \frac{1}{h}\text{dist}\,(\xi + hf(\tau, \xi); K)$$
$$+\frac{1}{h}\int_\tau^{\tau+h} \|f(\theta, \xi) - f(\tau, \xi)\|\, d\theta$$

and thus f satisfies (3.5.3). The proof is complete.

Problem 3.5.2. Let $(\tau, \xi) \in I \times K$ be arbitrary and choose $\rho > 0$ and $T \in I$, $T > \tau$, such that $D(\xi, \rho) \cap K$ is closed and there exists $M > 0$ such that the continuous extension $F : I \times X \to X$ of f satisfies

$$\|F(t, u)\| \le M \tag{14.3.4}$$

for each $t \in [\,\tau, T\,]$ and every $u \in D(\xi, \rho)$. This is always possible because K is locally closed and F continuous.

Next, diminishing $T > \tau$, if necessary, we may assume that

$$(T - \tau)(M + 1) \le \rho. \tag{14.3.5}$$

We first prove that the conclusion of Lemma 3.5.1 remains true if we replace T as above with a possible smaller number $\mu \in (\tau, T\,]$ which, at this stage, is allowed to depend on $\varepsilon \in (0, 1)$. Then, this being done, by using the Brezis-Browder Ordering Principle, we will prove that we can take $\mu = T$, independent of $\varepsilon \in (0, 1)$.

In view of Problem 3.5.1, there exist $\delta \in (0, \varepsilon)$ and $p \in X$, with $\|p\| \leq \varepsilon$, such that

$$\xi + \int_\tau^{\tau+\delta} f(\theta, \xi)\, d\theta + \delta p \in K.$$

Set $g(\theta) = f(\theta, \xi)$ and $r(\theta) = p$, for $\theta \in [\,\tau, \tau + \delta]$, and let us define u by

$$u(t) = \xi + \int_\tau^t f(\theta, \xi)\, d\theta + (t - \tau)p,$$

for each $t \in [\,\tau, \tau + \delta\,]$. Diminishing $\delta > 0$ if necessary, we may assume that

$$\sup_{s \in [\,\tau, \tau+\delta\,]} \|F(s, u(s)) - f(\tau, \xi)\| \leq \varepsilon.$$

Thus, the family $\mathcal{P}_{\tau+\delta} = \{[\,\tau, \tau + \delta)\}$ (in this case $\Gamma = \{1\}$, $t_1 = \tau$ and $s_1 = \tau + \delta$) and the functions r and u satisfy (i)-(vi) with T substituted by $\tau + \delta$. From now on the proof is similar with that one of Lemma 3.3.1.

Problem 3.5.3. To simplify presentation, we say that a triple $(\mathcal{P}_T, r, u)$ as in Lemma 3.5.1 is called an *ε-approximate solution* defined on $[\,\tau, T\,]$ for the Cauchy problem

$$\begin{cases} u'(t) = f(t, u(t)) \\ u(\tau) = \xi. \end{cases} \tag{14.3.6}$$

Let $(\varepsilon_n)_n$ be a sequence in $(0, 1)$, decreasing to 0, and let $((\mathcal{P}_T^n, r_n, u_n))_n$ be a sequence of ε_n-approximate solutions of (14.3.6) defined on $[\,\tau, T\,]$.

We will first show that $(u_n)_n$ is a Cauchy sequence in the sup-norm. Let $n \in \mathbb{N}$ and let us define $\sigma_n(t) = t_m^n$ for each $t \in [\,t_m^n, s_m^n)$ and $\sigma_n(T) = T$. We have

$$u_n(t) = \xi + \int_\tau^t f(s, u_n(\sigma_n(s)))\, ds + \int_\tau^t r_n(s)\, ds$$

and therefore, by (v) in Lemma 3.5.1, we conclude

$$u_n(t) = \xi + \int_\tau^t F(s, u_n(s))\, ds + \int_\tau^t G_n(s)\, ds, \tag{14.3.7}$$

where

$$G_n(s) = f(s, u_n(\sigma_n(s))) - F(s, u_n(s)) + r_n(s).$$

In view of (iv) and (vi) in Lemma 3.5.1, we have

$$\int_\tau^T \|f(s, u_n(\sigma_n(s))) - F(s, u_n(s)) + r_n(s)\|\, ds \leq 2(T - \tau)\varepsilon_n.$$

For a.a. $t \in [\,\tau, T\,]$ and all $n, k \in \mathbb{N}$, we have

$$u_n'(t) - u_k'(t) = F(t, u_n(t)) - F(t, u_k(t)) + G_n(t) - G_k(t).$$

From (ix), (ii) and (vi) in Exercise 1.6.1, we deduce

$$\begin{aligned} \frac{d^+}{dt}(\|u_n(t) - u_k(t)\|) &= [\,u_n(t) - u_k(t), u_n'(t) - u_k'(t)\,]_+ \\ &\leq [\,u_n(t) - u_k(t), F(t, u_n(t)) - F(t, u_k(t))\,]_+ \\ &\quad + [\,u_n(t) - u_k(t), G_n(t) - G_k(t)\,]_+ \end{aligned}$$

$$\leq \omega(t, \|u_n(t) - u_k(t)\|) + \|G_n(t) - G_k(t)\|.$$

Thus

$$\|u_n(t) - u_k(t)\| \leq \int_\tau^t \omega(s, \|u_n(s) - u_k(s)\|)\, ds + 2(T - \tau)(\varepsilon_n + \varepsilon_k).$$

Recalling that ω is a Carathéodory uniqueness function, we conclude that $(u_n)_n$ is a Cauchy sequence. Indeed, let us denote by $x_{n,k}(t) = \|u_n(t) - u_k(t)\|$ and by $\gamma_{n,k} = 2(T - \tau)(\varepsilon_n + \varepsilon_k)$. The last inequality rewrites

$$x_{n,k}(t) \leq \gamma_{n,k} + \int_\tau^t \omega(s, x_{n,k}(s))\, ds$$

for each $n, k = 1, 2, \ldots$ and $t \in [\,\tau, T\,]$. From Lemma 1.8.3, diminishing $T > \tau$ if necessary, we deduce that $\lim_{n,k} x_{n,k}(t) = 0$. Equivalently, we have

$$\lim_{n,k} \|u_n(t) - u_k(t)\| = 0,$$

uniformly for $t \in [\,\tau, T\,]$. So, there exists $u \in C([\,\tau, T\,]; X)$ such that

$$\lim_n u_n(t) = u(t)$$

uniformly for $t \in [\,\tau, T\,]$. Since $u_n(\sigma_n(t)) \in D(\xi, \rho) \cap K$ for each $n \in \mathbb{N}$, $D(\xi, \rho) \cap K$ is closed, $\lim_n \sigma_n(t) = t$ uniformly for $t \in [\,\tau, T\,]$, it follows that $u(t) \in D(\xi, \rho) \cap K$ and so $F(t, u(t)) = f(t, u(t))$ for each $t \in [\,\tau, T\,]$. Passing to the limit for $n \to \infty$ in the approximate equations (14.3.7), we get

$$u(t) = \xi + \int_\tau^t f(s, u(s))\, ds,$$

and this completes the proof.

Problem 3.5.4. The problem reduces to the nonautonomous case considered in Problem 3.4.1 by introducing the new unknown function $z = (t, u)$ and the new right hand side $F(z) = (1, f(z))$.

Problem 3.5.5. We introduce the new unknown function $z = (t, u)$ and the new right hand side $F(z) = (1, f(z))$. From now on we are in the autonomous case and we can use Problem 3.4.2.

Solutions to Chapter 4

Problem 4.1.1. Let $u : [\,\tau, c\,] \to D$ be a solution of (4.1.1). Let

$$T_m = \sup\{T \in (\tau, c\,];\ u(t) \in K, \text{ for each } t \in [\,\tau, T\,]\}.$$

Since K is closed and u is continuous, it follows that $u(T_m) \in K$. Now, if $T_m < c$, using once again the local invariance of K with respect to f, we conclude that there exists $d \in (T_m, c\,]$ such that $u(t) \in K$ for each $t \in [\,T_m, d\,]$, thereby contradicting the maximality of T_m. This contradiction can be eliminated only if $T_m = c$, as claimed.

Exercise 4.1.1. Let $u, v : [\tau, T] \to D$ be two solutions of (4.1.1). Arguing as in the proof of Theorem 4.1.2, we deduce that $[D_+g](t) \le \omega(t, g(t))$ for each $t \in [\tau, T)$, where $g(t) = \|u(t) - v(t)\|$ for $t \in [\tau, T]$. Thus $g \equiv 0$, which means that $u \equiv v$, as claimed.

Solutions to Chapter 5

Problem 5.3.1. Let $\mathcal{Z}$ be the negligible set given by Theorem 2.8.2. Then, for each $(t, \xi) \in (I \setminus \mathcal{Z}) \times K$ and $h > 0$ with $t + h \in I$, we have

$$\frac{1}{h}\text{dist}\,(\xi + hf(t,\xi); K) \le \frac{1}{h}\text{dist}\left(\xi + \int_t^{t+h} f(\theta,\xi)\,d\theta; K\right)$$

$$+\frac{1}{h}\left\| hf(t,\xi) - \int_t^{t+h} f(\theta,\xi)\,d\theta\right\|$$

$$\le \frac{1}{h}\text{dist}\left(\xi + \int_t^{t+h} f(\theta,\xi)\,d\theta; K\right) + \frac{1}{h}\int_t^{t+h} \|f(\theta,\xi) - f(t,\xi)\|\,d\theta.$$

From Theorem 2.8.2, it follows that the last term in the inequality above tends to 0 when $h \downarrow 0$. So, if f satisfies (5.3.1) at (t, ξ) then it satisfies (5.2.1). Conversely, if f satisfies (5.2.1), a similar argument shows that

$$\frac{1}{h}\text{dist}\left(\xi + \int_t^{t+h} f(\theta,\xi)\,d\theta; K\right) \le \frac{1}{h}\text{dist}\,(\xi + hf(t,\xi); K)$$

$$+\frac{1}{h}\int_t^{t+h} \|f(\theta,\xi) - f(t,\xi)\|\,d\theta$$

and thus f satisfies (5.3.1). The proof is complete.

Problem 5.4.1. With the notations in Section 5.4, we will show that the sequence $(u_n)_n$ given by (5.4.2) is fundamental. We have

$$\|u_n(t) - u_m(t)\| \le \int_\tau^t \|g_n(s) - g_m(s)\|\,ds + \int_\tau^t \|r_n(s) - r_m(s)\|\,ds$$

for $n, m = 1, 2, \ldots$. A standard calculation involving Lemma 5.3.1 shows that there exists a double-indexed sequence $(\gamma_{n,m})_{n,m}$, with $\lim_{n,m} \gamma_{n,m} = 0$, and such that

$$\|u_n(t) - u_m(t)\| \le \gamma_{n,m} + \int_\tau^t \mathfrak{L}(s)\|u_n(s) - u_m(s)\|\,ds,$$

for $n, m = 1, 2, \ldots$, where $\mathfrak{L}$ is given by Definition 5.2.1. The conclusion follows from Gronwall Lemma 1.8.4.

Problem 5.4.2. With the notations in Section 5.4, show that the family $\{u_n;\ n = 1, 2, \ldots\}$, where u_n is given by (5.4.2), satisfies the hypotheses of Theorem 1.3.6.

Problem 5.4.3. Just proceed as in the case of Problem 5.4.2, taking advantage of Lemma 1.3.1

Solutions to Chapter 6

Problem 6.1.1. Let $\xi \in K$. Let $u : [\,0, T\,] \to K$ be an exact solution of (6.1.1). Then, in view of Theorem 6.1.1, we have $u'(t) \in F(u(t)) \cap \mathcal{T}_K(u(t))$ at each point of differentiability of u. Thus, there exists at least one sequence $(t_k)_k$ of differentiability points of u such that $\lim_k t_k = 0$. Consequently, we have $\lim_k u(t_k) = \xi$ and this completes the proof.

Problem 6.1.2. Clearly each quasi-weakly compact set is weakly sequentially closed, wherefrom the conclusion.

Problem 6.1.3. Just repeat the proof of Theorem 6.1.2 by recalling that, in view of (iv) in Problem 2.3.5, we have $F(\xi) \in \mathcal{TS}_K(\xi)$ if and only if $\overline{F(\xi)} \in \mathcal{TS}_K(\xi)$.

Problem 6.5.1. Suppose that $(1, y) \in \mathcal{T}_{\mathcal{C}}(\tau, \xi)$. It follows that there exist two sequences $(h_n)_n$ and $(\theta_n)_n$ in $\mathbb{R}$, with $h_n \downarrow 0$, $\lim_n \theta_n = 0$, and $(q_n)_n$ in X, with $\lim_n q_n = 0$, and such that

$$(\tau, \xi) + h_n(1, y) + h_n(\theta_n, q_n) \in \mathcal{C}.$$

Clearly, $\tau_n = \tau + h_n + h_n\theta_n$ and $\xi_n = \xi + h_n y + h_n q_n$ satisfy the desired condition. The proof of the other implication is similar.

Problem 6.7.1. The proof is identical with that one of Problem 4.1.1.

Solutions to Chapter 7

Problem 7.2.1. It is convenient to use the characterization of tangency by sequences, given by Corollary 2.4.1. So, assuming that

$$(1, \omega(t, v(t))) \in \mathcal{T}_{\text{epi}(v)}(t, v(t)),$$

there exist $(h_m)_m$ in $\mathbb{R}_+$, and $(p_m)_m$, $(q_m)_m$ in $\mathbb{R}$, with $h_m \downarrow 0$ and $\lim_m p_m = 0$, $\lim_m q_m = 0$, and such that

$$(t, v(t)) + h_m(1, \omega(t, v(t))) + h_m(p_m, q_m) \in \text{epi}(v).$$

This means that

$$v(t + h_m + h_m p_m) \le v(t) + h_m\omega(t, v(t)) + h_m q_m.$$

Clearly this implies

$$v(t + h_m + h_m p_m) \le v(t) + \lambda + h_m\omega(t, v(t) + \lambda) + h_m q_m,$$

wherefrom the conclusion.

Problem 7.2.2. Let $K = \{(t, x); 0 \le t < T, f(t) \le x\}$. By Theorem 7.2.1 and Remark 7.2.1, K is viable with respect to the function $(t, y) \mapsto (1, M)$. Therefore, for $0 \le s \le t < T$ we have $f(s) + M(t - s) \ge f(t)$. Since f is lower semicontinuous, the inequality above is verified by $t = T$, too. Therefore, $f(t) - f(s) \le M(t - s)$

for $0 \leq s \leq t \leq T$. On the other hand, since $t \mapsto t + f(t)$ is increasing, it follows that $f(t) - f(s) \geq -(t - s)$. So, f is Lipschitz with constant $M + 1$.

Problem 7.3.1. The condition in hypothesis implies that the set $I \times \text{epi}(w)$ is viable with respect to $(f, 0)$. Let us define the preorder $\mathcal{P} : \text{epi}(w) \rightsquigarrow \text{epi}(w)$ by

$$\mathcal{P}(x, y) = \{(u, v); w(u) - v \leq w(x) - y\}.$$

Since w is continuous, $\mathcal{P}$ is closed. The conclusion follows by Lemma 7.3.1.

Exercise 7.9.1. Let $(\eta, t) \in \text{epi}\ (\underline{D}w(\xi))$ and $\varepsilon > 0$. We have $\underline{D}w(\xi)(\eta) \leq t$, so there exist $s \in (0, \varepsilon)$ and $p \in D(0, \varepsilon)$ such that $w(\xi + s(\eta + p)) - w(\xi) \leq s(t + \varepsilon)$. It follows that $(\xi + s(\eta + p), w(\xi) + s(t + \varepsilon)) \in \text{epi}(w)$. Proposition 2.4.2 implies that $(\eta, t) \in \mathcal{T}_{\text{epi}(w)}(\xi, w(\xi))$.

Conversely, let $(\eta, t) \in \mathcal{T}_{\text{epi}(w)}(\xi, w(\xi))$. This means that for every $\varepsilon > 0$, there exist $s \in (0, \varepsilon)$, $p \in D(0, \varepsilon)$ and $h \in (-\varepsilon, \varepsilon)$ such that $(\xi + t(\eta + p), w(\xi) + s(t + h)) \in \text{epi}(w)$, i.e., $w(\xi + s(\eta + p)) - w(\xi) \leq s(t + h)$. We get $(1/s)(w(\xi + s(\eta + p)) - w(\xi)) \leq t + \varepsilon$. This clearly implies that

$$\liminf_{\substack{s \downarrow 0 \\ p \to 0}} \frac{1}{s}[\, w(\xi + s(\eta + p)) - w(\xi)\,] \leq t,$$

i.e., $(\eta, t) \in \text{epi}\ (\underline{D}w(\xi))$.

The proof of the second equality is similar and is left to the reader.

Problem 7.9.1. The proof is similar to that one of Problem 7.3.1. Namely, let us consider the preorder $\mathcal{P} : \text{epi}(w) \rightsquigarrow \text{epi}(w)$ defined by

$$\mathcal{P}(x, y) = \{(u, v); w(u) - v \leq w(x) - y\},$$

and apply Lemma 7.3.2.

Exercise 7.9.2. The characteristics system is $v'(s) = 1$, $u'(s) = v(s)$ and has the solutions $v(s) = v(0) + s$, $u(s) = u(0) + sv(0) + s^2/2$. The function w does not satisfy the condition: *for every* $\xi \in \mathbb{R}$ *and for every* $s \in (0, +\infty)$, $w(\xi) + s \leq w(\xi + sw(\xi) + s^2/2)$ because if we take $\xi = 0$ and $s \in (0, 1)$, we have $w(0) + s = s < 1 = w(0) + sw(0) + s^2/2$.

Exercise 7.9.3. The characteristics system is $v'(s) = -1$ $u'(s) = v(s)$ and has the solutions $v(s) = v(0) - s$, $u(s) = u(0) + sv(0) - s^2/2$. The function w does not satisfy the condition: *for every* $\xi \in \mathbb{R}$ *and for every* $s \in (0, +\infty)$, $w(\xi) - s \leq w(\xi + sw(\xi) - s^2/2)$ because, if we take $\xi = 0$ and $s \in (0, 1)$, we have $w(0) - s = -s > -1 = w(0) + sw(0) - s^2/2$.

Solutions to Chapter 8

Problem 8.5.1. In this specific case, the tangency condition (8.5.3), i.e.,

$$\liminf_{h \downarrow 0} \frac{1}{h} \text{dist}\,((\tau, S(h)\xi) + h(1, f(\tau, \xi)); I \times K) = 0,$$

holds if and only if

$$\liminf_{h\downarrow 0}\frac{1}{h}\text{dist}\,(S(h)\xi+hf(\tau,\xi);K)=0,$$

since in our case, i.e., I open (to the right), we have $\liminf_{h\downarrow 0}\frac{1}{h}\text{dist}\,(\tau+h;I)=0$.

Solutions to Chapter 9

Problem 9.1.1. As in the case of Proposition 9.1.1, except for (vi), the remaining properties are direct consequences of the simple remark that $E\in\mathcal{TS}_K^A(\xi)$ if and only if there exist three sequences, $(h_n)_n$ in $\mathbb{R}_+$, with $h_n\downarrow 0$, $(\eta_n)_n$ in E and $(p_n)_n$ in X with $\lim_n p_n=0$ and such that

$$S(h_n)\xi+\int_0^{h_n}S(h_n-s)\eta_n\,ds+h_np_n\in K$$

for $n=1,2,\ldots$. To prove (vi), let $(\eta_n)_n$ in E, $h_n\downarrow 0$ and $(p_n)_n$ with $\lim_n p_n=0$ as above. Since E is compact, we may assume with no loss of generality that there exists $\eta\in E$ such that $\lim_n\eta_n=\eta$. Since

$$S(h_n)\xi+\int_0^{h_n}S(h_n-s)\eta\,ds+h_nq_n\in K,$$

where $q_n=\frac{1}{h_n}\int_0^{h_n}S(h_n-s)[\,\eta_n-\eta\,]\,ds+p_n$ for $n=1,2,\ldots$, and

$$\lim\frac{1}{h_n}\int_0^{h_n}S(h_n-s)[\,\eta_n-\eta\,]\,ds=0,$$

we get the conclusion.

Problem 9.1.2. In view of (iv) in Proposition 9.1.1, we have $E\in\mathcal{QTS}_K^A(\xi)$ if and only if $\overline{E}\in\mathcal{QTS}_K^A(\xi)$. So, to get the conclusion of Theorem 9.1.1, it suffices to show that $\overline{F(\xi)}\in\mathcal{QTS}_K^A(\xi)$. From now on the proof is similar to the one of Theorem 9.1.1.

Solutions to Chapter 10

Problem 10.1.1. Let $\xi\in D(A)$, let $\eta\in\mathcal{T}_K^A(\xi)$ and let us define the operator $A_\eta: D(A_\eta)\subseteq X\to X$ by $D(A_\eta)=D(A)$ and $A_\eta x=Ax+\eta$ for each $x\in D(A)$. Then $u(h,0,\xi,\eta)=S_\eta(h)\xi$ for each $h>0$, where $S_\eta(t):\overline{D(A_\eta)}\to\overline{D(A_\eta)}$ is the semigroup of nonlinear contractions generated by A_η. Since $\xi\in D(A_\eta)$, we have

$$\lim_{h\downarrow 0}\frac{1}{h}(u(h,0,\xi,\eta)-\xi)=A\xi+\eta.$$

But this shows that

$$\liminf_{h\downarrow 0}\frac{1}{h}\text{dist}\,(u(h,0,\xi,\eta);K)=\liminf_{h\downarrow 0}\frac{1}{h}\text{dist}\,(\xi+h(A\xi+\eta);K)$$

and this completes the proof.

Problem 10.1.2. If A is linear, the C^0-solution $u(\cdot, 0, \xi, \eta)$ is in fact a mild solution, i.e., it is given by

$$u(h, 0, \xi, \eta) = S(h)\xi + \int_0^h S(h-s)\eta\, ds$$

for each $h > 0$, wherefrom the conclusion.

Problem 10.5.1. This follows from the simple observation that, for $\zeta = (\tau, \xi)$, the solution z of the Cauchy problem

$$\begin{cases} z'(s) \in \mathcal{A}z(s) + \mathcal{F}(\zeta) \\ z(0) = \zeta, \end{cases}$$

is given by $z(s) = (\tau + s, u(s, 0, \xi, f(\tau, \xi))) = (\tau + s, u(\tau + s, \tau, \xi, f(\tau, \xi)))$, for each $s \in \mathbb{R}_+$.

Problem 10.10.1. Just follow the very same lines as those in the proof of Problem 4.1.1.

Solutions to Chapter 11

Problem 11.1.1. Clearly, if there exists $\eta \in E$ with $\eta \in \mathcal{T}_K^A(\xi)$, we have $E \in \mathcal{TS}_K^A(\xi)$. To prove the converse statement, let us first observe that $E \in \mathcal{TS}_K^A(\xi) \cap \mathcal{B}(X)$ if and only if there exist three sequences $(h_n)_n$ in $\mathbb{R}_+$, $(\eta_n)_n$ in E and $(p_n)_n$ in X with $h_n \downarrow 0$, $\lim_n p_n = 0$ and such that

$$u(h_n, 0, \xi, \eta_n) + h_n p_n \in K$$

for $n = 1, 2, \ldots$. Since E is compact, we may assume that there exists $\eta \in E$ such that $\lim_n \eta_n = \eta$. Inasmuch as, by (1.6.5), we have

$$\|u(h_n, 0, \xi, \eta_n) - u(h_n, 0, \xi, \eta)\| \leq h_n \|\eta_n - \eta\|,$$

it follows that

$$u(h_n, 0, \xi, \eta) + h_n q_n \in K$$

for $n = 1, 2, \ldots$, where $q_n = \frac{1}{h_n}(u(h_n, 0, \xi, \eta_n) - u(h_n, 0, \xi, \eta)) + p_n$. Finally, since $\lim_n q_n = 0$, it follows that $\eta \in \mathcal{T}_K^A(\xi)$, as claimed.

Problem 11.1.2. From (ii) in Remark 11.1.1, we deduce that $E \in \mathcal{TS}_K^A(\xi)$ if and only if $\overline{E} \in \mathcal{TS}_K^A(\xi)$. So, to get the conclusion of Theorem 11.1.1, it suffices to show that $\overline{F(\xi)} \in \mathcal{TS}_K^A(\xi)$. From now on just repeat the proof of Theorem 11.1.1.

Problem 11.2.1. First let us observe that whenever A is linear, a function $u(\cdot, \tau, \xi, f)$ is a C^0-solution of

$$\begin{cases} u'(t) = Au(t) + f(t) \\ u(\tau) = \xi, \end{cases}$$

on $[\,\tau, T\,]$, if and only if it is a mild solution of the same problem, on $[\,\tau, T\,]$, i.e.,

$$u(t, \tau, \xi, f) = S(t-\tau)\xi + \int_\tau^t S(t-s)f(s)\, ds$$

for each $t \in [0, T]$. See Remark 1.6.1. Thus, in this case, the mild solution operator, $f \mapsto u(\cdot, \tau, \xi, f)$, is linear continuous from $L^1(\tau, T; X) \to C([\tau, T]; X)$ and therefore its graph is closed. But the graph is convex and hence it follows that it is weakly-strongly sequentially closed, as claimed.

Problem 11.2.2. In view of (i) in Exercise 1.6.1 and (i) in Proposition 1.6.1, we have $(x, y)_+ = (y, J(x))$ for each $x \in X$. Let $(f_n)_n$ be a sequence in $L^1(\tau, T; X)$ with $\lim_n f_n = f$ weakly in $L^1(\tau, T; X)$ and $\lim_n u(t, \tau, \xi, f_n) = \widetilde{u}(t)$ uniformly for $t \in [\tau, T]$. We denote by $u_n(t) = u(t, \tau, \xi, f_n)$, for $n = 1, 2, \ldots$. Then, by Theorem 1.6.3, we have

$$\|u_n(t) - x\|^2 \leq \|u_n(s) - x\|^2 + 2\int_s^t (f_n(\theta) + y, J(u_n(\theta) - x))\, d\theta$$

for each $x \in D(A)$, each $y \in Ax$ and each $\tau \leq s \leq t \leq T$. Passing to the limit for $n \to \infty$ and taking into account that, by Theorem 1.1.4, J is uniformly continuous on bounded subsets in X, we conclude that

$$\|\widetilde{u}(t) - x\|^2 \leq \|\widetilde{u}(s) - x\|^2 + 2\int_s^t (f(\theta) + y, J(\widetilde{u}(\theta) - x))\, d\theta$$

for each $x \in D(A)$, each $y \in Ax$ and each $\tau \leq s \leq t \leq T$. Using once again Theorem 1.6.3, we deduce that $\widetilde{u} = u(\cdot, \tau, \xi, f)$, as claimed.

Solutions to Chapter 12

Problem 12.1.1. From 1.6.5, we have

$$\|S_{f(\tau,\xi)}(h)\xi - u(\tau + h, \tau, \xi, f(\cdot, \xi))\| \leq \int_\tau^{\tau+h} \|f(s, \xi) - f(\tau, \xi)\|\, ds$$

for each $(\tau, \xi) \in I \times K$ and $h > 0$ with $\tau + h \in I$. The conclusion follows directly from Theorem 2.8.5.

Solutions to Chapter 13

Problem 13.1.1. In view of Lemma 13.1.1, we have $\mathcal{T}_K(\xi) \subseteq \mathcal{T}_K^A(\xi)$, and therefore it suffices to show that $\mathcal{T}_K^A(\xi) \subseteq \mathcal{T}_K(\xi)$. Since, for each $t \in \mathbb{R}$, $G(t)$ is an isometry, we have

$$\operatorname{dist}(\xi + h\eta; K) = \operatorname{dist}(G(h)\xi + hG(h)\eta; G(h)K) = \operatorname{dist}(G(h)\xi + hG(h)\eta; K).$$

The conclusion follows from Proposition 8.1.1.

Problem 13.1.2. If $S(t)K \subseteq K$ for each $t \geq 0$, we have

$$\liminf_{h \downarrow 0} \frac{1}{h} \operatorname{dist}(S(h)\xi; K) = 0$$

and thus $0 \in \mathcal{T}_K^A(\xi)$ for each $\xi \in K$. Conversely, if $0 \in \mathcal{T}_K^A(\xi)$ for each $\xi \in K$ then, in view of Theorems 8.2.6 and 8.8.2, for each $\xi \in K$, there exists at least one global mild solution $u : [0, \infty) \to K$ of the problem $u'(t) = Au(t)$, $u(0) = \xi$. Since the unique mild solution of this problem is $u(t) = S(t)\xi$, this completes the proof.

Problem 13.2.1. Let $(u, \lambda) \in \mathcal{T}^{\mathcal{A}}_{\text{epi}(V)}(\xi, V(\xi))$. Then, there exist three sequences, $(h_n)_n$ and $(\theta_n)_n$ in $\mathbb{R}$ and $(w_n)_n$ in X, with $h_n \downarrow 0$, $\lim_n \theta_n = 0$, $\lim_n w_n = 0$, and such that

$$V(S(h_n)\xi + h_n(u + w_n)) \leq V(\xi) + h_n(\lambda + \theta_n) \leq \mu + h_n(\lambda + \theta_n)$$

for $n = 1, 2, \ldots$. But this yields $(u, \lambda) \in \mathcal{T}^{\mathcal{A}}_{\text{epi}(V)}(\xi, \mu)$ whenever $V(\xi) \leq \mu$, as claimed.

Problem 13.2.2. Corresponding to the function $g : X \to (-\infty, +\infty]$ we introduce the sequence of functions $(g_n)_n$, for $n > 2^{p-1}C$, given by

$$g_n(x) = \inf_{y \in X} \{g(y) + n\|x - y\|^p\},$$

for all $x \in X$.

First, we remark that $g_n(x) \in \mathbb{R}$ for all $x \in X$. This follows easily from the property of g to be proper. We also notice that, for every $x \in X$, the sequence $(g_n(x))_n$ is nondecreasing in $\mathbb{R}$ and $g_n(x) \leq g(x)$. Moreover, $\lim_n g_n(x) = g(x)$. Indeed, using the lower semicontinuity of g, we find for each $x \in X$ and $\varepsilon > 0$ some $\delta > 0$ such that $g(y) \geq g(x) - \varepsilon$ whenever $\|y - x\| \leq \delta$. At this point, let us observe that (13.2.3) implies that, if $\|y - x\| > \delta$, we have

$$\begin{aligned} g(y) + n\|x - y\|^p &\geq -C(1 + \|y\|^p) + n\|x - y\|^p \\ &\geq -C(1 + 2^{p-1}\|x\|^p) + (n - 2^{p-1}C)\delta^p, \end{aligned}$$

since $n > 2^{p-1}C$. It follows that there exists $n(\varepsilon, x) \in \mathbb{N}$ such that

$$g(y) + n\|x - y\|^p \geq g(x) - \varepsilon,$$

for all $y \in X$ and for all $n \in \mathbb{N}$, $n \geq n(\varepsilon, x)$. Thus $g_n(x) \uparrow g(x)$ as $n \to \infty$.

To complete the proof it suffices to justify that each function $g_n : X \to \mathbb{R}$ is Lipschitz continuous on the bounded subsets of X. To this end, fix both $n \in \mathbb{N}$ and a bounded subset B of X. As a first step in showing that g_n is Lipschitz continuous on B, we prove that the set

$$\tilde{B} = \{z \in X;\ \exists\, v \in B \text{ such that } g(z) + n\|z - v\|^p \leq g_n(v) + 1\} \tag{14.13.1}$$

is bounded in X. By (13.2.3), we derive

$$\begin{aligned} -C - n\|v\|^p + \left(\frac{n}{2^{p-1}} - C\right)\|z\|^p &\leq g(z) + n\|z - v\|^p \\ &\leq g_n(v) + 1 \leq g(x_0) + n\|v - x_0\|^p + 1, \end{aligned}$$

for all $z \in \tilde{B}$, where the element $v \in B$ is related to $z \in \tilde{B}$ in the sense of (14.13.1), and choosing some fixed $x_0 \in D(g)$. Since the set B is bounded and $n > 2^{p-1}C$, we deduce the boundedness of $\tilde{B}$. The boundedness of the set $\tilde{B}$ permits to consider the Lipschitz constant L_B of the function $\|\cdot\|^p$ on the bounded subset $B - \tilde{B} = \{\xi - \eta;\ \xi \in B,\ \eta \in \tilde{B}\}$ of X. Let $x, y \in B$ and $\varepsilon \in (0, 1]$. The definition of $g_n(x)$ yields some $z = z(x) \in X$ such that $g(z)+n\|x-z\|^p \leq g_n(x)+\varepsilon$, which means that $z \in \tilde{B}$. Then, by the definition of $g_n(y)$, we get

$$g_n(y) \leq g(z) + n\|y - z\|^p \leq g_n(x) + \varepsilon + n(\|y - z\|^p - \|x - z\|^p)$$
$$\leq g_n(x) + \varepsilon + nL_B\|x - y\|.$$

Letting $\varepsilon \to 0$, we deduce that g_n is Lipschitz on B. The proof is complete.

Problem 13.2.3. We take $V : X \to \mathbb{R}$ by $V(x) = \|x\|$ for every $x \in X$, $g : X \to \mathbb{R}$ by $g(x) = 1$ if $\|x\| > r$ and $g(x) = \|x\|/r$ if $\|x\| \leq r$. We show that (13.2.4) is satisfied. To prove this fact, let us first take $\|\xi\| \geq r$. For sufficiently small h, we have

$$\frac{1}{h}\left[V\left(S(h)\xi + h\left(-\frac{\xi}{\|\xi\|}\right)\right) - V(\xi)\right]$$
$$\leq \frac{1}{h}\left[\|S(h)\xi\|\left(1 - \frac{h}{\|\xi\|}\right) + h\left\|S(h)\frac{\xi}{\|\xi\|} - \frac{\xi}{\|\xi\|}\right\| - \|\xi\|\right]$$
$$\leq -1 + \left\|S(h)\frac{\xi}{\|\xi\|} - \frac{\xi}{\|\xi\|}\right\|.$$

This shows that

$$\underline{D}V(\xi)(f(\xi)) + 1 \leq 0$$

for every $\xi \in X$ with $\|\xi\| > r$. By a similar reasoning, we get

$$\underline{D}V(\xi)(f(\xi)) + \frac{\|\xi\|}{r} \leq 0$$

for every $\xi \in X$ with $\|\xi\| \leq r$.

Applying Theorem 13.2.1, we conclude that the (V, g) is a Lyapunov pair for (13.2.1). While $\|u(t)\| \geq r$, we have

$$\|u(t)\| \leq \|\xi\| - t.$$

Clearly, $\|u(T)\| = r$ at least for $T = \|\xi\| - r$.

Problem 13.8.1. First, let us observe that f is continuous and $f(D(0, r))$ is relatively compact being bounded and uniformly 2-sumable. Let us assume by contradiction that there exists $T > 0$ such that $u'(t) = f(u(t))$ has a T-periodic solution $u : \mathbb{R} \to D(0, r)$. Then

$$0 = \int_0^T u_n'(s)\,ds = a_n c_n + b_n \int_0^T (\|(u_k(s))_k\| - r)^2\,ds$$

where

$$\sum_{k=1}^{\infty} c_k^2 = \sum_{k=1}^{\infty}\left(\int_0^T u_k(s)\,ds\right)^2 < \infty.$$

Hence $\|(u_k(t))_k\| = r$, for each $t \in [0, T]$, which is impossible.

Bibliographical notes and comments

Chapter 1.

The results in this chapter are by now classical ones, and can be found in several well-known monographs and treatises: Dunford–Schwartz [**90**], Edwards [**91**], Hille–Phillips [**107**], Lakshmikantham-Leela [**119**], Yosida [**185**]. We also refer to Vrabie [**173**], [**175**] and [**176**].

Section 1.1. The presentation follows Hille–Phillips [**107**] and Yosida [**185**]. Theorem 1.1.1 is due to Mazur. Theorem 1.1.3 is due to Milman [**127**] and to Pettis [**144**] and Theorem 1.1.4 was proved by Kato [**113**].

Section 1.2. The Lebesgue-type integral for vector-valued functions was introduced by Bochner [**20**]. Theorem 1.2.2 is due to Bochner [**20**] and Theorem 1.2.6 was proved by Clarkson [**69**]. Theorem 1.2.3, which is an extension to vector-valued functions of the Dominated Convergence Theorem due to Lebesgue [**120**], is from Bochner [**20**].

Section 1.3. The "compact" part of Theorem 1.3.2 is due to Mazur and the "weakly" compact part to Krein–Šmulyan. Theorem 1.3.4 is due to Eberlein–Šmulyan and Theorem 1.3.6 is an infinite-dimensional version of the well-known Arzelà–Ascoli Theorem. Theorem 1.3.7 is a simple consequence of the Dunford Theorem. See Diestel–Uhl [**83**], Theorem 15, p. 76 and Theorem 1, p. 101. Theorem 1.3.8 is due to Diestel [**82**].

Section 1.4. Theorem 1.4.5 is the famous Hille-Yosida Generation Theorem, Theorem 1.4.6 is due to Feller [**95**], Miyadera [**129**] and Phillips [**145**], while Theorem 1.4.8 was proved by Lumer-Phillips [**122**]. For detailed proofs see Vrabie [**175**].

Section 1.5. The notion of mild solution was introduced by Segal [**153**] and imposed under this name by Browder [**38**]. Theorem 1.5.2 was proved by Vrabie [**172**] in the general case of nonlinear m-dissipative operators. A linear variant of the latter can be found in Vrabie [**175**]. Theorem 1.5.3 is due to Baras-Hassan-Veron [**9**].

Section 1.6. Except for (x), Exercise 1.6.1 includes Lemma 1.2.2, p. 8 and Corollary 1.2.1, p. 12 in Lakshmikantham-Leela [**119**]. We also refer to Pavel [**138**]. Suggested by the linear case, the notion of C^0-solution was introduced by Crandall [**75**] under the name of *mild solution*. This was called a *DS-limit solution* by Kobayashi [**116**]. In order to make a net distinction between the linear and nonlinear case, in this book, we preferred the name of C^0*-solution* used by Showalter [**158**]. In the case of evolutions driven by m-dissipative operators this turned out to coincide with the one of *integral solution* as defined successively by Benilan-Brezis [**18**] in the Hilbert space frame, and by Benilan [**17**] in the general Banach space frame. Theorems 1.6.2 and 1.6.3 are from Benilan *loc. cit.*, Theorem 1.6.4 is from Vrabie [**172**] and Theorem 1.6.5 from Baras [**8**].

Section 1.7. The extremely short presentation of the spaces $W^{m,p}(\Omega)$ follows Vrabie [**175**]. For details see Adams [**1**]. Theorem 1.7.1 is the famous Sobolev–Rellich–Kodrachov Imbedding Theorem. See Adams [**1**], Theorem 5.4, p. 97 and Theorem 6.2, p. 144. Theorems 1.7.2 and 1.7.3 are well-known and can be found in Vrabie [**175**], Theorem 4.1.2, p. 79, Theorem 7.2.5, p. 157 Theorem 4.1.3, p. 81 and Remark 4.1.3, p. 82. Theorem 1.7.4 is a consequence of Theorem 6, p. 64, in Protter-Weinberger [**148**]. The first part of Theorem 1.7.7 is due to Brezis-Strauss [**37**], while the sufficient condition for the compactness of the generated semigroup is from Badii-Diaz-Tesei [**7**]. Theorem 1.7.9 is a simple variant of Corollary 1 in Diaz–Vrabie [**81**]. A sharper compactness result, allowing β to be merely strictly monotone, was proved by Diaz-Vrabie [**79**].

Section 1.8. Lemmas 1.8.1, 1.8.2 and 1.8.3 are almost for sure well-known. We included them here with complete proof simply because we did not find appropriate references to their specific forms needed throughout the book. Lemma 1.8.4 is a simple extension to measurable functions of the well-known Gronwall Inequality in Vrabie [**176**], Lemma 1.5.2, p. 46.

Chapter 2.

Section 2.1. Theorem 2.1.1 is due to Brezis–Browder [**36**]. As we have already seen, this is an ordering principle similar to Zorn's Lemma, but based on the Axiom of Dependent Choice which, as shown by Feferman [**94**], turns out to be strictly weaker than the Axiom of Choice.

We notice that, in its turn, the Axiom of Dependent Choice implies *the Axiom of Countable Choice* stated below, which is sufficient to prove that a lot of remarkable properties in Real Analysis can be described by means of sequences. We emphasize that the Axiom of Dependent Choice is "far enough" from the Axiom of Countable Choice stated below, as shown by Howard–Rubin [**110**].

The Axiom of Countable Choice. *Let $\mathcal{S}$ be a nonempty set and let $\mathcal{F} = \{\mathbb{F}_m\,;\ m \in \mathbb{N}\}$ be a countable family of nonempty subsets in $\mathcal{S}$. Then, there exists a sequence $(\xi_m)_m$ with the property that $\xi_m \in \mathbb{F}_m$ for each $m \in \mathbb{N}$.*

We notice that, whenever possible, this framework, based on the Axiom of Dependent Choice, ought to be preferred simply because the results based on this axiom remain true no matter which initial assumption we make, i.e., no matter if we assume that either the Axiom of Choice, or its negation, holds true.

In the original formulation of Brezis–Browder Ordering Principle, it is assumed that $\mathcal{N}$ is bounded from above. In order to handle a larger class of applications this condition has been dropped in Cârjă–Ursescu [**54**], by obtaining the very slight extension here presented. A simple inspection of the proof shows that the conclusion of Theorem 2.1.1 remains true if (i) is replaced by the weaker condition:

(j) *For any increasing sequence $(\xi_k)_k$ in $\mathcal{S}$ with the property that the sequence $(\mathcal{N}(\xi_k))_k$ is strictly increasing, there exists some $\eta \in \mathcal{S}$ such that $\xi_k \preceq \eta$ for all $k \in \mathbb{N}$.*

For other extensions allowing $\mathcal{N}$ to take values in an ordered structure $(P, \leq)$ whose chains enjoy countable regularity properties, see Turinici [**162**] and the references therein. The proof of Theorem 2.1.1, here included, revealing the deep implication of the Axiom of Countable Choice, is due to Turinici [**163**]. The solution of Problem 2.1.1 is from Brezis–Browder [**36**].

Section 2.2. The result in Problem 2.2.1 stating that a set K is locally closed if and only if it is closed relative to some open set D is from Yorke [**182**]. Lemma 2.2.1 is a slight extension to infinite dimensional spaces of a result due to Cârjă–Ursescu [**54**]. A broader class of proximal sets than that one offered by Lemma 2.2.1 are the so-called *φ-convex sets* studied in finite dimensions by Federer [**93**] under the name of "*sets with positive rich*" in connection with the local uniqueness of projection and the smoothness of the distance function. This class includes both weakly closed and convex sets. As proved in Colombo–Goncharov [**70**], in Hilbert spaces, the sets with positive rich are nothing but the proximal sets for which the projection Π_K is single-valued and continuous on V.

Section 2.3. In order to handle in the most appropriate way viability problems concerning differential inclusions, Cârjă–Necula–Vrabie [**52**] have introduced the notion of tangent set. We notice that Section 3 is inspired from Cârjă–Necula–Vrabie [**52**].

Section 2.4. The general concept of a tangent vector, as given by Definition 2.4.1, was introduced independently by Bouligand [**32**] and Severi [**155**] in the very same volume of the very same journal. Problem 2.4.1 is a classical one, Problem 2.4.2, Examples 2.4.1 and 2.4.2 and Problem 2.4.3 are from Cârjă–Necula–Vrabie [**52**]. The cone $\mathcal{T}_K(\xi)$ was introduced by Bouligand [**32**], the proof proposed in Problem 2.4.4 is new, Propositions 2.4.2, 2.4.3 and 2.4.4 are classical and Theorem 2.4.1 is a variant of a similar result obtained by Quincampoix [**149**], Corollary 2.3, in the case when K_1, K_2 are closed subsets in a normed vector space. The tangency concept in Definition 2.4.2 was defined by Federer [**93**]. See also Girsanov [**101**]. A similar tangency concept defined only with topological notions was introduced by Ursescu [**165**] in general topological vector spaces. This concept, in a Banach space X, reduces to the one of Federer [**93**]. For other kind of tangent or normal cones, see Mordukhovich [**131**].

Section 2.5. The notion of metric normal vector to K at $\xi \in K$ was defined by Bony [**24**], the Bony tangent cone to K at $\xi \in K$ was introduced by Ursescu [**169**] and the tangent vector in the sense of Definition 2.5.4 by Clarke [**65**]. Example 2.5.1 is from Cârjă–Vrabie [**58**]. In this general form, Lemma 2.5.1 is new. As far as we know, (2.5.2) it is also new and extends some previous similar results proved independently by Ursescu [**166**] and by Cornet [**71**] in $X = \mathbb{R}^n$. We notice that, in all the above mentioned results, a weaker inclusion, i.e.,

$$\liminf_{\substack{\xi \to \xi_0 \\ \xi \in K}} \mathcal{T}_K(\xi) \subseteq \mathcal{C}_K(\xi_0),$$

is obtained. See also Treiman [**160**] for the proof of the inclusion above in general Banach spaces. For a similar result see Mordukhovich [**131**], Theorem 1.9, p. 14.

We emphasize that there are examples showing that the converse inclusion does not hold in infinite dimensional Banach spaces. See for instance Treiman [**160**]. We notice that Ursescu [**166**] proves a characterization of $\mathcal{C}_K(\xi)$ in general Banach spaces, pertaining also an immediate proof of Treiman's main result in [**160**]. Problem 2.5.1 is inspired from Cornet [**71**] and Proposition 2.5.1 is essentially due to Ursescu [**169**].

Section 2.6. Theorem 2.6.1 is due to Michael [**125**], [**126**]. Lemma 2.6.1 is classic. Lemma 2.6.2 is a particular case of a general closed graph result due to Castaing–Valadier [**59**].

Section 2.7. The presentation of the measures of noncompactness α, of Kuratowski, and β, of Hausdorff, is inspired from Akhmerov–Kamenskii–Potapov–Rodkina–Sadovskii [**3**] and Deimling [**77**]. Lemmas 2.7.1 and 2.7.2 are due to Mönch [**130**].

Section 2.8. Theorem 2.8.1 is a specific form of a Lusin-type continuity result due to Scorza Dragoni [**152**]. For more general results see Berliocchi–Lasry [**19**] and Kucia [**117**]. Theorem 2.8.2 is from Cârjă–Monteiro Marques [**47**] and Theorems 2.8.4 and 2.8.5 are new.

Chapter 3.

What we are referring to now as *viability* is in fact the actual name of an old concept introduced for the first time by Nagumo [**132**] and baptized by him "*rechts zuläsig*" in German, whose English translation is "*right admissibility*". Although Nagumo *loc. cit* proved the complete characterization of the viability of a locally closed set with respect to a continuous function in a finite dimensional space, one should not forget the earlier notable contribution of Perron [**142**] on the subject. More precisely, as far as we know, the first result in the spirit of what we mean nowadays by viability is due to Perron *loc. cit.*. He considered the case

$$\mathcal{C} = \{(t,u);\ \omega_1(t) \le u \le \omega_2(t),\ t \in [\,a,b\,]\},$$

where $\omega_1, \omega_2 : [\,a,b\,] \to \mathbb{R}$ and $f : \mathcal{C} \to \mathbb{R}$ are continuous, and $\omega_1(t) \le \omega_2(t)$ for each $t \in [\,a,b\,]$. In fact Perron *loc. cit.* proved nothing but Nagumo Viability Theorem 3.5.5 in this specific case, i.e., $X = \mathbb{R}$ and $\mathcal{C}$ as above. As far as we know, Nagumo's result (or variants of it) has been independently rediscovered several times in the seventies by Bony [**24**], Brezis [**35**], Crandall [**74**], Hartman [**106**], Martin [**124**] and Yorke [**182**], [**183**].

Section 3.1. Example 3.1.1 is from Dieudonné [**84**], Theorem 3.1.1 is a simple extension to general Banach spaces of the necessity part of the main result in Nagumo [**132**], while Theorem 3.1.2 is a slight extension of the former.

Section 3.2. Theorem 3.2.1, which is an easy extension of Theorem 3.2.2, is new. Theorem 3.2.2 is essentially due to Volkmann [**170**], who considered the specific case $\omega(r) = kr$, for each $r \in \mathbb{R}_+$, where $k \ge 0$ is fixed. Theorem 3.2.3 was proved by Brezis [**35**], Theorem 3.2.4 by Crandall [**74**] and Theorem 3.2.6 is due to Nagumo [**132**]. Finally, Theorem 3.2.5 is due to Ursescu [**169**].

Section 3.3. The idea of using a Zorn's Lemma-type argument to get approximate solutions in viability appeared first in Gautier [**99**] who also was the first to approach an infinite dimensional viability problem. The fact that Brezis–Browder Theorem 2.1.1 can produce the same effect as Zorn's Lemma was observed by Cârjă–Ursescu [**54**]. Lemma 3.3.1 is inspired from Deimling [**76**] and Bothe [**26**].

Section 3.4. The compactness argument used to prove the convergence of ε-approximate solutions is essentially inspired from Deimling [**76**] and Bothe [**29**].

Section 3.5. Theorem 3.5.3 is the infinite dimensional variant of the celebrated Nagumo [**132**] Viability Theorem 3.5.5, Theorem 3.5.4 is a consequence of Theorem 3.2.3 and Theorem 3.5.7 is a variant of a viability result of Martin [**124**]. A very general result of this type is due to Turinici [**161**], where the "dissipativity" condition in Definition 3.5.4 is satisfied only on $I \times K$ and ω is merely continuous and $\omega(t, 0) = 0$ for each $t \in I$.

Section 3.6. Theorem 3.6.1 is a non-open variant of a well known result in Ordinary Differential Equations. See for instance Vrabie [**176**]. Theorem 3.6.2 is new but not surprising. Definition 3.6.1 is adapted from Definition 3.2.5, p. 95 in Vrabie [**173**]. Theorem 3.6.3 and Corollary 3.6.1 are new.

Chapter 4.

The results in this chapter are mainly from Cârjă-Necula-Vrabie [**51**].

Section 4.1. The presentation follows both Cârjă–Necula–Vrabie *loc. cit.* and Cârjă–Vrabie [**58**]. Theorems 4.1.2 and 4.1.3 are new.

Section 4.2. The exterior tangency condition (4.2.1) was introduced in Cârjă–Necula–Vrabie [**51**], although some particular choices for the comparison function ω were considered earlier in Aubin [**5**]. Theorem 4.2.1 is from Cârjă–Necula–Vrabie *loc. cit.*.

Section 4.3. The two concepts of (D, K)-Lipschitz and of (D, K)-dissipative functions were introduced by Cârjă–Necula–Vrabie [**51**]. Theorems 4.3.1 and 4.3.2 are also from Cârjă–Necula–Vrabie *loc. cit.* We notice that a condition similar to (4.3.1), with ξ replaced by ξ_1, $\pi_K(\xi)$ replaced by ξ_2 and $\xi_1, \xi_2 \in V$, was used previously by Kenmochi–Takahashi [**114**].

Section 4.4. Theorems 4.4.1 and 4.4.2 are from Cârjă–Necula–Vrabie [**51**]. Example 4.4.1 is adapted from Aubin-Cellina [**6**], p. 203.

Section 4.5. Proposition 4.5.1 and Corollaries 4.5.1, 4.5.2 are new.

Chapter 5.

The first viability result in the case of a single-valued Carathéodory right-hand side is due to Ursescu [**167**].

Section 5.1. Theorem 5.1.1 is an infinite dimensional version of the necessity part of the main result in Ursescu [**167**] and Theorem 5.1.2 is new. Example 5.1.1 is from Bothe [**25**].

Section 5.2. Theorems 5.2.1, 5.2.2 and 5.2.3 extend the results in Section 3.5 to Carathéodory functions defined on cylindrical domains. Theorem 5.2.4 is due to Ursescu [**167**]. Theorem 5.2.5 is an extension of Ursescu *loc. cit.* which contains only the equivalence between (ii) and (iv) in the above mentioned theorem. We notice that, as we can see from Cârjă–Monteiro Marques [**48**], in the case of a Carathéodory function, the fact that the tangency condition (5.2.1) holds for each $(t, \xi) \in (I \setminus \mathcal{Z}) \times K$, with $\mathcal{Z}$ independent of $\xi \in K$, is equivalent to the fact that (5.2.1) holds for each $\xi \in K$ and each $t \in I \setminus \mathcal{Z}_\xi$, where $\mathcal{Z}_\xi$ is a negligible set depending on $\xi \in K$.

Section 5.3. Proposition 5.3.1 is from Cârjă–Vrabie [**58**], Lemma 5.3.1 is new and inspired from Cârjă–Vrabie *loc. cit.*

Section 5.4. The proof of Theorem 5.2.1 is a refinement of the proof of Theorem 3.2.2.

Section 5.5. Theorems 5.5.1 and 5.5.2 are new and extend their counterparts in Cârjă–Vrabie [**58**] referring to the continuous case.

Chapter 6.

The first viability result for the multi-valued case is due to Bebernes–Schuur [**12**]. We mention that, in 1936, Zaremba [**187**] proved that if $F : D \rightsquigarrow \mathbb{R}^n$ is u.s.c. with nonempty compact and convex values and D is open then, for each $\xi \in D$, the differential inclusion (6.1.1) has at least one solution $u : [\,0, T\,] \to D$ satisfying $u(0) = \xi$. It should be noticed that the concept of solution used by Zaremba *loc. cit.* is in the sense of the *contingent derivative*. More precisely, if $u : [\,0, T\,] \to \mathbb{R}^n$ is continuous and $t \in [\,0, T)$, the set

$$Du(t) = \left\{ \lim_m \frac{u(t + t_m) - u(t)}{t_m}\,;\; t_m \downarrow 0 \right\}$$

is called the *contingent derivative* of u at t. We say that $u : [\,0, T\,] \to D$ is a *contingent solution* of (6.1.1) if

$$\emptyset \neq Du(t) \subseteq F(u(t)) \tag{15.6.2}$$

for each $t \in [\,0, T)$. In 1961, Ważewski [**181**] proved that, if F is u.s.c. with nonempty, compact and convex values, u is a contingent solution to (6.1.1) if and only if u is an exact solution to (6.1.1). So, Zaremba's existence result is nothing more than the multi-valued counterpart of the Peano [**141**] Local Existence Theorem. In the same spirit, the viability result of Bebernes–Schuur *loc. cit.* is the multi-valued version of the Nagumo's Viability Theorem 3.5.5. The tangency condition used by Bebernes–Schuur *loc. cit.* is equivalent to (6.5.4). See Problem 6.5.1. Proposition 15.6.1 below, due to Ważewski *loc. cit.*, is in fact a finite dimensional variant of Theorem 6.1.2.

Proposition 15.6.1. *Let $K \subseteq \mathbb{R}^n$ be nonempty and locally closed and let $F : K \rightsquigarrow \mathbb{R}^n$ be u.s.c. with nonempty, convex, compact values. Then, for every $\xi \in K$, and every solution $u : [\,0, T\,] \to K$ to (6.1.1), with $u(0) = \xi$, there exist $\eta \in F(\xi)$ and a sequence $(t_m)_m$ in $(0, T)$ convergent to 0 such that the sequence $\left(\frac{1}{t_m}(u(t_m) - \xi)\right)_m$ converges to η.*

It is interesting to notice that, by using the viability theory developed in Section 6.5 for the locally closed set $K = \{(t, u(t))\,;\ t \in [0, T)\}$ and the multi-function $\{1\} \times F$, Cârjă–Monteiro Marques [**48**] proved that the condition (15.6.2) is also equivalent to each one of the following:

(i) $Du(t) \cap F(u(t)) \neq \emptyset$ for each $t \in [0, T)$, or
(ii) $\overline{\text{conv}} Du(t) \cap F(u(t)) \neq \emptyset$ for each $t \in [0, T)$.

In 1981, Haddad [**104**] obtained the first result on viability of preorders. In fact, Haddad adapted the proof of viability of sets in order to obtain viability of preorders. Cârjă–Ursescu [**54**] showed that the viability, as well as the invariance of preorders can be completely described in terms of viability, or invariance of sets. The structure of the set of all solutions of a differential inclusion on a subset K, for which Π_K has continuous selections, was studied by Plaskacz [**146**].

As far as the infinite dimensional case is concerned, i.e., the case in which instead of $\mathbb{R}^n$ we are considering an infinite dimensional Banach space X, we mention the pioneering contribution of Gautier [**99**]. Firstly, he used Zorn's Lemma in order to get approximate solutions defined on an a priori given interval. Secondly, he used a sufficient weak tangency condition of the form: *for each* $\xi \in K$, *there exist* $\eta \in F(\xi)$, *a sequence* $(h_m)_m$ *decreasing to* 0 *and a sequence* $(q_m)_m$ *weakly convergent to* 0 *satisfying* $\|q_m + \eta\| \leq 2\|\eta\|$ *and* $\xi + h_m(\eta + q_m) \in K$ *for each* $m \in \mathbb{N}$. This tangency is far from being necessary for the viability of K. A necessary and sufficient condition for the viability of K in a more general setting has been obtained by Cârjă–Vrabie [**55**] by means of the so-called "bounded weak tangency condition".

Although not presented here, the Carathéodory case for differential inclusions is well developed and there exists a rather large literature on the subject. Among the first notable results in this direction we mention those of Tallos [**159**], Ledyaev [**121**], Frankowska–Plaskacz–Rzeżuchowski [**97**]. In all these papers, theorems of Scorza Dragoni type are the main tools. Results of the same kind, can be found in Cârjă–Monteiro Marques [**48**] in the finite dimensional case, and in Cârjă–Monteiro Marques [**49**] in the infinite dimensional setting. There, a technique of approximation of the multi-function through the Aumann integral mean is used. The case in which F is measurable with respect to t and (strongly-weakly) u.s.c. with respect to u can be carried out with the help of a Measurable Selection Theorem, as for instance Kuratowski–Ryll-Nardzewski Theorem 3.1.1, p. 86 in Vrabie [**173**] combined with the special techniques used in Chapter 5.

Section 6.1. Theorem 6.1.1, Corollary 6.1.1 and Lemma 6.1.1 are from Cârjă–Necula–Vrabie [**52**].

Section 6.2. Theorem 6.2.1 is from Cârjă–Necula–Vrabie [**52**], Theorem 6.2.2 is a local, more precise (in the sense that here we deal with exact solutions instead of a.e. solutions) and more general form (we do not impose any growth condition) of a result due to Deimling [**78**], Theorem 6.2.3 is an autonomous version of a viability result in Bebernes–Schuur [**12**]. Example 6.2.1 is from Aubin–Cellina [**6**], p. 202. The lack of convexity of the values of F can be, however, compensated by an l.s.c. extra-assumption combined with a stronger tangency condition. More precisely, we have

Theorem 15.6.1. *Let K be locally closed and let $F : K \rightsquigarrow \mathbb{R}^n$ be both u.s.c. and l.s.c. with nonempty and closed values. If $F(\xi) \subseteq \mathcal{T}_K(\xi)$ for each $\xi \in K$ then K is viable with respect to F. Moreover, if u is a solution of* (6.1.1) *then u' is a regulated function*[15].

See Aubin–Cellina [**6**], p. 198.

Section 6.3. In this form, Lemma 6.3.1 is from Cârjă–Necula–Vrabie [**52**]. Its proof, using both the set tangency condition: $F(\xi) \in \mathcal{TS}_K(\xi)$ for each $\xi \in K$, and Brezis–Browder Theorem 2.1.1, is completely new.

Section 6.4. The proof of the convergence of a sequence of ε-approximate exact solutions is new but standard.

Section 6.5. The extension to the nonautonomous case is also from Cârjă–Necula–Vrabie [**52**]. We notice that we confined ourselves here merely to the case of multi-functions which are jointly (strongly-weakly) u.s.c. with respect to all variables, i.e., with respect to both t and u. For other viability results under some mixed hypotheses allowing the "nonautonomous" F to be l.s.c. on some set and u.s.c. on the complementary set, see Donchev [**87**].

Section 6.6. The results in this section are from Cârjă–Necula–Vrabie [**52**] too. Theorem 6.6.1 is standard and Theorem 6.6.2 has its roots in Theorem 3.2.3, p. 96 in Vrabie [**173**].

Section 6.7. A condition similar to that one in Definition 6.7.2, with ξ replaced by ξ_1 and $\pi_K(\xi)$ replaced by ξ_2 with $\xi_1, \xi_2 \in V$, has been used previously by Cârjă–Ursescu [**54**]. A condition, related to the latter, was introduced by Donchev [**86**]. This condition is known under the name of *one-sided Lipschitz condition.* As in the single-valued case, here (6.7.3) is also automatically satisfied for each $\xi \in K$, and therefore, in Definition 6.7.2, we have only to assume that (6.7.3) holds for each $\xi \in V \setminus K$.

A strictly more restrictive Lipschitz condition than the one in Definition 6.7.3, with ξ replaced by ξ_1 and $\pi_K(\xi)$ by ξ_2, with ξ_1, ξ_2 belonging to D, has been first considered by Filippov [**96**]. In the same spirit as Filippov *loc. cit.*, Kobayashi [**116**] has used a dissipative type condition even more restrictive. It is easy to see that if F is either (D, K)-Lipschitz, or (D, K)-dissipative, then it has the comparison property with respect to (D, K). We notice that there are examples showing that there exist multi-functions F which, although neither (D, K)-Lipschitz, nor (D, K)-dissipative, have the comparison property. See for instance the "single-valued" Examples 4.3.1 and 4.3.2. For other invariance results in the case $X = \mathbb{R}^n$

[15]A function is *regulated* if it is uniform limit of step functions.

and $F : I \times X \rightsquigarrow X$ satisfying an one-sided Lipschitz condition, see Donchev–Rios–Wolenski [**88**].

Chapter 7.

Section 7.1. Proposition 7.1.1 is due to Roxin [**151**].

Section 7.2. Theorem 7.2.1 was called to our attention by Ursescu [**169**] and provides a characterization of the viability of an epigraph of a certain function in the terms of a differential inequality. Similar results can be found in Clarke-Ledyaev-Stern [**66**], p. 266. Problem 7.2.2 is from Cannarsa–Frankowska–Sinestrari [**43**].

Section 7.3. Lemma 7.3.1 and Theorem 7.3.1 are slight extension of some results in Cârjă–Ursescu [**54**], while Lemma 7.3.2, which is a simple generalization of Lemma 7.3.1, is new.

Section 7.4. Theorem 7.4.1 is inspired from Browder [**39**] and Crandall [**73**].

Section 7.5. Theorem 7.5.1 is from Vrabie [**178**]. For some extensions see Vrabie [**177**], [**179**] and [**180**].

Section 7.6. The presentation follows Yorke [**182**]. Propositions 7.6.1∼7.6.4, are from Yorke [**182**], Theorem 7.6.1 is from Fukuhara [**98**][16] but the proof here presented is due to Yorke [**182**]. Theorems 7.6.2 and 7.6.3 are from Yorke [**182**].

Section 7.7. Theorem 7.7.1 was established by Kneser [**115**] but the proof herein is from Yorke [**182**].

Section 7.8. Theorem 7.8.1 is due to Yoshizawa [**184**], Theorem 7.8.2 is due to Yorke [**182**].

Section 7.9. The differentiability concept D was introduced by Severi [**156**]. See also Ursescu [**164**].

For specific cases of (7.9.1) as quasilinear, first order partial differential equations see Goursat [**103**], Perron [**143**], Kamke [**112**], Carathéodory [**44**], Courant–Hilbert [**72**]. For Bellman equations which are also particular cases of (7.9.1) see Bellman [**14**], [**15**], [**16**], Pontryagin–Boltyanskii– Gamkrelidze– Mishchenko [**147**], Boltyanskii [**22**], [**23**], Gonzalez [**102**], Hájek [**105**], Cesari [**60**], Clarke–Vinter [**68**]. Among other particular but important instances of (7.9.1) we also mention the *eikonal equation*. See Ishii [**111**].

As far as we know, the differentiability concept of Severi is the least restrictive for which the characterization in Theorem 7.9.1 holds true. Every other differentiability concepts used in the literature devoted to particular cases of equation (7.9.1) implies the classical Fréchet, better referred to as the Stolz–Young–Fréchet–Hadamard, differentiability. We emphasize that w is Fréchet differentiable at x if and only if both w is Severi differentiable at x, and the Severi differential $Dw(x)$ is linear on $\mathbb{R}^n$.

The property (C_1) is related to the so-called *weakly decreasing systems* discussed in Clarke–Ledyaev–Stern–Wolenski [**67**], p. 211, which in turn are related to the Lyapunov theory of stabilization. See also Clarke–Ledyaev–Stern–Wolenski *loc.*

[16] The old English transliteration of the Japanese name Hukuhara.

cit., p. 208. The examples in Exercises 7.9.2 and 7.9.3 are from Cârjă–Ursescu [**54**]. For an extension of the characteristics method to mild solutions generated by C_0-semigroups see Ursescu [**168**].

Chapter 8.

If $A \neq 0$, in order to handle also points $\xi \in K$ which do not belong to $D(A)$, Pavel [**137**] introduced the following tangency concept: by definition $\eta \in \mathcal{F}_K^A(\xi)$ if

$$\lim_{h \downarrow 0} \frac{1}{h} \text{dist}\,(S(h)\xi + h\eta; K) = 0. \tag{15.8.3}$$

One may easily see that $\mathcal{F}_K^A(\xi)$ is in fact an "A-variant" of the tangent cone of Federer [**93**] and $\mathcal{F}_K^A(\xi) \subseteq \mathcal{T}_K^A(\xi)$. More precisely, Pavel [**137**] shows that, whenever K is locally closed, A generates a compact C_0-semigroup and f is continuous on $I \times K$, a necessary and sufficient condition for viability is

$$\lim_{h \downarrow 0} \frac{1}{h} \text{dist}\,(S(h)\xi + hf(\tau, \xi); K) = 0 \tag{15.8.4}$$

for each $(\tau, \xi) \in I \times K$. We notice that, whenever $\xi \in K \cap D(A)$, the tangency condition above is equivalent to

$$\lim_{h \downarrow 0} \frac{1}{h} \text{dist}\,(\xi + h(A\xi + f(\tau, \xi)); K) = 0$$

which is nothing else than a stronger form of the Nagumo's tangency condition, i.e., $A\xi + f(\tau, \xi) \in \mathcal{F}_K(\xi)$ for each $\xi \in K \cap D(A)$. However, there are cases in which K is not included in $D(A)$, or even $K \cap D(A)$ is empty and in these cases we can use only (15.8.4). This happens for instance if K is the trajectory of a nowhere differentiable C^0-solution of (8.1.1).

Concerning $\mathcal{T}_K^A(\xi)$ (see Definition 8.1.3), as we have already noticed, if $A \equiv 0$, it is nothing but the contingent cone at $\xi \in K$ in the sense of Bouligand [**32**].

A necessary and sufficient condition for the existence of monotone solutions of semilinear differential inclusions has been obtained by Chiş-Şter [**61**].

Section 8.1. Proposition 8.1.2 and Theorem 8.1.1 are due to Pavel [**137**], [**138**], while Theorem 8.1.2 is nothing but a more precise reformulation of Theorem 8.1.1.

Section 8.2. Theorem 8.2.1 is from Burlică–Roşu [**41**] and handles into a unitary frame several previous viability results of different topological nature. Namely, it includes results referring to both β-compact or Lipschitz perturbations of infinitesimal generators of arbitrary C_0-semigroups, as Theorem 8.2.2, and to continuous perturbations of infinitesimal generators of compact C_0-semigroups, as Theorem 8.2.3, which is essentially due to Pavel [**137**]. See also Burlică–Roşu [**42**]. Theorems 8.2.4, 8.2.5 and 8.2.6 are immediate consequences of Theorem 8.2.2 but we were not able to establish their origins.

Section 8.3. Lemma 8.3.1 is from Cârjă–Vrabie [**55**].

Section 8.4. The proof of the convergence of a suitably chosen sequence of ε-approximate mild solutions in the case of Theorem 8.2.1 follows Burlică–Roşu [**42**].

Section 8.5. The results in this section, based on the classical trick of reducing the nonautonomous case to the autonomous one by introducing t as an extra unknown, are simple consequences of the main theorems in Sections 8.1 and 8.2.

Section 8.6. Theorems 8.6.1 and 8.6.2 are due to Burlică-Roşu [**42**] and extend some earlier existence results in Burlică-Roşu [**40**] referring to the simpler case in which the domain of definition of the perturbation is open.

Section 8.7. The proof of Theorem 8.6.2 follows Burlică-Roşu [**42**].

Section 8.8. Theorems 8.8.1 and 8.8.2 are simple rephrases of some classic results, while Theorem 8.8.3 is from Burlică-Roşu [**42**], but essentially inspired by Theorem 3.2.3, p. 96 in Vrabie [**173**].

Chapter 9.

For the semilinear multi-valued case we mention the pioneering works of Pavel-Vrabie [**139**] and [**140**]. Further developments are due to Shi [**157**], Lupulescu–Necula [**123**] and Cârjă-Vrabie [**55**]. See also the references therein.

Section 9.1. The notions of both A-quasi tangent and A-tangent set to a given set at a given point in Definitions 9.1.3 and 9.1.4 were introduced by Cârjă–Necula–Vrabie [**52**]. Proposition 9.1.1, Theorems 9.1.1 and 9.1.2 are also from Cârjă–Necula–Vrabie *loc. cit.*. It should be noted that Theorems 9.1.1 and 9.1.2 are the first general necessary conditions for mild viability avoiding both the compactness of the values of F and the use of a weak tangency condition as in Cârjă–Vrabie [**55**].

Section 9.2. Theorems 9.2.1, 9.2.2, 9.2.3 and 9.2.4 are also from Cârjă–Necula–Vrabie [**52**] and can be compared with previous results in Shi [**157**], where the viability problem deals with strong instead of mild solutions. A first result in Shi *loc. cit.* states that if X is reflexive, K is compact, A generates a compact differentiable semigroup and F is u.s.c. with nonempty, compact, convex values, then the tangency condition $F(\xi) \cap \mathcal{T}_K^A(\xi) \neq \emptyset$ for each $\xi \in K$ is necessary and sufficient for the viability of K with respect to $A+F$. If X, K and A are as above but F is strongly-weakly u.s.c. with bounded closed and convex values, then the viability of K with respect to $A+F$ is equivalent to the tangency condition

$$S(t)F(\xi) \cap \mathcal{T}^A_{S(t)K}(S(t)\xi) \neq \emptyset$$

for each $t > 0$ and $\xi \in K$, condition which clearly is implied by (9.2.1) but it is strictly stronger than the latter. Coming back to Theorems 9.2.1~9.2.4, it should be noticed firstly that we do not assume that X is reflexive. Secondly, we do not assume that A generates a differentiable semigroup and thirdly, we do not ask F to have compact values. In addition, in Theorems 9.2.1, 9.2.2 and 9.2.3 we do not ask K to be compact. Of course, the price (not at all high) to be payed is that the solutions are merely mild and not strong.

Section 9.3. Lemma 9.3.1 is from Cârjă–Necula–Vrabie [**52**] and is an "A-variant" of Lemma 6.3.1.

Sections 9.4~9.6. The proof of the convergence of a suitably chosen sequence of ε-approximate mild solutions, in the case of Theorems 9.2.1, 9.2.3 and 9.2.4, is also from Cârjă–Necula–Vrabie [**52**].

Section 9.7. By using the well-known trick of introducing t as a new unknown, we reduce the quasi-autonomous case to the autonomous one and we take advantage of the results in Sections 9.1 and 9.2 in order to get necessary and sufficient conditions for mild viability applicable to quasi-autonomous semilinear evolution inclusions.

Section 9.8. The results concerning the existence of noncontinuable and global solutions are also from Cârjă–Necula–Vrabie [**52**] and are simple extensions to the semilinear evolution inclusions of the corresponding results in Section 6.6.

Chapter 10.

The viability problem for fully nonlinear evolution equations, i.e. the case in which both A and f are nonlinear, with A unbounded but f still continuous, has been considered for the first time by Vrabie [**171**]. We notice that Vrabie *loc. cit.* introduced the suitable tangency condition to apply also for points of K which do not belong to $D(A)$. Namely, the tangency condition introduced in Vrabie *loc. cit.* is

$$\lim_{h \downarrow 0} \frac{1}{h} \text{dist}\,(u(\tau+h,\tau,\xi,f(\tau,\xi));K)=0, \tag{15.10.5}$$

where $u(\cdot,\tau,\xi,y)=v(\cdot)$ is the unique C^0-solution of the Cauchy problem

$$\begin{cases} v'(s) \in Av(s)+y \\ v(\tau)=\xi. \end{cases}$$

More precisely, Vrabie *loc. cit.* proved that if A is the generator of a compact semigroup of nonexpansive operators and (15.10.5) holds *uniformly* with respect to $(\tau,\xi) \in I \times K$, then $I \times K$ is C^0-viable with respect to $A+f$. We emphasize that, whenever A is linear, (15.10.5) is equivalent to

$$\lim_{h \downarrow 0} \frac{1}{h} \text{dist}\left(S(h)\xi + \int_\tau^{\tau+h} S(\tau+h-s)f(\tau,\xi)\,ds;K\right)=0$$

which, in view of Proposition 8.1.2, reduces to the tangency condition (15.8.3) introduced by Pavel. Subsequent contributions in this context are due to Bothe [**27**] who allowed K to depend on t as well. In particular, in the case when K is independent of t, Bothe [**27**] showed that (15.10.5) with "lim inf" instead of "lim" is necessary and sufficient for viability.

The case of an evolution equation driven by a single-valued perturbation of am m-dissipative operator has also been analyzed by Chiş-Şter [**63**], [**62**], who proved sufficient conditions in order for the corresponding evolution equation to have at least one monotone solution.

Section 10.1. Theorems 10.1.1, 10.1.3 and 10.1.4 are from Cârjă-Vrabie [**57**]. Theorem 10.1.2 is from Bothe [**27**], where it is mentioned without proof.

Section 10.2. Lemma 10.2.1 is a variant of a result in Cârjă-Vrabie [**57**].

Sections 10.3∼10.4. The proof of the convergence of a suitably chosen sequence of ε-approximate C^0-solutions in the case of Theorems 10.1.1 and 10.1.3 is from Cârjă-Vrabie [**57**], while that one in the case of Theorem 10.1.2 is new.

Section 10.5. Theorems 10.5.1 and 10.5.2 are from Bothe [**27**] although the first one has its roots in Vrabie [**171**]. Theorem 10.5.3 is a simple extension to the quasi-autonomous case of Theorem 10.1.3.

Section 10.6. Theorem 10.6.1 is from Cârjă–Vrabie [**57**], while Theorem 10.6.2 is new and uses some ideas from Vrabie [**173**].

Section 10.7∼10.9. Theorems 10.7.1∼10.7.5 as well as their proofs are from Necula–Vrabie [**135**].

Section 10.10. Theorems 10.10.1 and 10.10.2 are new and extend some previous results in Cârjă–Vrabie [**57**].

Other viability results referring to continuous perturbations of dissipative (but not necessarily m-dissipative) operators may be found in Bothe [**31**]. There, the author assumes that $K \subseteq \overline{D(A)}$ is closed, $f : I \times K \to X$ is continuous from the left with respect to $t \in I$ and continuous with respect to $x \in K$, the semigroup generated by A is compact and both the range condition

$$\lim_{h \downarrow 0} \frac{1}{h} \text{dist}\,(\xi; R(I - hA)) = 0, \tag{15.10.6}$$

for each $\xi \in K$ and the tangency condition

$$\liminf_{h \downarrow 0} \frac{1}{h} \text{dist}\,(\xi + hf(t,\xi); R(I - hA_{|K \cap D(A)})) = 0 \tag{15.10.7}$$

for each $(t,\xi) \in I \times K$ hold, and concludes that $I \times K$ is viable with respect to $A + f$.

Chapter 11.

Concerning the nonlinear multi-valued perturbed case, we mention the works of Bothe [**27**], Bressan–Staicu [**34**], Cârjă–Vrabie [**56**] and the references therein. Bothe [**27**] considers the case of a time-dependent set K and a tangency condition inspired from Vrabie [**171**] and which reduces to $F(\xi) \in \mathcal{T}_K^A(\xi)$ whenever K is constant and F is single valued. Bressan–Staicu [**34**] consider a tangency condition also inspired from Vrabie [**171**] and which reduces to the latter in many important specific cases as for instance when F is single valued and continuous. Cârjă–Vrabie [**56**] allow F to be strongly-weakly upper semicontinuous but use a tangency condition expressed in the terms of the weak topology on X.

In the case of a nonlinear m-dissipative operator A, the notions of both A-quasi tangent and A-tangent set to K at $\xi \in K$ were introduced by Cârjă–Necula–Vrabie [**53**].

Section 11.1. Theorems 11.1.1 and 11.1.2 are from Cârjă–Necula–Vrabie [**53**]. We note that, as far as we know, Theorems 11.1.1 and 11.1.2 are the first general necessary condition for C^0-viability avoiding the compactness of the values of F.

Section 11.2. Theorem 11.2.1∼11.2.4 are also from Cârjă–Necula–Vrabie [**53**].

Section 11.3∼11.5. Lemma 11.3.1 and the proofs of both Theorems 11.2.1 and 11.2.4 are also inspired from Cârjă–Necula–Vrabie [**53**].

Section 11.6. The extension to the quasi autonomous case is standard. Theorems 11.6.1 and 11.6.2 are immediate corollaries of Theorem 11.1.2, while Theorems 11.6.3∼11.6.5 are direct consequences of Theorems 11.2.2∼11.2.5.

Section 11.7. Theorems 11.7.1 and 11.7.2 are simple extensions to the fully nonlinear multi-valued case of Theorems 6.6.1 and 6.6.2.

Chapter 12.

Section 12.1. Theorems 12.1.1∼12.1.4 are from Cârjă–Vrabie [**57**]. The sufficiency parts in both Theorems 12.1.1 and 12.1.2 have been later proved by Bothe [**30**].

Section 12.2. Lemma 12.2.1 and its proof are from Cârjă–Vrabie [**57**].

Sections 12.3∼12.4. The proof of the convergence in the case of Theorems 12.1.1∼12.1.3 follows also Cârjă–Vrabie [**57**].

Section 12.5. Theorems 12.5.1 and 12.5.2 are simple extensions to the fully nonlinear case of Theorems 5.5.1 and 5.5.2.

Other viability results referring to Carathéodory perturbations of dissipative operators may be found in Bothe [**31**]. There, the author uses a stronger tangency condition which is only sufficient but he considers only dissipative operators which may lack m-dissipativity. Namely, Bothe *loc. cit.* assumes that $K \subseteq \overline{D(A)}$ is closed, $f : I \times K \to X$ is *almost continuous*[17], the semigroup generated by A is compact and both (15.10.6) and (15.10.7) are satisfied for each $(t,\xi) \in I \times K$. Then, he concludes that $I \times K$ is viable with respect to $A + f$.

As we have already pointed out, in the Carathéodory case, it is not possible to get viability for general locally closed noncylindrical sets $\mathcal{C}$. Therefore, in this case, we have to impose some extra-conditions on the set $\mathcal{C}$. One way to handle this problem, observed by Yorke [**182**] in the continuous case, is to identify $\mathcal{C}$ with the graph of a multi-function $K : D(K) \subseteq \mathbb{R} \rightsquigarrow X$, defined by

$$\begin{cases} D(K) = \{t \in \mathbb{R};\ \text{there exists } x \in X \text{ with } (t,x) \in \mathcal{C}\} \\ K(t) = \{x \in X;\ (t,x) \in \mathcal{C}\} \quad \text{for } \ t \in D(K). \end{cases}$$

Under these circumstances, the tangency condition

$$\mathcal{F}(\tau,\xi) \in \mathcal{T}^{\mathcal{A}}_{\mathcal{C}}(\tau,\xi),$$

with $\mathcal{F} = (1,f)$, $\mathcal{A} = (0,A)$ and $\mathcal{C} = \operatorname{graph}(K)$, takes the equivalent form

$$\liminf_{h \downarrow 0} \frac{1}{h} \operatorname{dist}\left(u(\tau+h,\tau,\xi,f(\tau,\xi)); K(\tau+h)\right) = 0.$$

Necessary and sufficient conditions for the viability of such kind of sets, under additional assumptions on the multi-function K, have been obtained by Bothe [**27**] and Necula [**133**], for the semilinear Carathéodory case, and by Necula [**134**], for

[17] We recall that f is *almost continuous* if for each $\varepsilon > 0$ there exists a closed set $I_\varepsilon \subseteq I$ with $\mu(I \setminus I_\varepsilon) \leq \varepsilon$ and such that $f_{|I_\varepsilon \times K}$ is continuous.

the fully nonlinear Carathéodory case. For the case when f is multi-valued, see Bothe [**26**].

Chapter 13.

Section 13.1. Lemma 13.1.1 and Theorems 13.1.1 and 13.1.2 are well-known but we do not know their origins. Theorems 13.1.3 and 13.1.4 are new and appear for the first time here.

Section 13.2. Lemma 13.2.3 is from Cârjă–Motreanu [**50**]. For a proof in the specific case $p = 2$, see Attouch–Azé [**4**]. Lemma 13.2.2 and Theorem 13.2.1 are also from Cârjă–Motreanu *loc. cit.*

Section 13.3. Theorem 13.3.1, although new, is a simple consequence of the main result in Burlică–Roşu [**42**].

Section 13.4. Theorem 13.4.1 is from Burlică-Roşu [**42**]. We notice that a similar result holds true if the homogeneous Dirichlet boundary conditions are replaced with homogeneous Neumann boundary conditions. Also, the comparison result can be reformulated and proved in the space $C_0(\overline{\Omega}) \times C_0(\overline{\Omega})$, where $C_0(\overline{\Omega})$ is the space of all continuous functions from $\overline{\Omega}$ to $\mathbb{R}$, vanishing on the boundary. For related results we refer to Ladde–Lakshmikantham–Vatsala [**118**]. For more details on predator-pray systems see Ainseba–Aniţa [**2**] and the references therein. Similar viability arguments for cooperative semilinear quadratic perturbed systems were used previously by Bohl–Marek [**21**].

Section 13.5. As far as we know, Lemma 13.5.1 is due to Benilan [**17**]. Theorem 13.5.1 is from Cârjă–Necula–Vrabie [**53**].

Section 13.6. Theorem 13.6.1 is from Necula–Vrabie [**135**] and extends Theorem 13.4.1 in two directions. First, it allows the diffusion operators to be nonlinear and second it allows one of them to be degenerate. For related comparison results see Ladde–Lakshmikantham–Vatsala [**118**].

Section 13.7. Theorem 13.7.1 and Corollary 13.7.1, as well as their proofs, based on viability techniques, are from Cârjă–Necula–Vrabie [**52**]. Similar results, in the case in which f is Lipschitz but A generates an arbitrary C_0-semigroup, were obtained recently by Cârjă [**46**], by using completely different arguments.

Section 13.8. Theorem 13.8.1 is due to Paicu [**136**] and is an abstract Banach space setting variant of an existence result proved in Caşcaval-Vrabie [**45**]. See also Vrabie [**174**] and Hirano–Shioji [**108**]. Problem 13.8.1 is from Deimling [**77**], Exercise 6, p. 85.

Applications of viability and invariance techniques to reaction diffusion systems and core-shell reaction diffusion systems can be found in Bothe [**28**] and [**31**]. For another concrete application, i.e., the take-of problem of an aircraft in the presence of windshear see Seube–Moitie–Leitmann [**154**]. There, the authors formulate the problem as a differential game and, using viability theory, they show how, in the given circumstances, to determine safe flight domains for the aircraft. The flow-invariance of controlled flux sets, as the Enstrophy and Helicity sets, for the Navier-Stokes equations, has been deducted by Barbu–Pavel [**11**], working with strong solutions and by using viability and invariance methods. For a very elegant and effective way of obtaining existence and uniqueness, via similar methods, for the Euler equation for a homogeneous, inviscid, incompressible fluid, see Bourguignon–Brezis [**33**].

Bibliography

[1] R. A. Adams, *Sobolev spaces*, Academic Press, Boston San Diego New York London Sidney Tokyo Toronto, 1978.

[2] B. Ainseba and S. Aniţa, Internal stabilization for reaction-diffusion problem modeling a predator–prey system, *Nonlinear Anal.*, **61**(2005), 491–501.

[3] R. R. Akhmerov, M. I. Kamenskii, A. S. Potapov, A. E. Rodkina and B. N. Sadovskii, *Measure of noncompactness and condensing operators*, Operator Theory: Advances and Applications Vol. **55**, Birkhäuser Verlag, Basel Boston Berlin, 1992.

[4] H. Attouch and D. Azé, Approximation and regularization of arbitrary functions in Hilbert spaces by the Lasry-Lions method, *Ann. Inst. H. Poincaré Anal. Non Linéaire*, **10** (1993), 289–312.

[5] J. P. Aubin, *Viability theory*, Birkhäuser, Boston–Basel–Berlin, 1991.

[6] J. P. Aubin and A. Cellina, *Differential inclusions*, Springer Verlag, Berlin–Heidelberg–New York–Tokyo, 1984.

[7] M. Badii, J. I. Diaz and A. Tesei, Existence and attractivity results for a class of degenerate functional parabolic problems, *Rend. Sem. Mat. Univ. Padova*, **78**(1987), 109–124.

[8] P. Baras, Compacité de l'opérateur definissant la solution d'une équation d'évolution non linéaire $(du/dt) + Au \ni f$, *C. R. Acad. Sci. Sér. I Math.*, **286**(1978), 1113-1116.

[9] P. Baras, J. C. Hassan and L. Veron, Compacité de l'opérateur definissant la solution d'une équation d'évolution non homogène, *C. R. Acad. Sci. Paris Sér. I Math.*, **284**(1977), 779-802.

[10] V. Barbu, *Nonlinear semigroups and differential equations in Banach spaces*, Noordhoff, Leyden, 1976.

[11] V. Barbu and N. H. Pavel, Flow-invariant closed sets with respect to nonlinear semigroup flows, *NoDEA Nonlinear Differential Equations Appl.*, **10**(2003), 57–72.

[12] W. Bebernes and I. D. Schuur, The Ważewski topological method for contingent equations, *Ann. Mat. Pura Appl.*, **87**(1970), 271–278.

[13] G. Beer, *Topologies on closed and closed convex sets*, Kluwer Academic Publishers, Mahematics and Its Applications, **268**, 1993.

[14] R. Bellman, Bottleneck problems and dynamic programming, *Proc. Nat. Acad. Sci. U.S.A.*, **39** (1953), 947–951.

[15] R. Bellman, Bottlenek problems, functional equations and dynamic programming, *Econometrica*, **23** (1955), 73–87.

[16] R. Bellman, *Dynamic programming,* Princeton University Press, Princeton, New Jersey, 1957.

[17] P. Benilan, Equations d'évolution dans un espace de Banach quelconque et applications, Thèse, Orsay (1972).

[18] P. Benilan and H. Brezis, Solutions faibles d'équations d'évolution dans les espaces de Hilbert, *Ann. Inst. Fourier* (*Grenoble*), **22**(1972), 311–329.
[19] H. Berliocchi and J. M. Lasry, Intégrandes normales et mesures paramétrées en calcul des variations, *Bull. Soc. Math. France*, **101**(1973), 129–184.
[20] S. Bochner, Integration von Funktionen, deren Werte die Elemente eines Vektoraumes sind, *Fund. Math.*, **20**(1933), 262–276.
[21] E. Bohl and I. Marek, Existence and uniqueness results for nonlinear cooperative systems, *Linear operators and matrices*, Oper. Theory Adv. Appl., **130**, Birkhäuser, Basel, 2002, 153–170.
[22] V. G. Boltyanskii, Sufficient optimality conditions, (Russian), *Dokl. Akad. Nauk SSSR*, **140** (1961), 994–997.
[23] V. G. Boltyanskii, Sufficient conditions for optimality and the justification of the programming method, (Russian) *Izv. Akad. Nauk SSSR Ser. Mat.*, **28** (1964), 481–514.
[24] J. M. Bony, Principe du maximum, inégalité de Harnack et unicité du problème de Cauchy pour les opérateurs elliptiques dégénérés, *Ann. Inst. Fourier* (*Grenoble*), **19**(1969), 277–304.
[25] D. Bothe, Multivalued differential equations on graphs, *Nonlinear Anal.*, **18**(1992), 245–252.
[26] D. Bothe, Multivalued differential equations on graphs and applications, Ph. D. Thesis, Paderborn, 1992.
[27] D. Bothe, Flow invariance for perturbed nonlinear evolution equations, *Abstr. Appl. Anal.*, **1**(1996), 417–433.
[28] D. Bothe, Reaction diffusion systems with discontinuities. A viability approach, *Nonlinear Anal.*, **30**(1997), 677–686.
[29] D. Bothe, Multivalued perturbations of m-accretive differential inclusions, *Israel J. Math.*, **108**(1998), 109–138.
[30] D. Bothe, Nonlinear evolutions with Carathéodory forcing, *J. Evol. Equ.*, **3** (2003), 375–394.
[31] D. Bothe, Flow invariance for nonlinear accretive evolutions under range conditions, *J. Evol. Equ.*, **5** (2005), 227–252.
[32] H. Bouligand, Sur les surfaces dépourvues de points hyperlimités, *Ann. Soc. Polon. Math.*, **9**(1930), 32–41.
[33] J. P. Bourguignon and H. Brezis, Remarks on the Euler equation, *J. Functional Analysis*, **15**(1974), 341–363.
[34] A. Bressan and V. Staicu, On nonconvex perturbations of maximal monotone differential inclusions, *Set-Valued Anal.*, **2**(1994), 415–437.
[35] H. Brezis, On a characterization of flow-invariant sets, *Comm. Pure Appl. Math.*, **23**(1970), 261–263.
[36] H. Brezis and F. E. Browder, A general principle on ordered sets in nonlinear functional analysis, *Adv. in Mathematics*, **21**(1976), 355–364.
[37] H. Brezis and W. A. Strauss, Semi-linear second-order elliptic equations in L^1, *J. Math. Soc. Japan*, **25**(1973), 565–590.
[38] F. E. Browder, Nonlinear equations of evolution, *Ann. of Math.*, **80**(1964), 485–523.
[39] F. E. Browder, *Nonlinear operators and nonlinear equations of evolution in Banach spaces*, Proc. Sympos. Pure Math., **Vol. 18 II**, Amer. Math. Soc., 1976.

[40] M. Burlică and G. Roşu, The initial value and the periodic problems for a class of reaction-diffusion systems, *Dyn. Contin. Discrete Impuls. Syst. Ser. A Math. Anal.*, in print.
[41] M. Burlică and D. Roşu, A viability result for semilinear reaction-diffusion systems, Proceedings of the International Conference of Applied Analysis and Differential Equations, 4–9 September 2006, Iaşi, Romania, O. Cârjă and I. I. Vrabie Editors, World Scientific, 2007, 31–44.
[42] M. Burlică and D. Roşu, Some viability results for a class of reaction-diffusion systems, to appear.
[43] P. Cannarsa, H. Frankowska and C. Sinestrari, Optimality conditions for the minimum time problem, *Set-Valued Anal.*, **8**(2000), 127–148.
[44] C. Carathéodory, *Variationsrechnung und partielle Differentialgleichungen erster Ordnung,* Teubner, Leipzig und Berlin, 1935.
[45] R. Caşcaval and I. I. Vrabie, Existence of periodic solutions for a class of nonlinear evolution equations, *Rev. Mat. Univ. Complut. Madrid*, **7**(1994), 325–338.
[46] O. Cârjă, On the minimal time function and the minimum energy problem: a nonlinear case, *Systems Control Lett.*, **55** (2006), 543–548.
[47] O. Cârjă and M. D. P. Monteiro Marques, Viability for nonautonomous semilinear differential equations, *J. Differential Equations*, **166**(2000), 328–346.
[48] O. Cârjă and M. D. P. Monteiro Marques, Viability results for nonautonomous differential inclusions, *J. Convex Anal.*, **7**(2000), 437–443.
[49] O. Cârjă and M. D. P. Monteiro Marques, Weak tangency, weak invariance and Carathéodory mappings, *J. Dynam. Control Systems*, **8**(2002), 445–461.
[50] O. Cârjă and D. Motreanu, Flow-invariance and Lyapunov pairs, *Dyn. Contin. Discrete Impuls. Syst. Ser. A Math. Anal.*, **13B**(2006), suppl., 185–198.
[51] O. Cârjă, M. Necula and I. I. Vrabie, Local invariance via comparison functions, *Electron. J. Differential Equations*, 2004, No. 50, 1–14.
[52] O. Cârjă, M. Necula and I. I. Vrabie, Necessary and sufficient conditions for viability for semilinear differential inclusions, paper in preparation.
[53] O. Cârjă, M. Necula and I. I. Vrabie, Necessary and sufficient conditions for viability for nonlinear evolution inclusions, paper in preparation.
[54] O. Cârjă and C. Ursescu, The characteristics method for a first order partial differential equation, *An. Ştiin. Univ. Al. I. Cuza Iaşi, Secţ. I a Mat.*, **39**(1993), 367–396.
[55] O. Cârjă and I. I. Vrabie, Some new viability results for semilinear differential inclusions, *NoDEA Nonlinear Differential Equations Appl.*, **4**(1997), 401–424.
[56] O. Cârjă and I. I. Vrabie, Viability results for nonlinear perturbed differential inclusions, *Panamer. Math. J.*, **9**, No. 1, (1999), 63–74.
[57] O. Cârjă and I. I. Vrabie, Viable domains for differential equations governed by Carathéodory perturbations of nonlinear m-accretive operators, *Differential Equations and Control Theory,* S. Aizicovici and N. H. Pavel Editors, Lecture Notes in Pure and Applied Mathematics, **225**, Marcel Dekker Inc. New York-Basel, 2002, 109–130.
[58] O. Cârjă and I. I. Vrabie, *Differential equations on closed sets*, Chapter 3 in Handbook of Differential Equations, Ordinary Differential Equations, volume 2, Edited by A. Cañada, P. Drábek and A. Fonda, Elsevier B.V. 2005, 147–238.
[59] C. Castaing and M. Valadier, *Convex analysis and measurable multifunctions*, Lecture Notes in Mathematics, **580**, Springer-Verlag, 1977.
[60] L. Cesari, *Optimization–theory and applications,* Springer, New York, 1983.

[61] I. Chiş-Şter, Existence of monotone solutions for semilinear differential inclusions, *NoDEA Nonlinear Differential Equations Appl.*, **6**(1999), 63–78.

[62] I. Chiş-Şter, Existence of monotone solutions for nonlinear perturbed differential inclusions, *Rev. R. Acad. Cienc. Exactas Fis. Nat.*, **93**(1999), 189–194.

[63] I. Chiş-Şter, Monotone solutions for single-valued perturbed nonlinear evolution equations, *Commun. Appl. Anal.*, **5**(2001), 295–306.

[64] I. Ciorănescu, *Aplicaţii de dualitate în analiza funcţională neliniară*, Editura Academiei, Bucureşti, 1974.

[65] F. H. Clarke, Generalized gradients and applications, *Trans. Amer. Math. Soc.*, **205**(1975), 247–262.

[66] F. H. Clarke, Yu. S. Ledyaev and R. J. Stern, Invariance, monotonicity and applications, *Nonlinear Analysis, Differential Equations and Control*, F. H. Clarke and R. J. Stern Editors, Kluwer Academic Publishers, 1999, 207-305.

[67] F. H. Clarke, Yu. S. Ledyaev, R. J. Stern and P. R. Wolenski, *Nonsmooth analysis and control theory*, Springer Verlag, 1998.

[68] F. H. Clarke and R. B. Vinter, Local optimality conditions and Lipschitzian solutions to the Hamilton–Jacobi equation, *SIAM J. Control Optim.*, **21** (1983), 856–870.

[69] J. A. Clarkson, Uniformly convex spaces, *Trans. Amer. Math. Soc.*, **40**(1936), 396–414.

[70] G. Colombo and V. V. Goncharov, Variational inequalities and regularity properties of closed sets in Hilbert spaces, *J. Convex Anal.*, **8**(2001), 197–221.

[71] B. Cornet, Regularity properties of open tangent cones. *Nonlinear analysis and optimization* (Louvain-la-Neuve, 1983). *Math. Programming Stud.* **30**(1987), 17–33.

[72] R. Courant and D. Hilbert, *Methoden der Mathematischen Physik,* Zweiter Band. Berlin, Julius Springer, 1937.

[73] M. G. Crandall, Differential equations on closed sets, *J. Math. Soc. Japan*, **22**(1970), 443-455.

[74] M. G. Crandall, A generalization of Peano's existence theorem and flow-invariance, *Proc. Amer. Math. Soc.*, **36**(1972), 151–155.

[75] M. G. Crandall, Nonlinear semigroups and evolution governed by accretive operators, in Nonlinear Functional Analysis and its Applications, F.E. Browder Editor, *Proc. Sympos. Pure. Math.*, **45-1**, 1986, 305–337.

[76] K. Deimling, *Ordinary differential equations in Banach spaces*, Lecture Notes in Mathematics 596, Springer Verlag Berlin Heidelberg New York, 1977.

[77] K. Deimling, *Nonlinear functional analysis*, Springer Verlag, Berlin Heidelberg New York Tokio, 1985.

[78] K. Deimling, Multivalued differential equations on closed sets, *Differential Integral Equations*, **1**(1988), 23–30.

[79] J. I. Diaz and I. I. Vrabie, Propriétes de compacité de l'opérateur de Green généralisé pour l'équation des millieux poreux, *C. R. Acad. Sci. Paris Sér. I Math.*, **309**(1989), 221–223.

[80] J. I. Diaz and I. I. Vrabie, Existence for reaction diffusion systems: A compactness method approach, *J. Math. Anal. Appl.*, **188**(1994), 521–540.

[81] J. I. Diaz and I. I. Vrabie, Compactness of the Green Operator of nonlinear diffusion equations: application to Boussinesq type systems in fluid mechanics, *Topol. Methods in Nonlinear Anal.*, **4**(1994), 399–416.

[82] J. Diestel, Remarks on weak compactness in $L_1(\mu; X)$, *Glasg. Math. J.*, **18**(1977), 87–01.

[83] J. Diestel and J. J. Uhl Jr., *Vector measures*, Mathematical Surveys, **15**, 1977.
[84] J. Dieudonné, Deux examples singuliers d'équations différentielles, *Acta Sci. Math. Szeged*, **12**(1950), 38–40.
[85] N. Dinculeanu, *Vector measures*, Pergamon Press, New York, 1967.
[86] T. Donchev, Functional differential inclusions with monotone right hand side, *Nonlinear Anal.*, **16**(1991), 533–542.
[87] T. Donchev, Mixed type semicontinuous differential inclusions in Banach spaces, *Ann. Polon. Math.*, **77**(2001), 245–259.
[88] T. Donchev, V. Rios and P. Wolenski, Strong invariance and one-sided Lipschitz multifunctions, *Nonlinear Anal.*, **60**(2005), 849–862.
[89] J. Dugundji, An extension of Tietze's theorem, *Pacific J. Math.*, **1**(1951), 353–367.
[90] N. Dunford and J. T. Schwartz, *Linear operators Part I: General theory*, Interscience Publishers, Inc. New York, 1958.
[91] R. E. Edwards, *Functional analysis theory and applications*, Holt, Rinehart and Winston, New York Chicago San Francisco Toronto London, 1965.
[92] R. Engelking, *Outline of general topology*, North-Holland Publishing Company Amsterdam, PWM – Polish Scientific Publishers Warszawa, 1968.
[93] H. Federer, Curvature measures, *Trans. Amer. Math. Soc.*, **93**(1959), 418–491.
[94] S. Feferman, Independence of the axiom of choice from the axiom of independent choices, *J. Sym. Logic*, **29**(1967), 226–226.
[95] W. Feller, On the generation of unbounded semi-groups of bounded linear operators, *Ann. of Math.*, **58**(1953), 166–174.
[96] A. F. Filippov, Classical solutions of differential equations with multi-valued right-hand side, *SIAM J. Control*, **5** (1967), 609–621.
[97] H. Frankowska, S. Plaskacz and T. Rzeżuchowski, Measurable viability theorems and Hamilton–Jacobi–Bellman equations, *J. Differential Equations*, **116**(1995), 265–305.
[98] M. Fukuhara[18], Sur les systèmes des équations différentielles ordinaires II, *Japan J. Math.*, **6**(1929), 269–299.
[99] S. Gautier, Equations differentielles multivoques sur un fermé, *Publications de l'Université de Pau*, (1973).
[100] B. R. Gelbaum and J. M. H. Olmsted, *Counterexamples in analysis*, Holden Day Inc., San Francisco, London, Amsterdam, 1964.
[101] I. V. Girsanov, *Lekcii po matematičeskoi teorii èkstremal'nyh zadač*, Moscow University, Moscow, 1970, i.e., *Lecture on mathematical theory of extremum problems* Lectures Bibliographical notes and comments in Economics and Mathematical Systems 67, Springer Verlag, New York, 1972.
[102] R. Gonzalez, Sur l'existence d'une solution maximale de l'équation de Hamilton - Jacobi, *C. R. Acad. Sci. Paris Sér. A–B*, **282** (1976), A1287–A1290.
[103] E. Goursat, *Lecons sur l'integration des équation partielles du premier ordre,* Hermann, Paris, 1891.
[104] G. Haddad, Monotone trajectories of differential inclusions and functional differential inclusions with memory, *Israel J. Math.*, **39**(1981), 83–100.
[105] O. Hájek, On differentiability of the minimal time function, *Funkcial. Ekvac.*, **20** (1977), 97–114.
[106] P. Hartman, On invariant sets and on a theorem of Ważewski, *Proc. Amer. Math. Soc.*, **32**(1972), 511–520.

[18]The old English transliteration of the Japanese name Hukuhara.

[107] E. Hille and R. S. Phillips, *Functional analysis and semi-groups*, Amer. Math. Soc. Colloquium Publications, **31**, Fourth Printing of Revised Edition, 1981.

[108] N. Hirano and N. Shioji, Invariant sets for nonlinear evolution equations, Cauchy problems and periodic problems, *Abstr. Applied Anal.*, No. 3, (2004), 183–203.

[109] E. W. Hobson, *The theory of functions of a real variable and the theory of Fourier's series*, Vol. 1, Cambridge University Press, Cambridge, 3rd edition, 1927.

[110] P. Howard and J. E. Rubin, The Boolean prime ideal theorem plus countable choice do [does] not imply dependent choice, *Math. Log. Q.*, **42**(1996), 410–420.

[111] H. Ishii, A simple, direct proof of uniqueness for solutions of the Hamilton - Jacobi equations of eikonal type, *Proc. Amer. Math. Soc.*, **100** (1987), 247–251.

[112] E. Kamke, *Differentialgleichungen reeler Funktionen,* Akademische Verlagsgesellschaft, Leipzig, 1930.

[113] T. Kato, Nonlinear semigroups and evolution equations, *J. Math. Soc. Japan.*, **19**(1967), 508–520.

[114] N. Kenmochi and T. Takahashi, Nonautonomous differential equations in Banach spaces, *Nonlinear Anal.*, **4**(1980), 1109–1121.

[115] H. Kneser, Über die Lösungen eines Systems gewöhnlicher Differentialgleichungen, das der Lipschitzschen Bedingung nicht genügt. *Berl. Ber.* (1923), 171–174.

[116] Y. Kobayashi, Difference approximation of Cauchy problems for quasi-dissipative operators and generation of nonlinear semigroups, *J. Math. Soc. Japan*, **27** (1975), 640–665.

[117] A. Kucia, Scorza Dragoni type theorems, *Fund. Math.*, **138**(1991), 197–203.

[118] G. S. Ladde, V. Lakshmikantham and A. S. Vatsala, *Monotone iterative techniques for nonlinear differential equations*, Monographs, Advanced Texts and Surveys in Pure and Applied Mathematics, **27**, Pitman 1985.

[119] V. Lakshmikantham and S. Leela, *Nonlinear differential equations in abstract spaces,* International Series in Nonlinear Mathematics, **2**, Pergamon Press, 1981.

[120] H. Lebesgue, Sur les intégrales singuliéres, *Ann. Fac. Sci. Univ. Toulouse, Math.-Phys.*, **1**(1909), 25–117.

[121] Yu. S. Ledyaev, Criteria for viability of trajectories of nonautonomous differential inclusions and their applications, *J. Math. Anal. Appl.*, **182**(1994), 165–188.

[122] G. Lumer, and R. S. Phillips, Dissipative operators in a Banach space, *Pacific J. Math.*, **11**(1961), 679-698.

[123] V. Lupulescu and M. Necula, Viability and local invariance for non-convex semilinear differential inclusions, *Nonlinear Funct. Anal. Appl.* **9**(2004), No. 3, 495–512.

[124] R. H. Martin Jr., Differential equations on closed subsets of a Banach space, *Trans. Amer. Math. Soc.*, **179**(1973), 399–414.

[125] E. Michael, Continuous selections I, *Ann. Math.*, **63**(1956), 361–382.

[126] E. Michael, Continuous selections II, *Ann. Math.*, **64**(1956), 562–580.

[127] D. Milman, On some criteria for regularity of spaces of type (B), *C. R. (Dokl.) Acad. Sci. URSS*, **20**(1938), 243–246.

[128] I. Miyadera, Generation of a strongly continuous semi-group operations, *Tôhoku Math. J.* (2) *Ser.*, **4**(1952), 109-114.

[129] I. Miyadera, *Nonlinear semigroups*, Translation of Mathematical Monographs **109**, American Mathematical Society, Providence, RI, 1992.

[130] H. Mönch, Boundary value problems for nonlinear ordinary differential equations of second order in Banach spaces, *Nonlinear Anal. T.M.A.*, **4**(1980), 985–999.

[131] B. S. Mordukhovich, *Variational analysis and generalized differentiation I Basic Theory*, Grundlehrer der Mathematisches Wissenschaften, Volume **330**, Springer, 2006.

[132] M. Nagumo, Über die Lage der Integralkurven gewöhnlicher Differentialgleichungen, *Proc. Phys.-Math. Soc. Japan*, **24**(1942), 551–559.

[133] M. Necula, Viability of variable domains for nonautonomous semilinear differential equations, *Panamer. Math. J.*, **11**(2001), 65–80.

[134] M. Necula, Viability of variable domains for differential equations governed by Carathéodory perturbations of nonlinear m-accretive operators, *An. Ştiinţ. Univ. Al. I. Cuza Iaşi, Sect. I a Mat.*, **48**(2002), 41–60.

[135] M. Necula and I. I. Vrabie, A viability result for a class of fully nonlinear reaction-diffusion systems, submitted.

[136] A. Paicu, Periodic solutions for a class of nonlinear evolution equations in Banach spaces, paper in preparation.

[137] N. H. Pavel, Invariant sets for a class of semi-linear equations of evolution, *Nonlinear Anal.*, **1**(1976/1977), 187–196.

[138] N. H. Pavel, *Differential equations, flow invariance and applications*, Research in Mathematics 113, Pitman, Boston, MA, 1984.

[139] N. H. Pavel and I. I. Vrabie, Equations d'évolution multivoques dans des espaces de Banach, *C. R. Acad. Sci. Paris Sér. A–B*, **287**(1978), A315–A317.

[140] N. H. Pavel and I. I. Vrabie, Semilinear evolution equations with multivalued right-hand side in Banach spaces, *An. Ştiinţ. Univ. Al. I. Cuza Iaşi Secţ. I a Mat.*, **25**(1979), 137–157.

[141] G. Peano, Démonstration de l'intégrabilité des équations différentielles ordinaires, *Math. Ann.*, **37**(1890), 182–228.

[142] O. Perron, Ein neuer Existenzbeweis für Integrale der Differentialgleichung, $y' = f(x,y)$, *Math. Ann.*, **76**(1915), 471–484.

[143] O. Perron, Partieller Differentialgleichungssysteme im reellen Gebiet, *Math. Zeischrift*, **27** (1928), 549–564.

[144] B. J. Pettis, A proof that every uniformly convex space is reflexive, *Duke Math. J.*, **5**(1939), 249–253.

[145] R. S. Phillips, On the generation of semi-groups of linear operators, *Pacific J. Math.*, **2**(1952), 343-369.

[146] S. Plaskacz, On the solution sets for differential inclusions, *Boll. Un. Mat. Ital. A*, (7) **6**(1992), 387–394.

[147] L. S. Pontryagin, V. G. Boltyanskii, R. V. Gamkrelidze and E. F. Mishchenko, *The mathematical theory of optimal processes,* (Russian), GIF-ML, Moscow, 1961.

[148] M. H. Protter and H. F. Weinberger, *Maximum principles in differential equations,* Springer Verlag New York Berlin Heidelberg Tokyo, 1984.

[149] M. Quincampoix, Differential inclusions and target problems, *SIAM J. Control Optim.*, **30**(1992), 324–335.

[150] T. Roubíček, *Nonlinear partial differential equations with applications*, International Series of Numerical Mathematics Volume **153**, Birkhäuser Verlag Basel-Boston-Berlin, 2005.

[151] E. Roxin, Stability in general control systems, *J. Differential Equations*, **1**(1965), 115–150.

[152] G. Scorza Dragoni, Un teorema sulle funzioni continue rispetto ad una e misurabili rispetto ad un'altra variabile, *Rend. Sem. Mat. Univ. Padova*, **17** (1948), 102–106.

[153] I. Segal, Non-linear semi-groups, *Ann. of Math.*, **78**(1963), 339–364.
[154] N. Seube, R. Moitie and G. Leitmann, Aircraft take-off windshear: a viability approach, *Set-Valued Analysis in Control Theory. Set-Valued Anal.*, **8**(2000), 163–180.
[155] F. Severi, Su alcune questioni di topologia infinitesimale, *Annales Soc. Polonaise*, **9**(1931), 97–108.
[156] F. Severi, Sulla differenziabilità totale delle funzioni di più variabili reali, *Ann. Mat. Pura Appl. IV. Ser.*, **13** (1934), 1–35.
[157] S. Z. Shi, Viability theorems for a class of differential operator inclusions, *J. Differential Equations*, **79**(1989), 232–257.
[158] R. E. Showalter, *Montone operators in Banach spaces and nonlinear partial differential equations*, Mathematical Surveys and Monographs, American Mathematical Society, Providence, RI, **49**, 1997.
[159] P. Tallos, Viability problems for nonautonomous differential inclusions, *SIAM J. Control. Optim.*, **29**(1991), 253–263.
[160] J. S. Treiman, Characterization of Clarke's tangent and normal cones in finite and infinite dimensions, *Nonlinear Anal.*, **7**(1983), 771–783.
[161] M. Turinici, Differential equations on closed sets under generalized dissipativity conditions, *An. Ştiint. Univ. Al. I. Cuza Iaşi, Mat.* (N.S.), **44**(1998), suppl., 611–637.
[162] M. Turinici, Brezis–Browder principles in Separable Ordered Sets, *Libertas Math.*, **26**(2006), 15–30.
[163] M. Turinici, Personal communication.
[164] C. Ursescu, Sur une généralisation de la notion de différentiabilité, *Atti Accad. Naz. Lincei Rend. Cl. Sci. Fis. Mat. Natur.*, (8) **54** (1973), 199–204.
[165] C. Ursescu, Tangent sets and differentiable functions, *Math. Control Theory Proc. Conf. Zakopane,* 1974, Banach Center Publications (1976), 151–155.
[166] C. Ursescu, Probleme de tangenţă cu aplicaţii în teoria optimizării, Internal Report, Grant 11996/25.09.1980, Iaşi 1981.
[167] C. Ursescu, Carathéodory solutions of ordinary differential equations on locally closed sets in finite-dimensional spaces, *Math. Japon.*, **31**(1986), 483–491.
[168] C. Ursescu, C_0-semigroups, characteristics method, and generic differential equations, *An. Univ. Timişoara Ser. Mat.-Inform.*, **34**(1996), 283–290.
[169] C. Ursescu, Personal communication.
[170] P. Volkmann, Über die Existenz von Lösungen der Differenialgleichungen $u' = f(u)$ in einer abgeschlossenen Menge, wenn f eine k-Mengenkontraktion ist, *Ordinary and Partial Differential Equations* Proc. of the Fourth Conf., Univ. Dundee 1976, Lecture Notes in Mathematics **564** Springer Verlag, 496–503.
[171] I. I. Vrabie, Compactness methods and flow-invariance for perturbed nonlinear semigroups, *An. Ştiinţ. Univ. "Al. I. Cuza" Iaşi Secţ. I a Mat.*, **27** (1981), 117–125.
[172] I. I. Vrabie, A compactness criterion in $C(0,T;X)$ for subsets of solutions of nonlinear evolution equations governed by accretive operators, *Rend. Sem. Mat. Univ. Politec. Torino*, **43** (1985), 149–157.
[173] I. I. Vrabie, *Compactness methods for nonlinear evolutions*, Second Edition, Pitman Monographs and Surveys in Pure and Applied Mathematics **75**, Longman 1995.
[174] I. I. Vrabie, Periodic solutions for nonlinear evolution equations in a Banach space, *Proc. Amer. Math. Soc.*, **109(3)**(1990), 653-661.
[175] I. I. Vrabie, C_0*-semigroups and applications*, North-Holland Publishing Co. Amsterdam, 2003.

[176] I. I. Vrabie, *Differential equations. An introduction to basic concepts, results and applications*, World Scientific, Publishing Co., Inc., River Edge NJ, 2004.

[177] I. I. Vrabie, A Nagumo type viability theorem, *An. Ştiinţ. Univ. Al. I. Cuza Iaşi*, (N.S.), **51**(2005), 293–308.

[178] I. I. Vrabie, Nagumo viability theorem. Revisited, *Nonlinear Anal.*, **64**(2006), 2043–2052.

[179] I. I. Vrabie, Viability under Carathéodory conditions, *Dyn. Contin. Discrete Impuls. Syst. Ser. A Math. Anal.*, **13B**(2006), suppl., 171–184.

[180] I. I. Vrabie, A viability result for a class of ordinary differential equations in Banach spaces, *Mathematical Analysis and Applications*, 143–157, AIP Conference Proceedings, Volume **835**, AIP Conf. Proc., **835** *Amer. Inst. Phys., Melville NY*, 2006.

[181] T. Ważewski, Sur un condition équivalent à l'équation au contingent, *Bull. Acad. Polon. Sci. Sér. Sci. Math. Astronom. et Phys.*, **9**(1961), 865–867.

[182] J. A. Yorke, Invariance for ordinary differential equations, *Math. Systems Theory*, **1**(1967), 353–372.

[183] J. A. Yorke, "Invariance for ordinary differential equations": Correction, *Math. Systems Theory*, **2**(1968), 381.

[184] T. Yoshizawa, Stability theory by Liapunov's second method, *Publications of the Mathematical Society of Japan*, **9**, 1966.

[185] K. Yosida, *Functional Analysis*, 3rd Edition, Springer Verlag, 1971.

[186] A. C. Zaanen and W. A. Luxemburg, Advanced Problems and Solutions: Solutions: 5029, *Amer. Math. Monthly*, **70**(1963), 674.

[187] S. C. Zaremba, Sur les équations au paratingent, *Bull. Sci. Math.*, (2) **60**(1936), 139–160.

Name Index

Subject Index

Notation Index

— $\alpha(B) = \inf\left\{\varepsilon > 0; \exists B_1, B_2, \ldots, B_{n(\varepsilon)} \in \mathcal{B}_\varepsilon(X),\ B \subseteq \bigcup_{i=1}^{n(\varepsilon)} B_i\right\}$

— $a\nabla w = \sum_{i=1}^{n} a_i \frac{\partial w}{\partial x_i}$ in the sense of distributions over $\mathbb{R}^n$

— $\mathcal{B}(X)$ the class of all bounded subsets in X

— $\mathcal{B}_\varepsilon(X)$ the class of all subsets in X whose diameters do not exceed ε

— $\beta(B) = \inf\left\{\varepsilon > 0; \exists x_1, x_2, \ldots, x_{n(\varepsilon)} \in X,\ B \subseteq \bigcup_{i=1}^{n(\varepsilon)} D(x_i, \varepsilon)\right\}$

— $C([\tau, T]; X) = \{u : [\tau, T] \to X\ u \text{ continuous}\}$

— $\overline{\text{conv}}\, C$ – the closed convex hull of C

— $\text{conv}\, C$ – the convex hull of C

— $C_{ub}(\mathbb{R}^n) = \{u : \mathbb{R}^n \to \mathbb{R};\ u \text{ uniformly continuous and bounded}\}$

— $\text{diam}\,(B) = \sup\{\|x - y\|;\ x, y \in B\}$

— $\text{dist}\,(C; D) = \inf_{x \in C, y \in D} \|x - y\|$

— $\text{dist}\,(x; D) = \inf_{y \in D} \|x - y\|$

— $\text{dist}_{HP}(B, C) = \max\{e(B; C), e(C; B)\}$

— $\text{dist}\,(x; y) = \|x - y\|$

— $[D_+ x](t) = \liminf_{h \downarrow 0} \frac{x(t+h) - x(t)}{h}$ the Dini right lower derivative of x at t

— $\mathcal{D}^\kappa u = \frac{\partial^{\kappa_1 + \kappa_2 + \cdots + \kappa_n} u}{\partial x_1^{\kappa_1} \partial x_2^{\kappa_2} \ldots \partial x_n^{\kappa_n}}$

— $\mathcal{D}_+ u(t)$ the set of all limit points of $h \mapsto h^{-1}(u(t+h) - u(t))$ for $h \downarrow 0$

— $D(\xi, \rho)$ – the closed ball with center ξ and radius ρ

— $\mathcal{E} = \{f \in L^1(\mathbb{R}_+; X);\ f(s) \in E \text{ a.e. for } s \in \mathbb{R}_+\}$

— $\text{epi}\,(g) = \{(\xi, t);\ g(\xi) \leq t,\ \xi \in D\}$, for $g : D \to \mathbb{R}$

— $e(B; C) = \sup_{x \in B} \text{dist}\,(x; C)$

— $F : K \rightsquigarrow X$ – the multi-function $F : K \to 2^X$

— $\text{Fin}\,(X^*)$ – the class of all finite subsets in X^*

— $\partial F_{\tau,\xi}(t)$ – the boundary of the set $F_{\tau,\xi}(t)$

— $\mathcal{F}_K(\xi)$ – set of all tangent vectors to K at ξ in the sense of Federer

— $H^1(\Omega) = W^{1,2}(\Omega)$

— $H^2(\Omega) = W^{2,2}(\Omega)$

— $\text{hyp}\,(g) = \{(\xi, t);\ g(\xi) \geq t,\ \xi \in D\}$, for $g : D \to \mathbb{R}$

— $H_0^1(\Omega) = W_0^{1,2}(\Omega)$

— $J(x) = \{x^* \in X^*;\ (x, x^*) = \|x\|^2 = \|x^*\|^2\}$ – the duality on X

— $\overline{K}$– the closure of the set K

— $|\kappa| = \kappa_1 + \kappa_2 + \cdots + \kappa_n$, for $\kappa = (\kappa_1, \kappa_2, \ldots, \kappa_n) \in \mathbb{N}^n$

— ∂K – the boundary of the set K

— $L^1_{\text{loc}}([0, T); X) = \{u : [0, T) \to X;\ u \in L^1(0, \widetilde{T}; X) \text{ for each } 0 < \widetilde{T} < T\}$

— $L^1(\tau, T) = L^1(\tau, T; \mathbb{R})$

— $L^\infty_{\text{loc}}(I) = \{u : I \to \mathbb{R};\ \forall [\tau, T] \subseteq I,\ u \in L^\infty(\tau, T)\}$

— $L^1(\tau, T; X) = L^1([\tau, T]; X)$

— $L^p(\Omega) = L^p(\Omega; \mathbb{R})$

— $L^p(\Omega; X) = L^p(\Omega, \mu; X)$ if μ is the Lebesgue measure

— $\mathcal{L}^p(\Omega, \mu; X)$ – the set of all measurable $f : \Omega \to X$ with $\|f\|^p$ μ-integrable

— $\mathcal{L}(X)$ – the space of all linear bounded operators from X to X

— $\mathbb{N}$ the set of positive integers

— $\|U\|_{\mathcal{L}(X)} = \sup_{\|x\|\leq 1} \|Ux\|$
— $\mathbb{N}^*$ the set of positive integers without 0
— (Ω, Σ, μ) – a measure space
— $\mathcal{QTS}_K^A(\xi) = \{E \subseteq X;\ (\forall \rho > 0)\ \liminf_{h\downarrow 0} \frac{1}{h} \operatorname{dist}(S_{\mathcal{E}}(h)\xi; K \cap D(\xi,\rho)) = 0\}$[19]
— $\mathcal{QTS}_K^A(\xi) = \{E \subseteq X;\ (\forall \rho > 0)\ \liminf_{h\downarrow 0} \frac{1}{h} \operatorname{dist}(S_{\mathcal{E}}(h)\xi; K \cap D(\xi,\rho)) = 0\}$[20]
— $\mathbb{R}$ the set of real numbers
— $\mathbb{R}_+$ the set of positive real numbers
— $\{S_\eta(t) : \overline{D(A)} \to \overline{D(A)};\ t \geq 0\}$ the semigroup generated by $A_\eta = A + \eta$
— $S_E(h)\xi = \{S(h)\xi + \int_0^h S(h-s)\eta\, ds;\ \eta \in E\}$[21]
— $S_E(h)\xi = \{u(h,0,\xi,\eta);\ \eta \in E\}$[22]
— $S_{\mathcal{E}}(h)\xi = \{S(h)\xi + \int_0^h S(h-s)f(s)\, ds;\ f \in \mathcal{E}\}$[23]
— $S_{\mathcal{E}}(h)\xi = \{u(h,0,\xi,f);\ f \in \mathcal{E}\}$[24]
— $\mathcal{S}_K^A(\xi) = \{\eta \in X;\ \lim_{h\downarrow 0} \frac{1}{h} \operatorname{dist}(u(h,0,\xi,\eta); K) = 0\}$
— $T_{\mathcal{C}} = \sup\{t \in \mathbb{R};$ there exists $\eta \in X,$ with $(t,\eta) \in \mathcal{C}\}$
— $\mathcal{T}_K(\xi) = \{\eta \in X;\ \liminf_{h\downarrow 0} \frac{1}{h} \operatorname{dist}(\xi + h\eta; K) = 0\}$
— $\mathcal{T}_K^A(\xi) = \{\eta \in X;\ \liminf_{h\downarrow 0} \frac{1}{h} \operatorname{dist}(S_\eta(h)\xi; K) = 0\}$
— $\mathcal{TS}_K(\xi) = \{E \subseteq X;\ (\forall \rho > 0)\ \liminf_{h\downarrow 0} \frac{1}{h} \operatorname{dist}(\xi + hE; K \cap D(\xi,\rho)) = 0\}$
— $\mathcal{TS}_K^A(\xi) = \{E \subseteq X;\ (\forall \rho > 0)\ \liminf_{h\downarrow 0} \frac{1}{h} \operatorname{dist}(S_E(h)\xi; K \cap D(\xi,\rho)) = 0\}$
— $u(\cdot,\tau,\xi,g(\cdot))$ the unique C^0-solution of $u'(t) \in Au(t) + g(t)$, $u(\tau) = \xi$
— $W^{m,p}(\Omega) = \{u : \Omega \to \mathbb{R};\ \mathcal{D}^\alpha u \in L^p(\Omega),$ for $|\alpha| \leq m\}$
— $W_0^{m,p}(\Omega)$ – the closure of $C_0^\infty(\Omega)$ in $W^{m,p}(\Omega)$
— $W^{1,1}_{\text{loc}}((\tau,T]; X) = \{u \in L^1([\delta,T]; X);\ u \in W^{1,1}([\delta,T]; X)$ for $\delta \in (\tau,T)\}$
— $(x,x^*) = x^*(x)$
— $x + C = \{y \in X;$ there exists $z \in C$ such that $y = x + z\}$
— X^* – the topological dual of X
— $[x,y]_- = \lim_{h\uparrow 0} \frac{\|x+hy\| - \|x\|}{h}$
— $(x,y)_- = \lim_{h\uparrow 0} \frac{\|x+hy\|^2 - \|x\|^2}{2h}$
— $[x,y]_+ = \lim_{h\downarrow 0} \frac{\|x+hy\| - \|x\|}{h}$
— $(x,y)_+ = \lim_{h\downarrow 0} \frac{\|x+hy\|^2 - \|x\|^2}{2h}$

[19]For A linear.
[20]For A nonlinear.
[21]For A linear.
[22]For A nonlinear.
[23]For A linear.
[24]For A nonlinear.

www.ingramcontent.com/pod-product-compliance
Ingram Content Group UK Ltd.
Pitfield, Milton Keynes, MK11 3LW, UK
UKHW020235290926
15470UKWH00002B/562

9 780444 527615